STUDY GUIDE AND FULL SOLUTIONS MANUAL
Susan McMurry

Fundamentals of General, Organic, and Biological

CHEMISTRY

EIGHTH EDITION

MCMURRY   BALLANTINE   HOEGER   PETERSON

D0162174

PEARSON

Editor-in-Chief: Jeanne Zalesky
Acquisitions Editor: Scott Dustan
Product Marketing Manager: Elizabeth Ellsworth
Program Manager: Sarah Shefveland
Project Manager: Beth Sweeten
Program Management Team Lead: Kristen Flatham
Project Management Team Lead: David Zielonka
Production Management: : Andrea Stefanowicz, Lumina Datamatics
Cover Designer: Seventeenth Street Studios
Operations Specialist: Maura Zaldivar-Garcia
Cover Photo Credit: Sebastian Kaulitzki/Spl/Age Fotostock

1 2 3 4 5 6 7 8 9 10—**V001**—19 18 17 16

www.pearsonhighered.com   ISBN-10: 0-13-426137-2; ISBN-13: 978-0-13-426137-9

Contents

Preface

How is food digested? What is DNA fingerprinting? How do anesthetics work?

For all of these questions, chemistry provides an answer. Chemistry, the study of matter, is essential for understanding the physical world, from acid rain to gene therapy to climate change. Chemical laboratory tests are a routine part of medical care. The functioning of our bodies is a result of thousands of biochemical reactions. To study life, you must first study chemistry.

Both the textbook and this *Study Guide and Solutions Manual* are designed to be as helpful as possible to you in your chemistry course. The textbook contains numerous solved problems that show you techniques for solving specific types of chemistry problems. In addition, the text includes applications of the material in each chapter to contemporary science problems. This *Study Guide* consists of several types of study aids: detailed solutions to all textbook problems, outlines of the chapters, and self-tests.

The following suggestions may help you:

1. *Attend all classes*. Although it is possible to learn chemistry just by reading the textbook and study guide, a lecturer makes difficult topics much easier to understand and ties together seemingly unrelated topics.

2. *Read the textbook*. The best way to study is to skim the chapter to be covered before class, and then to read it more thoroughly after the lecturer has discussed the material. Use the textbook to help you understand subjects you find especially difficult. It is also helpful to use the chapter outlines in the *Study Guide* after reading the textbook.

3. *Work the problems*. Start by reading carefully the Solved Problems in the textbook. Next, try to work the Practice Problems, using the steps outlined in the Solved Problems. Finally, attempt the Additional Problems at the end of the chapter. Although this *Study Guide and Solutions Manual* has detailed solutions to all text problems, it's better to use the solutions in this book as a last resort. You will learn more if you first struggle a bit with a problem before checking its answer than if you look up the answer right away.

4. Use the supplementary material in the *Study Guide* to help you review each chapter. The chapter outlines organize and restate both the major themes of each chapter and the detailed contents. Go over the list of key terms in the textbook to make sure that you know the meaning of each important term. Test yourself with the self-tests, which include multiple choice, sentence completions, matching, and true/false questions.

Acknowledgments

I would like to thank the following people for their assistance with this book: John McMurry (of course), David Ballantine, Carl Hoeger, Virginia Peterson, Andrea Stefanowicz, and Robert Dixon. I am grateful to my entire family for their patience and support.

Chapter 1 Matter and Measurements

Chapter Outline

I. Matter and its properties (Sections 1.1–1.2).
   A. Matter is anything that is physically real (Section 1.1).
   B. Properties are the characteristics of matter.
      1. Physical properties are those properties that can be measured without altering the identity of a substance. Examples include mass, melting point, boiling point, and color.
      2. Chemical properties must be determined by changing the identity of the substance. Examples include rusting, combustion, and chemical reactivity.
      3. A physical change does not alter the identity of a substance.
      4. A chemical change is a process in which one or more substances undergoes a change in identity.
   C. States of matter (Section 1.2).
      1. The states of matter are solid, liquid, and gas.
         a. Solids have definite volume and definite shape.
         b. Liquids have definite volume and indefinite shape.
         c. Gases have indefinite volume and indefinite shape.
      2. Changes of state are melting, boiling, condensing, and freezing.
II. Classification of matter (Sections 1.3–1.5).
   A. Mixtures (Section 1.3).
      1. Mixtures vary in composition and properties.
         a. A homogeneous mixture has a uniform composition at the molecular level.
         b. A heterogeneous mixture is a blend of two or more substances having nonuniform composition.
      2. Mixtures can be separated by physical methods.
   B. Pure substances.
      1. Pure substances include chemical compounds and chemical elements.
      2. Pure substances don't vary in composition and properties.
      3. Chemical compounds can be broken down to elements by chemical change.
      4. Chemical elements can't be broken down.
   C. Chemical elements and symbols (Sections 1.4).
      1. There are 118 elements; 91 of them occur naturally (Section 1.4).
      2. Elements are represented by one- or two-letter symbols.
      3. The symbols for elements can be combined to produce chemical formulas.
   D. Chemical reactions (Section 1.5).
         a. A chemical reaction represents a chemical change between pure substances.
         b. In a chemical reaction, the reactants are written on the left, the products are written on the right, and an arrow connects them.
III. Measurements (Sections 1.6–1.7).
   A. Physical quantities and scientific notation (Section 1.6).
      1. All physical quantities consist of a number plus a unit.
         a. SI units, the standard units for scientists, are the kilogram, the meter, and the kelvin.
         b. Metric units are the gram, the meter, the liter, and the degree Celsius.
         c. Other units, such as those for speed and concentration, can be derived from SI and metric units.

2. Prefixes are used with units to indicate multiples of 10.
  B. Scientific notation.
    1. In scientific notation, a number is written as the product of a number between 1 and 10 times 10 raised to a power.
      a. For numbers greater than 10, the power of 10 is positive.
      b. For numbers less than 1, the power of 10 is negative.
  C. Measuring mass, length, and volume (Section 1.7).
    1. Measuring mass.
      a. Mass is the amount of matter in a substance.
      b. Mass differs from weight (a measure of the gravitational pull exerted on an object).
      c. The SI unit of mass is the kilogram, but in chemistry the gram and milligram are more often used.
    2. Measuring length.
      a. The meter is the standard unit for length.
    3. Measuring volume.
      a. Volume is the amount of space that a substance occupies.
      b. Units for volume are the liter (L) and the cubic meter (m^3)—the SI unit.
IV. Numbers in measurement (Sections 1.8–1.10).
  A. Significant figures (Section 1.8).
    1. All measurements have a degree of uncertainty.
    2. For any measurement, the number of digits known with certainty, plus one digit considered uncertain, is known as the number of significant figures.
    3. Rules for significant figures:
      a. Zeroes in the middle of a number are always significant.
      b. Zeroes at the beginning of a number are never significant.
      c. Zeroes at the end of a number but after a decimal point are significant.
      d. Zeroes at the end of a number but before an implied decimal point may or may not be significant.
    4. Some numbers are exact and have an unlimited number of significant figures.
    5. Scientific notation is helpful in indicating the number of significant figures in a number.
  C. Rounding off numbers (Section 1.9).
    1. Numbers must be rounded off if they contain more digits than are significant.
    2. Rounding in calculations.
      a. In multiplication or division, the result cannot have more significant figures than any of the original numbers.
      b. In addition or subtraction, the result can't have more digits to the right of the decimal point than any of the original numbers.
    3. Rules of rounding:
      a. If the digit to be removed is 4 or less, drop it and remove all following digits.
      b. If the digit to be removed is 5 or greater, add 1 to the digit to the left of the digit you drop.
      c. Round off after all steps in a calculation have been carried out.
  D. Problem solving (Section 1.10).
    1. Converting a quantity from one unit to another. In the *factor-label* method:
      a. (quantity in old units) x (conversion factor) = (quantity in new units).
        i.   The conversion factor is a fraction that converts one unit to another.
        ii. All conversion factors are equal to 1.
      b. Units are treated as numbers.
      c. All unwanted units cancel.

2. Problem-solving techniques.
    a. Identify the information known.
    b. Identify the information needed in the answer.
    c. Use conversion factors to convert the given information to the answer.
    d. Make a ballpark estimate of the answer and compare it to the calculated answer.
V. Temperature, heat, and energy (Section 1.11).
    A. Measuring temperature.
        1. Units of temperature are the degree Fahrenheit, the degree Celsius, and the kelvin.
        2. Temperature in °C = Temperature in K − 273.15°.
        3. $°F = \left(\dfrac{9\,°F}{5\,°C} \times °C\right) + 32\ °F.$

        4. $°C = \dfrac{5\,°C}{9\,°F} \times (°F - 32\ °F).$
    B. Energy and heat.
        1. All chemical reactions are accompanied by a change in energy.
            a. Potential energy is stored energy.
            b. Kinetic energy is the energy of motion.
        2. Heat is the energy transferred from a hotter object to a cooler object when the two are in contact.
        3. Units of energy are the joule (SI) and the calorie.
        4. $\text{Specific heat} = \dfrac{\text{calories}}{\text{grams} \times °C}$
VI. Density and specific gravity (Section 1.12).
    A. Density.
        1. Density = Mass (g)/Volume (mL or cm^3).
        2. Density is temperature dependent.
    B. Specific gravity.
        1. $\text{Specific gravity} = \dfrac{\text{density of substance (g/mL)}}{\text{density of water (1 g/mL)}}.$
        2. Specific gravity has no units.

Solutions to Chapter 1 Problems

HOC 1.1 These are examples of chemical changes, since new substances are formed.

1.1   At 10 °C, acetic acid is a solid.

1.2   *Mixtures:* (a) concrete (heterogeneous);   (d) wood (heterogeneous)
         Pure substances: (b) helium (element);   (c) a lead weight (element)

1.3   *Physical changes:* (a) dissolving sugar in water
         Chemical change: (b) production of carbon dioxide by heating limestone; (c) frying an egg; (d) conversion of salicylic acid to acetylsalicylic acid

1.4   The process is a chemical change because the product is different in composition from the reactant. On the left is a pure substance, and on the right is a mixture of pure substances.

CIA 1.1   ASA is a solid at room temperature.

CIA 1.2   The change from SA to ASA is a chemical change because the chemical makeup of the product is different from the chemical makeup of the starting material.

1.5   (a) 2 (Na); (b) 1 (W); (c) 6 (Sr); (d) 5 (Ti); (e) 4 (F); (f) 3 (Sn).

1.6   (a) Ammonia (NH_3) contains one nitrogen atom and three hydrogen atoms.
(b) Sodium bicarbonate ($NaHCO_3$) contains one sodium atom, one hydrogen atom, one carbon atom, and three oxygen atoms.
(c) Octane (C_8H_{18}) contains eight carbon atoms and eighteen hydrogen atoms.
(d) Vitamin C ($C_6H_8O_6$) contains six carbon atoms, eight hydrogen atoms, and six oxygen atoms.

CIA 1.3 Only soluble forms of mercury are toxic. Hg_2Cl_2 passes through the body before it can be chemically converted to a soluble form, but methyl mercury chloride is soluble.

1.7   (a) 1 cm = 1 centimeter = 0.01 m   (b) 1 dg = 1 decigram = 0.1 g
(c) 1 km = 1 kilometer = 1000 m   (d) 1 μs = microsecond = 0.000 001 s
(e) 1 ng = 1 nanogram = 0.000 000 001 g

HOC 1.2
(a) The pennies sound different.
(b) The pre-1982 pennies weigh more.
(c) The post-1982 pennies are easier to cut. The pre-1982 pennies are made of solid copper.

1.8
Number	*Significant Figures*	*Reason*
(a) 3.45 m	3	
(b) 0.1400 kg	4	Rule 3
(c) 10.003 L	5	Rule 1
(d) 35 cents	Exact	

1.9   32.3 °C. The answer has three significant figures.

1.10 In scientific notation, a number is written as the product of a number between 1 and 10 times 10 raised to a power. In (a), 58 g = 5.8 x 10^1 g.

Value	*Scientific Notation*
(a) 0.058 g	5.8 x 10^{-2} g
(b) 46,792 m	4.6792 x 10^4 m
(c) 0.006 072 cm	6.072 x 10^{-3} cm
(d) 345.3 kg	3.453 x 10^2 kg

1.11
Value in Scientific Notation	*Value in Standard Notation*
(a) 4.885 x 10^4 mg	48,850 mg
(b) 8.3 x 10^{-6} m	0.000 0083 m
(c) 4.00 x 10^{-2} m	0.0400 m

1.12 (a) 6.3000 x 10^5 (b) 1.30 x 10^3 (c) 7.942 x 10^{11}

1.13 (a) 2.30 g   (b) 188.38 mL   (c) 0.009 L   (d) 1.000 kg

1.14 Remember:
(1) The sum or difference of two numbers can't have more digits to the right of the decimal point than either of the two numbers.
(2) The product or quotient of two numbers can't have more significant figures than either of the two numbers.

Calculation	*Rounded to:*
(a) 4.87 mL + 46.0 mL = 50.87 mL	50.9 mL
(b) 3.4 x 0.023 g = 0.0782 g	0.078 g
(c) 19.333 m − 7.4 m = 11.933 m	11.9 m
(d) 55 mg − 4.671 mg + 0.894 mg = 51.223 mg	51 mg
(e) 62,911 ÷ 611 = 102.96399	103

1.15

(a) $16.0 \text{ oz} \times \dfrac{28.35 \text{ g}}{1 \text{ oz}} = 454 \text{ g}$ (b) $2500 \text{ mL} \times \dfrac{1 \text{ L}}{1000 \text{ mL}} = 2.5 \text{ L}$

(c) $99.0 \text{ L} \times \dfrac{1 \text{ qt}}{0.9464 \text{ L}} = 105 \text{ qt}$

1.16

$0.840 \text{ quart} \times \dfrac{1 \text{ L}}{1.057 \text{ quart}} \times \dfrac{1000 \text{ mL}}{1 \text{ L}} = 795 \text{ mL}$

1.17

$20 \text{ mg methimazole} \times \dfrac{1 \text{ mL}}{8 \text{ mg methimazole}} = 2.5 \text{ mL}$

1.18

$2 \times 0.324 \text{ g} \times \dfrac{1000 \text{ mg}}{1 \text{ g}} \times \dfrac{1}{135 \text{ lb}} \times \dfrac{2.205 \text{ lb}}{1 \text{ kg}} = 10.6 \dfrac{\text{mg}}{\text{kg}}$

$2 \times 0.324 \text{ g} \times \dfrac{1000 \text{ mg}}{1 \text{ g}} \times \dfrac{1}{40 \text{ lb}} \times \dfrac{2.205 \text{ lb}}{1 \text{ kg}} = 36 \dfrac{\text{mg}}{\text{kg}}$

CIA 1.4

$(28 \,°\text{F} - 32 \,°\text{F}) \times \dfrac{5 \,°\text{C}}{9 \,°\text{F}} = -2.2 \,°\text{C} = 271 \text{ K}$

CIA 1.5

$°\text{F} = \left(\dfrac{9 \,°\text{F}}{5 \,°\text{C}} \times °\text{C} \right) + 32 \,°\text{F}$

For °C = 37 °C, °F = 99 °F.   For °C = 47 °C, °F = 117 °F

1.19   From Section 1.11, we find the formula:

$$(°F - 32 °F) \times \frac{5 °C}{9 °F} = °C$$

Substituting 136°F into the preceding formula:

$$(136 °F - 32 °F) \times \frac{5 °C}{9 °F} = 57.8 °C$$

To find the temperature in kelvins, add the temperature in degrees Celsius to 273.15 since the size of a degree is the same in both kelvins and degrees Celsius.

$$(273.15 + 57.8) K = 331.0 K$$

1.20 $39 °C = 102 °F$

1.21

$$\text{Heat (cal)} = \text{mass (g)} \times \text{temperature change (°C)} \times \text{specific heat}\left(\frac{cal}{g \cdot °C}\right)$$
$$= 350 \text{ g} \times (25 °C - 3 °C) \times \frac{1.0 \text{ cal}}{g \cdot °C}$$
$$= 7700 \text{ cal} = 7.7 \times 10^3 \text{ cal}$$

Ballpark check: The temperature change is approximately 20 °C. The mass of Coca-Cola times the temperature change equals 7000, a number close to the exact solution.

1.22

$$\text{Specific heat} = \frac{calories}{grams \times °C}$$

$$\text{Specific heat} = \frac{161 \text{ cal}}{75 \text{ g} \times 10.0 °C} = \frac{0.21 \text{ cal}}{g \cdot °C}; \quad \frac{674 \text{ J}}{75 \text{ g} \times 10.0 °C} = \frac{0.90 \text{ J}}{g \cdot °C}$$

1.23

$$\text{Density} = \frac{\text{Mass}}{\text{Volume}} = \frac{17.4 \text{ g}}{27.3 \text{ cm}^3} = \frac{0.637 \text{ g}}{\text{cm}^3}$$
The sample will float since it is less dense than water.

1.24

$$12.37 \text{ g} \times \frac{1 \text{ mL}}{1.474 \text{ g}} = 8.392 \text{ mL}$$

Ballpark check: Since the density of chloroform is approximately 1.5, the volume needed is about two-thirds of the mass needed.

1.25   Battery acid is more dense than water.

CIA 1.6 (a) BMI = 29   (b) BMI = 24   (c) BMI = 24

Individual (a), with a body mass index that borders on obese, is most likely to have increased health risks.

CIA 1.7

$$5.0 \text{ lb} \times \frac{454 \text{ g}}{1 \text{ lb}} \times \frac{1 \text{ mL}}{0.94 \text{ g}} \times \frac{1 \text{ L}}{10^3 \text{ mL}} = 2.4 \text{ L}$$

Understanding Key Concepts

1.26 Blue (from top to bottom): helium (He), neon (Ne), argon (Ar), krypton (Kr), xenon (Xe), radon (Rn).
Red (from top to bottom): copper (Cu), silver (Ag), gold (Au).

1.27

Element	*Sources*	*Uses*
Green: boron (B)	borate minerals	glass, ceramics, cleaning products
Blue: bromine (Br)	brine	flame retardants, pesticides
Red: vanadium (V)	bauxite, minerals	metal alloys, superconducting magnets

1.28 The element is americium (Am).

1.29 The specific gravity is found by reading the liquid level on the scale of the hydrometer.
(a) The specific gravity of this solution is 0.978.
(b) The answer has three significant figures.
(c) The solution is less dense than water.

1.30

The smaller graduated cylinder is more precise for two reasons. (1) The distance between two gradations represents a smaller volume, making it easier to measure volumes more accurately. (2) The percent error is greater when a small volume is measured in a large graduated cylinder.

1.31 The length of the pencil is either 8.0 cm or 3 1/8 in.

1.32 The liquid level is 0.11 mL before dispensing the sample and 0.25 mL after dispensing, indicating a sample size of 0.14 mL.

1.33 If two identical hydrometers are placed in ethanol and in chloroform, the hydrometer in chloroform will float higher. Since a hydrometer bulb sinks until it displaces a volume of liquid equal to its mass, it displaces a smaller volume of chloroform, the denser liquid, and the bulb floats higher.

Additional Section Problems

Chemistry and the Properties of Matter

1.34 A physical change doesn't alter the identity of a substance; it alters only its physical state. A chemical change results in a change in chemical identity of a substance.

1.35 *Physical changes:* (a) boiling water; (d) breaking of glass
Chemical changes: (b) decomposing water by passing a current through it; (c) exploding of potassium when placed in water

1.36 *Physical changes:* (a) making lemonade; (d) whipping cream
Chemical changes: (b) frying eggs; (c) burning a candle; (e) leaves changing color

States and Classification of Matter

1.37 A *gas* is a substance that has no definite shape or volume.
A *liquid* has no definite shape but has a definite volume.
A *solid* has a definite volume and a definite shape.

1.38 Melting, boiling, condensation, and freezing are all changes of state. *Melting* occurs when a solid is heated and becomes a liquid. *Boiling* occurs when a liquid is heated and becomes a gas. *Condensation* occurs when a gas is cooled and becomes a liquid. *Freezing* occurs when a liquid is cooled and becomes a solid. All changes of state are physical changes.

1.39 Sulfur dioxide is a gas at 298 K (25 °C).

1.40 No. At 25 °F (–4 °C), butane is a liquid.

1.41 *Mixtures:* (a) pea soup; (b) seawater; (d) urine; (f) multivitamin tablet
Pure substances: (c) propane (C_3H_8); (e) lead

1.42 (a) gasoline—(i) mixture; (iii) liquid
(b) iodine—(ii) solid; (v) chemical element
(c) water—(iii) liquid; (vi) chemical compound
(d) air—(i) mixture; (iv) gas
(e) blood—(i) mixture; (iii) liquid
(f) sodium bicarbonate—(ii) solid; (vi) chemical compound
(g) gaseous ammonia—(iv) gas; (vi) chemical compound
(h) silicon—(ii) solid); (v) chemical element

1.43   *Reactant*   *Products*
Hydrogen peroxide $\longrightarrow$ Water   +   Oxygen
chemical compound   *chemical compound*   *element*

1.44   *Reactants*   *Products*

Sodium   +   Water $\longrightarrow$ Hydrogen + Sodium hydroxide
element   chemical compound   *element   chemical compound*

Elements and Their Symbols

1.45 Oxygen (O) is the most abundant element in both the Earth's crust and the human body.

1.46 (a) Iodine (I); used as a disinfectant and in analytical chemistry.
(b) Chromium (Cr); used in metal alloys, dyes, and as a wood preservative.
(c) Technetium (Tc); radioactive: used in nuclear medicine.
(d) Arsenic (As); used in insecticides and in alloys.
(e) Barium (Ba); used in alloys, X-ray imaging, and fireworks.

1.47 (a) Nitrogen   (b) Potassium   (c) Chlorine   (d) Ca   (e) P   (f) Mn

1.48 (a) The symbol for bromine is Br (the second letter in a symbol is never capitalized).
(b) The symbol for manganese is Mn (Mg is the symbol for magnesium).
(c) The symbol for carbon is C (Ca is the symbol for calcium).
(d) The symbol for potassium is K (Po is the symbol for polonium).

1.49 (a) Carbon dioxide has the formula CO_2 (the number 2 is a subscript).
(b) The O in carbon dioxide must be capitalized to show that it is composed of carbon and oxygen.
(c) Table salt is composed of sodium (Na) and chlorine (Cl).

1.50 Carbon, hydrogen, nitrogen, and oxygen are present in glycine. The formula $C_2H_5NO_2$ represents 10 atoms.

1.51 Carbon (6 atoms), hydrogen (12 atoms), and oxygen (6 atoms) are present in glucose.

1.52 Ibuprofen: $C_{13}H_{18}O_2$

1.53 The atmosphere, a heterogeneous mixture, is composed of the elements oxygen, nitrogen, and argon and the compounds carbon dioxide and water.

Physical Quantities: Definitions and Units

1.54 A *physical quantity* is a physical property that can be measured; it consists of a number plus a unit.

1.55

Quantity	*SI Unit*	*Metric Unit*
Mass	kilogram (kg)	gram (g)
Volume	cubic meter (m^3)	liter (L)
Length	meter (m)	meter (m)
Temperature	kelvin (K)	degree celsius (°C)

1.56 (a) cubic centimeter (b) decimeter (c) millimeter (d) nanoliter
(e) milligram (f) cubic meter

1.57 (a) ng (b) cm (c) μL (d) μm (e) mg

1.58

$$1 \text{ mg } \times \frac{10^{-3} \text{ g}}{1 \text{ mg}} \times \frac{10^{12} \text{ pg}}{1 \text{ g}} = 10^9 \text{ pg}$$

$$35 \text{ ng } \times \frac{10^{-9} \text{ g}}{1 \text{ ng}} \times \frac{10^{12} \text{ pg}}{1 \text{ g}} = 3.5 \times 10^4 \text{ pg}$$

1.59

$$1 \text{ L } \times \frac{10^6 \text{ } \mu\text{L}}{1 \text{ L}} = 10^6 \text{ } \mu\text{L} \text{ ; } \quad 20 \text{ mL } \times \frac{10^3 \text{ } \mu\text{L}}{1 \text{ mL}} = 2 \times 10^4 \text{ } \mu\text{L}$$

Scientific Notation and Significant Figures

1.60 (a) 9.457×10^3 (b) 7×10^{-5} (c) 2.000×10^{10}
(d) 1.2345×10^{-2} (e) 6.5238×10^2

1.61 (a) 5280 (b) 0.082 05
(c) 0.000 018 4 (d) 63,700

1.62 (a) six (b) three (c) three (d) four
(e) 1–5 (f) 2–3

1.63 (a) five (b) three (c) four (d) one (e) three

1.64 (a) 7926 mi; 7900 mi; 7926.38 mi (b) $7.926\ 381 \times 10^3$ mi

1.65 (a) 1.4×10^4 km (b) 2.6×10^1 °C (c) 1.1×10^3 °C
(d) 1×10^{-5} mm (e) 1.1×10^{-1} nm

1.66 (a) 12.1 g (b) 96.19 cm (c) 263 mL (d) 20.9 mg

1.67 (a) 3.3×10^4 ft (b) 15 m^2 (c) 0.30 cm^3 (d) 81 cm^3

Unit Conversions and Problem Solving

1.68 (a) 0.3614 cg (b) 0.0120 ML (c) 0.0144 mm (d) 60.3 ng (e) 1.745 dL
(f) 1.5×10^3 cm

1.69

(a) $56.4 \text{ mi } \times \dfrac{1.609 \text{ km}}{1 \text{ mi}} = 90.7 \text{ km}$

$56.4 \text{ mi } \times \dfrac{1.609 \text{ km}}{1 \text{ mi}} \times \dfrac{1 \text{ Mm}}{10^3 \text{ km}} = 9.07 \times 10^{-2} \text{ Mm}$

(b) $2.0 \text{ L} \times \dfrac{1.057 \text{ qt}}{1 \text{ L}} = 2.1 \text{ qt}$

$2.0 \text{ L} \times \dfrac{10^3 \text{ mL}}{1 \text{ L}} \times \dfrac{1 \text{ fl oz}}{29.57 \text{ mL}} = 68 \text{ fl oz}$

(c) $7.0 \text{ ft} \times \dfrac{12 \text{ in.}}{1 \text{ ft}} = 84.0 \text{ in.}; \; 84.0 \text{ in.} + 2.0 \text{ in} = 86.0 \text{ in.}$

$86.0 \text{ in.} \times \dfrac{1 \text{ cm}}{0.3937 \text{ in.}} = 218 \text{ cm}; \quad 218 \text{ cm} \times \dfrac{1 \text{ m}}{100 \text{ cm}} = 2.18 \text{ m}$

(d) $1.35 \text{ lb} \times \dfrac{1 \text{ kg}}{2.205 \text{ lb}} = 0.612 \text{ kg}; \quad 0.612 \text{ kg} \times \dfrac{10^4 \text{ dg}}{1 \text{ kg}} = 6.12 \times 10^3 \text{ dg}$

1.70   (a) 97.8 kg   (b) 0.133 mL   (c) 0.46 ng   (d) 2.99 Mm

1.71   (a) $1.25 \times 10^5 \text{ m}$ (b) $6.285 \times 10^{-3} \text{ kg}$ (c) $4.735 \times 10^2 \text{ mL}$ (d) $6.74 \times 10^{-4} \text{ km}$

1.72

(a) $\dfrac{100 \text{ km}}{1 \text{ hr}} \times \dfrac{0.6214 \text{ mi}}{1 \text{ km}} = \dfrac{62.1 \text{ mi}}{1 \text{ hr}}$

(b) $\dfrac{62.1 \text{ mi}}{1 \text{ hr}} \times \dfrac{5280 \text{ ft}}{1 \text{ mi}} \times \dfrac{1 \text{ hr}}{60 \text{ min}} \times \dfrac{1 \text{ min}}{60 \text{ s}} = \dfrac{91.1 \text{ ft}}{1 \text{ s}}$

1.73

(a) $\dfrac{1200 \text{ ft}}{1 \text{ s}} \times \dfrac{1 \text{ mi}}{5280 \text{ ft}} \times \dfrac{60 \text{ s}}{1 \text{ min}} \times \dfrac{60 \text{ min}}{1 \text{ hr}} = \dfrac{820 \text{ mi}}{1 \text{ hr}}$

(b) $\dfrac{1200 \text{ ft}}{1 \text{ s}} \times \dfrac{0.3048 \text{ m}}{1 \text{ft}} = \dfrac{370 \text{ m}}{1 \text{ s}}$

1.74

(a) $6 \times 10^{-6} \text{ m} \times \dfrac{10^2 \text{ cm}}{1 \text{ m}} = 6 \times 10^{-4} \text{ cm}$

(b) $1 \text{ cm} \times \dfrac{1 \text{ cell}}{6 \times 10^{-4} \text{ cm}} = 2 \times 10^3 \text{ cells}$

$1 \text{ in.} \times \dfrac{1 \text{ cell}}{6 \times 10^{-6} \text{ m}} \times \dfrac{1 \text{ m}}{39.37 \text{ in}} = 4 \times 10^3 \text{ cells}$

1.75 $1 \text{ ft}^2 = (0.3048 \text{ m})^2 = 0.09290 \text{ m}^2$

$418{,}000 \text{ m}^2 \times \dfrac{1 \text{ ft}^2}{0.0929 \text{ m}^2} = 4.50 \times 10^6 \text{ ft}^2$

1.76

$\dfrac{200 \text{ mg}}{1 \text{ dL}} \times \dfrac{10 \text{ dL}}{1 \text{ L}} \times 5 \text{ L} = 10^4 \text{ mg}; \; 10^4 \text{ mg} \times \dfrac{1 \text{ g}}{10^3 \text{ mg}} = 10 \text{ g cholesterol}$

1.77

$$1200 \text{ mg} \times \frac{1.0 \text{ cup}}{290 \text{ mg}} = 4.1 \text{ cups}$$

1.78

$$\frac{1.2 \times 10^4 \text{ cells}}{1 \text{ mm}^3} \times \frac{10^6 \text{ mm}^3}{1 \text{ L}} \times 5 \text{ L} = 6 \times 10^{10} \text{ cells}$$

Energy, Heat, and Temperature

1.79 273.15 K – 195.8 = 77.4 K

$$°F = \left(\frac{9 \, °F}{5 \, °C} \times °C\right) + 32 \, °F = \left(\frac{9 \, °F}{5 \, °C} \times (-195.8 \, °C)\right) + 32 \, °F = -352.4 \, °F + 32 \, °F = -320.4 \, °F$$

1.80

$$\text{Heat (cal)} = \text{mass (g)} \times \text{temperature change (°C)} \times \text{specific heat}\left(\frac{\text{cal}}{\text{g} \cdot °C}\right)$$

$$= 30.0 \text{ g} \times (30.0 \, °C - 10.0 \, °C) \times \frac{0.895 \text{ cal}}{\text{g} \cdot °C} = 537 \text{ cal} = 0.537 \text{ kcal}$$

$$537 \text{ cal} \times \frac{4.18 \text{ J}}{1 \text{ cal}} = 2240 \text{ J} = 2.24 \text{ kJ}$$

1.81

$$\text{Temperature change (°C)} = \frac{\text{heat (cal)}}{\text{mass (g)} \times \text{specific heat}\left(\frac{\text{cal}}{\text{g} \cdot °C}\right)}$$

$$\text{Temperature change (°C)} = \frac{25.7 \text{ cal}}{18.4 \text{ g} \times 0.215 \left(\frac{\text{cal}}{\text{g} \cdot °C}\right)} = 6.50 \, °C$$

Final temperature = 20.0 °C + 6.5 °C = 26.5 °C

1.82

$$\text{Specific heat} = \frac{\text{cal}}{\text{g} \cdot °C} = \frac{23 \text{ cal}}{5.0 \text{ g} \times 50 \, °C} = 0.092 \, \frac{\text{cal}}{\text{g} \cdot °C} = 0.38 \, \frac{\text{J}}{\text{g} \cdot °C}$$

1.83

$$\text{Heat (cal)} = \text{mass (g)} \times \text{temperature change (°C)} \times \text{specific heat}\left(\frac{\text{cal}}{\text{g} \cdot °C}\right)$$

$$\text{Mass of fat} = \frac{0.94 \text{ g}}{1 \text{ cm}^3} \times 10 \text{ cm}^3 = 9.4 \text{ g}$$

$$\text{Heat (cal)} = 9.4 \text{ g} \times 10 \, °C \times 0.45 \left(\frac{\text{cal}}{\text{g} \cdot °C}\right) = 42 \text{ cal (180 J)}$$

1.84

For mercury: Temperature change (°C) = $\dfrac{250 \text{ cal}}{150 \text{ g} \times 0.033 \dfrac{\text{cal}}{\text{g} \cdot \text{°C}}}$ = 51 °C

Final temp = 25 °C + 51 °C = 76 °C

For iron: Temperature change (°C) = $\dfrac{250 \text{ cal}}{150 \text{ g} \times 0.106 \dfrac{\text{cal}}{\text{g} \cdot \text{°C}}}$ = 15.7 °C

Final temp = 25.0 °C + 15.7 °C = 40.7 °C

1.85

Specific heat = $\dfrac{\text{cal}}{\text{g} \cdot \text{°C}}$ = $\dfrac{100 \text{ cal}}{125 \text{ g} \times 28 \text{ °C}}$ = 0.029 $\dfrac{\text{cal}}{\text{g} \cdot \text{°C}}$ (0.12 $\dfrac{\text{J}}{\text{g} \cdot \text{°C}}$)

The calculated value for specific heat, 0.029 cal/g · °C, is closest in value to the specific heat of gold, 0.031 cal/g · °C.

Density and Specific Gravity

1.86

250 mg $\times \dfrac{1 \text{ g}}{10^3 \text{ mg}} \times \dfrac{1 \text{ cm}^3}{1.40 \text{ g}}$ = 0.179 cm³

1.87 From the expression for density, we know that 1 L of hydrogen has a mass of 0.0899 g. We need to find the number of liters that have a mass of 1.0078 g.

1.0078 g $\times \dfrac{1 \text{ L}}{0.0899 \text{ g}}$ = 11.2 L

1.88 To find the density of lead, divide the mass of lead by the volume of the bar in cm³. 0.500 cm $\times$ 1.55 cm $\times$ 25.00 cm = 19.38 cm³.

$\dfrac{220.9 \text{ g}}{19.38 \text{ cm}^3}$ = 11.4 $\dfrac{\text{g}}{\text{cm}^3}$

1.89 The volume of lithium is 0.82 cm $\times$ 1.45 cm $\times$ 1.25 cm = 1.5 cm³.

$\dfrac{0.794 \text{ g}}{1.5 \text{ cm}^3}$ = 0.53 $\dfrac{\text{g}}{\text{cm}^3}$

1.90 The density of ethanol at 25 °C is equal to its specific gravity at 25 °C times the density of water at 25 °C.

0.787 $\times \dfrac{0.997 \text{ g}}{1 \text{ mL}}$ = 0.785 $\dfrac{\text{g}}{\text{mL}}$; 125 g $\times \dfrac{1 \text{ mL}}{0.785 \text{ g}}$ = 159 mL

1.91   Use the same reasoning shown in the previous problem.

$$\text{density} = 1.1088 \text{ x } \frac{0.997 \text{ g}}{1 \text{ mL}} = \frac{1.11 \text{ g}}{1 \text{ mL}}$$

$$1.00 \text{ L } \text{ x } \frac{1000 \text{ mL}}{1 \text{ L}} \text{ x } \frac{1.11 \text{ g}}{1 \text{ mL}} = 1.11 \text{ x } 10^3 \text{ g } = 1.11 \text{ kg}$$

Conceptual Problems

1.92   Conversions between the Rankine scale and the Fahrenheit scale are easy, since the size of a degree is the same for both scales.

If -459.67 °F = 0 °R; 32 °F = 459.67 °F + 32 °F = 491.67 °R. Water melts at 491.67 °R

Water boils at 459.67 °F + 212 °F = 671.67 °R

1.93   The white solid is a chemical compound, and the brown gas and molten metal are elements.

1.94

$$3.125 \text{ in. } \text{ x } \frac{2.54 \text{ cm}}{1 \text{ in.}} = 7.94 \text{ cm}$$

The difference in the calculated length (7.94 cm) and the measured length (8.0 cm) is due to the lack of precision of the ruler and to rounding.

1.95   One carat = 200 mg = 0.200 g

$$\frac{0.200 \text{ g}}{1 \text{ carat}} \text{ x } 44.4 \text{ carat } = 8.88 \text{ g}$$

1.96

(a) $350 \text{ kcal } \text{ x } \frac{1 \text{ x } 10^3 \text{ cal}}{1 \text{ kcal}} = 3.50 \text{ x } 10^5 \text{ cal}; 3.50 \text{ x } 10^5 \text{ cal } \text{ x } \frac{4.18 \text{ J}}{\text{cal}} = 1.46 \text{ x } 10^6 \text{ J}$

(b) $\text{Temperature change (°C)} = \dfrac{\text{heat (cal)}}{\text{mass (g) x specific heat}\left(\dfrac{\text{cal}}{\text{g} \cdot \text{°C}}\right)}$

mass = 35.5 kg = $3.55 \text{ x } 10^4$ g; heat = $3.50 \text{ x } 10^5$ cal

$\text{Temperature change (°C)} = \dfrac{3.50 \text{ x } 10^5 \text{ cal}}{3.55 \text{ x } 10^4 \text{ g } \text{ x } 1.00\left(\dfrac{\text{cal}}{\text{g} \cdot \text{°C}}\right)} = 9.86 \text{ °C}$

1.97

(a) $\dfrac{9.0 \text{ mg}}{1 \text{ kg}} \text{ x } 130 \text{ lb } \text{ x } \dfrac{1 \text{ kg}}{2.21 \text{ lb}} = 530 \text{ mg}$

(b) $\dfrac{9.0 \text{ mg}}{1 \text{ kg}} \text{ x } 40 \text{ lb } \text{ x } \dfrac{1 \text{ kg}}{2.21 \text{ lb}} = 163 \text{ mg}$

To receive the recommended dose, the child would thus need about 1.3 of the 125 mg tablets.

1.98 3.9×10^{-2} g/dL iron, 8.3×10^{-3} g/dL calcium, 2.24×10^{-1} g/dL cholesterol

1.99

(a) 2.027×10^5 ft^3 x $\dfrac{0.0283 \text{ m}^3}{\text{ft}^3}$ x $\dfrac{1 \times 10^3 \text{ L}}{\text{m}^3}$ = 5.74×10^6 L

(b) 5.74×10^6 L x $\dfrac{0.174 \text{ g}}{1 \text{ L}}$ x $\dfrac{1 \text{ kg}}{1 \times 10^3 \text{ g}}$ = 9.99×10^2 kg

(c) 5.74×10^6 L x $\dfrac{1.20 \text{ g}}{1 \text{ L}}$ x $\dfrac{1 \text{ kg}}{1 \times 10^3 \text{ g}}$ = 6.89×10^3 kg

1.100

$75 \dfrac{\text{mL}}{\text{beat}}$ x $72 \dfrac{\text{beats}}{\text{min}}$ x $60 \dfrac{\text{min}}{\text{hr}}$ x $24 \dfrac{\text{hr}}{\text{day}}$ = $7.8 \times 10^6 \dfrac{\text{mL}}{\text{day}}$

1.101

15 g x $\dfrac{1000.0 \text{ mL}}{50.00 \text{ g}}$ = 300 mL

1.102

0.14 mL x $\dfrac{0.963 \text{ g}}{1 \text{ mL}}$ = 0.13 g

1.103 (a) The effective range of an alcohol thermometer in °C is 115 °C + 78.5 °C = 193.5 °C. Since a Fahrenheit degree is 9/5 of a Celsius degree, the range in °F is 9/5 x 193.5 = 348 °F

(b) The thermometer can contain either 0.79 g alcohol or 13.6 g mercury.

1.104

$\dfrac{85 \text{ mg}}{100 \text{ mL}}$ x $\dfrac{1 \text{ g}}{10^3 \text{mg}}$ x $\dfrac{10^3 \text{ mL}}{1 \text{ L}}$ x $\dfrac{0.9464 \text{ L}}{1 \text{ qt}}$ x $\dfrac{1 \text{ qt}}{2 \text{ pt}}$ x 11 pt = 4.4 g

4.4 g x $\dfrac{1 \text{ lb}}{454 \text{ g}}$ = 0.0097 lb

1.105

$\dfrac{3000 \text{ mL}}{1 \text{ day}}$ x $\dfrac{5 \text{ g}}{100 \text{ mL}}$ x $\dfrac{4 \text{ kcal}}{1 \text{ g}}$ = 600 kcal/day

1.106

$\left(\dfrac{100 \text{ mL}}{1 \text{ kg}} \text{ x } 10 \text{ kg}\right)$ + $\left(\dfrac{50 \text{ mL}}{1 \text{ kg}} \text{ x } 10 \text{ kg}\right)$ + $\left(\dfrac{20 \text{ mL}}{1 \text{ kg}} \text{ x } 35 \text{ kg}\right)$ = 2200 mL

1.107

7.5 grains x $\dfrac{1 \text{ fluidram}}{10 \text{ grains}}$ x $\dfrac{3.72 \text{ mL}}{1 \text{ fluidram}}$ = 2.8 mL

1.108

$$\text{Heat (cal)} = \text{mass (g)} \times \text{temperature change (°C)} \times \text{specific heat} \left(\frac{\text{cal}}{\text{g} \cdot °\text{C}}\right)$$

For water: specific heat = 1.00 cal /(g·°C);

$$\text{mass} = 3.00 \text{ L} \times \frac{10^3 \text{ mL}}{1 \text{ L}} \times \frac{1.00 \text{ g}}{1 \text{ mL}} = 3.00 \times 10^3 \text{ g};$$

$$\text{temperature change} = 90.0 \text{ °C} - 18.0 \text{ °C} = 72.0 \text{ °C}$$

$$\text{Calories needed} = 3.00 \times 10^3 \text{ g} \times 72.0 \text{ °C} \times \frac{1.00 \text{ cal}}{\text{g} \cdot °\text{C}} = 2.16 \times 10^5 \text{ cal} = 216 \text{ kcal}$$

$$216 \text{ kcal} \times \frac{1.0 \text{ tbsp}}{100 \text{ kcal}} = 2.2 \text{ tbsp butter}$$

1.109

$$\text{Specific heat} = \frac{\text{cal}}{\text{g} \cdot °\text{C}} = \frac{1350 \text{ cal}}{1620 \text{ g} \times 7.8 \text{ °C}} = 0.107 \frac{\text{cal}}{\text{g} \cdot °\text{C}} = 0.448 \frac{\text{J}}{\text{g} \cdot °\text{C}}$$

The calculated specific heat, 0.107 cal/g · °C, is closer to the value for iron than for gold.

1.110

$$\frac{1620 \text{ g}}{205 \text{ mL}} = 7.90 \frac{\text{g}}{\text{mL}}$$

The value calculated for density very nearly agrees with the density of iron (7.86 g/mL).

1.111 Use the dimensions of the cork and of the lead to find their respective volumes. Multiply the volume of each substance by its density to arrive at the mass of each.

Volume of cork: 1.30 cm x 5.50 cm x 3.00 cm = 21.5 cm³
Volume of lead: (1.15 cm)³ = 1.52 cm³

Mass of cork: Mass of lead:

$$21.5 \text{ cm}^3 \times \frac{0.235 \text{ g}}{1 \text{ cm}^3} = 5.05 \text{ g cork} \qquad 1.52 \text{ cm}^3 \times \frac{11.35 \text{ g}}{1 \text{ cm}^3} = 17.3 \text{ g lead}$$

Add the two masses, and divide by the sum of the two volumes to find the density of the combination.

$$\text{Density} = \frac{5.05 \text{ g} + 17.3 \text{ g}}{21.5 \text{ cm}^3 + 1.52 \text{ cm}^3} = \frac{22.40 \text{ g}}{23.0 \text{ cm}^3} = 0.974 \text{ g/cm}^3$$

The combination will float because its density is less than the density of water.

1.112 At the crossover point, °F = °C.

$$°\text{F} = \left(\frac{1.8 \text{ °F}}{°\text{C}} \times °\text{C}\right) + 32 \text{ °F} \quad \text{If °C = °F, °F} = 1.8 \text{ °F} + 32 °; -0.8 \text{ °F} = 32 °$$

The crossover temperature is °F = °C = –40°.

Group Problems

1.113 Using conversion rates from November 2015:

Currency per Dollar		per $500
Euro:	0.9186	459.3 euros
British pound:	0.6575	328.75 British pounds
Indian rupee:	65.881	32,9940 Indian rupees
Canadian $	1.3170	658.5 Canadian $

1.114 The formula of chloral hydrate, $C_2H_3Cl_3O_2$, indicates four elements and ten total atoms – two carbons, three hydrogens, three chlorines, and two oxygens.

1.115

Beverage	Alcohol Content	Specific Gravity
80 proof whiskey	40%	0.94
red wine	13%	0.99
domestic beer	5.3%	1.01

Specific gravity may be greater than 1 because of other components of beer.

1.116 A metric tonne = 1000 kg = 1×10^3 kg

230 million metric tonnes = 230×10^6 metric tonnes = 2.3×10^8 metric tonnes

$$2.30 \times 10^8 \text{ metric tonnes} \times \frac{1 \times 10^3 \text{kg}}{1 \text{ metric tonne}} = 2.30 \times 10^{11} \text{ kg}$$

$$2.30 \times 10^{11} \text{ kg} \times \frac{1 \text{ mL}}{1.83 \text{ g}} \times \frac{1 \text{ L}}{1000 \text{ mL}} \times \frac{10^3 \text{ g}}{1 \text{ kg}} = 1.26 \times 10^{11} \text{ L}$$

Sulfuric acid is used for the production of other chemicals, in fertilizers, in lead-acid batteries, and as a detergent.

Self-Test for Chapter 1

1. Write the full name of these units:
   (a) μm   (b) dL   (c) Mg   (d) L   (e) ng

2. Write the abbreviation for each of the following units:
   (a) kiloliter   (b) picogram   (c) centimeter   (d) hectoliter

3.

Quantity	Significant Figures?
(a) 1.0037 g	
(b) 0.0080 L	
(c) 0.008 L	
(d) 2 aspirin	
(e) 273,000 mi	

4. Express the following in scientific notation:
    (a) 0.000 070 3 g   (b) 137,100 m   (c) 0.011 L   (d) 18,371,008 mm
    How many significant figures do each of these quantities have?

5. Round the following to three significant figures:
    (a) 807.3 L   (b) 4,773,112 people   (c) 0.00127 g   (d) 10370 μm
    Express each of these quantities in scientific notation.

6. Convert the following quantities:
    (a) 256 g = _____ lb   (b) 417 mm = _____ m
    (c) 2.0 gallons = _____ L   (d) 2.17 m = _____ inches
    (e) 35 °C = _____ °F   (f) 298 K = _____ °C
    (g) 175 mL = _____ fl oz   (h) 175 mg = _____ oz

7. If the specific heat of gold is 0.031 cal/g °C, how many calories does it take to heat 10 g of gold from 0 °C to 100 °C?

8. If the density of ethanol is 0.7893 g/mL, how many grams does 275 mL of ethanol weigh?

Multiple Choice

1. Which of the following is not an SI unit?
    (a) kg (b) L (c) m (d) K

2. How many significant figures does the number 4500 have?
    (a) 2 (b) 3 (c) 4 (d) any of the above

3. Which of the following quantities is larger than a gram?
    (a) 1 nanogram (b) 1 dekagram (c) 1 centigram (d) 1 microgram

4. Which of the following conversion factors do you need for converting 3.2 lb/qt to kg/L?
    (a) $\dfrac{1 \text{ kg}}{2.2 \text{ lb}}$ x $\dfrac{1 \text{ qt}}{0.95 \text{ L}}$ (b) $\dfrac{2.2 \text{ lb}}{1 \text{ kg}}$ x $\dfrac{1 \text{ qt}}{0.95 \text{ L}}$ (c) $\dfrac{2.2 \text{ lb}}{1 \text{ kg}}$ x $\dfrac{0.95 \text{ L}}{1 \text{ qt}}$ (d) $\dfrac{1 \text{ kg}}{2.2 \text{ lb}}$ x $\dfrac{0.95 \text{ L}}{1 \text{ qt}}$

5. When written in scientific notation, the exponent in the number 0.000 007 316 is:
    (a) 10^{-6} (b) 6 (c) –6 (d) 10^{6}

6. The change of state that occurs when a gas is cooled to a liquid is called:
    (a) boiling (b) melting (c) evaporation (d) condensation

7. How many zeros are significant in the number 0.007 006?
    (a) 1 (b) 2 (c) 3 (d) 4

8. All of the following have uniform composition except:
    (a) element (b) solution (c) pure substance (d) chemical compound

9. To measure the amount of heat needed to raise the temperature of a given substance, you need to know all of the following except:
    (a) the mass of the substance (b) the specific heat of the substance
    (c) the density of the substance (d) the initial and final temperature

10. For which of the following is specific gravity a useful measure?
(a) to describe the amount of solids in urine   (b) to indicate the amount of heat necessary to raise the temperature of 1 g of a substance by 1 °C   (c) to determine if an object will float on water

Sentence Completion

1. The fundamental SI units are _____, _____, _____, and _____.

2. Physical quantities are described by a _____ and a _____.

3. The amount of heat necessary to raise the temperature of 1 g of a substance by 1 °C is the substance's _____ _____.

4. _____ is the symbol for the element bismuth.

5. The number 0.003 06 has _____ significant figures.

6. To convert from grams to pounds, use the conversion factor _____.

7. A _____ substance can be beaten or rolled into different shapes.

8. The method used for converting units is called the _____ _____ method.

9. _____ _____ is the density of a substance divided by the density of water at the same temperature.

10. The size of a degree is the same in both _____ and _____ units.

True or False

1. The units of specific gravity are g/mL.

2. The number 0.07350 has four significant figures.

3. The sum of 57.35 and 1.3 has four significant figures.

4. The conversion factor 1.057 quarts/liter is used to convert quarts into liters.

5. Rust formation is a chemical change.

6. The temperature in °C is always a larger number than the temperature in K.

7. Raising the temperature of 10 g of water by 10 °C takes less heat than raising the temperature of 10 g of gold by 10 °C.

8. Mass measures the amount of matter in an object.

9. Ice is more dense than water.

10. A nanogram is larger than a picogram.

Match each entry on the left with its partner on the right.

1.   50037

2.   1 centimeter

3. $\dfrac{2.205 \text{ lb}}{1 \text{ kg}}$

4.   263 K

5.   1 dekagram

6.   0.048

7.   1 m^3

8.   1 liter

9.   1 decimeter

10. $\dfrac{1 \text{ kg}}{2.205 \text{ lb}}$

11.   0 °F

12.   1 decigram

(a) Converts pounds to kilograms

(b) 0.1 grams

(c) Larger than one inch

(d) −17.8 °C

(e) Five significant figures

(f) SI unit of volume measure

(g) Converts kilograms to pounds

(h) 10 grams

(i) −10 °C

(j) Smaller than one inch

(k) Metric unit of volume measure

(l) Two significant figures

Chapter 2   Atoms and the Periodic Table

Chapter Outline

I. Atomic Theory (Sections 2.1–2.3).
   A. Fundamental assumptions about atoms (Section 2.1).
      1. All matter is composed of atoms.
      2. The atoms of each element are different from the atoms of all other elements.
      3. Chemical compounds consist of elements combined in definite proportions.
      4. Chemical reactions only change the way that atoms are combined in compounds; the atoms themselves are unchanged.
   B. Nature of the atom (Sections 2.1).
      1. Atoms are very small (~10^{-11} m in diameter).
      2. Atoms consist of subatomic particles.
         a. A proton is positively charged and has a mass of 1.6726×10^{-24} g.
         b. A neutron has no charge and has a mass of 1.6749×10^{-24} g.
         c. An electron is negatively charged and has a mass of 9.1093×10^{-28} g.
      3. The masses of atoms are expressed in relative terms.
         a. Under this system, a carbon atom with six protons and six neutrons is given a mass of exactly 12 atomic mass units (amu).
         b. Consequently, a proton and a neutron each have a mass of approximately 1 amu.
      4. Atoms are held together by the interplay of attractive and repulsive forces of positively charged protons and negatively charged electrons.
      5. Protons and neutrons are located in the nucleus of an atom, and electrons move about the nucleus.
      6. The size of the nucleus is small compared to the size of the atom.
   C. Composition of atoms (Section 2.2).
      1. Atoms of different elements differ from each other according to how many protons they contain.
      2. Z stands for the number of protons an atom has and is known as the atomic number.
      3. The number of electrons in an atom is the same as the number of protons.
      4. The mass number A stands for the number of protons plus the number of neutrons.
      5. Atoms are represented by showing the mass number as a superscript on the left side of the symbol for the chemical element; the atomic number is shown as a subscript on the left side.
   D. Isotopes and atomic weight (Section 2.3).
      1. Isotopes are atoms of the same element that differ only in the number of neutrons.
      2. A specific isotope is represented in the same way as an atom is represented.
      3. Most elements occur in nature as a mixture of isotopes.
      4. The atomic weight of an element can be calculated if the percent of contributing isotopes is known.
II. The Periodic Table (Sections 2.4–2.5).
   A. The periodic table is a classification of elements according to their properties (Section 2.4).
      1. Elements can be classified as metals, nonmetals, or metalloids.
         a. Metals are solids that are lustrous, brittle, malleable, and good conductors of heat and electricity.
         b. Nonmetals may be solid, liquid, or gas and are poor conductors.
         c. Metalloids have characteristics intermediate between metals and nonmetals.
      2. Elements are arranged by increasing atomic number in seven rows called periods.

3. Elements are also arranged in 18 vertical columns called groups.
    a. Main group elements occur on the left and right of the periodic table.
    b. Transition metal groups occur in the middle of the periodic table.
    c. Inner transition metal groups are shown separately at the bottom.
    d. The elements in each group have similar chemical properties.
    e. Groups can be numbered in two different ways.
B. Chemical characteristics of groups of elements (Section 2.5).
    1. Group 1A—Alkali metals (Li, Na, K, Rb, Cs, Fr).
        a. Shiny, soft, low-melting.
        b. React violently with water.
    2. Group 2A—Alkaline earth metals (Be, Mg, Ca, Sr, Ba, Ra).
        a. Lustrous, shiny metals.
        b. Less reactive than metals in group 1A.
    3. Group 7A—Halogens (F, Cl, Br, I, At).
        a. Corrosive, nonmetals (except At).
        b. Found in nature only in combination with other elements.
    4. Group 8A—Noble gases (He, Ne, Ar, Kr, Xe, Rn). Extremely unreactive.
C. Neighboring groups have similar behaviors.
    1. Metals.
        a. Found on the left side of the periodic table.
        b. Silvery, ductile, good conductors.
    2. Nonmetals.
        a. Found on the right side of the periodic table.
        b. Eleven of the 17 nonmetals are gases.
        c. Solid nonmetals are brittle and are poor conductors of electricity.
    3. Metalloids.
        a. Occur on the boundary between metals and nonmetals.
        b. Have intermediate chemical behavior.
III. Electrons (Sections 2.6–2.9).
    A. The properties of the elements are due to the distribution of electrons in their atoms (Section 2.6).
    B. Location of electrons.
        1. Electrons are located in specific regions about the nucleus.
        2. The energies of electrons are quantized.
        3. The locations of electrons are described by shells, subshells, and orbitals.
            a. Shells.
                i. Shells describe the energy level of an electron.
                ii. Shells are related to an electron's distance from the nucleus.
            b. Subshells.
                i. Subshells describe the energy levels of electrons within each shell.
                ii. The four types of subshell are *s, p, d,* and *f.*
            c. Orbitals.
                i. Orbitals are the regions of subshells in which electrons of a specific energy can be found.
                ii. An *s* subshell contains one orbital, a *p* subshell contains three orbitals, a *d* subshell contains five orbitals, and an *f* subshell contains seven orbitals.
                iii. Each orbital can hold two electrons, and they must be of opposite spin.
                iv. An *s* orbital is spherical, and a *p* orbital is dumbbell-shaped.
        4. Since the exact position of electrons can't be specified, orbitals are often referred to as electron clouds.

C. Electron configurations (Section 2.7).
   1. The specific arrangement of electrons in an atom's shells and subshells is known as its electron configuration.
   2. This arrangement can be predicted by using three rules.
      a. Electrons occupy the lowest orbitals available within each subshell.
         i. Within each shell, the subshell energy levels increase in the order *s, p, d, f.*
      b. Each orbital can hold only two electrons, and they must be of opposite spin.
      c. If two or more orbitals have the same energy, each orbital is half-filled before any orbital is completely filled.
   3. The number of electrons in each subshell is given by a superscript.
D. Electron configuration and the periodic table (Sections 2.8–2.9).
   1. Properties of elements are determined by their location in the periodic table (Section 2.8).
   2. The periodic table is divided into four regions.
      a. Elements in groups 1A and 2A are *s*-block elements because they result from filling *s* orbitals.
      b. Groups 3A–8A are *p*-block elements.
      c. Transition metals are *d*-block elements.
      d. Inner transition metals are *f*-block elements.
   3. The periodic table can be used as a reminder of the order of orbital filling.
   4. Elements within a group of the periodic table have similar electronic configurations of their valence electronic shells and thus similar chemical behavior.
   5. The valence electrons in a compound can be represented by electron dots (Section 2.9).

Solutions to Chapter 2 Problems

CIA 2.1 A normal light microscope can't reach the degree of precision of a scanning tunneling microscope.

CIA 2.2 (a) The character is 12 iron atoms wide.

$$(b) \ 12 \ atoms \ \times \ \frac{126 \ pm}{1 \ atom} \ \times \ \frac{1 \ cm}{1 \times 10^{10} \ pm} \ = \ 1.51 \times 10^{-7} \ cm$$

2.1 (a) Re   (b) Sr   (c) Te

2.2 Contribution from ^{39}K: 93.12% of 38.96 amu = 36.28 amu
Contribution from ^{41}K: 6.88% of 40.96 amu =   2.82 amu
——————————————————————
Atomic weight = 39.10 amu

The calculated value agrees exactly with the reported value.

2.3   Both bromine isotopes have 35 protons. The isotope with mass number 79 has 44 neutrons, and the isotope with mass number 81 has 46 neutrons.

The atomic number is written at the lower left of the element symbol, and the mass number is written at the upper left.

(a) The two isotopes of bromine:

$$^{79}_{35}Br \quad \text{and} \quad ^{81}_{35}Br$$

(b) Contribution from ^{79}Br: 50.69% of 78.92 amu $= 40.00$ amu

     Contribution from ^{81}Br: 49.31% of 80.91 amu $= 39.90$ amu

————————————————————————————————————

                 Average atomic weight $= 79.90$ amu

The values agree.

HOC 2.1

(a) Each pile has the same number of "protons"; each pile has a different number of "neutrons".

(b) The three piles represent the three isotopes of carbon.

$$^{12}_{6}C \quad ^{13}_{6}C \quad ^{14}_{6}C$$

(c) The answer (12.011) is identical.

2.4   The element is chlorine.

$$^{35}_{17}Cl \quad ^{37}_{17}Cl$$

2.5   Aluminum is in group 3A (or 13) and period 3.

2.6   The group 1B element in period 5 is silver (Ag).

The group 2A element in period 4 is calcium (Ca).

2.7

Group 5A Element	Period
Nitrogen (N)	2
Phosphorus (P)	3
Arsenic (As)	4
Antimony (Sb)	5
Bismuth (Bi)	6

2.8

Element	Name	Number in Periodic Table
(a) B	boron	5
(b) Si	silicon	14
(c) Ge	germanium	32
(d) As	arsenic	33
(e) Sb	antimony	51
(f) Te	tellurium	52

The metalloids appear on a diagonal boundary between metals and nonmetals.

2.9   Metals: titanium (Ti), scandium (Sc)

Nonmetals: selenium (Se), argon (Ar), astatine (At)

Metalloids: tellurium (Te)

2.10 (a) Krypton: (ii) nonmetal; (iv) main group element; (v) noble gas
(b) Strontium: (i) metal; (iv) main group element
(c) Nitrogen: (ii) nonmetal; (iv) main group element
(d) Cobalt: (i) metal; (iii) transition element

2.11 Use Figure 2.3 to arrive at an answer. Elements are listed in order of increasing atomic radius.
(a) Li, Na, K, Rb
(b) O, C, F, Li
(c) F, Cl, Br, I

2.12 Since Figure 2.4 is only partially helpful, look each element up online. Elements are listed in order of increasing melting point. Melting points are listed in kelvins.
(a) Rb (312 K), K (336 K), Na (371 K), Li (453 K)
(b) F (53 K), O (54 K), Li (453 K), C (3915 K– sublimes)
(c) F (53 K), Cl (171 K), Br (266 K), I (387 K)

2.13 (a) Sodium; Group 1A; Period 3; metal   (b) Oxygen; Group 6A; Period 2; nonmetal

CIA 2.3 (a) phosphorus (P)   (b) selenium (Se)

CIA 2.4 Fe – Group 8B (8)   Cr – Group 6B (6)   Co – Group 8B (9)
Cu – Group 1B (11)   Mo – Group 6B (6)   Mn – Group 7B (7)
Zn – Group 2B (12)

2.14 Twelve electrons are present in this atom, which is magnesium.

HOC 2.2
(a) N/A
(b) The number of layers depends on the size of the onion. Each layer represents a shell in an atom.
(c) As the layers get farther from the nucleus, the attractive forces decrease.
(d) The outermost layers are much larger than the inner layers and can hold more electrons.

2.15 Sulfur is a main group element (Group 6A) that is a nonmetal. The last electron is found in a $3p$ orbital.

2.16 To find the electron configuration of an atom, first find its atomic number. For C (carbon), the atomic number is 6; thus, carbon has 6 protons and 6 electrons. Then assign electrons to the proper orbitals. For carbon, the electron configuration is $1s^2\,2s^2\,2p^2$.

Element	Atomic Number	Electron Configuration
(a) C (carbon)	6	$1s^2\,2s^2\,2p^2$
(b) P (phosphorus)	15	$1s^2\,2s^2\,2p^6\,3s^2\,3p^3$
(c) Cl (chlorine)	17	$1s^2\,2s^2\,2p^6\,3s^2\,3p^5$
(d) K (potassium)	19	$1s^2\,2s^2\,2p^6\,3s^2\,3p^6\,4s^1$

2.17 Element 33 (As): $1s^2\,2s^2\,2p^6\,3s^2\,3p^6\,4s^2\,3d^{10}\,4p^3$. The 4p subshell is incompletely filled. Notice that only one electron occupies each orbital of the $4p$ subshell.

$\uparrow\ \uparrow\ \uparrow\quad 4p^3$

2.18   The atom has 31 electrons (and 31 protons) and can be identified as gallium.

2.19   (a) F: $1s^2\,2s^2\,2p^5$; [He] $\mathbf{2s^2\,2p^5}$ (b) Al: $1s^2\,2s^2\,2p^6\,3s^2\,3p^1$; [Ne] $\mathbf{3s^2\,3p^1}$

(c) As: $1s^2\,2s^2\,2p^6\,3s^2\,3p^6\,4s^2\,3d^{10}\,4p^3$; [Ar] $\mathbf{4s^2}\,3d^{10}\,\mathbf{4p^3}$
(Valence electrons are bold.)

2.20   In group 2A, all elements have the outer-shell configuration ns^2.

2.21   Chlorine, in group 7A (17), has 17 electrons. Its electron configuration is $1s^2\,2s^2\,2p^6\,3s^2$ $3p^5$. Two electrons are in shell 1, eight electrons are in shell 2, and seven electrons are in shell 3. The outer-shell configuration is $3s^2\,3p^5$.

2.22   The elements form group 6A (16) and have the general valence-shell configuration ns^2np^4.

2.23

$\cdot\overset{\displaystyle\cdot}{\text{X}}\cdot$

2.24

$:\!\overset{\displaystyle\cdot\cdot}{\underset{\displaystyle\cdot\cdot}{\text{Rn}}}\!:$ $\cdot\overset{\displaystyle\cdot}{\text{Pb}}\cdot$ $:\!\overset{\displaystyle\cdot\cdot}{\text{Xe}}\!:$ $\cdot\text{Ra}\cdot$

2.25   The wavelength of red light lies between 700–780 nm, and the wavelength of blue light lies between 400–480 nm. Blue light is associated with higher energy.

CIA 2.5 *Higher Energy*   *Lower Energy*

(a) ultraviolet   infrared
(b) gamma waves   microwaves
(c) X rays   visible light

CIA 2.6   Ultraviolet rays are more damaging to the skin because they are of higher energy than visible light.

Understanding Key Concepts

2.26

2.27   The element marked in red is a gas (fluorine—group 7A).
The element marked in blue has atomic number 79 (gold).
All elements in group 2A have chemical behavior similar to the element marked in green (calcium). These include beryllium, magnesium, strontium, barium, and radium.

2.28

2.29   Selenium (Z = 34) is the element whose orbital-filling diagram is shown. Notice that it is possible to identify the element by counting its electrons, which equal the number of protons and give the value of Z.

2.30   As: $1s^2\,2s^2\,2p^6\,3s^2\,3p^6$ $\underset{4s}{\uparrow\downarrow}$ $\underset{3d}{\uparrow\downarrow\ \uparrow\downarrow\ \uparrow\downarrow\ \uparrow\downarrow\ \uparrow\downarrow}$ $\underset{4p}{\uparrow\ \uparrow\ \uparrow}$

Additional Problems

Atomic Theory and the Composition of Atoms

2.31   (1) All matter is composed of atoms.
     (2) The atoms of a given element differ from the atoms of all other elements.
     (3) Chemical compounds consist of atoms combined in specific proportions.
     (4) Chemical reactions change only the way that atoms are combined in compounds; the atoms themselves are unchanged.

2.32   Atoms of different elements differ in the number of protons and electrons they have.

2.33

(a) Bi: $208.9804 \text{ amu} \times \dfrac{1.660\ 539 \times 10^{-24} \text{ g}}{1 \text{ amu}} = 3.470201 \times 10^{-22} \text{ g}$

(b) Xe: $131.29 \text{ amu} \times \dfrac{1.660\ 539 \times 10^{-24} \text{ g}}{1 \text{ amu}} = 2.1801 \times 10^{-22} \text{ g}$

(c) He: $4.0026 \text{ amu} \times \dfrac{1.660\ 539 \times 10^{-24} \text{ g}}{1 \text{ amu}} = 6.6465 \times 10^{-24} \text{ g}$

2.34

(a) $\dfrac{2.66 \times 10^{-23} \text{ g}}{1 \text{ atom O}} \times \dfrac{1 \text{ amu}}{1.660\ 539 \times 10^{-24} \text{ g}} = 16.0 \text{ amu}$

(b) $\dfrac{1.31 \times 10^{-22} \text{ g}}{1 \text{ atom Br}} \times \dfrac{1 \text{ amu}}{1.660\ 539 \times 10^{-24} \text{ g}} = 78.9 \text{ amu}$

2.35

$6.022 \times 10^{23} \text{ atoms} \times \dfrac{1.660\ 539 \times 10^{-24} \text{ g}}{1 \text{ amu}} \times \dfrac{14.01 \text{ amu}}{1 \text{ atom}} = 14.01 \text{ g}$

2.36

$6.022 \times 10^{23} \text{ atoms} \times \dfrac{1.660\ 539 \times 10^{-24} \text{ g}}{1 \text{ amu}} \times \dfrac{16.00 \text{ amu}}{1 \text{ atom}} = 16.00 \text{ g}$

2.37

$15.99 \text{ g} \times \dfrac{1 \text{ amu}}{1.660\ 539 \times 10^{-24} \text{ g}} \times \dfrac{1 \text{ atom}}{15.99 \text{ amu}} = 6.022 \times 10^{23} \text{ atoms}$

2.38

$12.00 \text{ g} \times \dfrac{1 \text{ amu}}{1.660\ 539 \times 10^{-24} \text{ g}} \times \dfrac{1 \text{ atom}}{12.00 \text{ amu}} = 6.022 \times 10^{23} \text{ atoms}$

2.39

Particle	*Mass in amu*	*Charge*
Proton	1.007 276	+1
Neutron	1.008 665	0
Electron	$5.485\ 799 \times 10^{-4}$	−1

2.40 Protons and neutrons are found in a dense central region called the nucleus. Electrons move about the nucleus in large, specifically defined regions called orbitals.

2.41

Isotope	*Argon-36*	*Argon-38*	*Argon-40*
Number of neutrons	18	20	22

2.42

Isotope	(a) $^{27}_{13}\text{Al}$	(b) $^{28}_{14}\text{Si}$	(c) $^{11}_{5}\text{B}$	(d) $^{115}_{47}\text{Ag}$
Number of protons	13	14	5	47
Number of neutrons	14	14	6	68
Number of electrons	13	14	5	47

2.43   Symbols (a) and (c) represent isotopes because they have the same atomic number but different mass numbers.

2.44   (a) fluorine-19   (b) neon-19   (c) fluorine-21   (d) magnesium-21

2.45

    (a) $^{14}_{6}C$ (b) $^{39}_{19}K$ (c) $^{20}_{10}Ne$

2.46

    (a) $^{120}_{50}Sn$ (b) $^{56}_{26}Fe$ (c) $^{226}_{88}Ra$

2.47

    $^{99}_{43}Tc$

2.48   Contribution from ^{63}Cu: 69.17% of 62.93 amu   =   43.53 amu
           Contribution from ^{65}Cu: 30.83% of 64.93 amu   =   20.02 amu
        —
                        Average atomic weight   =   63.55 amu

2.49   Contribution from ^{7}Li: 92.58% of 7.016 amu   =   6.495 amu
           Contribution from ^{6}Li: 7.42% of 6.015 amu   =   0.446 amu
        —
                        Average atomic weight   =   6.941 amu

The Periodic Table

2.50   The third period in the periodic table contains 8 elements because 8 electrons are needed to fill the *s* subshell (2 electrons) and *p* subshell (6 electrons) of the third shell.

2.51   The fourth period contains 18 elements because 18 electrons are needed to fill the *s* subshell (2 electrons) and *p* subshell (6 electrons) of the fourth shell, plus the *d* subshell (10 electrons) of the third shell.

2.52   Americium (Am; atomic number 95) is a metal.

2.53   The *p* subshell is being filled for the metalloid elements.

2.54   (a) They are metals.
         (b) They are transition metals.
         (c) The 3*d* subshell is being filled.

2.55 (a) They are metals.
(b) They are lanthanides, which are a subgroup of the inner transition elements.
(c) The 4*f* subshell is being filled.

2.56 (a) Rubidium: (i) metal; (v) main group element; (vii) alkali metal
(b) Tungsten: (i) metal; (iv) transition element
(c) Germanium: (iii) metalloid; (v) main group element
(d) Krypton: (ii) nonmetal; (v) main group element; (vi) noble gas

2.57 (a) Calcium: (i) metal; (v) main group element; (viii) alkaline earth metal
(b) Palladium: (i) metal; (iv) transition element
(c) Carbon: (ii) nonmetal; (v) main group element
(d) Radon: (ii) nonmetal; (v) main group element; (vi) noble gas

2.58 Selenium is chemically similar to sulfur.

2.59 Any element in Group 1A other than hydrogen is chemically similar to potassium.

2.60 The alkali metal family is composed of lithium, sodium, potassium, rubidium, cesium, and francium.

2.61 Fluorine, chlorine, bromine, iodine, and astatine make up the halogen family.

Electron Configuration

2.62 A maximum of two electrons can go into an orbital.

2.63 An *s* orbital is spherical and centered on the nucleus. A *p* orbital is dumbbell-shaped, and the three *p* orbitals extend out from the nucleus at 90° angles to each other.

2.64 First shell—2 electrons
Second shell—8 electrons
Third shell—18 electrons

2.65 The third shell contains nine orbitals: one *s* orbital, three *p* orbitals, and five *d* orbitals. The fourth shell contains sixteen orbitals: one *s* orbital, three *p* orbitals, five *d* orbitals, and seven *f* orbitals.

2.66 The third shell contains three subshells: 3*s*, 3*p*, 3*d*. The fourth shell contains four subshells: 4*s*, 4*p*, 4*d*, 4*f*. The fifth shell contains five subshells: 5*s*, 5*p*, 5*d*, 5*f*, and an additional subshell.

2.67 The last subshell of the fifth shell would contain nine orbitals and would be filled by 18 electrons.

2.68 Ten electrons are present; the element is neon.

2.69 Twenty-two electrons are present; the element is titanium.

2.70 (a) Sulfur   (b) Bromine   (c) Silicon

$\underset{3p^4}{\uparrow\downarrow\ \uparrow\ \uparrow}$ $\underset{4p^5}{\uparrow\downarrow\ \uparrow\downarrow\ \uparrow}$ $\underset{3p^2}{\uparrow\ \uparrow\ \_}$

2.71 (a) Rubidium   (b) Niobium   (c) Rhodium

$\underset{5s}{\uparrow}$ $\underset{5s}{\uparrow\downarrow}\quad\underset{4d}{\uparrow\ \uparrow\ \uparrow\ \_\ \_}$ $\underset{5s}{\uparrow\downarrow}\quad\underset{4d}{\uparrow\downarrow\ \uparrow\downarrow\ \uparrow\ \uparrow\ \uparrow}$

2.72 *Element:*   Sulfur   Bromine   Silicon   Rubidium   Niobium   Rhodium

Number of Unpaired Electrons:	Sulfur	Bromine	Silicon	Rubidium	Niobium	Rhodium
	2	1	2	1	3	3

2.73 *Element*   *Electron Configuration*

(a) Ti $1s^2\ 2s^2\ 2p^6\ 3s^2\ 3p^6\ 4s^2\ 3d^2$
(b) P $1s^2\ 2s^2\ 2p^6\ 3s^2\ 3p^3$
(c) Ar $1s^2\ 2s^2\ 2p^5\ 3s^2\ 3p^6$
(d) La $1s^2\ 2s^2\ 2p^6\ 3s^2\ 3p^6\ 4s^2\ 3d^{10}\ 4p^6\ 5s^2\ 4d^{10}\ 5p^6\ 6s^2\ 5d^1$

2.74 The element with atomic number 12 (Mg) has two electrons in its valence shell.

$$\cdot\text{Mg}\cdot$$

2.75 The number of valence electrons for elements in a main group is the same as the group number. Thus, group 4A elements have four valence electrons.

2.76 Beryllium: $2s$ Arsenic: $4p$

2.77 Group 5A(15) has the configuration $ns^2\ np^3$.

2.78 *Element:*   (a) Kr (b) C (c) Ca (d) K (e) B (f) Cl

Number of Valence-Shell Electrons:	(a) Kr	(b) C	(c) Ca	(d) K	(e) B	(f) Cl
	8	4	2	1	3	7

$$:\!\ddot{\text{Kr}}\!:\qquad \cdot\dot{\text{C}}\cdot\qquad \cdot\text{Ca}\cdot\qquad \cdot\text{K}\qquad \cdot\dot{\text{B}}\cdot\qquad :\!\ddot{\text{Cl}}\cdot$$

2.79 Group 6A: $ns^2\ np^4$ Group 2A: ns^2

Conceptual Problems

2.80 Helium, neon, argon, krypton, xenon, and radon make up the noble gas family.

2.81 Hydrogen can be placed in group 1A because it has one electron in its outermost (only) electron shell. Hydrogen can be placed in group 7A because only one electron is needed to fill its outermost (only) electron shell.

2.82 The undiscovered element beneath francium has the atomic number 119.

2.83 Pb: $1s^2\,2s^2\,2p^6\,3s^2\,3p^6\,4s^2\,3d^{10}\,4p^6\,5s^2\,4d^{10}\,5p^6\,6s^2\,4f^{14}\,5d^{10}\,6p^2$
Shell 1:   2 electrons   Shell 2:   8 electrons   Shell 3: 18 electrons
Shell 4: 32 electrons   Shell 5: 18 electrons   Shell 6:   4 electrons

2.84 Highest-energy occupied subshell: (a) I – $5p$; (b) Sc – $3d$; (c) As – $4p$; (d) Al – $3p$

2.85 (a) One atom of carbon has a mass of 12 amu. In grams:

$$12\ \text{amu}\ \times\ \frac{1.660\ 539\ \times\ 10^{-24}\ \text{g}}{1\ \text{amu}}\ =\ 1.99\ \times\ 10^{-23}\ \text{g}$$

(b) $$\frac{12\ \text{amu}}{1\ \text{atom}}\ \times\ \frac{1.660\ 539\ \times\ 10^{-24}\ \text{g}}{1\ \text{amu}}\ \times\ 6.02\ \times\ 10^{23}\ \text{atoms}\ =\ 12\ \text{g}$$

(c) Based on the answer to part (b), we predict that 6.02×10^{23} sodium atoms weigh about 23 grams.

2.86 The unidentified element is strontium, which occurs directly below calcium in group 2A and thus has similar chemical behavior. Strontium is a metal, has 38 protons, and is in the fifth period.

$$\cdot\,\text{Sr}\,\cdot$$

2.87 (a) Ge: $1s^2\,2s^2\,2p^6\,3s^2\,3p^6\,4s^2\,3d^{10}\,4p^2$; [Ar] $4s^2\,3d^{10}\,4p^2$
(b) The outer-shell electrons of germanium are in the $4s$ and $4p$ orbitals.

2.88 Tin, a metal, has an electron configuration by shell of 2 8 18 18 4.

2.89

$$8.6\ \text{mg}\ \times\ \frac{1\ \text{g}}{10^3\ \text{mg}}\ \times\ \frac{1\ \text{amu}}{1.660\ 539\ \times\ 10^{-24}\ \text{g}}\ \times\ \frac{1\ \text{atom}}{40.08\ \text{amu}}\ =\ 1.3\ \times\ 10^{20}\ \text{atoms}$$

2.90 (a) Electrons must fill the $4s$ subshell before entering the $3d$ subshell. The correct configuration:

$$1s^2\ 2s^2\ 2p^6\ 3s^2\ 3p^6\ 4s^2\ 3d^8$$

(b) Electrons must fill the $2s$ subshell before entering the $2p$ subshell. The correct configuration:

$$1s^2\ 2s^2\ 2p^3$$

(c) Silicon has 14 electrons. The p orbitals must be half-filled before any one orbital is completely filled. The correct configuration:

$$1s^2\ 2s^2\ 2p^6\ 3s^2\ \underset{3p}{\uparrow\ \uparrow\ —}$$

(d) The $3s$ electrons must have opposite spins. The correct configuration:

$$1s^2\ 2s^2\ 2p^6\ \underset{3s}{\uparrow\downarrow}$$

2.91 Count the electrons in each example to arrive at the atomic number. Then look up the answer in the periodic table.
(a) Cr   (b) Cu   (c) Mo   (d) Ag

2.92 An electron will fill or half-fill a d subshell instead of filling an s subshell of a higher shell.

2.93 Au: $1s^2\ 2s^2\ 2p^6\ 3s^2\ 3p^6\ 4s^2\ 3d^{10}\ 4p^6\ 5s^2\ 4d^{10}\ 5p^6\ 6s^1\ 4f^{14}\ 5d^{10}$

2.94 The last orbital filled in element 117 is $7p$.

Group Problems

2.95 J. J. Thomson aimed a cathode ray at a charged detector. Although he knew that the ray was composed of negatively charged particles, he noticed that the degree of bending of the particle wave by the detector indicated that the particles were much smaller than he would have expected and were later identified as electrons.

Ernest Rutherford fired alpha particles (helium nuclei) at a piece of gold foil a few atoms thick. He noticed that a few of the alpha particles were deflected, which indicated a collision with a mass that was later identified as a gold nucleus that was positively charged.

2.96 (a) Isotopes of several elements are used in cancer treatments. Among them are:
$$^{131}_{53}\text{I}\qquad ^{31}_{15}\text{P}\qquad ^{60}_{27}\text{Co}\qquad ^{177}_{62}\text{Sm}$$

Iodine-131 has 53 protons, 53 electrons, and 78 neutrons.
Phosphorus-31 has 15 protons, 15 electrons, and 16 neutrons.
Cobalt-60 has 27 protons, 27 electrons, and 33 neutrons.
Samarium-177 has 62 protons, 62 electrons, and 115 electrons.

(b) Osmium is the densest element. Its most abundant isotope, osmium-192 has 76 protons, 76 electrons, and 116 neutrons.
(c) Technetium (Z = 43) is not found in nature. It has 43 protons, 43 electrons, and 55 neutrons in its most common form.

2.97   Tellurium has a greater atomic weight because its nuclei contain, on the average, more neutrons than an iodine nucleus. Other pairs of elements in which the element with smaller Z has a larger value of A include argon/potassium, cobalt/nickel, tellurium/iodine, thorium/protactinium, and uranium/neptunium.

2.98   (a) In the graph of atomic radius versus atomic number (Figure 2.3), the elements on the left of the periodic table have the greatest atomic radius, and the elements on the right have the smallest radius. In Figure 2.4, the elements that are found in the middle of the periodic Table have the highest melting point, and the noble gases have the lowest boiling point.
(b) Electron affinity and ionization energy also show periodic trends.

Self-Test for Chapter 2

Multiple Choice

1. Which of the following is a metalloid?
    (a) carbon   (b) aluminum   (c) silicon   (d) phosphorus

2. Which of the following has a partially filled d subshell?
    (a) calcium   (b) vanadium   (c) zinc   (d) arsenic

3. Which of the following is not a part of atomic theory?
    (a) All metal is composed of atoms.   (b) The atoms of each element are different from the atoms of other elements.   (c) In chemical compounds, atoms are combined in specific proportions.   (d) Chemical reactions only change the way that atoms are combined.

4. How many isotopes of hydrogen are there?
    (a) one   (b) two   (c) three   (d) can't be determined

5. What holds protons and neutrons together in the nucleus?
    (a) attraction   (b) repulsion   (c) electrons   (d) internuclear forces

6. Which of the following has a mass number of 33?
    (a) $^{74}_{33}As$ (b) $^{35}_{17}Cl$ (c) $^{32}_{16}S$ (d) $^{33}_{16}S$

7. In which order are subshells usually filled?
    (a) s, p, d, f (b) d, f, p, s (c) s, p, f, d (d) s, d, p, f

8. The element with atomic number 38 is:
    (a) an alkali metal   (b) an alkaline earth metal   (c) a transition metal   (d) a metalloid

9. The element that has atomic weight = 91 and has 40 electrons is:
    (a) protactinium   (b) niobium   (c) antimony   (d) zirconium

10. The element that has electron configuration $1s^2\,2s^2\,2p^6\,3s^2\,3p^6\,4s^2\,3d^7$ is:
    (a) copper   (b) rhodium   (c) arsenic   (d) cobalt

11. An element that has electrons in its f subshell is:
    (a) europium   (b) technetium   (c) antimony   (d) xenon

12. $^{195}_{78}X$ is the symbol for:
    (a) gold   (b) platinum   (c) iridium   (d) iron

Sentence Completion

1. A _____ orbital is dumbbell-shaped.

2. An atomic mass unit is also known as a _____.

3. The nucleus of an atom is made up of _____ and _____.

4. The _____ _____ indicates the number of protons in an atom.

5. The electrons in an atom are grouped by energy into _____.

6. Elements belonging to the same _____ have similar chemical properties.

7. The atomic number of aluminum is _____.

8. Protons, neutrons, and electrons are known as _____ _____.

9. Atoms having the same number of protons but different numbers of neutrons are called _____.

10. The third shell contains _____ electrons.

11. The word _____ means that electrons can have certain energy values and no others.

True or False

1. Bismuth is a metal.

2. The mass of an electron is approximately 1 amu.

3. A $4s$ electron is higher in energy than a $3d$ electron.

4. Isotopes have the same number of protons but different numbers of neutrons.

5. An atom's atomic number indicates the number of protons and neutrons the atom has.

6. Elements in group 2A are more reactive than elements in group 1A.

7.   Elements in the same period have similar chemical properties.

8.   More elements are metals than are nonmetals.

9.   All compounds consist of atoms combined in specific proportions.

10.   A subshell contains only two electrons.

11.   An element can have the same number of protons, neutrons, and electrons.

12.   Atomic weight always increases with atomic number.

Match each entry on the left with its partner on the right.

1. $1s^2\ 2s^2$ (a) Mendeleev

2.   Group   (b) Number of protons in an element

3.   Atomic mass unit   (c) Reactive metals

4.   Formulated the periodic   (d) Average mass of a large number of an
      table   element's atoms

5.   Neutron   (e) Column in the periodic table

6. $^{28}_{14}Si$ (f) Row in the periodic table

7.   Group 1A   (g) Element with atomic mass of 14 amu

8.   Atomic number   (h) Electron configuration of beryllium

9. $^{29}_{14}Si$ (i) Element having 15 neutrons

10.   Atomic weight   (j) Subatomic particle with zero charge

11. $^{14}_{7}N$ (k) Dalton

12.   Period   (l) Element with atomic number 14

Chapter Outline

I. Ions (Sections 3.1–3.6).
   A. Ions are formed when a neutral atom either loses one or more electrons or gains one or
      more electrons (Section 3.1).
      1. Loss of one or more electrons forms a cation.
      2. Gain of one or more electrons forms an anion.
   B. The octet rule (Section 3.2).
      1. Main group elements undergo reactions that leave them with a noble gas configuration
         – an octet.
   C. Ions of some common elements (Section 3.3).
      1. Cations.

         a. Group 1A $M \cdot \longrightarrow M^+ + e^-$

         b. Group 2A $\cdot M \cdot \longrightarrow M^{2+} + 2e^-$

         c. Group 3A Al^{3+} is the only common cation.
         d. Transition metals form cations, but the charge on these cations is not as predictable
            as it is for main group elements.
      2. Anions.

         a. Group 6A $:\!\overset{\bullet}{X}\!: + 2e^- \longrightarrow :\!\overset{\bullet\bullet}{\underset{\bullet\bullet}{X}}\!:^{2-}$

         b. Group 7A $:\!\overset{\bullet}{X}\!: + e^- \longrightarrow :\!\overset{\bullet\bullet}{\underset{\bullet\bullet}{X}}\!:^-$

      3. Group 4A, group 5A, and group 8A elements don't usually form cations or anions.
   D. Periodic properties and ion formation (Section 3.4).
      1. Ionization energy measures the ease with which an atom gives up an electron.
         a. Energy must be supplied in order to remove an electron.
         b. Elements on the far left of the periodic table have smaller ionization energies and
            lose electrons more easily.
         c. Elements on the far right of the periodic table have larger ionization energies and
            lose electrons with difficulty.
      2. Electron affinity measures the ease with which an atom gains an electron.
         a. Energy is released when an atom gains an electron.
         b. Elements on the far right of the periodic table have larger electron affinities and gain
            electrons easily.
         c. Elements on the far left of the periodic table have smaller electron affinities and gain
            electrons less easily.
      3. Main group elements in the middle of the periodic table neither lose or gain electrons
         easily.
   E. Naming ions (Sections 3.5–3.6).
      1. Monatomic ions (Section 3.5).
         a. Main group cations are named by identifying the metal and then adding the word
            ion.
         b. Transition metal cations are named by identifying the metal, specifying the charge,
            and adding the word *ion*.

    c. Anions are named by replacing the end of the name of the element with *-ide* and adding the word *ion*.

  2. Polyatomic ions (Section 3.6).

    a. Polyatomic ions are composed of more than one atom.

    b. Subscripts in polyatomic ions indicate how many of each ion are present in the formula unit (no subscripts are necessary if only one ion is present).

    c. The names and structures of polyatomic ions should be memorized.

II. Ionic compounds (Sections 3.7–3.11).

  A. Formation of ionic compounds (Section 3.7).

    1. Compounds formed between an element on the far left side of the periodic table and an element on the far right of the periodic table are electronically neutral.

      a. These compounds consist of a large number of cations and anions packed together in a regular arrangement in a crystal.

      b. The bonds between ions in the crystal are known as ionic bonds.

      c. The crystal is known as an ionic solid.

  B. Formulas of ionic compounds (Section 3.8).

    1. Formulas are written so that the number of positive charges equals the number of negative charges.

      a. Cations are listed first, anions second.

      b. It is not necessary to write the charges of the ions.

      c. Use parentheses around a polyatomic ion if it has a subscript.

    2. A formula unit shows the simplest neutral unit of an ionic compound.

  C. Naming ionic compounds (Section 3.9).

    1. Ionic compounds are named by citing the cation and then the anion, with a space between the two words.

    2. If the cation can exhibit more than one charge, the charge must be specified.

  D. Properties of ionic compounds (Section 3.10).

    1. Ionic compounds are crystalline.

    2. Solutions of ionic compounds conduct electricity (if the compounds are soluble).

    3. Ionic compounds are high-melting.

    4. Many, but not all, ionic compounds are water-soluble.

  E. Acids and bases (Section 3.11).

    1. Acids are compounds that provide H^+ ions in solution.

    2. Bases are compounds that provide OH^- ions in solution.

    3. Some acids and bases can each provide more than one H^+ or OH^- ion in solution.

Solutions to Chapter 3 Problems

3.1   The Mg^{2+} ion is a cation.

3.2   The S^{2-} ion is an anion.

3.3   The ion shown, O^{-2}, is an anion because it has two more negative charges than positive charges.

3.4   Potassium (atomic number 19): $1s^2\, 2s^2\, 2p^6\, 3s^2\, 3p^6\, 4s^1$
       Argon (atomic number 18): $1s^2\, 2s^2\, 2p^6\, 3s^2\, 3p^6$
       Potassium can attain the noble gas configuration of argon by losing an electron from its $4s$ subshell, forming the K^+ cation.

3.5 Aluminum (atomic number 13): $1s^2\, 2s^2\, 2p^6\, 3s^2\, 3p^1$
Neon (atomic number 10): $1s^2\, 2s^2\, 2p^6$
Aluminum can attain the noble gas configuration of neon by losing three electrons, two from its $3s$ subshell and one from its $3p$ subshell, resulting in the formation of the Al^{3+} ion.

3.6 Y gains electrons, and X loses electrons.

$$X\!: \;+\; :\!\overset{\displaystyle\cdot}{\underset{\displaystyle\cdot}{Y}}\!: \;\longrightarrow\; X^{2+} \;+\; :\!\overset{\displaystyle\cdot\cdot}{\underset{\displaystyle\cdot\cdot}{Y}}\!:^{\,2-}$$

3.7 Although iron forms the ions Fe^{2+} and Fe^{3+}, Fe^{2+} is found in hemoglobin.
The electron configuration of Fe^{2+}: $1s^2\, 2s^2\, 2p^6\, 3s^2\, 3p^6\, 3d^6$

3.8

(a) $:\!\overset{\displaystyle\cdot}{\underset{\displaystyle\cdot\cdot}{Se}}\!\cdot \;+\; 2\,e^- \;\longrightarrow\; :\!\overset{\displaystyle\cdot\cdot}{\underset{\displaystyle\cdot\cdot}{Se}}\!:^{\,2-};\quad Se \;+\; 2\,e^- \;\longrightarrow\; Se^{2-}$

(b) $\cdot Ba\cdot \;\longrightarrow\; Ba^{2+} \;+\; 2\,e^-;\quad Ba \;\longrightarrow\; Ba^{2+} \;+\; 2\,e^-$

(c) $:\!\overset{\displaystyle\cdot\cdot}{Br}\!\cdot \;+\; e^- \;\longrightarrow\; :\!\overset{\displaystyle\cdot\cdot}{\underset{\displaystyle\cdot\cdot}{Br}}\!:^{\,-};\quad Br \;+\; e^- \;\longrightarrow\; Br^-$

3.9 For Na^+:
$$\frac{1\ mL}{3.2\ mg}\; x\; \frac{1000\ mg}{1\ g} = 310\ mL$$
For Cl^-:
$$\frac{1\ mL}{3.5\ mg}\; x\; \frac{1000\ mg}{1\ g} = 290\ mL$$

3.10 In line with the trend shown in Figure 3.2, you would expect the ionization energies of Rb and Cs to be even smaller than those for the other Group 1A elements. The actual values, 403 kJ/mol for Rb and 375 kJ/mol for Cs, confirm the expectation.

3.11 (a) According to Figure 3.2, B ($Z = 5$) loses an electron more easily than Be ($Z = 4$).
(b) Ca ($Z = 20$) loses an electron more easily than Co ($Z = 27$).
(c) Sc ($Z = 21$) loses an electron more easily than Se ($Z = 34$).

3.12 (a) According to Figure 3.2, H gains an electron more easily than He. Because the electron affinity for He is zero, it does not accept an electron.
(b) S gains an electron more easily than Si.
(c) Cr gains an electron more easily than Mn.

CIA 3.1 The RDI for sodium is 2300 mg, which is equivalent to 4 g of table salt.

CIA 3.2 Iodized salt contains 0.01% KI. The RDI for iodide is 150 μg/day.

3.13 (a) Cu^{2+} copper(II) ion (cupric ion) (b) F^- fluoride ion
(c) Mg^{2+} magnesium ion (d) S^{2-} sulfide ion

3.14 (a) Ag^+ (b) Fe^{2+} (c) Cu^+ (d) Te^{2-}

3.15 Na^+ sodium ion; K^+ potassium ion;
           Ca^{2+} calcium ion; Cl^- chloride ion

3.16   (a) NO_3^- nitrate ion   (b) CN^- cyanide ion
           (c) OH^- hydroxide ion   (d) HPO_4^{2-} hydrogen phosphate ion

3.17   *Group 1A:* Na^+, K^+ *Group 2A:* Mg^{2+}, Ca^{2+}
           Transition metal: Fe^{2+} *Halogen:* Cl^-

CIA 3.3   Most of the calcium present in the body is found in bones and teeth.

CIA 3.4   Sodium protects against fluid loss and is necessary for muscle contraction and
              transmission of nerve impulses.

3.18   (a) Two iodide ions, each with a charge of –1, will balance the +2 charge of one
           magnesium ion. The correct formula: MgI_2.
           (b) It takes three O^{2-} to balance two aluminum ions, each with charge of +3. The correct
           formula: Al_2O_3
           (c) $Fe_3(PO_4)_2$ (d) $Cr_2(SO_4)_3$

3.19 $(NH_4)_2CO_3$

3.20 $Al_2(SO_4)_3$ $Al(CH_3CO_2)_3$

3.21   Use the periodic table to identify the elements present and use the rules described in
           Section 3.9 to write the formulas. The formulas are K_2S (blue), $BaBr_2$ (red), and Al_2O_3
           (green).

HOC 3.1 no answer

3.22 Ag_2S—silver(I) sulfide. The charge on silver is +1.

3.23   (a) SnO_2 tin(IV) oxide   (b) $Ca(CN)_2$ calcium cyanide
           (c) Na_2CO_3 sodium carbonate   (d) Cu_2SO_4 copper(I) sulfate
           (e) $Ba(OH)_2$ barium hydroxide   (f) $Fe(NO_3)_2$ iron(II) nitrate

3.24   (a) Li_3PO_4 (b) $CuCO_3$ (c) $Al_2(SO_3)_3$ (d) CuF (e) $Fe_2(SO_4)_3$ (f) NH_4Cl

3.25   Since there are three calcium ions for every two nitride ions, the formula is Ca_3N_2.
           Calcium ion has a +2 charge, and nitride ion has a –3 charge.

HOC 3.2 (a) $CrPO_4$ (b) $FeCl_2$ (c) $Fe(NO_3)_3$ (d) $(NH_4)_2SO_3$

3.26   According to the listed melting points, NaCl has the strongest ionic bonds and RbCl has
           the weakest bonds. The trend shows that the largest cation forms compounds with the
           weakest ionic bonds.

CIA 3.5   The bulky cations can't pack in an orderly way and thus do not crystallize at room
              temperature.

CIA 3.6 They are nonvolatile (important in a very low-pressure environment), viscous (allowing them to be coated with a thin metallic film to form a parabolic reflective surface), and inexpensive (relative to the cost of a conventional lens).

3.27 Acids provide H^+ ions when dissolved in water.
Bases provide OH^- ions when dissolved in water.

Acids: HF $\xrightarrow{\text{dissolve in water}}$ H^+ + F^-

            HCN $\xrightarrow{\text{dissolve in water}}$ H^+ + CN^-

Bases: $Ca(OH)_2$ $\xrightarrow{\text{dissolve in water}}$ Ca^{2+} + $2\ OH^-$

            LiOH $\xrightarrow{\text{dissolve in water}}$ Li^+ + OH^-

3.28 Solution (a) represents HCl; the drawing shows that there are equal numbers of cations and anions in solution. Solution (b) represents H_2SO_4; two cations are present for each anion.

CIA 3.7

Ion	Name	Charge	Total Charge
Ca^{2+}	calcium ion	2+	10 x (2+) = 20+
PO_4^{3-}	phosphate ion	3–	6 x (3–) = 18–
OH^-	hydroxide ion	1–	2 x (1–) = 2–

The formula correctly represents a neutral compound because the number of positive charges equals the number of negative charges.

Understanding Key Concepts

3.29

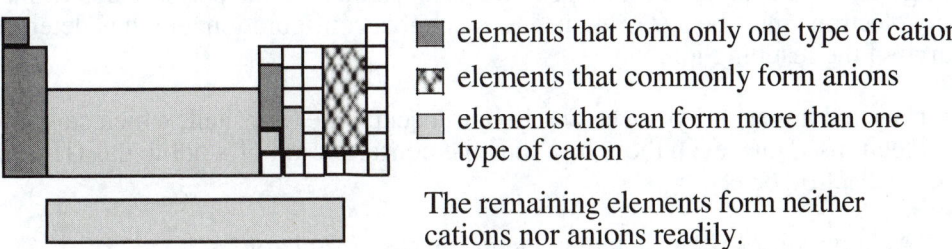

- ▨ elements that form only one type of cation
- ▨ elements that commonly form anions
- ▨ elements that can form more than one type of cation

The remaining elements form neither cations nor anions readily.

3.30

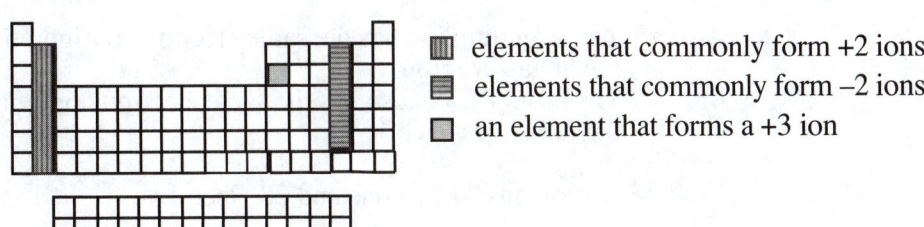

- ▦ elements that commonly form +2 ions
- ▦ elements that commonly form –2 ions
- ▦ an element that forms a +3 ion

3.31 Drawing (a) shows an ion that has two more electrons than protons and thus represents an O^{2-} ion.
Drawing (b) represents Na^+, an ion with one more proton than electron.
(c) Ca^{2+} (d) Fe^{2+}

3.32 Drawing (a) represents a sodium atom, and drawing (b) represents an Na^+ ion. The Na atom is larger because its $3s$ electron lies in an orbital that is farther from the nucleus.

3.33 Drawing (a) represents a chlorine atom, and drawing (b) represents a Cl^- anion. The anion is larger because its extra electron is not as tightly held by the positively charged nucleus as are the electrons of the atom.

3.34 $FeCl_2$: iron(II) chloride or ferrous chloride $FeCl_3$: iron(III) chloride or ferric chloride
FeO: iron(II) oxide or ferrous oxide Fe_2O_3: iron(III) oxide or ferric oxide
$PbCl_2$: lead(II) chloride $PbCl_4$: lead(IV) chloride
PbO: lead(II) oxide PbO_2: lead(IV) oxide

3.35 Use the ratios of atoms to assign answers. Picture (a) represents ZnS, because one cation is present for each anion. Picture (b) represents $PbBr_2$, (c) represents CrF_3, and (d) represents Al_2O_3.

3.36 Cr_2O_3 chromium(III) oxide

Additional Problems

Ions and the Octet Rule

3.37 The *octet rule* states that main group atoms undergo reactions in order to achieve a noble gas electron configuration with eight outer-shell electrons.

3.38 The octet rule states that atoms lose or gain electrons in order to attain the electron configuration of a noble gas. The position of an element in the periodic determines how many electrons must be lost or gained to reach this configuration and thus determines the charge of the resulting ion.

3.39 Electrons belonging to hydrogen and helium occupy the first shell, which can hold only two electrons. Thus, even though H^- has the configuration of a noble gas (He), it can't have an electron octet.

3.40 Se^{2-} An ion with 34 protons and 36 electrons has a –2 charge.

3.41 Sc^{2+} An ion with 21 protons and 19 electrons has a +2 charge.

3.42 (a) $X^{2+} = Sr^{2+}$; X = Sr   The strontium ion has the same electron configuration as the noble gas krypton.
(b) $X^- = Br^-$; X = Br   The bromide ion also has the same electron configuration as the noble gas krypton.

3.43 $Z^{3+} = Ga^{3+}$; Z = Ga. The Z^{3+} cation has 31 protons and 28 electrons.

3.44 (a) Rb^+ $1s^2\,2s^2\,2p^6\,3s^2\,3p^6\,4s^2\,3d^{10}\,4p^6$
(b) Br^- $1s^2\,2s^2\,2p^6\,3s^2\,3p^6\,4s^2\,3d^{10}\,4p^6$
(c) S^{2-} $1s^2\,2s^2\,2p^6\,3s^2\,3p^6$
(d) Ba^{2+} $1s^2\,2s^2\,2p^6\,3s^2\,3p^6\,4s^2\,3d^{10}\,4p^6\,5s^2\,4d^{10}\,5p^6$
(e) Al^{3+} $1s^2\,2s^2\,2p^6$

3.45 (a) Ca^{2+} (b) O^{2-} (c) Ti^{2+} (d) K^+ (e) Al^{3+}

Ions and Ionic Bonding

3.46 (a) $Ca \rightarrow Ca^{2+} + 2\,e^-$ (b) $Au \rightarrow Au^+ + e^-$
(c) $F + e^- \rightarrow F^-$ (d) $Cr \rightarrow Cr^{3+} + 3\,e^-$

3.47 (a) P^{3-} $1s^2\,2s^2\,2p^6\,3s^2\,3p^6$ (b) Li^+ $1s^2$
(c) Co^{2+} $1s^2\,2s^2\,2p^6\,3s^2\,3p^6\,4s^2\,3d^5$
(d) Tl^{3+} $1s^2\,2s^2\,2p^6\,3s^2\,3p^6\,4s^2\,3d^{10}\,4p^6\,5s^2\,4d^{10}\,5p^6\,4f^{14}\,5d^{10}$

3.48 (a) False. A cation is formed by *loss* of one or more electrons from an atom. (b), (c) False. Group 4A elements rarely form ions. (d) True.

3.49 (a) False. Ions have a regular arrangement in ionic solids.   (b) False. Ions vary in size. (c) True.   (d) False. Ionic solids have high melting points and boiling points.

Periodic Properties and Ion Formation

3.50 (a) O   (b) Li   (c) Zn   (d) N

3.51 (a) S   (b) I   (c) Br

3.52 None of the ions are stable because all lack eight outer-shell electrons.

3.53 (a) Magnesium forms only the Mg^{2+} cation.
(b) Tin forms both the Sn^{2+} and Sn^{4+} cations.
(c) Mercury forms Hg_2^{2+} and Hg^{2+} cations.
(d) Al forms the Al^{3+} cation.

3.54 Cr^{2+} $1s^2\,2s^2\,2p^6\,3s^2\,3p^6\,3d^4$
Cr^{3+} $1s^2\,2s^2\,2p^6\,3s^2\,3p^6\,3d^3$

3.55 Co $1s^2\,2s^2\,2p^6\,3s^2\,3p^6\,3d^7\,4s^2$
Co^{2+} $1s^2\,2s^2\,2p^6\,3s^2\,3p^6\,3d^7$
Co^{3+} $1s^2\,2s^2\,2p^6\,3s^2\,3p^6\,3d^6$

3.56 The ionization energy of Li^+ is much greater than that of Li. Li readily loses an electron to form an ion with eight outer-shell electrons, but Li^+ would need to lose one electron from a stable octet in order to form the Li^{2+} cation.

3.57 (a) K → K$^+$ + e^- loss of an electron by K
     K$^+$ + e^- → K   gain of an electron by K$^+$
(b)The second equation is the reverse of the first equation.
(c) Ionization energy of K = − (electron affinity of K$^+$)

Symbols, Formulas, and Names for Ions

3.58 (a) S^{2-} sulfide ion   (b) Sn^{2+} tin(II) ion   (c) Sr^{2+} strontium ion
(d) Mg^{2+} magnesium ion   (e) Au$^+$ gold(I) ion

3.59 (a) Cr^{2+} chromous ion   chromium(II) ion
(b) Fe^{3+} ferric ion   iron(III) ion
(c) Hg^{2+} mercuric ion   mercury(II) ion

3.60 (a) Se^{2-} (b) O^{2-} (c) Ag$^+$

3.61 (a) Fe^{2+} (b) Sn^{4+} (c) Pb^{2+} (d) Cr^{3+}

3.62 (a) OH$^-$ (b) HSO$_4^-$ (c) CH$_3$CO$_2^-$ (d) MnO$_4^-$ (e) OCl$^-$ (f) NO$_3^-$
(g) CO$_3^{2-}$ (h) Cr$_2$O$_7^{2-}$

3.63 (a) NO$_2^-$ nitrite ion   (b) CrO$_4^{2-}$ chromate ion
(c) NH$_4^+$ ammonium ion   (d) HPO$_4^{2-}$ hydrogen phosphate ion

Names and Formulas for Ionic Compounds

3.64 (a) NaHCO$_3$ (b) KNO$_3$ (c) CaCO$_3$ (d) NH$_4$NO$_3$

3.65 (a) Ca(OCl)$_2$ (b) CuSO$_4$ (c) Na$_3$PO$_4$

3.66

	S^{2-}	Cl$^-$	PO$_4^{3-}$	CO$_3^{2-}$
Copper(II)	CuS	CuCl$_2$	Cu$_3$(PO$_4$)$_2$	CuCO$_3$
Ca^{2+}	CaS	CaCl$_2$	Ca$_3$(PO$_4$)$_2$	CaCO$_3$
NH$_4^+$	(NH$_4$)$_2$S	NH$_4$Cl	(NH$_4$)$_3$PO$_4$	(NH$_4$)$_2$CO$_3$
Ferric ion	Fe$_2$S$_3$	FeCl$_3$	FePO$_4$	Fe$_2$(CO$_3$)$_3$

3.67

	O^{2-}	HSO$_4^-$	HPO$_4^{2-}$	C$_2$O$_4^{2-}$
K$^+$	K$_2$O	KHSO$_4$	K$_2$HPO$_4$	K$_2$C$_2$O$_4$
Ni^{2+}	NiO	Ni(HSO$_4$)$_2$	NiHPO$_4$	NiC$_2$O$_4$
NH$_4^+$	(NH$_4$)$_2$O	NH$_4$HSO$_4$	(NH$_4$)$_2$HPO$_4$	(NH$_4$)$_2$C$_2$O$_4$
Chromous	CrO	Cr(HSO$_4$)$_2$	CrHPO$_4$	CrC$_2$O$_4$

3.68 copper(II) sulfide   copper(II) chloride   copper(II) phosphate   copper(II) carbonate
calcium sulfide   calcium chloride   calcium phosphate   calcium carbonate
ammonium sulfide   ammonium chloride   ammonium phosphate   ammonium carbonate
ferric sulfide   ferric chloride   ferric phosphate   ferric carbonate

3.69   potassium oxide   potassium bisulfate   potassium hydrogen phosphate   potassium oxalate

nickel(II) oxide   nickel(II) bisulfate   nickel(II) hydrogen phosphate   nickel(II) oxalate

ammonium oxide   ammonium bisulfate   ammonium hydrogen phosphate   ammonium oxalate

chromous oxide   chromous bisulfate   chromous hydrogen phosphate   chromous oxalate

3.70 (a) $MgCO_3$ magnesium carbonate   (b) $Ca(CH_3CO_2)_2$ calcium acetate
          (c) $AgCN$ silver(I) cyanide   (d) $Na_2Cr_2O_7$ sodium dichromate

3.71 (a) $Fe(OH)_2$ iron(II) hydroxide   (b) $KMnO_4$ potassium permanganate
          (c) Na_2CrO_4 sodium chromate   (d) $Ba_3(PO_4)_2$ barium phosphate

3.72 $Ca_3(PO_4)_2$ (d) is the correct formula because the six positive charges from the three Ca^{2+} ions are balanced by the six negative charges of the two PO_4^{3-} ions.

3.73 (a) $Al_2(SO_4)_3$ (b) $(NH_4)_3PO_4$ (c) Rb_2SO_4

Acids and Bases

3.74 An acid provides H^+ ions when dissolved in water. A base provides OH^- ions when dissolved in water.

3.75 Acids: H_2CO_3, HCN
          Bases: $Mg(OH)_2$, KOH

3.76

(a) H_2CO_3 $\xrightarrow{\text{dissolve in water}}$ $2H^+$ + CO_3^{2-}

(b) HCN $\xrightarrow{\text{dissolve in water}}$ H^+ + CN^-

(c) $Mg(OH)_2$ $\xrightarrow{\text{dissolve in water}}$ Mg^{2+} + $2OH^-$

(d) KOH $\xrightarrow{\text{dissolve in water}}$ K^+ + OH^-

3.77 (a) carbonate anion   (b) cyanide anion

Conceptual Problems

3.78 The hydride ion has the same electron configuration ($1s^2$) as the noble gas helium.

3.79 The H^- ion has a stable noble gas configuration, but the Li^- ion doesn't and is thus likely to be unstable. (Notice in Figure 3.2 that the electron affinity of Li is very small, indicating that Li is unlikely to gain an electron to form Li^-.)

3.80 (a) CrO_3 (b) VCl_5 (c) MnO_2 (d) MoS_2

3.81 H_3AsO_4

3.82   (a) Cu_3PO_4 copper(I) phosphate   (b) Na_2SO_4 sodium sulfate
(c) MnO_2 manganese(IV) oxide   (d) $AuCl_3$ gold(III) chloride
(e) $Pb(CO_3)_2$ lead(IV) carbonate   (f) Ni_2S_3 nickel(III) sulfide

3.83   (a) $Co(CN)_2$ (b) UO_3 (c) $SnSO_4$ (d) MnO_2 (e) K_3PO_4 (f) Ca_3P_2
(g) $LiHSO_4$ (h) $Al(OH)_3$

3.84

Ion	*Protons*	*Electrons*	*Neutrons*
(a) $^{16}O^{2-}$	8	10	8
(b) $^{89}Y^{3+}$	39	36	50
(c) $^{133}Cs^+$	55	54	78
(d) $^{81}Br^-$	35	36	46

3.85   (a) X is likely to be a metal, because metals are more likely to form cations.
(b) Y is likely to be a nonmetal.
(c) The formula for the product is X_2Y_3.
(d) X is likely to be in Group 3A or Group 3B (or to be a transition metal) and Y is likely to be in Group 6A.

3.86   (a) Mn^{4+} (b) Cu^+ (c) Ti^{4+}

Group Questions

3.87   (a) The most commonly used alums are $KAl(SO_4)_2 \cdot 12\ H_2O$ and $NaAl(SO_4)_2 \cdot 12\ H_2O$. The water molecules are referred to as water of hydration.
(b) Uses of alums include water purification, as a flame retardant, and in deodorants.

3.88   The formula of calcium gluconate is $Ca(C_6H_{11}O_7)_2$ Calcium has two positive charges, and each gluconate anion has one negative charge. Calcium ion is monatomic, and gluconate is polyatomic because it contains more than one atom.

3.89   The formula of borax is $Na_2B_4O_7 \cdot 10\ H_2O$. The anion in borax, $B_4O_7^{2-}$ is named tetraborate.

3.90   The most common fluoride in toothpaste is SnF_2, and it is named either tin(II) fluoride or stannous fluoride.

Self-Test for Chapter 3

Multiple Choice

1. How many atoms does a formula unit of Li_2CO_3 contain?
    (a) 3   (b) 4   (c) 5   (d) 6

2. How many ions are produced when a formula unit of Li_2CO_3 is dissolved in water?
    (a) 3   (b) 4   (c) 5   (d) 6

3. Which of the following is not a property of ionic compounds?
    (a) crystalline   (b) conductor of electricity   (c) high melting   (d) 1:1 ratio of cations to anions

4. Which of the following is the correct name for $Fe(NO_3)_3$?
    (a) ferrous nitrate   (b) iron nitrate   (c) iron(III) nitrate   (d) iron(II) nitrate

5. Bromine has a:
    (a) large ionization energy and large electron affinity   (b) large ionization energy and small electron affinity   (c) small ionization energy and large electron affinity   (d) small ionization energy and small electron affinity

6. H_2CrO_4 is an acid that:
    (a) can provide one H^+ ion when dissolved   (b) can provide two H^+ ions when dissolved
    (c) can provide one OH^- when dissolved   (d) H_2CrO_4 is not an acid.

7. An element that loses three electrons to attain the electron configuration of argon is:
    (a) titanium   (b) scandium   (c) calcium   (d) potassium

8. The formula for gold(III) chloride is:
    (a) Au_3Cl (b) $AuCl$ (c) $AuCl_2$ (d) $AuCl_3$

9. Which of the following anions is not biologically important?
    (a) Cl^- (b) Br^- (c) HCO_3^- (d) HPO_4^-

10. The redness of rubies is due to which ion?
    (a) iron   (b) titanium   (c) aluminum   (d) chromium

Sentence Completion

1. _____ _____ measures the ease with which an atom gives up an electron.

2. The name of K_3PO_4 is _____ _____.

3. Radium (atomic number 88) loses _____ electrons to achieve a noble gas configuration.

4. NO_3^- is an example of a _____ ion.

5. The formulas of ionic compounds are _____ formulas.

6. Atoms of main group elements tend to combine in chemical compounds so that they attain _____ outer-shell electrons.

7.   A _____ provides OH⁻ ions in water.

8.   _____ is the principal mineral component of bone.

9.   Ionic compounds are usually _____ solids.

10.   The first three elements in group ____A form neither cations nor anions.

True or False

1.   Zinc can form ions with different charges.

2. Na^+ and F^- have the same electron configuration.

3.   Ionization energy measures the amount of energy released when an ion is formed from a neutral atom.

4. $Co(CO_3)_2$ is a possible compound.

5.   A solution of H_3PO_4 contains only H^+ and $H_2PO_4^-$ ions.

6.   Ionic crystals conduct electricity.

7.   Cuprous ion is the same as copper(II) ion.

8.   Many ionic compounds are not water-soluble.

9.   Group 5A consists of metals and nonmetals.

10.   Elements in group 7A have the largest ionization energies.

11.   The octet rule is limited to main group elements.

12.   Both men and women suffer the same percent bone loss over their lifetimes.

Match each entry on the left with its partner on the right.

1. $FeBr_2$ (a) Sulfite anion

2. S^{2-} (b) Electron-dot symbol

3. NH_4^+ (c) Alkali metal

4. SO_3^{2-} (d) Has the same electron configuration as Na^+

5. Ca   (e) Iron(II) bromide

6. $FeBr_3$ (f) Transition metal

7. Ar   (g) Sulfate anion

8. ·Be·   (h) Alkaline earth metal

9. Co   (i) Has the same electron configuration as Cl^-

10. SO_4^{2-} (j) Iron(III) bromide

11. Ne   (k) Polyatomic cation

12. K   (l) Sulfide anion

Chapter 4   Molecular Compounds

Chapter Outline

I. Covalent bonds (Sections 4.1–4.4).
   A Formation of covalent bonds (Section 4.1).
      1. Covalent bonds occur when two atoms share electrons.
      2. Electron sharing results from the overlap of orbitals of two atoms.
      3. The optimum distance between nuclei of two atoms is the bond length.
      4. Seven elements exist as diatomic molecules: H_2, N_2, O_2, F_2, Cl_2, Br_2, I_2.
   B. Covalent bonds and the periodic table (Section 4.2).
      1. Molecular compounds result when atoms form covalent bonds to another atom or to more than one atom.
      2 The octet rule states that each atom in a molecular compound shares the number of electrons necessary to achieve a noble gas configuration.
      3 Most main group elements form from one to four covalent bonds and obey the octet rule.
         a. Boron forms only three bonds because it has only three valence electrons.
         b. Sulfur and phosphorus may form five or six bonds if they use *d* orbitals.
   C. Multiple covalent bonds (Section 4.3).
      1. Some atoms can share more than one electron pair to form multiple bonds.
         a. A double bond is formed when two pairs are shared.
         b. A triple bond is formed when three pairs are shared.
         c. Carbon, nitrogen, and oxygen are the elements most often present in multiple bonds.
      2. Even when multiple bonds occur, the atoms still obey the octet rule.
   D. Coordinate bonds occur when one atom donates both of the shared electrons (Section 4.4).
II. Molecular compounds (Sections 4.5–4.8).
   A. Characteristics of molecular compounds (Section 4.5).
      1. Molecular compounds are electrically neutral and have no charged particles.
      2. Molecular compounds have low melting and boiling points.
      3. Molecular compounds may be solids, liquids, or gases.
      4. Molecular compounds are usually insoluble in water and do not conduct electricity.
   B. Molecular formulas (Section 4.6).
      1. Molecular formulas show the numbers and kinds of atoms in one molecule of a compound.
      2. Structural formulas show how atoms are connected.
      3. Lewis structures are structural formulas that show electron lone pairs (Section 4.7).
         a. One approach to drawing Lewis structures involves knowing common bonding patterns.
         b. The other approach is a general method.
            i. Find the number of valence electrons for all atoms.
            ii. Draw a line between each pair of connected atoms to represent an electron pair.
            iii. Place lone pairs around all peripheral atoms to give them octets.
            iv. Place remaining electrons around the central atom.
            v. If the central atom doesn't have an octet, use an electron pair from a neighboring atom to form a multiple bond to the central atom.
      4. Some larger organic molecules are written as condensed structures.

C. Molecular shapes can be predicted by using the VSEPR model (Section 4.8).
   1. Draw a Lewis structure of the molecule, and identify the atom whose geometry you want to know.
   2. Count the number of charge clouds around the atom.
   3. Predict shape by assuming that the charge clouds orient in space so that they are as far apart as possible.
      a. If there are two charge clouds, the geometry is linear.
      b. If there are three charge clouds, the geometry is trigonal planar or bent.
      c. If there are four charge clouds, the geometry is tetrahedral, trigonal pyramidal, or bent.
III. Polar covalent bonds (Sections 4.9–4.11).
   A. Polar covalent bonds occur when the electrons in a covalent bond are attracted more to one atom than another (Section 4.9).
      1. The ability of an atom to attract electrons is called electronegativity.
      2. Atoms with electronegativity differences between 0.5 and 1.9 form polar covalent bonds.
      3. Atoms with electronegativity differences >1.9 form ionic bonds.
   B. Molecules containing polar covalent bonds can be polar (Section 4.10).
      1. Molecular polarity depends on both the presence of polar covalent bonds and on molecular shape.
      2. Polarity has a dramatic effect on molecular properties.
   C. Naming binary molecular compounds (Section 4.11).
      1. The less electronegative element is always written first.
      2. Name the first element in the compound, using a prefix if necessary.
      3. Name the second element, using an *-ide* ending and using a prefix if necessary.

Solutions to Chapter 4 Problems

4.1

shared electron pair

Each iodine atom achieves the noble gas configuration of xenon.

4.2 (a) PH_3 hydrogen—one covalent bond; phosphorus—three covalent bonds
      (b) H_2Se hydrogen—one covalent bond; selenium—two covalent bonds
      (c) HCl   hydrogen—one covalent bond; chlorine—one covalent bond
      (d) SiF_4 fluorine—one covalent bond; silicon—four covalent bonds

4.3 (a) CH_2Cl_2. Carbon forms four covalent bonds. Two bonds form between carbon and hydrogen, and the other two form between carbon and chlorine.
      (b) BH_3. Boron forms three covalent bonds.
      (c) NI_3. Nitrogen forms three covalent bonds.
      (d) $SiCl_4$. Silicon, like carbon, forms four covalent bonds.

4.4 In acetic acid, all hydrogen atoms have two outer-shell electrons and all carbon and oxygen atoms have eight outer-shell electrons.

Acetic acid

4.5

4.6

$$H_3B + :NH_3 \longrightarrow H_3B:NH_3$$

The coordinate covalent bond forms between the lone pair of nitrogen and the vacant orbital of boron.

4.7 $AlCl_3$ is a covalent molecular compound, and Al_2O_3 is ionic.

4.8 C and N must be bonded to each other. After drawing the C—N bond and all bonds to hydrogens, two electrons remain; they are a lone pair on nitrogen. All atoms now have a noble gas configuration.

CH_5N: 14 valence electrons

4.9

(a)

(b)

(c)

4.10   (a) For phosgene, $COCl_2$:

> *Step 1:*   Total valence electrons
> $4 \; e^-$ (from C) $+ 6 \; e^-$ (from O) $+ 2 \times 7 \; e^-$ (from Cl) $= 24 \; e^-$

> *Step 2:*   Six electrons are involved in the covalent bonds.
>
> $$\text{Cl}-\text{C}-\text{Cl} \quad (\text{with } \text{O} \text{ above C})$$

> *Step 3:*   The other 18 electrons are placed in nine lone pairs.
>
> $$:\ddot{\text{Cl}}-\text{C}-\ddot{\text{Cl}}: \quad (\text{with } :\ddot{\text{O}}: \text{ above C})$$

> *Step 4:*   All electrons are used up in the preceding structure, but carbon does not have an electron octet, so one electron pair must be moved from oxygen to form a carbon–oxygen double bond.

> *Step 5:*   The 24 electrons have been used up, and all atoms have a complete octet.
>
> $$:\ddot{\text{Cl}}-\text{C}=\ddot{\text{Cl}}: \quad (\text{with } :\text{O}: \text{ double bonded above C})$$

(b) For OCl^-:

> *Step 1:*   Total valence electrons
> $6 \; e^-$ (from O) $+ 7 \; e^-$ (from Cl) $+ 1 \; e^-$ (negative charge) $= 14 \; e^-$

> *Step 2:*   Two electrons used. $\text{O}-\text{Cl}$

> *Step 3:*   Twelve additional electrons used. $:\ddot{\text{O}}-\ddot{\text{Cl}}:^-$

> *Step 4:*   The above structure uses 14 valence electrons, and all atoms have complete octets.

(c) For H_2O_2:

> *Step 1:* $2 \; e^-$ (from 2 H) $+ 2 \times 6 \; e^-$ (from O) $= 14$ electrons

> *Step 2:*   Six electrons are involved in covalent bonds. $\text{H}-\text{O}-\text{O}-\text{H}$

> *Step 3:*   The other eight electrons are placed in four lone pairs. $\text{H}-\ddot{\text{O}}-\ddot{\text{O}}-\text{H}$

> *Step 4:*   The 14 electrons have been used up, and all atoms have complete octets.

(d) For SCl_2: 20 electrons

$$:\ddot{\text{Cl}}-\ddot{\text{S}}-\ddot{\text{Cl}}:$$

4.11

HNO_3:   24 electrons

4.12   (a) The molecular formula for methyl methacrylate is $C_6H_{10}O_2$.
(b) Methyl methacrylate has two double bonds and four electron lone pairs.

Methyl methacrylate

CIA 4.1   Carbon monoxide forms a coordinate covalent bond with Fe^{2+} in hemoglobin.

CIA 4.2

CO is reactive because it can form coordinate covalent bonds with compounds that can accept an electron pair. NO is reactive because it has an unpaired electron.

4.13   Ozone is a bent molecule containing both positive and negative charges. It is a strong oxidant, and two ozones decompose to form three diatomic oxygens.

Ozone

CIA 4.3   Teflon is a polymer of tetrafluoroethylene. Its structure, $-CF_2CF_2CF_2CF_2-$ resembles that of polyethylene, with –F replacing –H.

CIA 4.4   The geometry at (a) is tetrahedral, and the geometry at (b) is trigonal planar.

4.14

tetrahedral

4.15 Carbon, the central atom, is surrounded by four bonds. Referring to Table 4.3, we see that chloroform has tetrahedral geometry.

Chloroform

Each carbon of dichloroethylene is surrounded by three charge clouds. Dichloroethylene is planar, with 120° bond angles.

Dichloroethylene

4.16 Both molecules are bent and have bond angles of approximately 90°.

4.17

Methionine

a. bent
b. tetrahedral
c. tetrahedral
d. trigonal planar
e. pyramidal

4.18 Use Figure 4.7 to predict electronegativity:

Least electronegative — — — — — — —> Most electronegative

H (2.1), P (2.1) < S (2.5) < N (3.0) < O (3.5)

4.19

Electronegativity	Difference	Type of Bond
(a) I (2.5), Cl (3.0)	0.5	polar covalent
(b) Li (1.0), O (3.5)	2.5	ionic
(c) Br (2.8), Br (2.8)	0	covalent
(d) P (2.1), Br (2.8)	0.7	polar covalent

(a) δ^+ δ^-
     I—Cl

(b)   Li$^+$ O$^-$

(d) δ^+ δ^-
     P—Br

The Br–Br bond (c) is nonpolar.

4.20

Formaldehyde is polar because of the polarity of the carbon–oxygen bond and because of the two electron lone pairs.

4.21   The —CH₃ portions of diethyl ether are tetrahedral. The C–O–C portion is bent and has a bond angle of 112°. The molecule has the indicated polarity because of the polar C–O bonds and because of the two lone pairs of electrons.

4.22

The difference in electronegativity between Li (1.0) and C (2.5) indicates that the bond between them is polar covalent, with carbon the more electronegative atom. The electrostatic potential map shows that carbon is electron-rich (red) and lithium is electron-poor (blue).

HOC 4.1
(a) The C–H covalent bonds are nonpolar.
(b) The C–Cl bonds are polar, with the –Cl end of the bond bearing the partial negative charge. The whole molecule is also polar, again with the partial negative charge being located between the two –Cl atoms.
(c) Although all of the C–Cl bonds are polar, the molecule itself is nonpolar.

CIA 4.5
Chemical names are complicated because the name of each of the 20 million known chemicals must be unique and must contain enough information for chemists to identify the composition and structure of each chemical.

CIA 4.6

$$CH_3C{=}CHCH_2CH_2C{=}CHCH_2OH \qquad \text{Geraniol} \qquad C_{10}H_{18}O$$

with CH_3 substituents above the two $=C$ carbons

4.23   (a) S₂Cl₂   Disulfur dichloride   (b) ICl   Iodine monochloride
(c) ICl₃   Iodine trichloride

4.24   (a) SeF₄   Selenium tetrafluoride   (b) P₂O₅   Diphosphorus pentoxide
(c) BrF₃   Bromine trifluoride

Understanding Key Concepts

4.25   (a) tetrahedral geometry   (b) pyramidal geometry   (c) trigonal planar geometry

4.26   All models except (c) represent a molecule with a tetrahedral central atom. In models (b) and (d), some atoms are hidden.

4.27

(a) (b)

Acetaminophen $C_8H_9NO_2$

(c) All carbons except for the starred carbon have trigonal planar geometry. The starred carbon has tetrahedral geometry. Nitrogen has pyramidal geometry.

4.28

Vitamin C

4.29

(a) (b)

Thalidomide $C_{13}H_{10}N_2O_4$

(c) All carbons except for the starred carbons have trigonal planar geometry. The starred carbons have tetrahedral geometry. The nitrogens have pyramidal geometry.

4.30

:O: ⟵ electron-rich

Acetamide

The electrostatic potential maps show that oxygen is the most electron-rich atom (red) in acetamide. The hydrogens are most electron-poor (blue).

Additional Problems

Covalent Bonds and Molecular Compounds

4.31   A covalent bond is a chemical bond in which two electrons are shared between two atoms. An ionic bond is formed by the attraction of a positively charged ion to a negatively charged ion. In an ionic bond, both electrons of the electron pair *belong* to the negatively charged ion.

4.32   In a covalent bond, each atom donates an electron to the bond. Both electrons in a coordinate covalent bond come from the same atom.

4.33   Multiple bonds between atoms form in order that each atom can achieve an outer-shell electron octet.

4.34   Form covalent bonds: (a) aluminum and bromine; (b) carbon and fluorine
Form ionic bonds: (c) cesium and iodine; (d) zinc and fluorine; (e) lithium and chlorine

4.35

4.36   Tellurium, a group 6A element, forms two covalent bonds, as do oxygen, sulfur, and other members of group 6A.

4.37   In the same group as nitrogen (5A), antimony is expected to form 3 covalent bonds. Thus, $SbCl_3$ is covalent, and $SbCl_5$ is ionic.

4.38   If the central atom is expected to form fewer bonds than appear in the formula, the compound probably contains a coordinate covalent bond.
Coordinate covalent bonds: (b) $Cu(NH_3)_4^{2+}$ (c) NH_4^+

4.39   Coordinate covalent bonds: (b) BF_4^- (c) H_3O^+

4.40   Since tin is a member of group 4A, it forms four covalent bonds. $SnCl_4$ is the most likely formula for a molecular compound of tin and chlorine.

4.41   A low-boiling, low-melting compound of gallium and chlorine is most likely covalent (ionic compounds are high-boiling and high-melting). Since gallium is a group 3A element, a likely formula is $GaCl_3$.

4.42   The indicated bond is coordinate covalent because both electrons in the bond come from nitrogen.

4.43 The indicated bond is coordinate covalent because both electrons in the bond come from sulfur.

$$:\overset{..}{\underset{}{O}}:$$

← coordinate covalent

$$:\overset{..}{\underset{..}{Cl}}\diagup\overset{S}{\diagdown}\overset{..}{\underset{..}{Cl}}:$$

Structural Formulas

4.44 (a) A *molecular formula* shows the numbers and kinds of atoms in a molecule; a *structural formula* shows how the atoms in a molecule are bonded to one another.
(b) A *structural formula* shows the bonds between atoms; a *condensed structure* shows central atoms and the atoms connected to them written as groups but does not show bonds.
(c) A *lone pair* of valence electrons is a pair that is not shared; a *shared pair* of electrons is shared between two atoms as a covalent bond.

4.45 Two possible methods to distinguish between the compounds:
(a) Take the melting point of the two compounds. The covalent solid melts at a lower temperature.
(b) Try to dissolve the two solids in water. The ionic solid is more likely to be water-soluble than the covalent solid.

4.46

(a) N_2	10 valence electrons	N_2 contains a triple bond.
(b) NOCl	18 valence electrons	N and O are joined by a double bond.
(c) CH_3CH_2CHO	24 valence electrons	C and O form a double bond.
(d) OF_2	20 valence electrons	

4.47

(a) $:C\equiv O:$ (b) $CH_3\overset{..}{\underset{..}{S}}H$ (c) $H-\overset{..+}{\underset{|}{O}}-H$ with H below   (d) $H_3C-\overset{..}{\underset{|}{N}}-CH_3$ with H below

4.48 A compound with the formula C_2H_8 has too many hydrogens and can't exist because any structure drawn would violate the rules of valence.

4.49 Structure (a) is reasonable. Structure (b) would be reasonable if there were a double bond between the two carbons bearing the –OH groups. Structure (c) would be reasonable if the hydrogen bonded to oxygen were moved to the carbon bearing the C=O double bond.

4.50

(a) $H-\overset{..}{\underset{..}{O}}-\overset{..}{N}=\overset{..}{\underset{..}{O}}$

(b) $H-\overset{\overset{H}{|}}{\underset{\underset{H}{|}}{C}}-C\equiv N:$

(c) $H-\overset{..}{\underset{..}{F}}:$

4.51

$$\left[\begin{array}{c} \text{:O:} \\ \| \\ \text{:O—N—O:} \end{array} \right]^{-}$$

Three oxygen atoms and one nitrogen atom yield a total of 23 valence electrons. An additional electron is added so that the octet rule is satisfied for all atoms.

4.52   (a) $CH_3CH_2CH_3$ (b) $H_2C=CHCH_3$ (c) CH_3CH_2Cl

4.53

(a)

(b)

(c)

4.54 CH_3COOH

Drawing Lewis Structures

4.55   Recall that in Section 4.2 it was stated that sulfur (and selenium) can form more than four covalent bonds.

(a)

(b)

(c)

(d)

(e)

(f)

4.56

(a)

(b)

(c)

In (b), sulfur can also use *d* orbitals to form three double bonds to the oxygens.

4.57

H—C—C—O—H   Ethanol

4.58

H—C—O—C—H   Dimethyl ether

(structure with H atoms above and below each C, and two lone pairs on O)

4.59

(structure of C=C with :Cl: groups) Tetrachloroethylene contains a double bond.

4.60

H—N—O—H   Hydroxylamine

(structure with H below N, lone pair on N and two lone pairs on O)

4.61 Each atom of carbonate ion has 2 inner-shell electrons, plus the 24 electrons pictured, for a total of 32 electrons. Together, carbon and the three oxygens have 30 protons. Since there are two more electrons than protons, carbonate ion has a charge of –2.

$$\left[\begin{array}{c} :O: \\ \| \\ :O-C-O: \end{array}\right]^{2-}$$

4.62

(a) $\left[\begin{array}{c} :O: \\ \| \\ H-C-O: \end{array}\right]^{-}$ (b) $\left[\begin{array}{c} :O: \\ | \\ :O-S-O: \end{array}\right]^{2-}$ (c) $\left[:S-C\equiv N:\right]^{-}$ (d) $\left[\begin{array}{c} :O: \\ | \\ :O-P-O: \\ | \\ :O: \end{array}\right]^{3-}$

(e) $\left[:O-Cl=O:\right]^{-}$

Molecular Geometry

4.63 Use Table 4.3 to predict molecular geometry. B equals the number of bonds, and E equals the number of electron lone pairs:

Molecule	Number of Bonds	Number of Lone Pairs	Shape	Bond Angle
(a) AB$_3$	3	0	trigonal planar	120°
(b) AB$_2$E	2	1	bent	120°

4.64 As in the previous problem, use Table 4.3.

Molecule	Number of Bonds	Number of Lone Pairs	Shape	Bond Angle
(a) AB_4	4	0	tetrahedral	109°
(b) AB_3E	3	1	pyramidal	109°
(c) AB_2E_2	2	2	bent	109°

4.65

(a)   (b)   (c)   (d)   (e)

4.66

Molecule	Number of Bonding Pairs Around Central Atom	Number of Lone Pairs	Shape	Bond Angle
(a) SiF_4	4	0	tetrahedral	109°
(b) CF_2Cl_2	4	0	tetrahedral	109°
(c) SO_3	3*	0	trigonal planar	120°
(d) BBr_3	3	0	trigonal planar	120°
(e) NF_3	3	1	pyramidal	109°

* For determining shape, the double bond counts as one bond pair, and the number of bonds is three.

4.67

tetrahedral →   ← trigonal planar
Alanine

4.68

Vinyl acetate

All carbons are trigonal planar, except for the starred carbon, which is tetrahedral.

Polarity of Bonds and Molecules

4.69 The most electronegative elements are found on the upper right side of the periodic table. The least electronegative elements are found on the left side of the periodic table.

4.70   Using the periodic table, count up from element 114. Element 119 should occur under francium and accordingly should be one of the least electronegative elements, with an electronegativity of around 0.8.

4.71   *Less electronegative* $----\!>$ *More electronegative*
              K < Be < Si < B < O

4.72   *More electronegative* $----\!>$ *Less electronegative*
              Cl > C > Cu > Ca > Cs

4.73

$$\overset{\delta+\ \ \delta-}{\text{(a) I—Br}}\quad\overset{\delta-\ \ \delta+}{\text{(b) O—H}}\quad\overset{\delta+\ \ \delta-}{\text{(c) C—F}}\quad\overset{\delta-\ \ \delta+}{\text{(d) N—C}}\quad\text{(e) nonpolar}$$

4.74   Bonds in (b), (c), and (d) are nonpolar.

$$\overset{\delta-\ \ \delta+}{\text{(a) O—Cl}}\quad\overset{\delta+\ \ \delta-}{\text{(e) C—O}}$$

4.75   *Electronegativity*   *Difference*   *Type of Bond*
          _____
          (a) Be (1.5), F (4.0)   2.5   ionic
          (b) Ca (1.0), Cl (3.0)   2.0   ionic
          (c) O (3.5), H (2.1)   1.4   polar covalent
          (d) Be (1.5), Br (2.8)   1.3   polar covalent

4.76   Use Figure 4.7 to determine bond polarities.

   Least polar bonds $----\!>$ *Most polar bonds*

   PH_3 < HCl < H_2O < CF_4

4.77   Both compounds have dipole moments due to the lone pair electrons of the central atom. The dipole moment of NH_3 is greater because of the bond polarities of the three N–H bonds; the P–H bonds are nonpolar.

4.78

(a)   (b)   (c)   (d) nonpolar

   H—Cl   H--P—H   H--O

                    H   H

   polar   polar   polar

4.79

:Ö=C=Ö: $^{+}$S=O
                          $^{-}$:O:

The bond polarities of the individual bonds cancel in CO_2, but bent geometry makes SO_2 polar.

4.80 Both molecules are polar because of the lone pair electrons. Water is more polar because the bonds between oxygen and hydrogen are polar covalent, but the bonds between sulfur and hydrogen are nonpolar.

Names and Formulas of Molecular Compounds

4.81 (a) PI_3 phosphorus triiodide   (b) $AsCl_3$ arsenic trichloride
     (c) P_4S_3 tetraphosphorus trisulfide   (d) Al_2F_6 dialuminum hexafluoride
     (e) N_2O_5 dinitrogen pentoxide   (f) $AsCl_5$ arsenic pentachloride

4.82 (a) SeO_2 selenium dioxide   (b) XeO_4 xenon tetroxide
     (c) N_2S_5 dinitrogen pentasulfide   (d) P_3Se_4 triphosphorus tetraselenide

4.83 (a) NO_2 nitrogen dioxide   (b) SF_6 sulfur hexafluoride
     (c) BrI_3 bromine triiodide   (d) N_2O_3 dinitrogen trioxide
     (e) NI_3 nitrogen triiodide   (f) IF_7 iodine heptafluoride

4.84 (a) $SiCl_4$ silicon tetrachloride   (b) NaH sodium hydride
     (c) SbF_5 antimony pentafluoride   (d) OsO_4 osmium tetroxide

Conceptual Problems

4.85 (a) It was thought that noble gases couldn't form bonds because they already had a full electron octet.

(b)

Xenon tetrafluoride

There are six electron clouds around the central atom.

(c) The Xe–F bonds are polar covalent. Each atom contributes one electron to the bond.

4.86

(a)

(b) The C=O carbon atoms have trigonal planar geometry, and the other carbons have tetrahedral geometry. (c) The C=O bonds are polar.

4.87

Both molecules are expected to be polar because of their C—O bonds.

4.88   Consult Figure 4.3 for help.

(a) Carbon forms four bonds. The correct formula is CCl_4.
(b) Nitrogen forms three bonds. The correct formula is N_2H_4.
(c) Sulfur forms two bonds. The correct formula is H_2S.
(d) C_2OS *could* actually be correct (S=C=C=O), but compounds with such adjacent double bonds are rare. More likely is the formula COS, a structural relative of carbon dioxide (S=C=O).

4.89   (a) $BaCl_2$ ionic bonds (b) $Ca(NO_3)_2$ ionic and covalent bonds (c) BCl_4^- covalent bonds and coordinate covalent bonds (d) $TiBr_4$ covalent bonds

4.90

(a)

$$\left[\begin{array}{c} H \\ | \\ H-P-H \\ | \\ H \end{array} \right]^+$$

(b) Phosphonium ion is tetrahedral.

(c) Phosphorus donates its lone pair of electrons to H^+ to form a coordinate covalent bond.
(d) Number of electrons: 15 (from P) + 3 (from 3 H) = 18
     Number of protons: 15 (from P) + 4 (from 4 H) = 19
PH_4^+ is positive because the number of protons exceeds the number of electrons by one.

4.91   Both figures show that the halogens and the group 6A elements are the most electronegative. It is surprising to see in Figure 3.2 that group 5A elements have small electron affinities, whereas their electronegativities given in Figure 4.7 are significant.

4.92   *Compound   Name*
_____

(a) $CaCl_2$ calcium chloride
(b) $TeCl_2$ tellurium dichloride
(c) BF_3 boron trifluoride
(d) $MgSO_4$ magnesium sulfate
(e) K_2O potassium oxide
(f) FeF_3 iron(III) fluoride
(g) PF_3 phosphorus trifluoride

4.93   The structures of sulfite anion and sulfur trioxide are shown below. It is also possible to draw a structure for SO_3 in which sulfur forms three double bonds to the three oxygens.

Sulfite anion has four charge clouds, including one electron lone pair, and has pyramidal geometry. Sulfur trioxide has three charge clouds and has planar geometry.

4.94

Chloral hydrate

4.95

Dichromate ion

4.96

Oxalic acid

4.97

(a)   O=Se=O

(b)

4.98

(a)

(b)

4.99   Ammonium ion and sulfate ion are tetrahedral. Phosphite ion is pyramidal.

Ammonium ion   Sulfate ion   Phosphite ion

Group Problems

4.100 (a) oxygen: (i) (iv)   (b) potassium: (iii)   (c) phosphorus: (ii)
          (d) iodine: (i) (iv)   (e) hydrogen: (i) (ii)   (f) cesium: (iii)

4.101

$$H-\overset{\overset{\displaystyle H}{|}}{N}-\overset{\overset{\displaystyle ..}{N}}{\underset{\displaystyle H}{|}}-H$$ Hydrazine

4.102

$$H-\overset{\overset{\displaystyle H}{|}}{C}-\overset{\overset{\displaystyle :\ddot{O}:}{|}}{S}-\overset{\overset{\displaystyle H}{|}}{C}-H$$ Dimethyl sulfoxide

4.103 $TiBr_4$ is a molecular compound, and TiO_2 is an ionic compound. The electronegativity difference between Ti and Br (approx. 1.4) indicates polar covalent bonds, whereas the electronegativity difference between Ti and O (approx. 2.1) indicates ionic bonds.

Self-Test for Chapter 4

Multiple Choice

1. Which of these diatomic molecules contains a double bond?
    (a) H_2 (b) I_2 (c) O_2 (d) N_2

2. Which of these molecules is pyramidal?
    (a) CBr_4 (b) $AlCl_3$ (c) SF_6 (d) PH_3

3. Choose the element that forms three covalent bonds and has one lone pair of electrons.
    (a) C   (b) N   (c) O   (d) F

4. How many charge clouds does the molecule H_2S have?
    (a) 2   (b) 3   (c) 4   (d) can't tell

5. Which of the following double bonds is not likely to be found in organic molecules?
    (a) O=O   (b) C=C   (c) C=O   (d) C=N

6. All of the following are true about the polyatomic ion BF_4^- except:
    (a) All atoms don't have electron octets.   (b) It has polar covalent bonds.
    (c) It contains a coordinate covalent bond.   (d) It has a tetrahedral shape.

7. One of these molecules doesn't contain a polar covalent bond. Which is it?
    (a) $BeBr_2$ (b) PCl_3 (c) CO_2 (d) CS_2

8. Which of the following compounds is polar?
   (a) $BeBr_2$ (b) PCl_3 (c) CO_2 (d) CS_2

9. A name for S_2F_2 is:
   (a) sulfur fluoride   (b) sulfur difluoride   (c) disulfur fluoride   (d) disulfur difluoride

10. Which of the following is not true for molecular compounds?
    (a) They are water-soluble.   (b) They can be solids, liquids, or gases.
    (c) They are composed of nonmetals.   (d) They have low melting points.

Sentence Completion

1. An _____ element strongly attracts electrons.

2. A molecular compound that occurs in living organisms is called a _____ .

3. In Lewis structures, a line represents a _____ bond.

4. _____ and _____ are the most common triple bonds in chemical compounds.

5. The formula X_2 represents a _____ molecule.

6. A molecule whose central atom forms three bonds and has no lone pairs has a _____ _____ shape.

7. _____ orbitals in sulfur and phosphorus can be used for covalent bonds.

8. N_2 contains a _____ bond.

9. _____ molecular compounds are formed from only two elements.

10. Many molecular compounds are soluble in _____ liquids.

11. Large organic molecules are often written as _____ structures.

True or False

1. Br_2 is a molecular compound.

2. Six electrons are used to form a triple bond.

3. Coordinate covalent bonds occur only in cations or anions, not in neutral compounds.

4. The first step in drawing a Lewis structure is to find the total number of electrons in the combined atoms.

5. $AlCl_3$ is a pyramidal molecule.

6. The shape of ionic compounds can be predicted by the VSEPR model.

7. Electronegativity decreases in going down the periodic table.

8. CI_4 contains polar covalent bonds.

9.   Both planar and bent molecules have bond angles of 120°.

10.   Group 7A elements can form more than one covalent bond.

11.   Bent compounds may have either one or two lone pairs.

Match each entry on the left with its partner on the right.

1. BCl_3 (a) Contains a pure covalent bond

2.   Planar molecule   (b) $BeCl_2$

3.   Bent molecule   (c) Central atom has more than an electron octet

4. Cl_2 (d) NH_3

5. BCl_4^- (e) Contains a triple bond

6.   NaI   (f) H_2O

7. $C{\equiv}O$ (g) Contains a double bond

8. SF_6 (h) Central atom doesn't have an electron octet

9.   Linear molecule   (i) Contains an ionic bond

10. $O{=}C{=}O$ (j) $AlCl_3$

11.   Pyramidal molecule   (k) Contains a coordinate covalent bond

12. CH_3Cl (l) Contains a polar covalent bond

Chapter 5   Classification and Balancing of Chemical Reactions

Chapter Outline

I. Chemical equations (Sections 5.1–5.2).
   A. Writing chemical equations (Section 5.1).
      1. A chemical equation describes a chemical reaction.
         a. The reactants are written on the left.
         b. The products are written on the right.
         c. An arrow goes between them to indicate the chemical change.
      2. The number and kinds of atoms must be the same on both sides of the equation.
         a. This is known as the law of conservation of mass.
      3. Numbers that are placed in front of formulas are called coefficients.
      4. States of matter (s), (g), (l) are often placed after chemical formulas.
   B. Balancing chemical equations (Section 5.2).
      1. Write an unbalanced equation using the correct formulas for all substances.
      2. Add appropriate coefficients to balance the atoms of each type, one at a time.
      3. Check to make sure that the numbers and kinds of atoms are balanced.
      4. Make sure all coefficients are reduced to their lowest whole-number values.
II. Chemical reactions (Sections 5.3–5.7).
   A. Classes of chemical reactions (Section 5.3).
      1. Precipitation reactions occur when an insoluble solid is formed.
      2. Acid–base reactions occur when an acid and a base react to yield water and a salt.
      3. Redox reactions occur when electrons are transferred between reaction partners.
   B. Precipitation reactions.
      1. To predict whether a precipitation reaction occurs, you must know the solubilities of the products.
      2. Table 5.1 gives general solubility rules to predict if a precipitation reaction will occur.
   C. Acid–base reactions (Section 5.4).
      1. Acid–base reactions are known as neutralization reactions because when the reaction is complete the solution is neither acidic nor basic.
      2. An example: $HA(aq) + MOH(aq) \longrightarrow H_2O(l) + MA(aq)$
      3. When a carbonate or bicarbonate is one of the reactants, CO_2 is also produced.
   D. Redox reactions (Sections 5.5–5.6).
      1. Definition of redox reactions (Section 5.5).
         a. A redox reaction occurs when electrons are transferred from one atom to another.
         b. The substance that loses electrons is oxidized and is known as the reducing agent.
         c. The substance that gains electrons is reduced and is known as the oxidizing agent.
         d. Oxidation and reduction always occur together.
            i. In the reaction of a metal with a nonmetal, the metal loses electrons, and the nonmetal gains electrons.
            ii. In the reaction of two nonmetals, the *more metallic* element loses electrons, and the *less metallic* element gains electrons.
         e. Redox reactions occur during corrosion, combustion, respiration, and photography.

2. Recognizing redox reactions (Section 5.6).
   a. With some substances, it is not obvious if a redox reaction has occurred.
   b. In a neutral compound, oxidation numbers are assigned to each element to indicate electron ownership.
   c. To assign oxidation numbers:
      i.   An atom in its elemental state has an oxidation number of zero.
      ii. A monatomic ion has an oxidation number equal to its charge.
      iii. In a molecular compound, an atom usually has the same oxidation number that it would have if it were a monatomic ion.
      iv. The sum of the oxidation numbers in a neutral compound is zero.
      v.   The oxidation number of oxygen is usually –2; the oxidation number of hydrogen is usually +1.
E. Net ionic equations (Section 5.7).
   1. A molecular equation shows reactants and products as molecules and doesn't indicate if they are ions.
   2. An ionic equation is written if ions are involved and shows all ionic reactants and products.
   3. A net ionic equation includes only the ions that undergo change and deletes spectator ions.

Solutions to Chapter 5 Problems

5.1 *Step 1:* Write the unbalanced equation.

$$O_2 \longrightarrow O_3$$

Step 2: Balance the atoms of each type, one by one.

$$3\,O_2 \longrightarrow 2\,O_3 \quad \text{The equation is balanced.}$$

5.2 (a) $Ca(OH)_2 + 2\,HCl \longrightarrow CaCl_2 + 2\,H_2O$

(b) $4\,Al + 3\,O_2 \longrightarrow 2\,Al_2O_3$

(c) $2\,CH_3CH_3 + 7\,O_2 \longrightarrow 4\,CO_2 + 6\,H_2O$

(d) $2\,AgNO_3 + MgCl_2 \longrightarrow 2\,AgCl + Mg(NO_3)_2$

5.3 *Step 1:* $A + B_2 \longrightarrow A_2B_2$

Step 2: $2\,A + B_2 \longrightarrow A_2B_2$ The equation is balanced.

HOC 5.1

1 cup butter + 1 cup sugar + 1 egg + 2 tbsp orange juice + 2 1/2 cup flour + 1 tbsp vanilla + 1 tsp baking powder = 36 butter cookies
Grams are the best units if all ingredients have to be measured in the same units.

5.4   Use Table 5.1 for solubility guidelines.

Insoluble
(a) $CdCO_3$ (most carbonates are insoluble)
(c) $PbSO_4$ (exception to the rule that sulfates are soluble)
(e) Hg_2Cl_2 (exception to the rule that chlorides are soluble)

Soluble
(b) Na_2S
(d) $(NH_4)_3PO_4$

5.5   As in Worked Example 5.4, identify the products, and use Table 5.1 to predict their solubility. If the products are insoluble, a precipitation reaction will occur.

(a) $NiCl_2(aq) + (NH_4)_2S(aq) \longrightarrow 2\,NH_4Cl(aq) + NiS(s)$
A precipitation reaction will occur.

(b) $2\,AgNO_3(aq) + CaBr_2(aq) \longrightarrow Ca(NO_3)_2(aq) + 2\,AgBr(s)$
A precipitation reaction will occur.

CIA 5.1 $CaCl_2(aq) + Na_2C_2O_4(aq) \longrightarrow 2\,NaCl(aq) + CaC_2O_4(s)$

CIA 5.2   (a) $2\,C_5H_4N_4 + 3\,O_2 \longrightarrow 2\,C_5H_4N_4O_3$
(b) This is a redox reaction in which purine is oxidized and O_2 is reduced.

5.6   (a) $2\,CsOH(aq) + H_2SO_4(aq) \longrightarrow Cs_2SO_4(aq) + 2\,H_2O(l)$

(b) $Ca(OH)_2(aq) + 2\,CH_3CO_2H(aq) \longrightarrow Ca(CH_3CO_2)_2(aq) + 2\,H_2O(l)$

(c) $NaHCO_3(aq) + HBr(aq) \longrightarrow NaBr(aq) + CO_2(g) + H_2O(l)$

HOC 5.2
1. The acetic acid smells pungent and irritating.
2. The solution bubbles, indicating the release of CO_2 and indicating a neutralization reaction. You may or may not still smell acetic acid. If you do smell it, add another antacid, and you may again see CO_2 bubbles.

5.7   (a) $AgNO_3(aq) + KCl(aq) \longrightarrow AgCl(s) + KNO_3(aq)$
Precipitation: Solid AgCl is formed.

(b) $2\,Al(s) + 3\,Br_2(l) \longrightarrow 2\,AlBr_3(s)$
Redox reaction: Al^{+3} is formed from Al, and Br^- is formed from Br_2.

(c) $Ca(OH)_2(aq) + 2\,HNO_3(aq) \longrightarrow 2\,H_2O(l) + Ca(NO_3)_2(aq)$
Acid–base neutralization.

5.8

Oxidized Reactant/ Reducing Agent	*Reduced Reactant/ Oxidizing Agent*
(a) Fe	Cu^{2+}
(b) Mg	Cl_2
(c) Al	Cr_2O_3

The oxidizing agent is the reduced reactant, and the reducing agent is the oxidized reactant.

5.9 $2 K(s)$ + $Br_2(l)$ $\longrightarrow$ $2 KBr(s)$
          Reducing agent   *Oxidizing agent*

CIA 5.3 Elemental cadmium is oxidized, and nickel [as NiO(OH)] is reduced.

CIA 5.4 Lithium is oxidized, and iodine is reduced.

5.10 *Compound*   *Oxidation Number of Metal*   *Name*

	Compound	Oxidation Number of Metal	Name
(a)	VCl_3	+3	vanadium(III) chloride
(b)	$SnCl_4$	+4	tin(IV) chloride
(c)	CrO_3	+6	chromium(VI) oxide
(d)	$Cu(NO_3)_2$	+2	copper(II) nitrate
(e)	$NiSO_4$	+2	nickel(II) sulfate

5.11 Oxidation numbers are written above the atoms.

       +1–2   +2–1   +1–1   +2–2
(a) $Na_2S(aq) + NiCl_2(aq)$ $\longrightarrow$ $2 NaCl(aq) + NiS(s)$
This reaction is not a redox reaction because no atoms change oxidation numbers.

       0   +1–2   +1–2+1   0
(b) $2 Na(s) + 2 H_2O(l)$ $\longrightarrow$ $2 NaOH(aq) + H_2(g)$
Sodium is oxidized and hydrogen is reduced in this reaction.

       0   0   +4–2
(c) $C(s) + O_2(g)$ $\longrightarrow$ $CO_2(g)$
Carbon is oxidized and oxygen is reduced in this redox reaction.

       +2-2   0   +4–2
(d) $2 CO(s) + O_2(g)$ $\longrightarrow$ $2 CO_2(g)$
Carbon is oxidized and oxygen is reduced in this redox reaction.

       +2–2   +1–1   +2–1   +1–2
(e) $CuO(s) + 2 HCl(aq)$ $\longrightarrow$ $CuCl_2(aq) + H_2O(l)$
This is not a redox reaction because no atoms change oxidation number.

       +7–2   +4–2   +1–2   +2   +6–2   +1
(f) $2 MnO_4^-(aq) + 5 SO_2(g) + 2 H_2O(l)$ $\longrightarrow$ $2 Mn^{2+}(aq) + 5 SO_4^{2-}(aq) + 4 H^+(aq)$
In this redox reaction, the oxidation number of Mn changes from +7 to +2 and the oxidation number of S changes from +4 to +6.

5.12 *Reducing Agent   Oxidizing Agent*

	Reducing Agent	Oxidizing Agent
(b)	Na	H_2O
(c)	C	O_2
(d)	CO	O_2
(f)	SO_2	MnO_4^-

5.13   (a) Write the equation, including all ions.

$$Zn(s) + Pb^{2+} + 2\,NO_3^- \longrightarrow Zn^{2+} + 2\,NO_3^- + Pb(s)$$

The nitrate ions on each side cancel.

$$Zn(s) + Pb^{2+}(aq) \longrightarrow Zn^{2+}(aq) + Pb(s)$$

(b) $OH^-(aq) + H^+(aq) \longrightarrow H_2O(l)$
K^+ and SO_4^{2-} cancel, and coefficients are reduced.

(c) $2\,Fe^{3+}(aq) + Sn^{2+}(aq) \longrightarrow 2\,Fe^{2+}(aq) + Sn^{4+}(aq)$
The chloride ions cancel.

5.14   Reactions (a) and (c) are redox reactions, and (b) is a neutralization reaction.

5.15
$$\overset{0}{(a)\ Zn}(s) + \overset{+2}{Pb^{2+}}(aq) \longrightarrow \overset{+2}{Zn^{2+}}(aq) + \overset{0}{Pb}(s)$$
Zinc is oxidized, and lead is reduced.

$$\overset{+3}{(c)\ 2\,Fe^{3+}}(aq) + \overset{+2}{Sn^{2+}}(aq) \longrightarrow \overset{+2}{2\,Fe^{2+}}(aq) + \overset{+3}{Sn^{4+}}(aq)$$
Tin is oxidized, and iron is reduced.

Understanding Key Concepts

5.16   Product mixture (d) is the only mixture that contains the same number of atoms as (a).

5.17   (c) $2\,A + B_2 \longrightarrow A_2B_2$

5.18   (d) $\longrightarrow$ (c)
reactants   products

5.19   (a) Mixing of sodium and carbonate ions produces outcome (1). No insoluble product is formed, and sodium and carbonate ions remain in solution.
(b) Mixing barium and chromate ions produces outcome (2). $BaCrO_4$ precipitate forms (one barium ion for each chromate ion), and the excess barium ions remain in solution.
(c) Mixing silver and sulfite ions yields outcome (3). Ag_2SO_3 precipitate forms (two silver ions for each sulfite ion), and no Ag^+ or SO_3^{2-} ions remain in solution.

5.20   The observed product is a precipitate that contains two cations per anion. From the list of cations, Ag^+ is the only one that forms precipitates and has a 2:1 cation/anion ratio with the anions listed. The anion can be either CO_3^{2-} or CrO_4^{2-} because both form precipitates with Ag^+. Thus, the possible products are Ag_2CO_3 and Ag_2CrO_4.

5.21   (a) $CaCl_2$ is dissolved in beaker A. (PbI_2 has the same ratio of cations to anions but is insoluble in water.)
(b) Na_2SO_4 is dissolved in beaker B.
(c) $CaSO_4$ is the precipitate, and Na^+ and Cl^- are the spectator ions.

Additional Problems

Balancing Chemical Equations

5.22 A balanced equation is an equation in which the number of atoms of each kind is the same on both sides of the reaction arrow.

5.23 Changing the subscripts on a substance to balance an equation changes the identity of the substance and makes the equation meaningless.

5.24 (a) $2\,HCl + CaCO_3 \longrightarrow CO_2 + H_2O + CaCl_2$

   (b) $2\,K(s) + Br_2(l) \longrightarrow 2\,KBr(s)$

   (c) $C_3H_8(g) + 5\,O_2(g) \longrightarrow 3\,CO_2(g) + 4\,H_2O(g)$

5.25 $4\,NH_3(g) + Cl_2(g) \longrightarrow N_2H_4(l) + 2\,NH_4Cl(s)$

5.26 (a) $2\,C_2H_6(g) + 7\,O_2(g) \longrightarrow 4\,CO_2(g) + 6\,H_2O(l)$

   (b) balanced

   (c) $2\,Mg(s) + O_2(g) \longrightarrow 2\,MgO(s)$

   (d) $2\,K(s) + 2\,H_2O(l) \longrightarrow 2\,KOH(aq) + H_2(g)$

5.27 (a) balanced

   (b) $C_2H_8N_2 + 2\,N_2O_4 \longrightarrow 3\,N_2 + 2\,CO_2 + 4\,H_2O$

   (c) balanced

   (d) $2\,N_2O \longrightarrow 2\,N_2 + O_2$

5.28 (a) $Hg(NO_3)_2(aq) + 2\,LiI(aq) \longrightarrow 2\,LiNO_3(aq) + HgI_2(s)$

   (b) $I_2(s) + 5\,Cl_2(g) \longrightarrow 2\,ICl_5(s)$

   (c) $4\,Al(s) + 3\,O_2(g) \longrightarrow 2\,Al_2O_3(s)$

   (d) $CuSO_4(aq) + 2\,AgNO_3(aq) \longrightarrow Ag_2SO_4(s) + Cu(NO_3)_2(aq)$

   (e) $2\,Mn(NO_3)_3(aq) + 3\,Na_2S(aq) \longrightarrow Mn_2S_3(s) + 6\,NaNO_3(aq)$

5.29 (a) $4\,NO_2(g) + O_2(g) \longrightarrow 2\,N_2O_5(g)$

   (b) $P_4O_{10}(s) + 6\,H_2O(l) \longrightarrow 4\,H_3PO_4(aq)$

   (c) $B_2H_6(l) + 3\,O_2(g) \longrightarrow B_2O_3(s) + 3\,H_2O(l)$

   (d) $Cr_2O_3(s) + 3\,CCl_4(l) \longrightarrow 2\,CrCl_3(s) + 3\,COCl_2(aq)$
   (e) $4\,Fe_3O_4(s) + O_2(g) \longrightarrow 6\,Fe_2O_3(s)$

5.30 (a) $2 C_4H_{10}(g) + 13 O_2(g) \longrightarrow 8 CO_2(g) + 10 H_2O(l)$

(b) $C_2H_6O(g) + 3 O_2(g) \longrightarrow 2 CO_2(g) + 3 H_2O(l)$

(c) $2 C_8H_{18}(g) + 25 O_2(g) \longrightarrow 16 CO_2(g) + 18 H_2O(l)$

5.31 (a) $2 C_4H_{10}(g) + 9 O_2(g) \longrightarrow 8 CO(g) + 10 H_2O(l)$

(b) $C_2H_6O(g) + 2 O_2(g) \longrightarrow 2 CO(g) + 3 H_2O(l)$

(c) $2 C_8H_{18}(g) + 17 O_2(g) \longrightarrow 16 CO(g) + 18 H_2O(l)$

5.32 $4 HF + SiO_2 \longrightarrow SiF_4 + 2 H_2O$

5.33 $Na_2CO_3(aq) + 2 HNO_3(aq) \longrightarrow CO_2(g) + 2 NaNO_3(aq) + H_2O(l)$

Types of Chemical Reactions

5.34 (a) $Mg(s) + 2 HCl(aq) \longrightarrow MgCl_2(aq) + H_2(g)$ redox reaction

(b) $KOH(aq) + HNO_3(aq) \longrightarrow KNO_3(aq) + H_2O(l)$ neutralization reaction

(c) $Pb(NO_3)_2(aq) + 2 HBr(aq) \longrightarrow PbBr_2(s) + 2 HNO_3(aq)$ precipitation reaction

(d) $Ca(OH)_2(aq) + 2 H_2SO_4(aq) \longrightarrow 2 H_2O(l) + CaSO_4(aq)$ neutralization reaction

5.35

(a) $2 H^+(aq) + SO_4^{2-}(aq) + 2 K^+(aq) + 2 OH^-(aq) \rightarrow 2 K^+(aq) + SO_4^{2-}(aq) + 2 H_2O(l)$
$H^+(aq) + OH^-(aq) \rightarrow H_2O(l)$

(b) $Mg^{2+}(aq) + 2 OH^-(aq) + 2 H^+(aq) + 2 Cl^-(aq) \rightarrow Mg^{2+}(aq) + 2 Cl^-(aq) + 2 H_2O(l)$
$H^+(aq) + OH^-(aq) \rightarrow H_2O(l)$

5.36

(a) $Ba^{2+}(aq) + 2 NO_3^-(aq) + 2 K^+(aq) + SO_4^{2-}(aq) \rightarrow BaSO_4(s) + 2 K^+(aq) + 2 NO_3^-(aq)$
$Ba^{2+}(aq) + SO_4^{2-}(aq) \rightarrow BaSO_4(s)$

(b) $Zn(s) + 2 H^+(aq) + SO_4^{2-}(aq) \rightarrow Zn^{2+}(aq) + H_2(g) + SO_4^{2-}(aq)$
$Zn(s) + 2 H^+(aq) \rightarrow Zn^{2+}(aq) + H_2(g)$

5.37 redox: (a), (c), (d); neutralization: (b)

5.38 redox: (b), (c); precipitation: (a), (d), (e)

5.39 Use Section 5.3 as a guide.
Only $ZnSO_4$ is soluble in water.

5.40 Use Section 5.3 as a guide.
Only $Ba(NO_3)_2$ is soluble in water.

5.41 A precipitation reaction occurs only in (b). A neutralization occurs with the reagents in (a).

    (b) $FeCl_2(aq) + 2\,KOH(aq) \longrightarrow Fe(OH)_2(s) + 2\,KCl(aq)$

5.42 If the products consist of soluble ions, no reaction will occur. No reaction occurs in (b) and (c).

    (a) $2\,NaBr(aq) + Hg_2(NO_3)_2(aq) \longrightarrow Hg_2Br_2(s) + 2\,NaNO_3(aq)$

    (d) $(NH_4)_2CO_3(aq) + CaCl_2(aq) \longrightarrow CaCO_3(s) + 2\,NH_4Cl(aq)$

    (e) $2\,KOH(aq) + MnBr_2(aq) \longrightarrow Mn(OH)_2(s) + 2\,KBr(aq)$

    (f) $3\,Na_2S(aq) + 2\,Al(NO_3)_3(aq) \longrightarrow Al_2S_3(s) + 6\,NaNO_3(aq)$

5.43 In net ionic reactions, no spectator ions appear. Otherwise, the equations are balanced for number of atoms and charge, and coefficients are reduced to their lowest common denominators.

    (a) $Mg(s) + Cu^{2+}(aq) \longrightarrow Mg^{2+}(aq) + Cu(s)$

    (b) $2\,Cl^-(aq) + Pb^{2+}(aq) \longrightarrow PbCl_2(s)$

    (c) $2\,Cr^{3+}(aq) + 3\,S^{2-}(aq) \longrightarrow Cr_2S_3(s)$

5.44 (a) $2\,Au^{3+}(aq) + 3\,Sn(s) \longrightarrow 3\,Sn^{2+}(aq) + 2\,Au(s)$

    (b) $2\,I^-(aq) + Br_2(l) \longrightarrow 2\,Br^-(aq) + I_2(s)$

    (c) $2\,Ag^+(aq) + Fe(s) \longrightarrow Fe^{2+}(aq) + 2\,Ag(s)$

5.45 (a) $FeSO_4(aq) + Sr(OH)_2(aq) \longrightarrow Fe(OH)_2(s) + SrSO_4(s)$

    (b) $Na_2S(aq) + ZnSO_4(aq) \longrightarrow Na_2SO_4(aq) + ZnS(s)$

5.46 (a) The net ionic equation is the same as 5.45 (a) because no ions are spectator ions.

    (b) $S^{2-}(aq) + Zn^{2+}(aq) \longrightarrow ZnS(s)$

Redox Reactions and Oxidation Numbers

5.47 In general, the best reducing agents are metals. The most reactive reducing agents are in groups 1A and 2A. The most reactive oxidizing agents are in groups 6A and 7A.

5.48 The most easily reduced elements are found in groups 6A and 7A. The most easily oxidized elements are in groups 1A and 2A.

5.49 *Gains electrons:* (a) oxidizing agent, (d) substance undergoing reduction
Loses electrons: (b) reducing agent, (c) substance undergoing oxidation

5.50 Oxidation number increases: (b) reducing agent, (c) substance undergoing oxidation
Oxidation number decreases: (a) oxidizing agent, (d) substance undergoing reduction

5.51   +5 -2   +4-2   0+1-2   +1+5-2
(a) N_2O_5 (b) SO_3^{2-} (c) CH_2O (d) $HClO_3$

5.52   (a) Co: +3   (b) Fe: +2   (c) U: +6   (d) Cu: +2   (e) Ti: +4   (f) Sn: +2

5.53   The reduced element gains electrons, and the oxidized element loses electrons.

	(a)	(b)	(c)
Oxidized	Si	Br	Sb
Reduced	Cl	Cl	Cl

5.54

	(a)	(b)	(c)	(d)
Oxidized	S	Na	Zn	Cl
Reduced	O	Cl	Cu	F

5.55   (a) $2\,Al(s) + 3\,H_2SO_4(aq) \longrightarrow Al_2(SO_4)_3(aq) + 3\,H_2(g)$
(b) $2\,Fe(s) + 3\,Cl_2(g) \longrightarrow 2\,FeCl_3(s)$
(c) $5\,CO(g) + I_2O_5(s) \longrightarrow I_2(s) + 5\,CO_2(g)$

5.56   (a) $N_2O_4(l) + 2\,N_2H_4(l) \longrightarrow 3\,N_2(g) + 4\,H_2O(g)$
(b) $CaH_2(s) + 2\,H_2O(l) \longrightarrow Ca(OH)_2(aq) + 2\,H_2(l)$
(c) $2\,Al(s) + 6\,H_2O(l) \longrightarrow 2\,Al(OH)_3(s) + 3\,H_2(g)$

5.57   *Reducing Agent*   *Oxidizing Agent*

(a) Al H_2SO_4
(b) Fe Cl_2
(c) CO I_2O_5

5.58   *Reducing Agent*   *Oxidizing Agent*

(a) N_2H_4 N_2O_4
(b) CaH_2 H_2O
(c) Al H_2O

Conceptual Problems

5.59   (a) $2\,Al(s) + Fe_2O_3(s) \longrightarrow Al_2O_3(l) + 2\,Fe(l)$

(b) $2\,NH_4NO_3(s) \longrightarrow 2\,N_2(g) + O_2(g) + 4\,H_2O(g)$

5.60   (a) $Li_2O(s) + H_2O(g) \longrightarrow 2\,LiOH(s)$
(b) This is not a redox reaction because the oxidation numbers of reactant and product atoms remain the same.

5.61 $CuCl_2(aq) + Na_2CO_3(aq) \longrightarrow CuCO_3(s) + 2\,NaCl(aq)$
A precipitate of $CuCO_3$ forms.

$Cu^{2+}(aq) + CO_3^{2-}(aq) \longrightarrow CuCO_3(s)$

5.62 (a) $Al(OH)_3(aq) + 3\,HNO_3(aq) \longrightarrow Al(NO_3)_3(aq) + 3\,H_2O(l)$ neutralization

     (b) $3\,AgNO_3(aq) + FeCl_3(aq) \longrightarrow 3\,AgCl(s) + Fe(NO_3)_3(aq)$ precipitation

     (c) $(NH_4)_2Cr_2O_7(s) \longrightarrow Cr_2O_3(s) + 4\,H_2O(g) + N_2(g)$ redox

     (d) $Mn_2(CO_3)_3(s) \longrightarrow Mn_2O_3(s) + 3\,CO_2(g)$ none

5.63 (a) $P_4 + 5\,O_2 \longrightarrow 2\,P_2O_5$

     (b) The reactants both have an oxidation number of zero. In the product, the oxidation number of oxygen is –2, and the oxidation number of phosphorus is +5. P_4 is the reducing agent, and O_2 is the oxidizing agent.

5.64

Compound	Oxidation Number of Metal	Compound	Oxidation Number of Metal
(a) MnO_2	+4	(b) CrO_2	+4
Mn_2O_3	+3	CrO_3	+6
$KMnO_4$	+7	Cr_2O_3	+3

5.65 (a) The oxidation number of chromium in dichromate is +6.
     (b) The oxidation number of carbon in C_2H_5OH is –2; the oxidation number of carbon in CO_2 is +4.
     (c) The oxidizing agent is dichromate, and the reducing agent is C_2H_5OH.

5.66 $Fe^{3+}(aq) + 3\,NaOH(aq) \longrightarrow Fe(OH)_3(s) + 3\,Na^+(aq)$

     $Fe^{3+}(aq) + 3\,^-OH(aq) \longrightarrow Fe(OH)_3(s)$

5.67 $Mg^{2+}(aq) + CO_3^{2-}(aq) \longrightarrow MgCO_3(s)$

     $Ca^{2+}(aq) + CO_3^{2-}(aq) \longrightarrow CaCO_3(s)$

5.68 $2\,Bi^{3+}(aq) + 3S^-(aq) \longrightarrow Bi_2S_3(s)$

5.69 (a) $Fe_2O_3(s) + 3\,CO(g) \longrightarrow 2\,Fe(s) + 3\,CO_2(g)$ (b) This is a redox reaction.

5.70 $CO_2(g) + 2\,NH_3(g) \longrightarrow NH_2CONH_2(s) + H_2O(l)$

5.71 (a) $MgCO_3(aq) + 2\,HCl(aq) \longrightarrow MgCl_2(aq) + CO_2(g) + H_2O(l)$

     $CO_3^{2-}(aq) + 2\,H^+(aq) \longrightarrow CO_2(g) + H_2O(l)$

     (b) This is a neutralization reaction.

5.72 (a) Reactants: $Mn = +4$, $I = -1$; Products: $Mn = +2$, $I = 0$
     (b) MnO_2 is the oxidizing agent, and I^- is the reducing agent.

Group Problems

5.73 (a) The principal products of high-temperature combustion of nitrogen and sulfur are NO_2 and SO_2.

$$S(s) + O_2(g) \longrightarrow SO_2(g)$$
$$N_2(g) + O_2(g) \longrightarrow 2\,NO(g)$$
$$2\,NO(g) + O_2(g) \longrightarrow 2\,NO_2(g)$$

(b) Sulfur dioxide is oxidized to sulfur trioxide, which reacts with water vapor to form sulfuric acid.

$$2\,SO_2(g) + O_2(g) \longrightarrow 2\,SO_3(g)$$
$$SO_3(g) + H_2O(g) \longrightarrow H_2SO_4(l)$$

Nitrogen dioxide reacts with water vapor to form nitric acid and nitrous acid

$$2\,NO_2(g) + H_2O(g) \longrightarrow HNO_3(g) + HNO_2(g)$$

(c) All of the reactions involving oxygen are oxidations, and the reactions involving water are hydrations.

5.74 (a) The active ingredient in milk of magnesia is $Mg(OH)_2$.

(b) $Mg(OH)_2(s) + 2\,HCl(aq) \longrightarrow MgCl_2(aq) + H_2O(l)$

5.75 HCl indicates the drug is the hydrochloride salt of an amine. Hydrochlorides are prepared by treating the drug with HCl in a neutralization reaction between an amine and HCl. Drugs are prepared as amine hydrochlorides to increase solubility.

Self-Test for Chapter 5

Multiple Choice

1. Which of the following salts is soluble in water?
   (a) $FeSO_4$ (b) $BaSO_4$ (c) $SrSO_4$ (d) $PbSO_4$

2. Which of the following compounds has sulfur in a +1 oxidation state?
   (a) S_2F_2 (b) H_2SO_4 (c) SO_2 (d) Na_2S

3. Which of the following is a treatment for gout?
   (a) oxalate (b) purine (c) allopurinol (d) urate

4. Which of the following is not a component of a dry cell battery?
   (a) MnO_2 (b) I_2 (c) NH_4Cl (d) Zn

5. When the equation $SiCl_4 + H_2O \longrightarrow SiO_2 + HCl$ is balanced, the coefficients are:
   (a) 1,1,1,1   (b) 1,1,1,2 (c) 1,2,1,2 (d) 1,2,1,4

6. Which of the following is not a redox process?
   (a) respiration   (b) corrosion   (c) neutralization   (d) combustion

7. If you wanted to remove Ba^{2+} from solution, which reagent would you add?
   (a) CH_3CO_2H (b) NaOH   (c) H_2SO_4 (d) HCl

8. In the reaction $2\,Ca + O_2 \longrightarrow 2\,CaO$, calcium is:
   (a) reduced/oxidizing agent   (b) oxidized/reducing agent   (c) reduced/reducing agent
   (d) oxidized/oxidizing agent

9. When HCl is added to a solution, bubbles of gas appear. Which compound is probably present
   in the solution?
   (a) $SrCO_3$ (b) $Ca(OH)_2$ (c) AgBr   (d) K_2SO_4

10. Which lead-containing compound is probably soluble in water?
    (a) PbS   (b) $PbSO_4$ (c) $PbCl_2$ (d) $Pb(NO_3)_2$

Sentence Completion

1. The numbers placed in front of formulas to balance equations are called _____.

2. A _____ is a solid that forms during a reaction.

3. Ions that appear on both sides of the reaction arrow are _____ ions.

4. The substances in a reaction can be solids, liquids, or gases, or they can be in _____

   solution.

5. When acid is added to a solution, bubbles of gas are evolved if either _____ or _____ ions

   are present.

6. _____ is the science of extracting and purifying metals.

7. In a chemical equation, _____ are shown on the left of the reaction arrow.

8. The oxidation state of gold in $AuCl_3$ is _____.

9. In the reaction $2\,Mg + O_2 \longrightarrow 2\,MgO$, Mg is the _____ agent.

10. A reaction between an acid and a base is a _____ reaction.

True or False

1.   Lithium and iodine are the redox components of a dry cell battery.

2.   If the chloride of a cation is water-insoluble, the bromide and iodide of that cation are also

insoluble.

3.   Limestone ($CaCO_3$) can be dissolved by treatment with HNO_3.

4.   In a redox equation, the oxidizing agent is oxidized and the reducing agent is reduced.

5.   If all coefficients in a chemical equation are even numbers, the equation is not properly

balanced.

6.   In a redox reaction, a reducing agent becomes more negative.

7.   In the reaction between K and Cl_2, K is the oxidizing agent.

8.   Increased urate concentration in urine leads to gout.

9.   Most sodium salts are soluble in water.

10.   The same compound can be both oxidized and reduced in a reaction.

Match each entry on the left with its partner on the right.

1. Zn	(a) Net ionic equation
2. $C_3H_8O + O_2 \longrightarrow C_3H_6O + H_2O$	(b) Deterioration of a metal by oxidation
3. Respiration	(c) Redox reaction
4. $H^+ + OH^- \longrightarrow H_2O$	(d) Precipitation reaction
5. $CoCO_3$	(e) Salt soluble in water
6. $S + O_2 \longrightarrow SO_2$	(f) Unbalanced equation
7. $HCl + NaOH \longrightarrow NaCl + H_2O$	(g) Manganese has an oxidation number of +7
8. Corrosion	(h) Oxidized in a battery
9. $KMnO_4$	(i) Salt insoluble in water
10. MnO_2	(j) Neutralization reaction
11. $AgNO_3 + NaCl \longrightarrow AgCl + NaNO_3$	(k) Biological redox reaction
12. $CoCl_2$	(l) Reduced in a battery

Chapter 6 Chemical Reactions: Mole and Mass Relationships

Chapter Outline

I. The mole (Section 6.1).
   A. Molecular weight is the sum of the atomic weights of atoms in a molecule.
   B. Formula weight is the sum of the atomic weights of atoms in a formula unit.
   C. One mole of any pure substance has a mass equal to its molecular or formula weight in grams.
      1. Avogadro's number (6.022×10^{23}), known as the mole, represents the number of formula units that has a mass in grams equal to its weight in amu.
II. Mole–mass relationships (Section 6.2–6.4).
   A. Gram–mole conversions (Section 6.2).
      1. Molar mass is the mass in grams of one mole of any substance.
      2. Molar mass is a conversion factor that allows calculation of moles from grams and of grams from moles.
   B. Mole relationships and chemical equations (Section 6.3).
      1. The coefficients in an equation tell how many moles of reactant or product are involved in the reaction.
      2. The coefficients can be put in a mole ratio, which can be used as a conversion factor.
   C. Mass relationships and chemical equations (Section 6.4).
      1. Mole–mole conversions are made by using mole ratios.
      2. Mole–mass conversions are made by using molar mass as a conversion factor.
      3. Mass–mass conversions can be made by a mass–mole conversion of one substance, mole ratios, and a mole–mass conversion of the other substance.
   D. A summary of calculations using mole–mass relationships:
      1. Write the balanced equation.
      2. Choose mole–mass relationships and mole ratios to calculate the desired quantity.
      3. Set up the factor-label method to calculate the answer.
      4. Check the answer with a ballpark solution.
III. Limiting reagent and percent yield (Section 6.5).
   A. In a reaction, the reactant that is consumed first is the limiting reagent.
   B. Percent yield = actual yield/theoretical yield x 100%.
   C. Theoretical yield is found by using a mass–mass calculation.

Solutions to Chapter 6 Problems

6.1   (a) For ibuprofen, $C_{13}H_{18}O_2$:

At. wt. of 13 C = 13 x 12.0 amu = 156.0 amu
At. wt. of 18 H = 18 x 1.0 amu = 18.0 amu
At. wt. of 2 O = 2 x 16.0 amu = 32.0 amu

MW of $C_{13}H_{18}O_2$ = 206.0 amu

(b) For phenobarbital, $C_{12}H_{12}N_2O_3$:

At. wt. of 12 C = 12 x 12.0 amu = 144.0 amu
At. wt. of 12 H = 12 x 1.0 amu = 12.0 amu
At. wt. of 2 N = 2 x 14.0 amu = 28.0 amu
At. wt. of 3 O = 3 x 16.0 amu = 48.0 amu

MW of $C_{12}H_{12}N_2O_3$ = 232.0 amu

6.2
At. wt. of 6 C = 6 x 12.0 amu = 72.0 amu
At. wt. of 8 H = 8 x 1.0 amu = 8.0 amu
At. wt. of 6 O = 6 x 16.0 amu = 96.0 amu

MW of $C_6H_8O_6$ = 176.0 amu

Since the molecular weight of ascorbic acid is 176.0 amu, 6.022×10^{23} molecules have a mass of 176 g.

$$500 \text{ mg ascorbic acid } \times \frac{1 \text{ g}}{10^3 \text{ mg}} \times \frac{6.022 \times 10^{23} \text{ molecules}}{176 \text{ g}} = 1.71 \times 10^{21} \text{ molecules}$$

A 500 mg tablet contains 1.71×10^{21} molecules of ascorbic acid.

6.3
At. wt. of 9 C = 9 x 12.0 amu = 108.0 amu
At. wt. of 8 H = 8 x 1.0 amu = 8.0 amu
At. wt. of 4 O = 4 x 16.0 amu = 64.0 amu

MW of $C_9H_8O_4$ = 180.0 amu

$$5.0 \times 10^{20} \text{ molecules } \times \frac{180.0 \text{ g}}{1 \text{ mol}} \times \frac{1 \text{ mol}}{6.022 \times 10^{23} \text{ molecules}} = 0.15 \text{ g}$$

5.0×10^{20} molecules of aspirin weigh 0.15 g.

6.4

Cytosine $C_4H_5N_3O$

At. wt. of 4 C = 4 x 12.0 amu = 48.0 amu
At. wt. of 5 H = 5 x 1.0 amu = 5.0 amu
At. wt. of 3 N = 3 x 14.0 amu = 42.0 amu
At. wt. of 1 O = 1 x 16.0 amu = 16.0 amu

MW of $C_4H_5N_3O$ = 111.0 amu

6.5 Molar mass of C_2H_6O = 46.0 g/mol

$$10.0 \text{ g } \times \frac{1 \text{ mol}}{46.0 \text{ g}} = 0.217 \text{ mol in a } 10.0 \text{ g sample}$$

Ballpark check: The sample size, 10.0 g, is about 1/5 mol, or about 0.2 mol.

$$0.10 \text{ mol } \times \frac{46.0 \text{ g}}{1 \text{ mol}} = 4.6 \text{ g in a } 0.10 \text{ mol sample}$$

6.6 Molar mass of acetaminophen = 151 g

$$0.0225 \text{ mol } \times \frac{151 \text{ g}}{1 \text{ mol}} = 3.40 \text{ g}$$

5.00 g acetaminophen weighs more than 0.0225 mol.

6.7 Molar mass of urea $(C_5H_4N_4O_3)$ = 168 g

$$0.50 \text{ g urea } \times \frac{1 \text{ mol}}{168 \text{ g urea}} \times \frac{10^6 \text{ } \mu mol}{1 \text{ mol}} = 3.0 \times 10^3 \text{ } \mu mol \text{ urea}$$

6.8 (a) $Ni(s) + 2 HCl(aq) \longrightarrow NiCl_2(aq) + H_2(g)$

$$9.81 \text{ mol HCl } \times \frac{1 \text{ mol Ni}}{2 \text{ mol HCl}} = 4.91 \text{ mol Ni}$$

(b) $6.00 \text{ mol Ni } \times \frac{1 \text{ mol NiCl}_2}{1 \text{ mol Ni}} = 6.00 \text{ mol NiCl}_2 \text{ from } 6.00 \text{ mol Ni}$

$$12.00 \text{ mol HCl } \times \frac{1 \text{ mol NiCl}_2}{2 \text{ mol HCl}} = 6.00 \text{ mol NiCl}_2 \text{ from } 12.00 \text{ mol HCl}$$

6.00 mol $NiCl_2$ can be formed from 6.00 mol Ni and 12.00 mol HCl.

6.9 $6 CO_2 + 6 H_2O \longrightarrow C_6H_{12}O_6 + 6 O_2$

$$15.0 \text{ mol glucose } \times \frac{6 \text{ mol CO}_2}{1 \text{ mol glucose}} = 90.0 \text{ mol CO}_2$$

86   **Chapter 6**

6.10   (a) This is a mole–to–mole problem.

$$9.90 \text{ mol SiO}_2 \text{ x } \frac{4 \text{ mol HF}}{1 \text{ mol SiO}_2} = 39.6 \text{ mol HF}$$

(b) This is a mass–to–mass problem.

$$23.0 \text{ g SiO}_2 \text{ x } \frac{1 \text{ mol SiO}_2}{60.1 \text{ g SiO}_2} \text{ x } \frac{2 \text{ mol H}_2\text{O}}{1 \text{ mol SiO}_2} \text{ x } \frac{18.0 \text{ g H}_2\text{O}}{1 \text{ mol H}_2\text{O}} = 13.8 \text{ g H}_2\text{O}$$

Ballpark check: About 1/3 mol SiO_2 is used to produce 2/3 mol H_2O. Since the molar mass of H_2O is 18 g, we expect about 12 g H_2O, which is reasonably close to the calculated answer.

6.11   (a)

$$5.00 \text{ g W } \text{ x } \frac{1 \text{ mol W}}{183.8 \text{ g W}} = 0.0272 \text{ mol W}$$

(b) Using the mole ratios in the given equation, we can see that 0.0272 moles of WO_3 and 3 x (0.0272 mol) = 0.0816 mol H_2 are required.

(c) For WO_3:

$$5.00 \text{ g W } \text{ x } \frac{1 \text{ mol W}}{183.8 \text{ g W}} \text{ x } \frac{1 \text{ mol WO}_3}{1 \text{ mol W}} \text{ x } \frac{231.8 \text{ g WO}_3}{1 \text{ mol WO}_3} = 6.31 \text{ g WO}_3$$

For H_2:

$$5.00 \text{ g W } \text{ x } \frac{1 \text{ mol W}}{183.8 \text{ g W}} \text{ x } \frac{3 \text{ mol H}_2}{1 \text{ mol W}} \text{ x } \frac{2.02 \text{ g H}_2}{1 \text{ mol H}_2} = 0.165 \text{ g H}_2$$

6.31 g WO_3 and 0.165 g H_2 are needed to produce 5.00 g W.
Ballpark check: About 3/4 of the mass of WO_3 is composed of W. Thus, 5.00 g W comes from about 4/3 x 5.00 g W = 6.33 g WO_3, a result almost identical to the calculated result.

6.12   First, find the limiting reagent.

$$19.4 \text{ g C}_2\text{H}_4 \text{ x } \frac{1 \text{ mol C}_2\text{H}_4}{28.0 \text{ g C}_2\text{H}_4} = 0.693 \text{ mol C}_2\text{H}_4; \text{ } 50 \text{ g HCl x } \frac{1 \text{ mol HCl}}{36.5 \text{ g HCl}} = 1.4 \text{ mol HCl}$$

Ethylene is the limiting reagent.

$$19.4 \text{ g ethylene } \text{ x } \frac{1 \text{ mol ethylene}}{28.0 \text{ g ethylene}} \text{ x } \frac{1 \text{ mol ethyl chloride}}{1 \text{ mol ethylene}} \text{ x } \frac{64.5 \text{ g ethyl chloride}}{1 \text{ mol ethyl chloride}}$$

$$= 44.7 \text{ g ethyl chloride}$$

$$\frac{25.5 \text{ g ethyl chloride actually formed}}{44.7 \text{ g theoretical yield of ethyl chloride}} \text{ x } 100\% = 57.0\%$$

Ballpark check: The amount of ethylene reactant corresponds to about 2/3 mol. The theoretical yield of product is about 2/3 mol ethyl chloride (somewhat more than 40 g). Since the actual yield is 25.5 g, the percent yield should be around 60%.

Copyright ©2017 Pearson Education, Inc.

6.13   Note: to save space, ethylene oxide will be represented as EO, and ethylene glycol as EG.

$$35.0 \text{ g EO} \times \frac{1 \text{ mol EO}}{44.0 \text{ g EO}} \times \frac{1 \text{ mol EG}}{1 \text{ mol EO}} \times \frac{62.0 \text{ g EG}}{1 \text{ mol EG}} = 49.3 \text{ g ethylene glycol}$$

This quantity represents the maximum possible amount of ethylene glycol that can be produced. If the reaction occurs in 96.0% yield, then 49.3 g EG x 0.960 = 47.3 g EG is actually formed.

Ballpark check: A bit more than 3/4 mol EO is used to produce the same number of mol of EG, corresponding to about 47 g EG, close to the calculated result. The percent yield is also in close agreement.

6.14   According to the equation $A_2 + 2 B_2 \longrightarrow 2 AB_2$, two B_2 are needed for each A_2. In the illustration, there are seven A_2 and 17 B_2. Since the seven A_2 can consume only 14 B_2, A_2 is the limiting reagent.

HOC 6.1
(a) The number of treats is determined by the number of crackers, bananas, or squares of chocolate you had at the beginning.
(b) If you started with 20 crackers and 12 banana slices, your actual yield would be 10 treats. Crackers would be the limiting reagent.

CIA 6.1

$$8 \text{ mg Fe} \times \frac{1 \text{ g}}{10^3 \text{ mg}} \times \frac{1 \text{ mol Fe}}{55.8 \text{ g Fe}} \times \frac{4 \text{ mol O}_2}{1 \text{ mol Fe}} = 6 \times 10^{-4} \text{ mol O}_2$$

CIA 6.2   The formula for ferrous sulfate is $FeSO_4$. Its molar mass is 151.9 g/mol.

$$250 \text{ mg FeSO}_4 \times \frac{55.8 \text{ g Fe}}{151.9 \text{ g FeSO}_4} = 91.8 \text{ mg Fe}$$

CIA 6.3

$$8.0 \text{ mg} \times \frac{1 \text{ g}}{10^3 \text{ mg}} \times \frac{1 \text{ mol}}{55.8 \text{ g}} = 1.4 \times 10^{-4} \text{ mol for adult males}$$

$$18 \text{ mg} \times \frac{1 \text{ g}}{10^3 \text{ mg}} \times \frac{1 \text{ mol}}{55.8 \text{ g}} = 3.2 \times 10^{-4} \text{ mol for premenopausal women}$$

Understanding Key Concepts

6.15   Methionine: $C_5H_{11}NO_2S$
At. wt of   5 C =   5 x 12.0 amu =   60.0 amu
At. wt of 11 H = 11 x 1.0 amu =   11.0 amu
At. wt of   N =   =   14.0 amu
At. wt of   2 O =   2 x 16.0 amu =   32.0 amu
At. wt of   S =   =   32.1 amu
MW of $C_5H_{11}NO_2S$ = 149.1 amu

6.16   (a) $A_2 + 3 B_2 \longrightarrow 2 AB_3$

(b) $1.0 \text{ mol } A_2 \text{ x } \dfrac{2 \text{ mol } AB_3}{1 \text{ mol } A_2} = 2.0 \text{ mol } AB_3$

$1.0 \text{ mol } B_2 \text{ x } \dfrac{2 \text{ mol } AB_3}{3 \text{ mol } B_2} = 0.67 \text{ mol } AB_3$

6.17   There are 10 molecules of A and 7 molecules of B_2. According to the equation, 10 molecules of AB can theoretically be formed, with two molecules of B_2 left over.

6.18 B_2 is the limiting reagent. Since there are 11 B_2 in (a) but only 8 AB_2 in (b), the percent yield is 8/11 x 100% = 73%.

6.19   Note: To save space, ethylene oxide will be represented as EO, and ethylene glycol as EG.

$9.0 \text{ g } H_2O \text{ x } \dfrac{1 \text{ mol } H_2O}{18.0 \text{ g } H_2O} \text{ x } \dfrac{1 \text{ mol } EO}{1 \text{ mol } H_2O} \text{ x } \dfrac{44.0 \text{ g } EO}{1 \text{ mol } EO} = 22 \text{ g ethylene oxide}$

$9.0 \text{ g } H_2O \text{ x } \dfrac{1 \text{ mol } H_2O}{18.0 \text{ g } H_2O} \text{ x } \dfrac{1 \text{ mol } EG}{1 \text{ mol } H_2O} \text{ x } \dfrac{62.0 \text{ g } EG}{1 \text{ mol } EG} = 31 \text{ g ethylene glycol}$

Additional Problems

Molar Masses and Moles

6.20   One mole of a substance is an amount equal to its formula weight in grams. One mole of a molecular compound contains 6.022×10^{23} molecules.

6.21   Molecular weight is the sum of the atomic weights of the individual atoms in a molecule. Formula weight is the sum of the atomic weights of the individual atoms in a formula unit of any compound, whether molecular or ionic.
Molar mass is the mass in grams of 6.022×10^{23} molecules or formula units of any substance.

6.22

$\dfrac{6.022 \times 10^{23} \text{ units } Na_2SO_4}{1 \text{ mol } Na_2SO_4} \text{ x } \dfrac{2 \text{ Na}^+ \text{ ions}}{1 \text{ unit } Na_2SO_4} = \dfrac{1.204 \times 10^{24} \text{ Na}^+ \text{ ions}}{1 \text{ mol } Na_2SO_4}$

$\dfrac{6.022 \times 10^{23} \text{ units } Na_2SO_4}{1 \text{ mol } Na_2SO_4} \text{ x } \dfrac{1 \text{ SO}_4{}^{2-} \text{ ion}}{1 \text{ unit } Na_2SO_4} = \dfrac{6.022 \times 10^{23} \text{ SO}_4{}^{2-} \text{ ions}}{1 \text{ mol } Na_2SO_4}$

6.23   Each formula unit of K_2SO_4 contains three ions—two K^+ ions and one $SO_4{}^{2-}$ ion. Thus, one mole of K_2SO_4 contains three moles of ions, and

$1.75 \text{ mol } K_2SO_4 \text{ x } \dfrac{3 \text{ mol ions}}{1 \text{ mol } K_2SO_4} = 5.25 \text{ mol ions}$

6.24

$$16.2 \text{ g Ca} \times \frac{1 \text{ mol}}{40.1 \text{ g Ca}} \times \frac{6.022 \times 10^{23} \text{ atoms}}{1 \text{ mol}} = 2.43 \times 10^{23} \text{ atoms}$$

6.25

$$2.68 \times 10^{22} \text{ atoms} \times \frac{1 \text{ mol}}{6.022 \times 10^{23} \text{ atoms}} \times \frac{238.0 \text{ g}}{1 \text{ mol}} = 10.6 \text{ g uranium}$$

6.26–6.28

Compound	Molar Mass	Moles C	Atoms C	Grams C
(a) $CaCO_3$	100.1 g	1	6.022×10^{23}	12.0
(b) $CO(NH_2)_2$	60.0	1	6.022×10^{23}	12.0
(c) $C_2H_6O_2$	62.0	2	1.204×10^{24}	24.0

6.29   Molar mass of caffeine = 194 g

$$125 \text{ mg caffeine} \times \frac{1 \text{ g}}{10^3 \text{ mg}} \times \frac{1 \text{ mol caffeine}}{194 \text{ g caffeine}} = 6.44 \times 10^{-4} \text{ mol caffeine}$$

6.30   Molar mass of aspirin = 180 g

$$500 \text{ mg aspirin} \times \frac{1 \text{ g}}{10^3 \text{ mg}} \times \frac{1 \text{ mol aspirin}}{180 \text{ g aspirin}} = 2.78 \times 10^{-3} \text{ mol aspirin}$$

6.31   Molar mass of $C_{16}H_{13}ClN_2O$:
$(16 \times 12.0 \text{ g}) + (13 \times 1.0 \text{ g}) + (35.5 \text{ g}) + (2 \times 14.0 \text{ g}) + 16.0 \text{ g} = 284.5 \text{ g/ mol Valium}$

6.32–6.34

Compound	Molar Mass	Number of Moles in 4.50 g	Number of Grams in 0.075 mol
(a) $Al_2(SO_4)_3$	342.3 g	0.0131 mol	26 g
(b) $NaHCO_3$	84.0 g	0.0536 mol	6.3 g
(c) $C_4H_{10}O$	74.0 g	0.0608 mol	5.6 g
(d) $C_{16}H_{18}N_2O_5S$	350.1 g	0.0129 mol	26 g

Note: Slightly different values for molar mass result when exact atomic weights are used.

6.35

$$8.5 \times 10^{20} \text{ formula units} \times \frac{1 \text{ mol}}{6.022 \times 10^{23} \text{ formula units}} = 1.4 \times 10^{-3} \text{ mol}$$

$$0.0014 \text{ mol CaC}_2O_4 \times \frac{128 \text{ g CaC}_2O_4}{1 \text{ mol CaC}_2O_4} = 0.18 \text{ g CaC}_2O_4$$

Mole and Mass Relationships from Chemical Equations

6.36 (a) $N_2(g) + O_2(g) \longrightarrow 2\,NO(g)$

(b) 7.50 mol of N_2 are needed to react with 7.50 mol of O_2.

(c) $3.81\ \text{mol } N_2 \ \times \ \dfrac{2\ \text{mol NO}}{1\ \text{mol } N_2} \ = \ 7.62\ \text{mol NO}$

(d) $0.250\ \text{mol NO} \ \times \ \dfrac{1\ \text{mol } O_2}{2\ \text{mol NO}} \ = \ 0.125\ \text{mol } O_2$

6.37 (a) $C_4H_8O_2(l) + 2\,H_2(g) \longrightarrow 2\,C_2H_6O(l)$

(b) 3.0 mol of ethyl alcohol are produced from 1.5 mol of ethyl acetate.

(c) $1.5\ \text{mol } C_4H_8O_2 \ \times \ \dfrac{2\ \text{mol } C_2H_6O}{1\ \text{mol } C_4H_8O_2} \ \times \ \dfrac{46.0\ \text{g } C_2H_6O}{1\ \text{mol } C_2H_6O} \ = \ 138\ \text{g } C_2H_6O$

(d) $12.0\ \text{g } C_4H_8O_2 \ \times \ \dfrac{1\ \text{mol } C_4H_8O_2}{88.0\ \text{g } C_4H_8O_2} \ \times \ \dfrac{2\ \text{mol } C_2H_6O}{1\ \text{mol } C_4H_8O_2} \ \times \ \dfrac{46.0\ \text{g } C_2H_6O}{1\ \text{mol } C_2H_6O}$

$= 12.5\ \text{g } C_2H_6O$

(e) $12.0\ \text{g } C_4H_8O_2 \ \times \ \dfrac{1\ \text{mol } C_4H_8O_2}{88.0\ \text{g } C_4H_8O_2} \ \times \ \dfrac{2\ \text{mol } H_2}{1\ \text{mol } C_4H_8O_2} \ \times \ \dfrac{2.02\ \text{g } H_2}{1\ \text{mol } H_2} \ = \ 0.551\ \text{g } H_2$

6.38 (a) Molar mass of $Mg(OH)_2$ = 24.3 g + (2 x 16.0 g) + (2 x 1.0 g) = 58.3 g

(b) $1.2\ \text{g } Mg(OH)_2 \times \dfrac{1\ \text{mol } Mg(OH)_2}{58.3\ \text{g } Mg(OH)_2} = 0.021\ \text{mol } Mg(OH)_2$

Since there are 3 teaspoons in a tablespoon, the dose per teaspoon is 0.0070 mol.

6.39 (a) $N_2(g) + 3\,H_2(g) \longrightarrow 2\,NH_3(g)$

(b) $16.0\ \text{g } NH_3 \ \times \ \dfrac{1\ \text{mol } NH_3}{17.0\ \text{g } NH_3} \ \times \ \dfrac{1\ \text{mol } N_2}{2\ \text{mol } NH_3} \ = \ 0.471\ \text{mol } N_2$

(c) $75.0\ \text{g } N_2 \ \times \ \dfrac{1\ \text{mol } N_2}{28.0\ \text{g } N_2} \ \times \ \dfrac{3\ \text{mol } H_2}{1\ \text{mol } N_2} \ \times \ \dfrac{2.02\ \text{g } H_2}{1\ \text{mol } H_2} \ = \ 16.2\ \text{g } H_2$

6.40 (a) $N_2H_4(l) + 3 O_2(g) \longrightarrow 2 NO_2(g) + 2 H_2O(g)$

(b) $165 \text{ g } N_2H_4 \times \dfrac{1 \text{ mol } N_2H_4}{32.0 \text{ g } N_2H_4} \times \dfrac{3 \text{ mol } O_2}{1 \text{ mol } N_2H_4} = 15.5 \text{ mol } O_2$

(c) $15.5 \text{ mol } O_2 \times \dfrac{32.0 \text{ g } O_2}{1 \text{ mol } O_2} = 496 \text{ g } O_2$

6.41 (a) $Fe_2O_3(s) + 3 CO(g) \longrightarrow 2 Fe(s) + 3 CO_2(g)$

(b) $3.02 \text{ g } Fe_2O_3 \times \dfrac{1 \text{ mol } Fe_2O_3}{159.7 \text{ g } Fe_2O_3} \times \dfrac{3 \text{ mol } CO}{1 \text{ mol } Fe_2O_3} \times \dfrac{28.0 \text{ g } CO}{1 \text{ mol } CO} = 1.59 \text{ g } CO$

(c) $1.68 \text{ mol } Fe_2O_3 \times \dfrac{3 \text{ mol } CO}{1 \text{ mol } Fe_2O_3} \times \dfrac{28.0 \text{ g } CO}{1 \text{ mol } CO} = 141 \text{ g } CO$

6.42 (a) $2 Mg(s) + O_2(g) \longrightarrow 2 MgO(s)$

(b) $25.0 \text{ g } Mg \times \dfrac{1 \text{ mol } Mg}{24.3 \text{ g } Mg} \times \dfrac{1 \text{ mol } O_2}{2 \text{ mol } Mg} \times \dfrac{32.0 \text{ g } O_2}{1 \text{ mol } O_2} = 16.5 \text{ g } O_2$

   41.5 g MgO will result from reaction of 25.0 g Mg and 16.5 g O_2.

(c) $25.0 \text{ g } O_2 \times \dfrac{1 \text{ mol } O_2}{32.0 \text{ g } O_2} \times \dfrac{2 \text{ mol } Mg}{1 \text{ mol } O_2} \times \dfrac{24.3 \text{ g } Mg}{1 \text{ mol } Mg} = 38.0 \text{ g } Mg$

   63.0 g MgO will result from reaction of 25.0 g O_2 and 38.0 g Mg.

6.43 (a) Add the two reactions:

$TiO_2(s) + 2 Cl_2(g) + 2 C(s) + 2 Mg(s) \longrightarrow Ti(s) + 2 CO(g) + 2 MgCl_2(s)$

You can see that the mole ratio of TiO_2:Ti is 1:1, and that the ratio of either titanium species to any other component is 1:2.

(b) One mole of TiO_2 forms one mole of Ti.

(c) $95 \text{ kg Ti} \times \dfrac{1 \text{ mol Ti}}{47.9 \text{ g Ti}} \times \dfrac{1 \text{ mol } TiO_2}{1 \text{ mol Ti}} \times \dfrac{79.9 \text{ g } TiO_2}{1 \text{ mol } TiO_2} = 158 \text{ kg } TiO_2$

6.44

$105 \text{ kg } Fe_2O_3 \times \dfrac{10^3 \text{ g}}{1 \text{ kg}} \times \dfrac{1 \text{ mol } Fe_2O_3}{159.7 \text{ g } Fe_2O_3} \times \dfrac{3 \text{ mol } CO}{1 \text{ mol } Fe_2O_3} = 1.97 \times 10^3 \text{ mol } CO$

6.45

$4 \times 10^8 \text{ kg } SO_2 \times \dfrac{10^3 \text{ g}}{1 \text{ kg}} \times \dfrac{1 \text{ mol}}{64 \text{ g}} = 6 \times 10^9 \text{ mol}$

$6 \times 10^9 \text{ mol } H_2SO_4 \times \dfrac{98 \text{ g}}{1 \text{ mol}} \times \dfrac{1 \text{ kg}}{1 \times 10^3 \text{ g}} = 6 \times 10^8 \text{ kg } H_2SO_4$

6.46

$$1.5 \text{ kg Fe}_2\text{O}_3 \ \text{x} \ \frac{10^3 \text{ g}}{1 \text{ kg}} \ \text{x} \ \frac{1 \text{ mol Fe}_2\text{O}_3}{159.7 \text{ g}} \ \text{x} \ \frac{2 \text{ mol Fe}}{1 \text{ mol Fe}_2\text{O}_3} = 19 \text{ mol Fe}$$

6.47 Each mole of *scrubber* removes one mole of CO_2.

$$1.0 \text{ kg NaOH x} \ \frac{1 \times 10^3 \text{g}}{1 \text{ kg}} \ \text{x} \ \frac{1 \text{ mol NaOH}}{40 \text{ g NaOH}} \ \text{x} \ \frac{1 \text{ mol CO}_2}{1 \text{ mol NaOH}} = 25 \text{ mol CO}_2$$

$$1.0 \text{ kg Ca(OH)}_2 \text{ x} \ \frac{1 \times 10^3 \text{g}}{1 \text{ kg}} \ \text{x} \ \frac{1 \text{ mol Ca(OH)}_2}{74 \text{ g Ca(OH)}_2} \ \text{x} \ \frac{1 \text{ mol CO}_2}{1 \text{ mol Ca(OH)}_2} = 14 \text{ mol CO}_2$$

One kilogram of NaOH removes more CO_2 than one kilogram of $Ca(OH)_2$.

6.48

$$7.5 \text{ mol O}_2 \ \text{x} \ \frac{1 \text{ mol B}_2\text{H}_6}{3 \text{ mol O}_2} \ \text{x} \ \frac{27.67 \text{ g B}_2\text{H}_6}{1 \text{ mol B}_2\text{H}_6} = 69 \text{ g B}_2\text{H}_6$$

Limiting Reagent and Percent Yield

6.49

(a) $25.0 \text{ g CO} \ \text{x} \ \dfrac{1 \text{ mol CO}}{28.0 \text{ g CO}} = 0.893 \text{ mol CO};$ $6.00 \text{ g H}_2 \ \text{x} \ \dfrac{1 \text{ mol H}_2}{2.02 \text{ g H}_2} = 2.97 \text{ mol H}_2$
Carbon monoxide is the limiting reagent.

(b) $10.0 \text{ g CO} \ \text{x} \ \dfrac{1 \text{ mol CO}}{28.0 \text{ g CO}} \ \text{x} \ \dfrac{1 \text{ mol CH}_3\text{OH}}{1 \text{ mol CO}} \ \text{x} \ \dfrac{32.0 \text{ g CH}_3\text{OH}}{1 \text{ mol CH}_3\text{OH}} = 11.4 \text{ g CH}_3\text{OH}$

(c) $\dfrac{9.55 \text{ g}}{11.4 \text{ g}} \ \text{x} \ 100\% = 83.8\% \text{ yield}$

6.50 The balanced equation appears in the answer to Problem 6.40.

(a) $75.0 \text{ kg N}_2\text{H}_4 \ \text{x} \ \dfrac{1 \text{ kmol N}_2\text{H}_4}{32.0 \text{ kg N}_2\text{H}_4} = 2.34 \text{ kmol N}_2\text{H}_4$

   $75.0 \text{ kg O}_2 \ \text{x} \ \dfrac{1 \text{ kmol O}_2}{32.0 \text{ kg O}_2} = 2.34 \text{ kmol O}_2$

Although the same number of moles of each are present, O_2 is the limiting reagent because three moles of O_2 are needed to react with each mole of hydrazine, according to the balanced equation.

(b) $75.0 \text{ kg O}_2 \ \text{x} \ \dfrac{1 \text{ kmol O}_2}{32.0 \text{ kg O}_2} \ \text{x} \ \dfrac{2 \text{ kmol NO}_2}{3 \text{ kmol O}_2} \ \text{x} \ \dfrac{46.0 \text{ kg NO}_2}{1 \text{ kmol NO}_2} = 71.9 \text{ kg NO}_2$

(c) $\dfrac{59.3 \text{ g actually formed}}{71.9 \text{ g theoretical yield}} \ \text{x} \ 100\% = 82.5\%$

6.51 (a) $CH_4(g) + 2 Cl_2(g) \longrightarrow CH_2Cl_2(l) + 2 HCl(g)$

(b) $50.0 \text{ g } CH_4 \times \dfrac{1 \text{ mol } CH_4}{16.0 \text{ g } CH_4} \times \dfrac{2 \text{ mol } Cl_2}{1 \text{ mol } CH_4} \times \dfrac{71.0 \text{ g } Cl_2}{1 \text{ mol } Cl_2} = 444 \text{ g } Cl_2$

(c) If the reaction occurred in 100% yield:

$50.0 \text{ g } CH_4 \times \dfrac{1 \text{ mol } CH_4}{16.0 \text{ g } CH_4} \times \dfrac{1 \text{ mol } CH_2Cl_2}{1 \text{ mol } CH_4} \times \dfrac{85.0 \text{ g } CH_2Cl_2}{1 \text{ mol } CH_2Cl_2} = 266 \text{ g } CH_2Cl_2$

Since the reaction occurs in 76% yield:

$266 \text{ g} \times 0.76 = 202 \text{ g } CH_2Cl_2$ are actually formed.

6.52

(a) $55.8 \text{ g } K_2PtCl_4 \times \dfrac{1 \text{ mol } K_2PtCl_4}{415.3 \text{ g } K_2PtCl_4} \times \dfrac{2 \text{ mol } NH_3}{1 \text{ mol } K_2PtCl_4} \times \dfrac{17.0 \text{ g } NH_3}{1 \text{ mol } NH_3} = 4.57 \text{ g } NH_3$

(b) If the reaction occurred in 100% yield:

$55.8 \text{ g } K_2PtCl_4 \times \dfrac{1 \text{ mol } K_2PtCl_4}{415.3 \text{ g } K_2PtCl_4} \times \dfrac{1 \text{ mol cisplatin}}{1 \text{ mol } K_2PtCl_4} \times \dfrac{300.1 \text{ g cisplatin}}{1 \text{ mol cisplatin}}$
$= 40.3 \text{ g cisplatin}$

Since the reaction occurs in 95% yield, $40.3 \text{ g} \times 0.95 = 38 \text{ g cisplatin}$ are actually formed.

6.53

(a) $75 \text{ g } C_6H_6 \times \dfrac{1 \text{ mol } C_6H_6}{78 \text{ g } C_6H_6} = 0.96 \text{ mol } C_6H_6$

$27.5 \text{ g } HNO_3 \times \dfrac{1 \text{ mol } HNO_3}{63.0 \text{ g } HNO_3} = 0.437 \text{ mol } HNO_3$

Nitric acid is the limiting reagent.

(b) $0.437 \text{ mol } HNO_3 \times \dfrac{1 \text{ mol } C_6H_5NO_2}{1 \text{ mol } HNO_3} \times \dfrac{123 \text{ g}}{1 \text{ mol } C_6H_5NO_2} = 53.8 \text{ g } C_6H_5NO_2$

6.54

$\dfrac{48.2 \text{ g}}{53.8 \text{ g}} \times 100\% = 89.6\%$

Conceptual Problems

6.55 Molar mass of batrachotoxin ($C_{31}H_{42}N_2O_6$) = 538 g/mol

$0.05 \text{ } \mu g \times \dfrac{1 \text{ g}}{10^6 \text{ } \mu g} \times \dfrac{1 \text{ mol}}{538 \text{ g}} \times \dfrac{6.022 \times 10^{23} \text{ molecules}}{1 \text{ mol}} = 6 \times 10^{13} \text{ molecules}$

6.56

(a) $15.0 \text{ g Zn} \times \dfrac{1 \text{ mol Zn}}{65.4 \text{ g Zn}} \times \dfrac{1 \text{ mol H}_2}{1 \text{ mol Zn}} \times \dfrac{2.02 \text{ g H}_2}{1 \text{ mol H}_2} = 0.463 \text{ g H}_2$

(b) In this redox reaction, H^+ is reduced (oxidizing agent) and Zn is oxidized (reducing agent).

6.57 (a) $C_{12}H_{22}O_{11}(s) \longrightarrow 12 \text{ C}(s) + 11 \text{ H}_2O(l)$

(b) $60.0 \text{ g sucrose} \times \dfrac{1 \text{ mol sucrose}}{342 \text{ g sucrose}} \times \dfrac{12 \text{ mol C}}{1 \text{ mol sucrose}} \times \dfrac{12.0 \text{ g C}}{1 \text{ mol C}} = 25.3 \text{ g carbon}$

(c) $6.50 \text{ g C} \times \dfrac{1 \text{ mol C}}{12.0 \text{ g C}} \times \dfrac{11 \text{ mol H}_2O}{12 \text{ mol C}} \times \dfrac{18.0 \text{ g H}_2O}{1 \text{ mol H}_2O} = 8.94 \text{ g H}_2O$

6.58 (a) $Cu(s) + 4 H^+(aq) + 2 NO_3^-(aq) \longrightarrow Cu^{2+}(aq) + 2 NO_2(g) + 2 H_2O(l)$

(b) $35.0 \text{ g HNO}_3 \times \dfrac{1 \text{ mol HNO}_3}{63.0 \text{ g HNO}_3} \times \dfrac{1 \text{ mol Cu}}{4 \text{ mol HNO}_3} \times \dfrac{63.5 \text{ g Cu}}{1 \text{ mol Cu}} = 8.82 \text{ g Cu}$

35.0 g HNO_3 is more than enough to dissolve 5.00 g Cu.

6.59

(a) $1.50 \text{ g C}_2H_6O \times \dfrac{1 \text{ mol C}_2H_6O}{46.0 \text{ g C}_2H_6O} \times \dfrac{2 \text{ mol K}_2Cr_2O_7}{3 \text{ mol C}_2H_6O} \times \dfrac{294.2 \text{ g K}_2Cr_2O_7}{1 \text{ mol K}_2Cr_2O_7}$

$= 6.40 \text{ g K}_2Cr_2O_7$

(b) $80.0 \text{ g C}_2H_6O \times \dfrac{1 \text{ mol C}_2H_6O}{46.0 \text{ g C}_2H_6O} \times \dfrac{3 \text{ mol C}_2H_4O_2}{3 \text{ mol C}_2H_6O} \times \dfrac{60.0 \text{ g C}_2H_4O_2}{1 \text{ mol C}_2H_4O_2}$

$= 104 \text{ g C}_2H_4O_2$

6.60

$100.0 \text{ lb C}_6H_{12}O_6 \times \dfrac{454 \text{ g}}{1 \text{ lb}} \times \dfrac{1 \text{ mol C}_6H_{12}O_6}{180.0 \text{ g C}_6H_{12}O_6} \times \dfrac{2 \text{ mol C}_2H_6O}{1 \text{ mol C}_6H_{12}O_6} \times \dfrac{46.0 \text{ g C}_2H_6O}{1 \text{ mol C}_2H_6O}$

$= 2.32 \times 10^4 \text{ g C}_2H_6O$

$2.32 \times 10^4 \text{ g ethanol} \times \dfrac{1 \text{ mL ethanol}}{0.789 \text{ g ethanol}} \times \dfrac{1 \text{ qt}}{946.4 \text{ mL}} = 31.1 \text{ qt ethanol}$

6.61 (a) $4 NH_3(g) + 5 O_2(g) \longrightarrow 4 NO(g) + 6 H_2O(g)$

(b) $17.0 \text{ g NH}_3 \times \dfrac{1 \text{ mol NH}_3}{17.0 \text{ g}} \times \dfrac{1 \text{ mol NO}}{1 \text{ mol NH}_3} \times \dfrac{30.0 \text{ g NO}}{1 \text{ mol NO}} = 30.0 \text{ g NO}$

6.62 $2\,NaOH(aq) + Cl_2(g) \longrightarrow NaOCl(aq) + NaCl(aq) + H_2O(l)$

$32.5\ g\ NaOH \times \dfrac{1\ mol\ NaOH}{40.0\ g\ NaOH} \times \dfrac{1\ mol\ NaOCl}{2\ mol\ NaOH} = 0.406\ mol\ NaOCl$

6.63   (a) $BaCl_2(aq) + Na_2SO_4(aq) \longrightarrow BaSO_4(s) + 2\,NaCl(aq)$

(b) $27.4\ g\ Na_2SO_4 \times \dfrac{1\ mol\ Na_2SO_4}{142\ g\ Na_2SO_4} \times \dfrac{1\ mol\ BaSO_4}{1\ mol\ Na_2SO_4} \times \dfrac{233\ g\ BaSO_4}{1\ mol\ BaSO_4}$

$= 45.0\ g\ BaSO_4$

6.64   (a) $3\,NO_2(g) + H_2O(l) \longrightarrow 2\,HNO_3(aq) + NO(g)$

(b) $65.0\ g\ NO_2 \times \dfrac{1\ mol\ NO_2}{46.0\ g\ NO_2} \times \dfrac{2\ mol\ HNO_3}{3\ mol\ NO_2} \times \dfrac{63.0\ g\ HNO_3}{1\ mol\ HNO_3}$

$= 59.3\ g\ HNO_3$

(c) $\dfrac{43.8\ g}{59.3\ g} \times 100\% = 73.9\%$

6.65

(a) $47\ g\ C_7H_6O_3 \times \dfrac{1\ mol\ C_7H_6O_3}{138\ g\ C_7H_6O_3} = 0.34\ mol\ C_7H_6O_3$

$25\ g\ C_4H_6O_3 \times \dfrac{1\ mol\ C_4H_6O_3}{102\ g\ C_4H_6O_3} = 0.25\ mol\ C_4H_6O_3$

Acetic anhydride is the limiting reagent.

$0.25\ mol\ C_4H_6O_3 \times \dfrac{1\ mol\ C_9H_8O_4}{1\ mol\ C_4H_6O_3} \times \dfrac{180\ g\ C_9H_8O_4}{1\ mol\ C_9H_8O_4} = 45\ g\ C_9H_8O_4$

(b) $\dfrac{35\ g}{45\ g} \times 100\% = 78\%$

6.66

$15.2\ g\ Ag \times \dfrac{1\ mol\ Ag}{108\ g\ Ag} \times \dfrac{1\ mol\ AgNO_3}{1\ mol\ Ag} \times \dfrac{170\ g\ AgNO_3}{1\ mol\ AgNO_3} = 23.9\ g\ AgNO_3$

6.67   (a) $P_4(s) + 10\,Cl_2(g) \longrightarrow 4\,PCl_5(s)$

(b) $15.2\ g\ P_4 \times \dfrac{1\ mol\ P_4}{124\ g\ P_4} \times \dfrac{4\ mol\ PCl_5}{1\ mol\ P_4} \times \dfrac{208\ g\ PCl_5}{1\ mol\ PCl_5} = 102\ g\ PCl_5$

6.68

$80.0\ kg\ H_2O \times \dfrac{1\ mol\ H_2O}{18.0\ g\ H_2O} \times \dfrac{1\ mol\ Li_2O}{1\ mol\ H_2O} \times \dfrac{29.8\ g\ Li_2O}{1\ mol\ Li_2O} = 132\ kg\ Li_2O$

6.69   (a) $6\,NH_4ClO_4(s) + 10\,Al(s) \longrightarrow 4\,Al_2O_3(s) + 2\,AlCl_3(s) + 12\,H_2O(g) + 3\,N_2(g)$

(b) $14.5 \times 10^3\ g\ NH_4ClO_4 \times \dfrac{1\ mol\ NH_4ClO_4}{117\ g\ NH_4ClO_4} \times \dfrac{15\ mol\ gases}{6\ mol\ NH_4ClO_4} = 310\ mol\ gases$

Remember that both water and nitrogen are gases in this problem.

Group Problems

6.70   (a) The recommended daily intake of calcium for adult men and premenopausal women is 1000 mg.

(b) Note: For simplicity, calcium citrate is indicated by *C cit*, and 1000 mg = 1.00 g

$$1.00 \text{ g Ca} \times \frac{1 \text{ mol Ca}}{40.1 \text{ g Ca}} \times \frac{1 \text{ mol C cit}}{3 \text{ mol Ca}} \times \frac{498.5 \text{ g C cit}}{1 \text{ mol C cit}} = 4.14 \text{ g calcium citrate}$$

6.71   (a) An aspirin tablet weighs 325 g.

$$325 \text{ g aspirin} \times \frac{1 \text{ mol aspirin}}{180 \text{ g aspirin}} = 1.81 \text{ mol aspirin}$$

$$1.81 \text{ mol} \times \frac{6.022 \times 10^{23} \text{ molecules}}{1 \text{ mol}} = 1.08 \times 10^{24} \text{ molecules aspirin}$$

6.72   (a) The formula for Lovastatin: $C_{24}H_{36}O_5$

24 C x (12.0 amu) + 36 H x (1.00 amu) + 5 O x (16 amu) = 404 amu
molar mass = 404 g/mol

(b) $$10 \text{ mg lovastatin} \times \frac{1 \text{ g}}{10^3 \text{ mg}} \times \frac{1 \text{ mol}}{404 \text{ g}} = 2.5 \times 10^{-5} \text{ mol lovastatin}$$

6.73   (a) Pyrite, FeS_2, has a molar mass of 119.7 g.

(b) Two moles of SO_2 are produced from each mole of pyrite.

$$1.0 \text{ kg FeS}_2 \times \frac{10^3 \text{ g}}{1 \text{ kg}} \times \frac{1 \text{ mol FeS}_2}{119.7 \text{ g}} \times \frac{2 \text{ mol SO}_2}{1 \text{ mol FeS}_2} = 17 \text{ mol SO}_2$$

6.74   In the United States, the recommended daily intake of Vitamin C is 75 mg for women and 90 mg for men.
(a) My multivitamin supplement contains 90 mg Vitamin C.
(b) This is 100% of the daily dosage for men and 120% of the dosage for women.

Self-Test for Chapter 6

Multiple Choice

1. What mass of $CaCO_3$ has the same number of molecules as 21 g of NaF?
    (a) 10 g   (b) 20 g   (c) 50 g   (d) 100 g

2. Consider the following equation: $LiOH + CO_2 \longrightarrow LiHCO_3$. In which situation is LiOH the limiting reagent?
    (a) 6 g LiOH + 11 g CO_2 (b) 10 g LiOH + 16 g CO_2
    (c) 4 g LiOH + 7 g CO_2 (d) 9 g LiOH + 17 g CO_2

3. In the equation $P_2O_5 + 3 H_2O \longrightarrow 2 H_3PO_4$ the mole ratio of product to H_2O is:
    (a) 2:1   (b) 2:3   (c) 3:2   (d) 1:2

4. In the reaction $2 Ca + O_2 \longrightarrow 2 CaO$, how many grams of CaO can be produced from 20 g of Ca? (a) 20 g   (b) 28 g   (c) 40 g   (d) 56 g

5. $3 NO_2 + H_2O \longrightarrow 2 HNO_3 + NO$

    All the following statements about this reaction are true except:
    (a) The starting material is both oxidized and reduced.
    (b) NO_2 in the atmosphere might contribute to acid rain.
    (c) At least 6 g of H_2O are needed to react completely with 46 g of NO_2.
    (d) 92 g of NO_2 produces 28 g of NO.

True or False

1. A mole of oxygen atoms has the same mass as a mole of nitrogen atoms.

2. The coefficients in chemical reactions show the relative numbers of moles of reactants and products.

3. Percent yield is the amount of a substance that can theoretically be produced divided by the amount actually produced.

4. Mole ratios are used to convert between moles and grams of a compound.

5. In the reaction $4 Al + 3 O_2 \longrightarrow 2 Al_2O_3$, the mole ratio of product to Al is 2.

Match each entry on the left with its partner on the right.

1. 42.0 amu   (a) Converts moles Na to grams Na

2. 23.0 g sodium / 1 mol   (b) Molar mass of NaF

3. 42.0 g   (c) Converts grams Na to moles Na

4. 1 mol / 23.0 g sodium   (d) Avogadro's number

5. 6.02×10^{23} (e) Formula weight of NaF

Chapter 7   Chemical Reactions: Energy, Rates, and Equilibrium

Chapter Outline

I. Energy (Sections 7.1–7.4).
  A. Energy and chemical bonds (Section 7.1).
    1. There are two kinds of energy.
      a. Potential energy is stored energy.
      b. Kinetic energy is the energy of motion.
    2. Chemical bonds are a form of potential energy.
    3. Whether a reaction occurs, and how much heat is associated, depends on the potential energy of reactants and products.
  B. Heat changes during chemical reactions (Section 7.2).
    1. Bond dissociation energies measure the strength of covalent bonds.
      a. Bond breaking requires heat and is endothermic.
      b. Bond formation releases heat and is exothermic.
    2. The reverse of an endothermic process is exothermic, and the reverse of an exothermic process is endothermic.
    3. The law of conservation of energy states that energy can neither be created nor destroyed during any physical or chemical change.
    4. The difference between the energy needed for breaking bonds and the energy released in forming bonds is the heat of reaction, also known as enthalpy (ΔH).
  C. Exothermic and endothermic reactions: ΔH (Section 7.3).
    1. In exothermic reactions, the bond dissociation energies of the products are greater than the bond dissociation energies of the reactants, and ΔH is negative.
    2. In endothermic reactions, the bond dissociation energies of the products are smaller than the bond dissociation energies of the reactants, and ΔH is positive.
    3. The amount of heat released in the reverse of a reaction is equal to that absorbed in the forward reaction, but ΔH has the opposite sign.
    4. The heat of reaction can be determined by the formula:
      $\Delta H = \Sigma$(bond dissociation energies)$_{\text{reactants}} - \Sigma$(bond dissociation energies)$_{\text{products}}$
  D. Free energy: ΔG (Section 7.4).
    1. Spontaneous processes.
      a. A spontaneous process proceeds without any external influence.
      b. A nonspontaneous process needs a constant external source of energy.
    2. Entropy (ΔS) measures the amount of disorder in a system.
      a. If disorder increases, ΔS is positive.
      b. If disorder decreases, ΔS is negative.
    3. The absorption of heat and the increase or decrease in disorder determine if a reaction is spontaneous.
      a. $\Delta G = \Delta H - T\Delta S$.
      b. If ΔG is negative, the process is spontaneous, and the reaction is exergonic.
      c. If ΔG is positive, the process is not spontaneous, and the reaction is endergonic.
    4. ΔG for the reverse of a reaction is equal in value to ΔG for the forward reaction, but the sign is changed.
    5. Some nonspontaneous processes become spontaneous when temperature increases.
II. Reaction rates (Sections 7.5–7.6).
  A. How reactions occur (Section 7.5).
    1. In addition to the value of ΔG, other factors determine if a reaction will occur.
      a. Reactants must collide in the correct orientation.
      b. The energy of collision must be great enough to cause bond breaking or forming.

      c. Many reactions with a favorable free energy don't occur at room temperature, and heat must be added to get them started.

   2. The energy changes during a reaction can be graphed on a reaction energy diagram.

   3. The amount of energy needed to produce favorable collisions is the activation energy E_{act}.

      a. E_{act} determines the reaction rate.

      b. The size of E_{act} is unrelated to the size of ΔH.

  B. Factors that affect reaction rate (Section 7.6).

   1. Increasing temperature increases reaction rate.

   2. Increasing the concentration of reactants increases reaction rate.

   3. Catalysts increase the reaction rate by lowering E_{act}.

III. Chemical equilibrium (Sections 7.7–7.9).

  A. Reversible reactions (Section 7.7).

   1. Some reactions proceed to virtual completion.

   2. Other reactions go to partial completion, at which point products begin to reform starting material.

      a. These reactions are reversible.

      b. The reaction from left to right is the forward reaction.

      c. The reaction from right to left is the reverse reaction.

   3. At some point the rate of the forward reaction equals the rate of the reverse reaction, and equilibrium is established.

      a. In biological systems, many reactions may be linked to establish an equilibrium called homeostasis.

   4. It is not necessary for the concentrations of products and reactants to be equal at equilibrium.

  B. Equilibrium (Section 7.8).

   1. Equilibrium equations.

      a. For the reaction $a\,A + b\,B \longrightarrow c\,C + d\,D$,

$$K = \frac{[C]^c[D]^d}{[A]^a[B]^b}$$

      b. K = equilibrium constant.

      c. Expression on the right = equilibrium constant expression.

      d. The concentrations of pure solids or liquids are omitted from the equilibrium expression.

   2. The value of K determines the position of equilibrium.

      a. When $K \gg 1$, reaction goes to completion.

      b. When $K > 1$, the forward reaction is favored.

      c. When K is of intermediate value, significant amounts of reactants and products are present at equilibrium.

      d. When $K < 1$, the reverse reaction is favored.

      e. When $K \ll 1$, there is essentially no reaction.

  C. Effect of changing reaction conditions (Section 7.9).

   1. Le Châtelier's principle: When a stress is applied to a system, the equilibrium shifts to remove the stress.

   2. Effect of changing concentration.

      a. Increasing the concentration of reactants favors the forward reaction.

      b. Increasing the concentration of products favors the reverse reaction.

   3. Effect of changing temperature.

      a. Decreasing temperature favors an exothermic reaction.

      b. Increasing temperature favors an endothermic reaction.

4. Effect of changing pressure.
   a. There is no effect unless one of the reactants or products is a gas.
   b. Increasing pressure shifts the equilibrium in the direction that produces fewer gas molecules.

Solutions to Chapter 7 Problems

7.1   Kinetic energy: (b) a bullet in flight; (d) wind; (f) spinning windmill blades
Potential energy: (a) gunpowder; (c) a cell phone battery; (e) candy bar

In (a), potential energy is converted to kinetic energy when gunpowder explodes.
In (c), potential energy is converted to electrical energy when the phone is used.
In (e) potential energy is converted to heat when a candy bar is eaten.

7.2   Use Table 7.1 to find bond energies. The larger the bond energy, the more stable the molecule.
(a) N_2 (226 kcal/mol; 946 kJ/mol) is more stable than O_2 (119 kcal/mol; 498 kJ/mol).
(b) CO_2 is more stable than CO because it has two carbon–oxygen bonds.

CIA 7.1   A gram of fat contains more energy (9 kcal/g or 38 kJ/g) than a gram of carbohydrate (4 kcal/g or 17 kJ/g).

CIA 7.2   The chips contain 22.5 g carbohydrate and 22.5 g fat.

$$22.5 \text{ g carbohydrate} \times \frac{4.0 \text{ kcal}}{1 \text{ g carbohydrate}} = 90 \text{ kcal}; \quad 22.5 \text{ g fat} \times \frac{9.0 \text{ kcal}}{1 \text{ g fat}} = 200 \text{ kcal}$$

A small bag of potato chips contains 290 Cal (kcal), or 1200 kJ.

7.3   (a) Use the following expression to calculate ΔH:

$\Delta H = \Sigma(\text{bond dissociation energies})_{\text{reactants}} - \Sigma(\text{bond dissociation energies})_{\text{products}}$

$$\begin{aligned}\Sigma(\text{bond dissociation energies})_{\text{reactants}} &= 12(\text{C=O}) + 12(\text{O–H}) \\ &= 12(191 \text{ kcal/mol}) + 12(112 \text{ kcal/mol}) \\ &= 3636 \text{ kcal/mol}\end{aligned}$$

$$\begin{aligned}\Sigma(\text{bond dissociation energies})_{\text{products}} &= 5(\text{C–C}) + 7(\text{C–H}) + 7(\text{C–O}) + 5(\text{O–H}) + 6(\text{O=O}) \\ &= [5(83) + 7(99) + 7(86) + 5(112) + 6(119)] \text{ kcal/mol} \\ &= 2984 \text{ kcal/mol}\end{aligned}$$

$\Delta H = 3636 \text{ kcal/mol} - 2984 \text{ kcal/mol} = +652 \text{ kcal/mol}$
652 kcal/mol x 4.18 kJ/kcal = 2730 kJ/mol

(b) The reaction is endothermic because ΔH is positive.

7.4   (a) The reaction is endothermic because ΔH is positive.

(b) $\dfrac{801 \text{ kcal}}{4 \text{ mol Al}} \times 1.00 \text{ mol Al} = 200 \text{ kcal } (836 \text{ kJ}) \text{ required}$

(c) $10.0 \text{ g Al} \times \dfrac{1 \text{ mol Al}}{27.0 \text{ g Al}} \times \dfrac{801 \text{ kcal}}{4 \text{ mol Al}} = 74.2 \text{ kcal } (310 \text{ kJ}) \text{ required}$

7.5

$$127 \text{ g NO } \times \frac{1 \text{ mol}}{30.0 \text{ g NO}} \times \frac{43 \text{ kcal}}{2 \text{ mol NO}} = 91 \text{ kcal (380 kJ) absorbed}$$

7.6 $\Delta H = \Sigma(\text{bond dissociation energies})_{\text{reactants}} - \Sigma(\text{bond dissociation energies})_{\text{products}}$

$\Sigma(\text{bond dissociation energies})_{\text{reactants}} = (\text{C–C}) + 5(\text{C–H}) + (\text{C–O}) + (\text{O–H}) + 3(\text{O=O})$
$= [83 + 5(99) + 86 + 112 + 3(119)] \text{ kcal/mol}$
$= 1133 \text{ kcal/mol}$
$\Sigma(\text{bond dissociation energies})_{\text{products}} = 4(\text{C=O}) + 6(\text{O–H})$
$= [4(191) + 6(112)] \text{ kcal/mol} = 1436 \text{ kcal/mol}$
$\Delta H = 1133 \text{ kcal/mol} - 1436 \text{ kcal/mol} = -303 \text{ kcal/mol}$

$$\frac{303 \text{ kcal}}{1 \text{ mol}} \times \frac{1 \text{ mol}}{46 \text{ g}} \times \frac{1 \text{ Cal}}{1 \text{ kcal}} = \frac{6.59 \text{ Cal}}{1 \text{ g}}$$

This value agrees fairly well with the caloric value of alcohol (7.1 Cal/g) shown in Table 7.1.

HOC 7.1
(a) For a representative energy bar, fat = 6 g, protein = 20 g, carbohydrate = 16 g
The calculated value: [(6 g x 9 Cal/g) + (20 g x 4 Cal/g) + (16 g x 4 Cal/g)] = 198 Cal
This figure is in very close agreement with the reported value, 200 Cal.

(b) For a representative candy bar, fat = 12 g, protein = 4 g, carbohydrate = 33 g
The calculated value: [(12 g x 9 Cal/g) + (4 g x 4 Cal/g) + (33 g x 4 Cal/g)] = 256 Cal
This figure is also close to the 250 Cal reported on the bar.

7.7   (a) Entropy increases because the number of molecules increases.
(b) Entropy decreases because disorder decreases when gases condense.
(c) Entropy decreases because two moles of gaseous product are formed from three moles of gaseous reactants.

7.8   (a) The sign of ΔS is positive because disorder increases from reactant to products.
(b) In the expression $\Delta G = \Delta H - T\Delta S$, ΔH is a negative number. Because ΔS is positive, $(-T\Delta S)$ is negative at all temperatures. Thus, ΔG is negative at all temperatures, and the reaction is spontaneous at all temperatures.

7.9   In the process of rust formation, ΔG is negative because it is stated that the process is spontaneous. Since rust formation is slow, E_{act} is large.

7.10   The rate of reaction can be increased by either increasing concentration of reactants, increasing the temperature, or adding a catalyst.

7.11

$$K = \frac{[\text{products}]}{[\text{reactants}]}$$

(a) $K = \dfrac{[NO_2]^2}{[N_2O_4]}$ (b) $K = \dfrac{[H_2O]^2}{[H_2S]^2[O_2]}$ (c) $K = \dfrac{[Br_2][F_2]^5}{[BrF_5]^2}$

Don't forget that the coefficients of the reactants and products appear as exponents in the equilibrium expression. In (b), the concentration of solid sulfur doesn't enter into the equation.

7.12 If K is greater than 1, the reaction favors products. If K is less than 1, the reaction favors reactants.
(a) Products are strongly favored.
(b) Reactants are strongly favored.
(c) Products are somewhat favored.

7.13 In the expression for the equilibrium constant, the concentration of the product HI is squared in the numerator. The concentrations of the reactants appear in the denominator. Substitute the given concentrations into the expression. The equilibrium favors products.

$$K = \frac{[HI]^2}{[H_2][I_2]} = \frac{[0.507]^2}{[0.0510][0.174]} = 29.0$$

7.14

(a) $K = \dfrac{[AB]^2}{[A_2][B_2]}$; $K = \dfrac{[AB]^2}{[A_2][B]^2}$

(b) $K = \dfrac{[1]^2}{[3][3]} = 0.11; K = \dfrac{[4]^2}{[2][3]^2} = \dfrac{16}{18} = 0.89$

CIA 7.3 Body temperature is regulated by the thyroid gland and by the hypothalamus region of the brain.

CIA 7.4 Dilation of the blood vessels cools the body by allowing more blood to flow close to the surface of the body.

7.15 High pressure favors the production of SO_3 because increasing the pressure shifts the equilibrium in the direction that decreases the number of molecules in the gas phase. Low temperature favors the production of SO_3 because exothermic reactions are favored by lower temperatures.

7.16 (a) Increasing the temperature shifts the equilibrium to the left, favoring reactants.
(b) Increasing the pressure shifts the equilibrium to the right, favoring product.
(c) Removing CH_4 from the reaction vessel causes more CH_4 to be formed and shifts the equilibrium toward the right, favoring product.

7.17

$$272 \text{ kcal} \times \frac{1 \text{ mol}}{678 \text{ kcal}} = 0.401 \text{ mol glucose}$$

Understanding Key Concepts

7.18 ΔH is positive because energy must be supplied in order to break the attractive forces between molecules of the crystal. ΔS is also positive because the molecules of gas are more disordered than the molecules of solid. ΔG is negative because it is stated in the problem that the reaction is spontaneous.

7.19 ΔH is negative because energy is released when a liquid condenses. ΔS is negative because disorder decreases when a gas condenses. ΔG is negative because it is stated in the problem that the reaction is spontaneous.

7.20 (a) $2\ A_2\ +\ B_2\ \longrightarrow\ 2\ A_2B$

(b) ΔG is negative because the reaction is spontaneous. ΔS is negative because the product mixture has fewer gas molecules and less disorder than the reactant mixture. ΔH must be negative in order for ΔG to be negative since $(-T\Delta S)$ is positive.

7.21 (a) The blue curve represents the faster reaction, since less energy is needed to surmount the energy barrier leading to formation of products.
(b) The red curve represents a spontaneous reaction since the energy of its products is less than the energy of its reactants: ΔG is negative.

7.22 (a) ΔS is positive because the product has more disorder than the reactants.
(b) At low temperatures, the reaction is nonspontaneous because ΔH is positive and $T\Delta S$ is small. As the temperature rises, a certain temperature is reached at which $\Delta H\ =\ T\Delta S$. At this temperature, $\Delta G\ =\ 0$. Above this temperature, ΔG is negative and the reaction is spontaneous.

Additional Problems

Enthalpy and Heat of Reaction

7.23 In an endothermic reaction, the total enthalpy of the reactants is greater than the total enthalpy of the products. In other words, the reactants have stronger bonds and are more stable, while the products have weaker bonds and are less stable. Energy must, therefore, be supplied for the reaction to take place.

7.24 In a chemical reaction, the difference in bond energies between products and reactants is known as the *heat of reaction,* or enthalpy change.

7.25 (a) ΔH is positive. $Br_2(l) + 7.4\ \text{kcal/mol} \longrightarrow Br_2(g)$

(b) $5.8\ \text{mol}\ Br_2\ \times\ \dfrac{7.4\ \text{kcal}}{1\ \text{mol}\ Br_2}\ =\ 43\ \text{kcal needed}$

(c) $82\ \text{g}\ Br_2\ \times\ \dfrac{1\ \text{mol}\ Br_2}{159.8\ \text{g}\ Br_2}\ \times\ \dfrac{31.0\ \text{kJ}}{1\ \text{mol}\ Br_2}\ =\ 15.9\ \text{kJ needed}$

7.26 (a) ΔH is negative. $H_2O(l) \longrightarrow H_2O(s) + 1.44$ kcal

(b) 2.5 mol H_2O x $\dfrac{-6.02 \text{ kJ}}{1 \text{ mol } H_2O}$ = -15.1 kJ

(c) 32 g H_2O x $\dfrac{1 \text{ mol } H_2O}{18 \text{ g } H_2O}$ x $\dfrac{-1.44 \text{ kcal}}{1 \text{ mol } H_2O}$ = -2.6 kcal

(d) +1.44 kcal/mol

7.27 (a) 2 $C_2H_2(g)$ + 5 $O_2(g)$ $\longrightarrow$ 4 $CO_2(g)$ + 2 $H_2O(g)$

(b) $\Delta H = \Sigma$(bond dissociation energies)$_{reactants} - \Sigma$(bond dissociation energies)$_{products}$

Σ(bond dissociation energies)$_{reactants}$ = 4(C–H) + 2(C≡C) + 5(O=O)
= [4(99) + 2(201) + 5(119)] kcal/mol
= 1393 kcal/mol (5820 kJ/mol)

Σ(bond dissociation energies)$_{products}$ = 8(C=O) + 4(O–H)
= [8(191) + 4(112)] kcal/mol
= 1976 kcal/mol (8260 kJ/mol)

ΔH = 1393 kcal/mol – 1976 kcal/mol = -583 kcal/mol = -2440 kJ/mol

(c)

$\dfrac{2440 \text{ kJ}}{1 \text{ mol}}$ x $\dfrac{1 \text{ mol}}{2 \text{ mol } C_2H_2}$ x $\dfrac{1 \text{ mol } C_2H_2}{26 \text{ g } C_2H_2}$ = $\dfrac{46.9 \text{ kJ}}{1 \text{ g } C_2H_2}$

This value is similar to the energy values for gasoline and natural gas and indicates that acetylene is one of the best fuels.

7.28

(a) :N≡N: + 2 :Ö=Ö: $\longrightarrow$ 2 $\cdot\overset{+}{N}=\overset{..}{\underset{..}{O}}$
$^-$:Ö:

(b) Σ(bond dissociation energies)$_{reactants}$ = N≡N + 2(O=O)
= 226 kcal/mol + 2(119 kcal/mol)
= 464 kcal/mol (1940 kJ/mol)
Σ(bond dissociation energies)$_{products}$ = 2(N=O) + 2(N–O)
= 2(145 kcal/mol) + 2(48 kcal/mol)
= 386 kcal/mol (1610 kJ/mol)
ΔH = 464 kcal/mol – 386 kcal/mol = +78 kcal/mol (330 kJ/mol)

7.29 (a) $C_6H_{12}O_6 + 6 O_2 \longrightarrow 6 CO_2 + 6 H_2O$

(b) $\dfrac{-3.8 \text{ kcal}}{1 \text{ g glucose}} \times \dfrac{180 \text{ g glucose}}{1 \text{ mol glucose}} \times 1.50 \text{ mol glucose} = -1.0 \times 10^3 \text{ kcal} = -4.2 \times 10^3 \text{ kJ}$

(c) The production of glucose from CO_2 and H_2O is an endothermic process.

$\dfrac{3.8 \text{ kcal}}{1 \text{ g glucose}} \times 15.0 \text{ g glucose} = 57 \text{ kcal (238 kJ) needed to produce 15 g glucose}$

7.30 (a) $2 C_8H_{18} + 25 O_2 \longrightarrow 16 CO_2 + 18 H_2O + \text{heat}$

(b) ΔH is negative, because energy is released.

(c) $\dfrac{1002 \text{ kJ}}{5.00 \text{ g } C_8H_{18}} \times \dfrac{114 \text{ g } C_8H_{18}}{1 \text{ mol } C_8H_{18}} = \dfrac{2.28 \times 10^4 \text{ kJ (5460 kcal)}}{1 \text{ mol } C_8H_{18}} \text{ released}$

(d) $450.0 \text{ kcal} \times \dfrac{1 \text{ mol } C_8H_{18}}{5460 \text{ kcal}} = 0.0824 \text{ mol } C_8H_{18}$

$0.0824 \text{ mol} \times \dfrac{114 \text{ g}}{1 \text{ mol}} = 9.39 \text{ g } C_8H_{18}$

(e) $17.0 \text{ g } C_8H_{18} \times \dfrac{1002 \text{ kJ}}{5.00 \text{ g } C_8H_{18}} = 3.40 \times 10^3 \text{ kJ (814 kcal) released}$

Entropy and Free Energy

7.31 Increase in entropy: (a)
Decrease in entropy: (b), (c)

7.32 (a) Entropy decreases because there is an increase in order.
(b) Entropy decreases because there are fewer moles of gaseous product than reactant.
(c), (f) Entropy decreases because formation of a precipitate decreases entropy.
(d), (e) Entropy increases because there are more moles of gaseous products than gaseous reactants.

7.33 The two factors that influence the spontaneity of a reaction are: (1) the release or absorption of heat, and (2) the increase or decrease in entropy.

7.34 An exothermic reaction releases heat (negative ΔH), whereas an exergonic reaction is spontaneous (negative ΔG).

7.35 The free-energy change (ΔG) of a chemical reaction shows whether or not a reaction is spontaneous. If the sign of ΔG is negative, the reaction is spontaneous. Of the two factors that contribute to ΔG (ΔH and ΔS), ΔH is usually larger at low temperatures. Thus, if a reaction is spontaneous (negative ΔG), it is usually exothermic (negative ΔH).

7.36 If an endothermic reaction is accompanied by a large increase in entropy, the ($-T\Delta S$) term in the expression for ΔG becomes large and negative and outweighs the positive enthalpy term, resulting in a negative value for ΔG and indicating an exergonic reaction.

7.37   (a) Dissolving NaCl is endothermic since ΔH is positive.
(b Entropy increases because disorder increases.
(c) Since the dissolution of NaCl is spontaneous, ΔG for the reaction must be negative. Because we already know that ΔH is positive, it must be true that $T\Delta S$ is the major contributor to ΔG.

7.38   (a) Entropy decreases because two atoms of liquid and one molecule of gas react to yield two molecules of solid.
(b) Because ΔS is negative, $-T\Delta S$ is positive. The reaction is spontaneous up to the temperature at which $\Delta H = T\Delta S$; above this temperature, the reaction no longer is spontaneous.

7.39   (a) $H_2(g) + Br_2(l) \longrightarrow 2\,HBr(g)$
(b) Entropy increases because the number of gaseous product molecules is greater than the number of gaseous reactant molecules, and thus the sign of ΔS is positive.
(c) The process is spontaneous at all temperatures because ΔH is negative and ΔS is positive.
(d) $\Delta G = \Delta H - T\Delta S$; $\Delta H = -17.4$ kcal/mol; $\Delta S = 27.2$ cal/(mol·K); $T = 300$ K

$$T\Delta S = 300\ \text{K} \times \frac{27.2\ \text{cal}}{\text{mol K}} \times \frac{1\ \text{kcal}}{10^3\ \text{cal}} = \frac{8.16\ \text{kcal}}{\text{mol}} \left(\frac{34.1\ \text{kJ}}{\text{mol}}\right)$$

$$\Delta G = -17.4\ \text{kcal/mol} - 8.16\ \text{kcal/mol} = -25.6\ \text{kcal/mol}\ (-107\ \text{kJ/mol})$$

7.40   (a) ΔS is negative because two molecules of a gas combine to form one molecule of liquid.
(b) This reaction is spontaneous only up to a certain temperature, above which $T\Delta S$ is greater than ΔH and the reaction becomes nonspontaneous.

Rates of Chemical Reactions

7.41   The *activation energy* of a reaction is the amount of energy needed for reactants to surmount the energy barrier to reaction.

7.42   A reaction with $E_{act} = +5$ kcal is faster than one with $E_{act} = +10$ kcal because less energy is needed to surmount the energy barrier.

7.43   The reaction rate for the reverse of an endergonic reaction is faster than the rate for the forward reaction. The reaction rate for the reverse of an exergonic reaction is slower than the rate for the forward reaction.

7.44   Increasing concentration increases the reaction rate because the increased crowding of reactants causes more collisions.

7.45   A catalyst is a substance that increases reaction rate by lowering the activation energy barrier, yet remains unchanged when the reaction is completed.

7.46   A catalyst that reduces the energy of activation of a forward reaction by 5.0 kcal/mol (from 28.0 kcal/mol to 23.0 kcal/mol) also reduces the energy of activation of the reverse reaction by 5.0 kcal/mol. Lowering the energy of activation of the reverse reaction, however, will not cause it to take place if it is not spontaneous.

7.47 (a)The negative value of ΔG indicates that diamonds spontaneously turn into graphite at 25 °C.
(b) Because this behavior is not observed, the activation energy of the reaction must be extremely high and the reaction rate must be extremely slow.

7.48 (a) The reaction of hydrogen with carbon is not spontaneous at 25 °C because ΔG is positive.
(b) It is not possible to find a catalyst for the reaction because it is not spontaneous at any temperature.

Chemical Equilibria

7.49 In a reversible reaction, chemical equilibrium is a state in which the rates of the forward reaction and the reverse reaction are equal. The amounts of reactants and products need not be equal at equilibrium.

7.50 Catalysts lower the height of the activation energy barrier for both forward and reverse reactions by the same amount but do not change the equilibrium constant.

7.51 Remember that solids and pure liquids do not appear in the equilibrium expression

(a) $K = \dfrac{[CO_2]^2}{[CO]^2[O_2]}$ (b) $K = \dfrac{[MgCl_2][H_2]}{[HCl]}$

(c) $K = \dfrac{[H_3O^+][F^-]}{[HF]}$ (d) $K = \dfrac{[SO_2]}{[O_2]}$

7.52

(a) $K = \dfrac{[H_2S]^2}{[S_2][H_2]^2}$ (b) $K = \dfrac{[HCl]^2}{[H_2S][Cl_2]}$

(c) $K = \dfrac{[BrCl]^2}{[Br_2][Cl_2]}$ (d) $K = \dfrac{[CO][H_2]}{[H_2O]}$

7.53
$$K = \frac{[NO_2]^2}{[N_2O_4]} = \frac{(0.0325 \text{ mol/L})^2}{(0.147 \text{ mol/L})} = 7.19 \times 10^{-3} \text{ Reactants are favored.}$$

7.54
$$K = \frac{[CO_2]^2}{[CO]^2[O_2]} = \frac{(0.11 \text{ mol/L})^2}{(0.025 \text{ mol/L})^2 (0.015 \text{ mol/L})} = 1.3 \times 10^3 \text{ Products are favored.}$$

7.55

(a) $K = 7.19 \times 10^{-3} = \dfrac{(0.0250 \text{ mol/L})^2}{[N_2O_4]}$; $[N_2O_4] = 0.0869 \text{ mol/L}$

(b) $K = 7.19 \times 10^{-3} = \dfrac{[NO_2]^2}{(0.0750 \text{ mol/L})}$; $[NO_2] = 0.0232 \text{ mol/L}$

7.56

(a) $1.3 \times 10^3 = \dfrac{(0.18 \text{ mol/L})^2}{(0.0200 \text{ mol/L})^2 [O_2]}$

$[O_2] = \dfrac{(0.18 \text{ mol/L})^2}{(0.0200 \text{ mol/L})^2 \times 1.3 \times 10^3} = 0.062 \text{ mol/L}$

(b) $[CO_2]^2 = K \times [CO]^2 \times [O_2] = 1.3 \times 10^3 \times (0.080 \text{ mol/L})^2 \times 0.520 \text{ mol/L}$

$= 4.3 \text{ (mol/L)}^2; [CO_2] = 2.1 \text{ mol/L}$

7.57   If the pressure is raised, relatively more reactants than products are found at equilibrium, because high pressure favors the reaction that produces fewer gas molecules.

7.58   If the pressure is lowered, a relatively greater concentration of reactants than products is found at equilibrium, because lower pressure favors the reaction that produces more gas molecules.

Le Châtelier's Principle

7.59   (a) The reaction is endothermic.
(b) Reactants are favored at equilibrium.
(c) (1) Increasing pressure favors formation of ozone, since increased pressure shifts the equilibrium in the direction of the reaction that produces fewer gas molecules.
(2) Increasing the O_2 concentration increases the amount of O_3 formed.
(3) Increasing the concentration of O_3 shifts the equilibrium to the left.
(4) A catalyst has no effect on the equilibrium.
(5) Increasing the temperature shifts the equilibrium to the right because the reaction is endothermic.

7.60   (a) This reaction is exothermic.
(b) Products are favored at equilibrium.
(c) (1) Increasing pressure has no effect on the equilibrium since there are the same number of reactant gas molecules and product gas molecules.
(2) Increasing the HCl concentration shifts the equilibrium to the left.
(3) Decreasing the Cl_2 concentration decreases the amount of HCl formed.
(4) Increasing the H_2 concentration increases the amount of HCl formed.
(5) A catalyst has no effect on the equilibrium.

7.61   According to Le Châtelier's principle, an increase in pressure shifts the equilibrium in the direction that decreases the number of molecules in the gas phase.
(a) The concentration of products decreases.
(b) The concentration of products remains the same.
(c) The concentration of products increases.

7.62   A decrease in volume increases the system pressure.
(a) The equilibrium shifts to favor reactants.
(b) The equilibrium shifts to favor products.
(c) The position of equilibrium remains the same.

7.63 $CO(g) + H_2O(g) \rightleftharpoons CO_2(g) + H_2(g)$ $\Delta H = -9.8$ kcal/mol (-41 kJ/mol)
Because it has a negative value of ΔH, the reaction is exothermic, and the equilibrium favors products. Decreasing the temperature increases the amount of H_2.

7.64 $3 O_2(g) \rightleftharpoons 2 O_3(g)$ $\Delta H = +68$ kcal/mol ($+285$ kJ/mol)
Increasing the temperature of the reaction increases the equilibrium constant.

7.65 $H_2(g) + I_2(g) \rightleftharpoons 2 HI(g)$ $\Delta H = -2.2$ kcal/mol (-9.2 kJ/mol)
(a) Adding I_2 to the reaction mix increases the equilibrium concentration of HI.
(b) Removing H_2 causes the equilibrium concentration of HI to decrease.
(c) Addition of a catalyst does not change the equilibrium concentration of HI.
(d) Increasing the temperature decreases the concentration of HI.

7.66 $Fe^{3+}(aq) + Cl^-(aq) \rightleftharpoons FeCl^{2+}(aq)$
(a) Addition of $Fe(NO_3)_3$ increases the equilibrium concentration of $FeCl^{2+}$.
(b) Removal of Cl^- by precipitation decreases the concentration of $FeCl^{2+}$.
(c) Increasing the temperature of the endothermic reaction increases the equilibrium concentration of $FeCl^{2+}$.
(d) Addition of a catalyst does not change the equilibrium concentration of $FeCl^{2+}$.

Conceptual Problems

7.67 (a) $C_2H_5OH + 3 O_2 \longrightarrow 2 CO_2 + 3 H_2O + 327$ kcal
(b) ΔH is negative because energy is released.

(c) 5.00 g $C_2H_5OH \times \dfrac{1 \text{ mol } C_2H_5OH}{46.0 \text{ g } C_2H_5OH} \times \dfrac{327 \text{ kcal}}{1 \text{ mol } C_2H_5OH} = 35.5$ kcal

(d) Calories needed = specific heat of H_2O x mass (g) x temperature change (°C)
Specific heat of water = 1.00 cal /(g · °C); mass of 500 mL H_2O = 500 g;
Temperature change = 80.0 °C

Calories = $\dfrac{1.00 \text{ cal}}{\text{g} \cdot °C} \times 500.0$ g $\times 80.0$ °C $= 4.00 \times 10^4$ cal $= 40.0$ kcal

40.0 kcal $\times \dfrac{5.00 \text{ g } C_2H_5OH}{35.5 \text{ kcal}} = 5.63$ g C_2H_5OH

(e) $\dfrac{35.5 \text{ kcal}}{5.00 \text{ g } C_2H_5OH} \times \dfrac{0.789 \text{ g } C_2H_5OH}{1 \text{ mL } C_2H_5OH} = 5.60$ kcal/mL $= 23.4$ kJ/mol

7.68 $N_2(g) + 3 H_2(g) \longrightarrow 2 NH_3(g)$ $\Delta H = -22$ kcal/mol (-92 kJ/mol)

(a) The production of ammonia from its elements is an exothermic process.

(b) $\dfrac{-22 \text{ kcal}}{2 \text{ mol } NH_3} \times 0.700$ mol $NH_3 = -7.7$ kcal (-32 kJ)

7.69 (a) $Fe_3O_4(s) + 4 H_2(g) \longrightarrow 3 Fe(s) + 4 H_2O(g)$ $\Delta H = +36$ kcal/mol (151 kJ/mol)

(b) 55 g Fe $\times \dfrac{1 \text{ mol Fe}}{55.8 \text{ g Fe}} \times \dfrac{36 \text{ kcal}}{1 \text{ mol Fe}_3\text{O}_4} \times \dfrac{1 \text{ mol Fe}_3\text{O}_4}{3 \text{ mol Fe}} = 12$ kcal (50 kJ)

(c) 75 g Fe $\times \dfrac{1 \text{ mol Fe}}{55.8 \text{ g Fe}} \times \dfrac{4 \text{ mol H}_2}{3 \text{ mol Fe}} \times \dfrac{2.0 \text{ g H}_2}{1 \text{ mol H}_2} = 3.6$ g H_2

(d) Reactants are favored in this reaction.

7.70 (a) CO removes Hb from the bloodstream because the reaction of CO with Hb is more favorable than the reaction of O_2 with Hb. Less Hb is available to react with O_2, and less HbO_2 is available to the tissues.

(b) Administering high doses of O_2 to a victim of CO poisoning shifts the equilibrium in the following reaction to the right.

$Hb(CO)(aq) + O_2(aq) \rightleftharpoons HbO_2(aq) + CO(aq)$

This shift results in displacement of CO from Hb(CO) and in formation of HbO_2 to replenish body tissues with O_2.

7.71

(a)

(b) Σ(bond dissociation energies)$_{\text{reactants}}$ = (C=O) + 2(C–N) + 4(N–H) + 2(O–H)
= [178 + 2(73) + 4(93) + 2(112)] kcal/mol
= 920 kcal/mol

Σ(bond dissociation energies)$_{\text{products}}$ = 2(C=O) + 6(N–H)
= [2(191) + 6(93)] kcal/mol = 940 kcal/mol

ΔH = 920 kcal/mol – 940 kcal/mol = –20 kcal/mol (–84 kJ/mol)

7.72

(a) 10.0 g $H_2O \times \dfrac{1 \text{ mol H}_2\text{O}}{18.0 \text{ g H}_2\text{O}} \times \dfrac{9.72 \text{ kcal}}{1 \text{ mol H}_2\text{O}} = 5.40$ kcal needed

(b) 22.6 kJ are released.

7.73 (a) $4 NH_3(g) + 5 O_2(g) \rightleftharpoons 4 NO(g) + 6 H_2O(g)$ + heat

(b) $K = \dfrac{[\text{NO}]^4 [\text{H}_2\text{O}]^6}{[\text{NH}_3]^4 [\text{O}_2]^5}$

(c) (1) Raising the pressure shifts the equilibrium to the left and favors reactants.
(2) Adding $NO(g)$ shifts the equilibrium to the left and favors reactants.
(3) Decreasing NH_3 shifts the equilibrium to the left and favors reactants.
(4) Lowering the temperature shifts the equilibrium to the right and favors products.

7.74   (a) $2\ CH_3OH(l)\ +\ 3\ O_2(g)\ \longrightarrow\ 2\ CO_2(g)\ +\ 4\ H_2O(g)$

      (b) $1.85\ mol\ CH_3OH\ \times\ \dfrac{-174\ kcal}{1\ mol\ CH_3OH}\ =\ -322\ kcal\ (-1340\ kJ)$

      (c) $50.0\ g\ CH_3OH\ \times\ \dfrac{1\ mol\ CH_3OH}{32.0\ g\ CH_3OH}\ \times\ \dfrac{-728\ kJ}{1\ mol\ CH_3OH}\ =\ -1140\ kJ\ (-272\ kcal)$

7.75

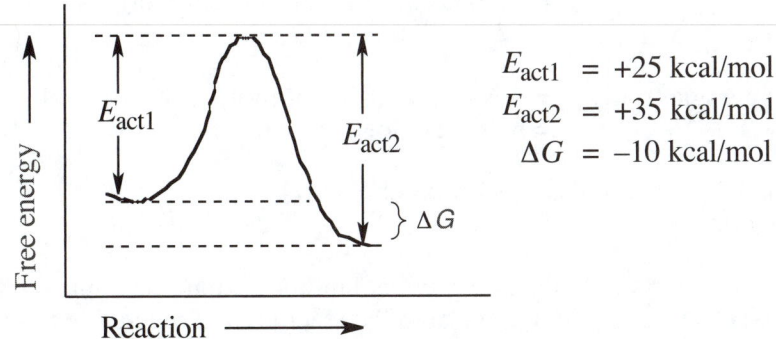

$$E_{act1}\ =\ +25\ kcal/mol$$
$$E_{act2}\ =\ +35\ kcal/mol$$
$$\Delta G\ =\ -10\ kcal/mol$$

      (a) The forward process is exergonic.

      (b) $\Delta G = -10\ kcal/mol\ (-4.2\ kJ/mol)$

7.76

      (a) $0.255\ mol\ \times\ \dfrac{-848.9\ kJ}{2\ mol\ Al}\ =\ -108\ kJ\ (-25.9\ kcal)$

      (b) $5.00\ g\ \times\ \dfrac{1\ mol\ Al}{27.0\ g\ Al}\ \times\ \dfrac{-202.9\ kcal}{2\ mol\ Al}\ =\ -18.8\ kcal$

      18.8 kcal of heat is released.

7.77

      $1.00\ g\ \times\ \dfrac{1\ mol\ Na}{23.0\ g}\ \times\ \dfrac{-88.0\ kcal}{2\ mol\ Na}\ =\ -1.91\ kcal\ (-7.99\ kJ/mol)$

      1.91 kcal of heat is evolved. The reaction is exothermic.

Group Problems

7.78   Let us use as an example a smallish bag of potato chips. The 77.9 oz. bag claims to contain 440 Cal. According to the nutrition information on the bag, the calories come from protein (5g), fat (29 g) and carbohydrate (42 g). Using the information in the Chemistry in Action feature, we calculate (5 g x 4 Cal/g = 20 Cal) + (29 g x 9 Cal/g = 260 Cal) + (42 g x 4 Cal/g = 170 Cal) = 450 Cal, which agrees pretty closely with the information on the bag.

      (a) A 160 pound person uses up approximately 290 Cal/hr in biking, 600 Cal/hr in moderate running, and 420 Cal/hr in swimming.

      (b) You would either have to bike for about 11/2 hours, run for 45 minutes, or swim for nearly an hour to burn the calories.

7.79   When blood glucose levels rise, insulin is released from the pancreas. Insulin sends a signal to cells to take up glucose, which is metabolized there. If blood glucose drops too low, glucagon, a second hormone, sends a signal to other tissues to convert other substances into glucose in order to restore the blood glucose level. This system obeys Le Châtelier's principle because, as glucose is consumed, more glucose is supplied to maintain blood glucose levels.

7.80   (a)

$$\Delta H = \Sigma(\text{bond dissociation energies})_{\text{reactants}} - \Sigma(\text{bond dissociation energies})_{\text{products}}$$

$$
\begin{aligned}
\Sigma(\text{bond dissociation energies})_{\text{reactants}} &= N\equiv N + 3(H-H) \\
&= (226 \text{ kcal/mol}) + 3(103 \text{ kcal/mol}) \\
&= 535 \text{ kcal/mol}
\end{aligned}
$$

$$
\begin{aligned}
\Sigma(\text{bond dissociation energies})_{\text{products}} &= 6(N-H) = 6(93 \text{ kcal/mol}) \\
&= 558 \text{ kcal/mol}
\end{aligned}
$$

$$\Delta H = 535 \text{ kcal/mol} - 558 \text{ kcal/mol} = -23 \text{ kcal/mol} = -96 \text{ kJ/mol}$$
The reaction is exothermic.

(b) According to Le Châtelier's principle, the yield of ammonia might be improved by adding more N_2 or NH_3, removing ammonia from the reaction mix, decreasing temperature, or by increasing pressure.

(c) Haber and Bosch tried all of the methods in (b). Lowering the temperature was unsuccessful because the resulting energy of activation was too low for the reaction to proceed, Although both other methods were successful, removing ammonia from the system was most practical.

Self-Test for Chapter 7

Multiple Choice

1. In which of the following processes does entropy decrease?
   (a) $C_6H_{12}O_6(s) + 6 \ O_2(g) \longrightarrow 6 \ CO_2(g) + 6 \ H_2O(l)$

   (b) $2 \ Na(s) + 2 \ H_2O(l) \longrightarrow 2 \ NaOH(aq) + H_2(g)$

   (c) $2 \ Mg(s) + O_2(g) \longrightarrow 2 \ MgO(s)$

   (d) $Zn(s) + CuSO_4(aq) \longrightarrow Cu(s) + ZnSO_4(aq)$

2. The reaction that will proceed at any temperature has:
   (a) negative ΔH, negative ΔS (b) negative ΔH, positive ΔS
   (c) positive ΔH, positive ΔS (d) positive ΔH, negative ΔS

3. Which of the conditions in Problem 2 indicates the reaction with the largest positive value of ΔG?

4. In the reaction $CH_3CO_2H + CH_3OH \rightleftharpoons CH_3CO_2CH_3 + H_2O$, the yield of $CH_3CO_2CH_3$ can be improved by:
   (a) distilling off the $CH_3CO_2CH_3$ (b) removing the water
   (c) using more CH_3CO_2H and CH_3OH (d) all of the above

5. The equilibrium expression for the reaction $C_6H_6 + 3\,H_2 \longrightarrow C_6H_{12}$ is:

(a) $\dfrac{[C_6H_{12}]}{[C_6H_6]\,[H_2]}$ (b) $\dfrac{[C_6H_{12}]}{[C_6H_6]\,[H_2]^3}$ (c) $\dfrac{[C_6H_{12}]}{[C_6H_6]\,[3\,H_2]}$ (d) $\dfrac{[C_6H_6]\,[H_2]^3}{[C_6H_{12}]}$

6. A reaction with which of the following K values is most likely to go to completion at room temperature?
(a) $K = 10^5$ (b) $K = 10^2$ (c) $K = 10^{-1}$ (d) $K = 10^{-4}$

7. In which of the following reactions will increasing the pressure decrease the yield of product?
(a) $2\,Mg(s) + O_2(g) \longrightarrow 2\,MgO(s)$
(b) $H_2C{=}CH_2(g) + H_2(g) \longrightarrow CH_3CH_3(g)$
(c) $C(s) + H_2O(g) \longrightarrow CO(g) + H_2(g)$
(d) $3\,O_2(g) \longrightarrow 2\,O_3(g)$

8. According to Le Châtelier's principle, which of the following changes also changes the value of K?
(a) changing temperature   (b) changing concentration
(c) changing pressure   (d) all of the above

9. All of the following statements are true about the reaction $3\,O_2(g) \longrightarrow 2\,O_3(g)$ (heat of formation = +34 kcal/mol at 25 °C) except:
(a) The reaction is endothermic.   (b) ΔS is negative.   (c) ΔG is positive.
(d) The rate of the reverse reaction increases with temperature.

10. In the reaction $Si + O_2 \longrightarrow SiO_2$ ($\Delta H = -218$ kcal/mol at 25 °C), what is ΔH in kcal when 80 g of O_2 reacts with 70 g of Si?
(a) 218 kcal   (b) –218 kcal   (c) –436 kcal   (d) –545 kcal

Sentence Completion

1. A reaction that has equal amounts of reactants and products at equilibrium has $K =$ ___.

2. The law of _____ ___ _____ states that energy can be neither created nor destroyed.

3. Addition of a _____ increases the rate of a reaction.

4. A reaction that easily proceeds in either direction is _____.

5. A reaction that absorbs heat from the surroundings is an _____ reaction.

6. An _____ _____ ____ gives the relationship between the concentrations of products and reactants at equilibrium.

7. The rate of a reaction can be increased by increasing _____ or _____ or by adding a _____.

8. The reactions that continually take place in the body are known as _____.

9. _____ ___ is unrelated to reaction spontaneity.

10. For a chemical reaction to occur, reactant molecules must _____.

True or False

1. The caloric value of food tells how much energy is absorbed when food is burned in oxygen.

2. In a chemical equilibrium, both product and reactant concentrations remain constant.

3. According to Le Châtelier's principle, changing pressure only affects equilibrium if gaseous reactants or products are involved.

4. In an exothermic reaction, the bond dissociation energies of the products are greater than the bond dissociation energies of the reactants.

5. A reaction with a large E_{act} will probably have a large heat of reaction.

6. Raising the temperature of a reaction always increases the rate of reaction.

7. A very unfavorable reaction may take place if the temperature is high enough.

8. Catalysts increase the rate of a reaction by increasing the number of collisions between reactants.

9. At chemical equilibrium, all chemical reaction stops.

10. At low temperatures, the spontaneity of a reaction is determined by ΔH.

Match each entry on the left with its partner on the right.

1.	Enzyme	(a) Shows energy relationships in a reaction
2.	Exergonic reaction	(b) A reaction that gives off heat
3.	K	(c) Reduces the size of E_{act}
4.	Enthalpy	(d) Energy needed for a reaction to occur
5.	Reaction energy diagram	(e) Measure of the amount of disorder in a reaction
6.	Free energy	(f) Reaction that is not spontaneous
7.	Exothermic reaction	(g) Reaction that can go in either direction
8.	Endergonic reaction	(h) Reaction that absorbs heat
9.	E_{act}	(i) Difference in energy of products and reactants
10.	Entropy	(j) Reaction that is spontaneous
11.	Reversible reaction	(k) Determines if a reaction is spontaneous
12.	Endothermic reaction	(l) Measures the ratio of products to reactants at equilibrium

Chapter 8   Gases, Liquids, and Solids

Chapter Outline

I. Introduction to gases, liquids, and solids (Section 8.1).
   A. Phases of matter are determined by the attractive forces between molecules.
      1. In gases, attractive forces are very weak.
      2. In liquids, attractive forces are strong.
      3. In solids, forces are so strong that atoms are held in place.
   B. During changes of phase, also known as changes of state, heat is either absorbed or released.
      1. Every change of state is characterized by a free-energy change, ΔG.
         a. ΔG consists of an enthalpy term and an entropy term.
         b. The enthalpy term is a measure of the heat absorbed or released during a phase change.
         c. The entropy change is a measure of the change in molecular disorder.
      2. At the temperature where a change of phase occurs, two states are in equilibrium.
      3. At the change from solid to liquid, two phases are at equilibrium at the melting point.
      4. At the change from liquid to gas, two phases are at equilibrium at the boiling point.
II. Intermolecular forces (Section 8.2).
   A. Intermolecular forces are the forces that act between molecules.
   B. In an ideal gas, intermolecular forces are unimportant.
   C. Three types of intermolecular forces are important in liquids and solids.
      1. London forces.
         a. Short-lived polarity in molecules causes a temporary attraction between molecules.
         b. London forces increase with increasing molecular weight and vary with molecular shape.
      2. Dipole–dipole forces occur when the positive end of a polar molecule is attracted to the negative end of a second molecule.
      3. Hydrogen bonding.
         a. Hydrogen bonds occur between a hydrogen atom bonded to an electronegative atom (O, N, F) and an unshared electron pair of a second electronegative atom.
         b. In hydrogen bonds, the hydrogen atom is partially bonded to two different electronegative atoms.
         c. Hydrogen bonds can be quite strong, and hydrogen bonding is responsible for elevated boiling points.
III. Gases (Sections 8.3–8.11).
   A. The kinetic–molecular theory explains the behavior of gases (Section 8.3).
      1. A gas consists of a great many molecules moving about with no attractive forces.
      2. The amount of space that molecules occupy is much smaller than the space between molecules.
      3. The energy of the molecules is related to Kelvin temperature.
      4. When molecules collide, their total kinetic energy is conserved.
      5. A gas that obeys all these behaviors is an ideal gas.
   B. Pressure (Sections 8.4).
      1. Pressure is defined as force per unit area.
      2. Units of pressure are mmHg, Pascal (in the SI system), atmosphere, and pounds per square inch (psi).
      3. Gas pressure can be measured by using a barometer or a manometer.

C. Gas laws (Sections 8.5–8.11).
  1. Boyle's law (for a fixed amount of gas at constant T) (Section 8.5).
    a. The pressure of a gas is inversely proportional to its volume.
    b. $P_1V_1 = P_2V_2$.
  2. Charles's law (for a fixed amount of gas at constant P) (Section 8.6).
    a. The volume of a gas is directly proportional to its temperature in K.
    b. $V_1/T_1 = V_2/T_2$.
  3. Gay-Lussac's law (for a fixed amount of gas at constant V) (Section 8.7).
    a. Pressure is directly proportional to temperature in K.
    b. $P_1/T_1 = P_2/T_2$.
  4. Combined gas law (Section 8.8).
    a. $P_1V_1/T_1 = P_2V_2/T_2$ for a fixed amount of gas.
  5. Avogadro's law (at constant T and P) (Section 8.9).
    a. The volume of a gas is directly proportional to its molar amount at constant T and P.
    b. $V_1/n_1 = V_2/n_2$.
    c. Standard temperature and pressure = 273.15 K and 1 atm.
    d. Standard molar volume of a gas = 22.4 L.
  6. Ideal gas law (Section 8.10).
    a. $PV = nRT$, where R is a gas constant.
  7. Dalton's law of partial pressure (Section 8.11).
    a. Mixtures of gases behave the same as a pure gas.
    b. The partial pressure of a gas in a mixture is the same as the gas would have if it were alone.
IV. Liquids and solids (Sections 8.12–8.14).
  A. Liquids (Section 8.12).
    1. Evaporation occurs when molecules near the surface of a liquid escape into the gaseous state.
      a. When molecules are in the gaseous state, they obey gas laws.
    2. The contribution of the partial pressure of the escaped gas to the total pressure is known as vapor pressure.
      a. Vapor pressure rises with increasing temperature.
    3. The normal boiling point of a liquid is at 760 mmHg.
      a. The boiling point rises or falls with atmospheric pressure.
    4. Viscosity and surface tension are properties of liquids.
      a. Viscosity is a liquid's resistance to flow.
      b. Surface tension is the resistance of a liquid to spread out.
  B. Solids (Section 8.13).
    1. A crystalline solid has atoms, molecules, or ions rigidly held in an orderly arrangement.
      a. Categories of crystalline solids include ionic solids, molecular solids, covalent network solids, and metallic solids.
    2. Particles in an amorphous solid do not have an orderly arrangement.
    3. Solid water is less dense than liquid water.
  C. Changes of phase (Section 8.14).
    1. The heat needed to completely melt a solid is the heat of fusion.
    2. The heat needed to completely vaporize a liquid is the heat of vaporization.

Solutions to Chapter 8 Problems

8.1 (a) The change of state from liquid to gas is unfavored by ΔH (positive sign) but is favored by ΔS (large positive value).
(b) For the change from gas to liquid, $\Delta H = -9.72$ kcal/mol (-40.1 kJ/mol) and $\Delta S = -26.1$ cal/mol·K (-109 kJ/mol·K).

8.2 Boiling points generally increase with increasing molecular (or atomic) weight because of London dispersion forces.
(a) Kr, Ar, Ne. This series is arranged in order of decreasing boiling point.
(b) Cl_2, Br_2, I_2. This series is arranged in order of increasing boiling point.

8.3 Methyl alcohol (a) and methylamine (c) are capable of hydrogen bonding because each contains a hydrogen atom bonded to an electronegative atom. Ethylene (b) does not form hydrogen bonds.

8.4 (a) *London forces* are the only intermolecular forces between nonpolar ethane molecules, and thus ethane has a low boiling point.
(b) The major intermolecular force between ethyl alcohol molecules is *hydrogen bonding,* which causes ethyl alcohol to be high boiling. Dipole–dipole interactions and London forces are also present but are weaker than hydrogen bonding.
(c) *Dipole–dipole* interactions are the principal forces between ethyl chloride molecules and cause the boiling point of ethyl chloride to be higher than that of ethane. London forces are also present.

8.5 Use the appropriate conversion factor from Section 8.4.

0.289 atm x $\dfrac{760 \text{ mmHg}}{1 \text{ atm}}$ = 220 mmHg

0.289 atm x $\dfrac{14.7 \text{ psi}}{1 \text{ atm}}$ = 4.25 psi

0.289 atm x $\dfrac{101,325 \text{ Pa}}{1 \text{ atm}}$ = 2.93 x 10^4 Pa

8.6

32 psi x $\dfrac{1 \text{ atm}}{14.7 \text{ psi}}$ = 2.2 atm

32 psi x $\dfrac{760 \text{ mmHg}}{14.7 \text{ psi}}$ = 1700 mmHg

32 psi x $\dfrac{101,325 \text{ Pa}}{14.7 \text{ psi}}$ = 2.2 x 10^5 Pa

CIA 8.1 Increased concentrations of CO_2 in the atmosphere, coupled with an increase in the average global temperature, are evidence for global warming.

CIA 8.2 The three most important greenhouse gases are carbon dioxide, water vapor, and methane.

8.7

$$29.5 \text{ inHg } \times \frac{25.4 \text{ mm}}{1 \text{ in}} = 749 \text{ mmHg} = 749 \text{ torr}$$

$$749 \text{ torr } \times \frac{1 \text{ atm}}{760 \text{ torr}} = 0.985 \text{ atm}$$

8.8   The pressure of the gas inside the manometer is greater than atmospheric pressure because the mercury level is higher in the open arm of the manometer than in the closed end. Since atmospheric pressure is 750 mmHg and since the difference in mercury levels in the manometer arm is 250 mmHg, the pressure of the gas inside the manometer is 750 mm + 250 mm = 1000 mmHg.

CIA 8.3   *Systolic pressure* (the higher number) is the maximum blood pressure developed in the artery just after contraction. *Diastolic pressure* (the lower number) is the minimum pressure that occurs at the end of the heartbeat cycle. A blood pressure reading of 180/110 is an indication of high blood pressure.

CIA 8.4

$$180 \text{ mmHg } \times \frac{1 \text{ atm}}{760 \text{ mmHg}} = 0.24 \text{ atm}; 110 \text{ mmHg } \times \frac{1 \text{ atm}}{760 \text{ mmHg}} = 0.14 \text{ atm}$$

Blood pressure would be reported as 0.24/0.14.

CIA 8.5

$$180 \text{ mmHg } \times \frac{1 \text{ in}}{25.4 \text{ mm}} = 7.48 \text{ inHg}; 110 \text{ mmHg } \times \frac{1 \text{ in}}{25.4 \text{ mm}} = 4.33 \text{ inHg}$$

The ambient barometric pressure is close to 30 mmHg. This indicates that systolic pressure is about 1/4 ambient barometric pressure.

8.9   If the variables in the problem are pressure and volume, Boyle's law must be used.

Solution: $P_1 \times V_1 = P_2 \times V_2$

$$V_2 = \frac{P_1 \times V_1}{P_2} = \frac{(90 \text{ atm})(5.0 \text{ L})}{(1.0 \text{ atm})} = 450 \text{ L}$$

Ballpark check: Since the pressure is reduced by almost 100 times, the volume must increase by almost 100 times, from 5 L to 500 L.

The ballpark solution and the detailed solution agree.

8.10   *Solution:* $P_1 \times V_1 = P_2 \times V_2$

$$V_2 = \frac{P_1 \times V_1}{P_2} = \frac{(4.0 \text{ atm})(3.2 \text{ L})}{(10.0 \text{ atm})} = 1.3 \text{ L}$$

Ballpark check: Since the pressure increases by three times, the volume must decrease by the same amount, from 4.0 L to 1.3 L.

At a pressure of 0.70 atm: $\qquad V_2 = \frac{P_1 \times V_1}{P_2} = \frac{(4.0 \text{ atm})(3.2 \text{ L})}{(0.70 \text{ atm})} = 18 \text{ L}$

8.11 Charles's law is used to calculate volume or temperature changes when pressure and quantity remain constant.

Solution:

$$\frac{V_1}{T_1} = \frac{V_2}{T_2}$$

$$T_2 = \frac{V_2 \times T_1}{V_1} = \frac{(1.0 \text{ L})(273 \text{ K})}{0.30 \text{ L}} = 910 \text{ K } (637 \text{ °C})$$

Ballpark check: Since the volume increases approximately threefold, the temperature must also increase threefold, from 273 K to around 900 K.

$$\text{If } V_2 = 0.20 \text{ L}: \quad T_2 = \frac{V_2 \times T_1}{V_1} = \frac{(0.20 \text{ L})(273 \text{ K})}{0.30 \text{ L}} = 182 \text{ K } (-91 \text{ °C})$$

HOC 8.1

According to Charles's Law, the volume of a gas is directly proportional to the temperature when pressure and number of moles are held constant. Since the pressure must remain constant, the volume must decrease when the temperature is lowered, and the volume must increase when the temperature is raised. This happens because the kinetic energy of the molecules is less when the temperature is lowered, and the volume must decrease in order to maintain constant pressure.

8.12 Use Gay-Lussac's law when pressure and temperature vary, and volume and quantity are unchanged. Don't forget to change temperature from °C to K.

Solution:

$$\frac{P_1}{T_1} = \frac{P_2}{T_2}$$

$$P_2 = \frac{P_1 \times T_2}{T_1} = 30 \text{ psi} \times \frac{318 \text{ K}}{288 \text{ K}} = 33 \text{ psi}$$

Ballpark check: Since the temperature increase is about 10%, the pressure increase is also expected to be approximately 10%.

8.13 In this problem, *P*, *V*, and *T* vary.

$$\frac{P_1 V_1}{T_1} = \frac{P_2 V_2}{T_2}; \quad P_1 = 752 \text{ mmHg}; \, T_1 = 295 \text{ K}; \, V_1 = 275 \text{ L}$$

$$P_2 = 480 \text{ mmHg}; \, T_2 = 241 \text{ K}; \, V_2 = ?$$

$$V_2 = \frac{P_1 V_1 T_2}{P_2 T_1} = \frac{(752 \text{ mmHg})(275 \text{ L})(241 \text{ K})}{(480 \text{ mmHg})(295 \text{ K})} = 352 \text{ L}$$

8.14 The temperature increase [from 18 °C (291 K) to 50 °C (323 K)] increases the volume by about 10%. The pressure increase (from 1 atm to 2 atm) decreases the volume by half. The resulting balloon volume should be somewhat more than half the original volume, as represented by balloon (a).

8.15   Use Avogadro's Law; the quantity on the left side of the equation is the standard molar volume of a gas, 22.4 L/mol.

$$\frac{V_1}{n_1} = \frac{V_2}{n_2}$$

$$n_2 = \frac{V_2 \times n_1}{V_1} = 1.00 \times 10^5 \text{ L CH}_4 \times \frac{1.0 \text{ mol}}{22.4 \text{ L}} = 4.46 \times 10^3 \text{ mol CH}_4$$

The same container could also hold 4.46×10^3 moles of CO_2.

$$4.46 \times 10^3 \text{ mol CH}_4 \times \frac{16.0 \text{ g CH}_4}{1 \text{ mol CH}_4} = 7.14 \times 10^4 \text{ g CH}_4$$

$$4.46 \times 10^3 \text{ mol CO}_2 \times \frac{44.0 \text{ g CO}_2}{1 \text{ mol CO}_2} = 1.96 \times 10^5 \text{ g CO}_2$$

Ballpark check: Since one mole of a gas occupies 22.4 L at STP, a 100,000 L container holds 100,000/22.4 moles, or about 4500 moles. Thus, the ballpark solution and the exact solution agree.

8.16 $PV = nRT; P = nRT/V$

$$n = 3.2 \text{ g} \times \frac{1 \text{ mol}}{44.0 \text{ g}} = 0.073 \text{ mol}; R = 0.0821 \frac{\text{L} \cdot \text{atm}}{\text{mol} \cdot \text{K}}$$

$$T = 20 \,°\text{C} = 293 \text{ K}; V = 350 \text{ mL} = 0.35 \text{ L}$$

$$P = \frac{0.073 \text{ mol} \times \dfrac{0.0821 \text{ L} \cdot \text{atm}}{\text{mol} \cdot \text{K}} \times 293 \text{ K}}{0.35 \text{ L}} = 5.0 \text{ atm}$$

8.17 $PV = nRT; n = PV/RT$

$$P = 150 \text{ atm}; V = 180 \text{ L He}; R = 0.0821 \frac{\text{L} \cdot \text{atm}}{\text{mol} \cdot \text{K}}; T = 25 \,°\text{C} = 298 \text{ K}$$

$$n = \frac{150 \text{ atm} \times 180 \text{ L He}}{0.0821 \dfrac{\text{L} \cdot \text{atm}}{\text{mol} \cdot \text{K}} \times 298 \text{ K}} = 1.1 \times 10^3 \text{ mol He}$$

$$1.1 \times 10^3 \text{ mol He} \times \frac{4.0 \text{ g}}{1 \text{ mol He}} = 4.4 \times 10^3 \text{ g He}$$

8.18 We can use the ideal gas law to compare the volumes of two gases: Since the piston is moveable, $P = 1$ atm in all cases.

(a) The increase in temperature increases the volume to 450/300 = 3/2 of the original volume, and the decrease in amount decreases the volume to 0.200/0.300 = 2/3 of the original value. Since 3/2 x 2/3 = 1, the two changes exactly cancel, and the final volume is the same as the original volume.

$T_1 = 300$ K; $n_1 = 0.300$ mol; $T_2 = 450$ K; $n_2 = 0.200$ mol; $V_1/V_2 = ?$

$$\frac{V_1}{V_2} = \frac{n_1 T_1 P_2}{n_2 T_2 P_1} = \frac{n_1 T_1}{n_2 T_2}; \quad \frac{V_1}{V_2} = \frac{(0.300 \text{ mol})(300 \text{ K})}{(0.200 \text{ mol})(450 \text{ K})} = 1$$

(b) The decrease in temperature decreases the volume to 200/300 = 2/3 of the original, and the increase in amount increases the volume to 0.400/0.300 = 4/3 of the original volume. Since 2/3 x 4/3 = 8/9, the final volume is a bit less than the original volume.

$T_1 = 300$ K; $n_1 = 0.300$ mol; $T_2 = 200$ K; $n_2 = 0.400$ mol; $V_1/V_2 = ?$

$$\frac{V_1}{V_2} = \frac{(0.300 \text{ mol})(300 \text{ K})}{(0.400 \text{ mol})(200 \text{ K})} = \frac{9}{8} \quad \text{The volume in (b) is 8/9 of the original volume.}$$

$P = 1$ atm   (a)   (b)

$T = 300$ K $T = 450$ K $T = 200$ K
$n = 0.300$ mol $n = 0.200$ mol $n = 0.400$ mol

8.19 0.98 x 9.5 atm = 9.3 atm He
0.020 x 9.5 atm = 0.19 atm O_2

The partial pressure of oxygen in diving gas (0.19 atm) is approximately equal to the partial pressure of oxygen in air (0.21 atm).

8.20

$$\frac{573 \text{ mmHg}}{760 \text{ mmHg}} \times 100\% = 75.4\% \text{ N}_2; \quad \frac{100 \text{ mmHg}}{760 \text{ mmHg}} \times 100\% = 13.2\% \text{ O}_2$$

$$\frac{40 \text{ mmHg}}{760 \text{ mmHg}} \times 100\% = 5.3\% \text{ CO}_2; \quad \frac{47 \text{ mmHg}}{760 \text{ mmHg}} \times 100\% = 6.2\% \text{ H}_2\text{O}$$

8.21 We know from the previous problem that the partial pressure of O_2 in the lungs at atmospheric pressure is 13.2%. Thus, at 265 mmHg,

265 mmHg x 0.132 = 35.0 mmHg

8.22 The ratio of He molecules/Xe molecules is 8/4 or 2/1. Since 2/3 of the molecules are He, P_{He} = 2/3(750 mmHg) = 500 mmHg. P_{Xe} = 250 mmHg.

8.23 (a) To melt isopropyl alcohol:

$$1.50 \text{ mol } \times \frac{60.0 \text{ g}}{1 \text{ mol}} \times \frac{21.4 \text{ cal}}{1 \text{ g}} \times \frac{1 \text{ kcal}}{10^3 \text{ cal}} = 1.93 \text{ kcal}$$

(b) To boil isopropyl alcohol:

$$1.50 \text{ mol } \times \frac{60.0 \text{ g}}{1 \text{ mol}} \times \frac{159 \text{ cal}}{1 \text{ g}} \times \frac{1 \text{ kcal}}{10^3 \text{ cal}} = 14.3 \text{ kcal}$$

8.24

$$2.50 \text{ mol } H_2O \times \frac{18.0 \text{ g}}{1 \text{ mol } H_2O} \times \frac{2260 \text{ J}}{1 \text{ g}} \times \frac{1 \text{ kJ}}{10^3 \text{ J}} = 102 \text{ kJ}$$

8.25 *Smallest* ΔH_{vap} –>*Largest* ΔH_{vap}
Ether (85.6 cal/g); Isopropyl alcohol (159 cal/g); Ammonia (327 cal/g); Water (540 cal/g)

The order is based on the strength of intermolecular forces, which range from dispersion forces (ether) to hydrogen bonding (ammonia and water).

CIA 8.6 The supercritical state of matter is a condition intermediate between liquid and gas, in which there is some space between molecules yet they are too close together to be truly a gas.

CIA 8.7 Supercritical CO_2 is nontoxic and nonflammable and can be continuously recycled.

CIA 8.8 At 25 °C and 50 atm, CO_2 is probably close to a gas/liquid phase equilibrium (solid line on graph).

Understanding Key Concepts

8.26

(a) According to Charles's law, the volume of a gas at fixed pressure is directly proportional to its temperature. Since the temperature increases by 50% (from 300 K to 450 K), volume also increases by 50%.
(b) Boyle's law states that the volume of a gas is inversely proportional to its pressure. Doubling the pressure halves the volume.
(c) For changes in both pressure and temperature, the combined gas law states that PV/T is constant. Thus, reducing both the pressure and the temperature by one-third leaves the volume unchanged.

8.27 Drawing (b) represents the situation at 150 K if the boiling point of the gas is 200 K; at this temperature, the gas is a liquid. Drawing (c) represents the gas in a sealed container after the temperature has been lowered from 350 K to 150 K if the boiling point is 100 K. The gas remains a gas, and its volume is unchanged; only pressure is reduced because the molecules are moving slower.

8.28 Drawing (a) shows the sample of water as a solid, its state at 200 K. At 300 K, water is a liquid with a very low vapor pressure, represented by (c).

8.29 At equilibrium, the gases are totally mixed.

8.30 Both sides of the manometer have equal heights because both sides have equal pressure.

8.31 The horizontal parts of the graph represent phase changes.
(a) Melting point: 10 °C
(b) Boiling point: 75 °C
(c) Heat of fusion: 1.3 kcal/mol (the distance between the two points of the curve at the melting point)
(d) Heat of vaporization: 7.5 kcal/mol (the distance between the two points of the curve at the boiling point)

8.32 Refer to Problem 8.18 for an approach to solving this problem. Note the quantities that change and those that remain constant; these differ in each part of the problem. Be sure to convert temperature to kelvins.

(a) $\dfrac{V_2}{V_1} = \dfrac{n_2 T_2 P_1}{n_1 T_1 P_2} = \dfrac{T_2}{T_1}; \quad \dfrac{V_2}{V_1} = \dfrac{(323\ \text{K})}{(298\ \text{K})} = 1.08$

(b) $\dfrac{V_2}{V_1} = \dfrac{n_2 T_2 P_1}{n_1 T_1 P_2} = \dfrac{T_2 P_1}{T_1 P_2}; \quad \dfrac{V_2}{V_1} = \dfrac{(448\ \text{K})(0.92\ \text{atm})}{(298\ \text{K})(2.7\ \text{atm})} = 0.51$

(c) $\dfrac{V_2}{V_1} = \dfrac{n_2 T_2 P_1}{n_1 T_1 P_2} = \dfrac{n_2 P_1}{n_1 P_2}; \quad \dfrac{V_2}{V_1} = \dfrac{(0.22\ \text{mol})(0.92\ \text{atm})}{(0.075\ \text{mol})(2.7\ \text{atm})} = 1.0$

(a)	(b)	(c)

$T = 298$ K $T = 323$ K $T = 448$ K $T = 298$ K

$n = 0.075$ mol $n = 0.075$ mol $n = 0.075$ mol $n = 0.22$ mol

$P = 0.92$ atm $P = 0.92$ atm $P = 2.7$ atm $P = 2.7$ atm

8.33 There are 12 molecules of gas in the container, and 4 are green (partial pressure = 240 mmHg.) The total pressure is 3 x 240 mmHg = 720 mmHg.
P_{red} + P_{yellow} + P_{green} = 720 mm:
    red = 6/12 = 1/2;   yellow = 2/12 = 1/6;   green = 4/12 = 1/3
P_{red} = 1/2 x 720 mmHg = 360 mmHg;
P_{yellow} = 1/6 x 720 mmHg = 120 mmHg;
P_{green} = 1/3 x 720 mmHg = 240 mmHg

Additional Problems

Intermolecular Forces

8.34 (a) All molecules exhibit London forces, which increase in strength with increasing molecular weight.
(b) Dipole–dipole interactions are important for molecules that are polar.
(c) Hydrogen bonding occurs between an unshared electron pair on an electronegative atom (O, N, or F) and a hydrogen atom bonded to a second electronegative atom (O, N, F).

8.35 Hydrogen bonding takes place between molecules having an electron lone pair and a hydrogen bonded to an electronegative atom, such as NH_3 (d) or CH_3CO_2H (f). Dipole–dipole interactions are the most important interactions between molecules that have polar covalent bonds, such as HCN (b) or CH_3Cl (e). London dispersion forces are most important for molecules that do not experience other types of forces, such as N_2 (a) or CCl_4 (c). Notice that CCl_4 contains polar covalent bonds, but their dipole moments cancel because of symmetry.

8.36 Ethanol is higher boiling than dimethyl ether because of hydrogen bonding. Since molecules of ethanol are strongly attracted to each other, the boiling point of ethanol is higher than that of dimethyl ether, whose molecules are held together by weaker dipole–dipole interactions.

8.37 London forces are the major attractive forces for Br_2 and I_2 molecules. Since London forces become stronger with increasing molar mass, the melting point of I_2 is higher than the melting point of Br_2.

Gases and Pressure

8.38 One atmosphere is the amount of pressure needed to hold a column of mercury 760 mm high.

8.39 Four common units for measuring pressure: atmosphere (atm), mmHg, pounds per square inch (psi), Pascal (Pa).

$$1 \text{ atm} = 760 \text{ mmHg} = 14.7 \text{ psi} = 101,325 \text{ Pa}$$

8.40 (1) A gas consists of many tiny particles moving about at random with no attractive forces between particles.
(2) The amount of space occupied by gas molecules is much smaller than the amount of space between molecules.
(3) The average kinetic energy of gas particles is proportional to Kelvin temperature.
(4) When molecules collide, they spring apart elastically, and their total energy is constant.

8.41 According to the kinetic–molecular theory of gases, gas pressure is due to collisions with the walls of the container. The more collisions, the higher the pressure.

8.42

(a) $1 \text{ atm} \times \dfrac{760 \text{ mmHg}}{1 \text{ atm}} = 760 \text{ mmHg}$

(b) $25.3 \text{ psi} \times \dfrac{760 \text{ mmHg}}{14.7 \text{ psi}} = 1310 \text{ mmHg}$

(c) $7.5 \text{ atm} \times \dfrac{760 \text{ mmHg}}{1 \text{ atm}} = 5.7 \times 10^3 \text{ mmHg}$

(d) $28.0 \text{ in.Hg} \times \dfrac{25.4 \text{mm}}{1 \text{ in.}} = 711 \text{ mmHg}$

(e) $41.8 \text{ Pa} \times \dfrac{760 \text{ mmHg}}{101,325 \text{ Pa}} = 0.314 \text{ mmHg}$

8.43

(a) $440 \text{ mmHg} \times \dfrac{1 \text{ atm}}{760 \text{ mmHg}} = 0.58 \text{ atm}$

(b) $440 \text{ mmHg} \times \dfrac{133.3224 \text{ Pa}}{1 \text{ mmHg}} = 59,000 \text{ Pa}$

8.44 When the level in the arm connected to the container is 176 mm (17.6 cm) lower than the level open to the atmosphere, the pressure in the gas container is greater than atmospheric pressure.

$$754.3 \text{ mmHg} + 176 \text{ mmHg} = 930 \text{ mmHg} = 1.22 \text{ atm}$$

8.45   When the level in the arm connected to the container is 283 mm (28.3 cm) higher than the level open to the atmosphere, the pressure in the gas container is less than atmospheric pressure.

$$283 \text{ mmHg} \times \frac{1 \text{ atm}}{760 \text{ mmHg}} = 0.372 \text{ atm} \qquad 1.021 \text{ atm} - 0.372 \text{ atm} = 0.649 \text{ atm}$$

$$0.649 \text{ atm} \times \frac{760 \text{ mmHg}}{1 \text{ atm}} = 493 \text{ mmHg}$$

Boyle's Law

8.46   Boyle's law: Volume varies inversely with pressure at constant temperature and number of moles, that is, $P_1V_1 = P_2V_2$.

8.47   Two assumptions of the KMT explain the behavior of gases described by Boyle's Law. (1) The amount of space occupied by gas particles is much smaller than the space between particles. This allows compression of a gas without interactions between gas molecules. (2) Collisions between gas molecules and between molecules and the wall of a container are elastic. This means that all of the kinetic energy results in gas pressure.

Boyle's law involves an inverse proportionality: Pressure and volume change in opposite directions.

8.48 $P_1V_1 = P_2V_2$; $P_1 = 65.0$ mmHg; $P_2 = 385$ mmHg; $V_1 = 600.0$ mL $= 0.600$ L; $V_2 = ?$

$$V_2 = \frac{P_1V_1}{P_2} = \frac{65.0 \text{ mmHg} \times 0.600 \text{ L}}{385 \text{ mmHg}} = 0.101 \text{ L} = 101 \text{ mL}$$

8.49 $P_1V_1 = P_2V_2$; $P_1 = 1.00$ atm; $P_2 = ?$; $V_1 = 2.85$ L; $V_2 = 1.70$ L

$$P_2 = \frac{P_1V_1}{V_2} = \frac{1.00 \text{ atm} \times 2.85 \text{ L}}{1.70 \text{ L}} = 1.68 \text{ atm}$$

8.50 $P_1V_1 = P_2V_2$; $P_1 = 5.0$ atm; $P_2 = 1.0$ atm; $V_1 = 0.350$ L; $V_2 = ?$

$$V_2 = \frac{P_1V_1}{P_2} = \frac{5.00 \text{ atm} \times 0.350 \text{ L}}{1.00 \text{ atm}} = 1.75 \text{ L}$$

8.51 $P_1V_1 = P_2V_2$; $P_1 = 1$ atm $= 760$ mmHg; $V_1 = 1.25$ L; $P_2 = 220$ mmHg $V_2 = ?$;

$$V_2 = \frac{P_1V_1}{P_2} = \frac{760 \text{ mmHg} \times 1.25 \text{ L}}{220 \text{ mmHg}} = 4.32 \text{ L}$$

Charles's Law

8.52   Charles's law: Volume varies directly with temperature at constant pressure and number of moles; that is, $V_1/T_1 = V_2/T_2$.

8.53 An important assumption of the KMT relates to the behavior of gases described by Charles's law: The average kinetic energy of gas particles is proportional to the Kelvin temperature. This means that the relationship of temperature and volume is linear.

Charles's law involves a direct proportionality: Volume and temperature change in the same direction.

8.54

$$\frac{V_1}{T_1} = \frac{V_2}{T_2}; \qquad V_1 = 960 \text{ L}; \ V_2 = 1200 \text{ L}; \ T_1 = 291 \text{ K}; \ T_2 = \ ?$$

$$T_2 = \frac{V_2 T_1}{V_1} = \frac{1200 \text{ L} \ \times \ 291 \text{ K}}{960 \text{ L}} = 364 \text{ K} = 91 \ ^\circ\text{C}$$

8.55

$$\frac{V_1}{T_1} = \frac{V_2}{T_2}; \qquad V_1 = 875 \text{ L}; \ V_2 = 955 \text{ L}; \ T_1 = \ ? \ ; \ T_2 = 56 \ ^\circ\text{C} = 329 \text{ K}$$

$$T_1 = \frac{V_1 T_2}{V_2} = \frac{875 \text{ L} \ \times \ 329 \text{ K}}{955 \text{ L}} = 301 \text{ K} = 28 \ ^\circ\text{C}$$

8.56

$$\frac{V_1}{T_1} = \frac{V_2}{T_2}; \qquad V_1 = 185 \text{ mL}; \ V_2 = \ ?; \ T_1 = 38 \ ^\circ\text{C} = 311 \text{ K}; \ T_2 = 97 \ ^\circ\text{C} = 370 \text{ K}$$

$$V_2 = \frac{V_1 T_2}{T_1} = \frac{185 \text{ mL} \ \times \ 370 \text{ K}}{311 \text{ K}} = 220 \text{ mL}$$

8.57

$$\frac{V_1}{T_1} = \frac{V_2}{T_2}; \qquad V_1 = 43.0 \text{ L}; \ V_2 = \ ?; \ T_1 = 25 \ ^\circ\text{C} = 298 \text{ K}; \ T_2 = 2.8 \ ^\circ\text{C} = 276 \text{ K}$$

$$V_2 = \frac{V_1 T_2}{T_1} = \frac{43.0 \text{ L} \ \times \ 276 \text{ K}}{298 \text{ K}} = 39.8 \text{ L}$$

Gay-Lussac's Law

8.58 Gay-Lussac's law: Pressure varies directly with temperature at constant volume and number of moles; that is, $P_1/T_1 = P_2/T_2$.

8.59 The assumptions explained for Boyle's Law and Charles's Law explain the behavior of gases under Gay-Lussac's law.
Gay-Lussac's law involves a direct proportionality: Pressure and temperature change in the same direction.

8.60

$$\frac{P_1}{T_1} = \frac{P_2}{T_2}; \qquad P_1 = 0.95 \text{ atm}; \ P_2 = \ ?; \ T_1 = 25 \ ^\circ\text{C} = 298 \text{ K}; \ T_2 = 117 \ ^\circ\text{C} = 390 \text{ K}$$

$$P_2 = \frac{P_1 T_2}{T_1} = \frac{0.95 \text{ atm} \ \times \ 390 \text{ K}}{298 \text{ K}} = 1.2 \text{ atm}$$

8.61

$$\frac{P_1}{T_1} = \frac{P_2}{T_2}; \qquad P_1 = 3.85 \text{ atm}; \ P_2 = 18.0 \text{ atm}; \ T_1 = 25\,^{\circ}\text{C} = 298 \text{ K}; \ T_2 = ?$$

$$T_2 = \frac{P_2 T_1}{P_1} = \frac{18.0 \text{ atm} \ \times \ 298 \text{ K}}{3.85 \text{ atm}} = 1390 \text{ K} = 1120\,^{\circ}\text{C}$$

Combined Gas Law

8.62

$$\frac{P_1 V_1}{T_1} = \frac{P_2 V_2}{T_2}; \qquad P_1 = 760 \text{ mmHg}; \ V_1 = 2.84 \text{ L}; \ T_1 = 273 \text{ K}$$
$$P_2 = 520 \text{ mmHg}; \ V_2 = 7.50 \text{ L}; \ T_2 = ?$$

$$T_2 = \frac{P_2 V_2 T_1}{P_1 V_1} = \frac{520 \text{ mmHg} \ \times \ 7.50 \text{ L} \ \times \ 273 \text{ K}}{760 \text{ mmHg} \ \times \ 2.84 \text{ L}} = 493 \text{ K} = 220\,^{\circ}\text{C}$$

8.63

$$\frac{P_1 V_1}{T_1} = \frac{P_2 V_2}{T_2}; \qquad P_1 = 120 \text{ atm}; \ V_1 = 6.80 \text{ L}; \ T_1 = 20\,^{\circ}\text{C} = 293 \text{ K}$$

$$P_2 = 1.00 \text{ atm}; \ V_2 = ?; \ T_2 = 273 \text{ K}$$

$$V_2 = \frac{P_1 V_1 T_2}{T_1 P_2} = \frac{120 \text{ atm} \ \times \ 6.80 \text{ L} \ \times \ 273 \text{ K}}{293 \text{ K} \ \times \ 1.00 \text{ atm}} = 760 \text{ L}$$

8.64

$$\frac{P_1 V_1}{T_1} = \frac{P_2 V_2}{T_2}; \qquad P_1 = 748 \text{ mmHg}; \ V_1 = 75.4 \text{ mL}; \ T_1 = 23\,^{\circ}\text{C} = 296 \text{ K}$$
$$P_2 = 760 \text{ mmHg}; \ V_2 = ?; \ T_2 = 0\,^{\circ}\text{C} = 273 \text{ K}$$

$$V_2 = \frac{P_1 V_1 T_2}{T_1 P_2} = \frac{748 \text{ mmHg} \ \times \ 75.4 \text{ mL} \ \times \ 273 \text{ K}}{296 \text{ K} \ \times \ 760 \text{ mmHg}} = 68.4 \text{ mL}$$

8.65 (a) For one mole of gas, the combined gas law states that V is proportional to T/P, or $V \propto T/P$. Thus, if the pressure is halved and the temperature is doubled, the new volume is four times greater than the original volume.

$$\frac{P_1 V_1}{T_1} = \frac{P_2 V_2}{T_2}; \qquad P_2 = 0.5 P_1; \ T_2 = 2 T_1; \ V_2 = ?$$

$$V_2 = \frac{P_1 V_1 T_2}{T_1 P_2} = \frac{P_1 V_1 \ \times \ 2 T_1}{T_1 \ \times \ 0.5 P_1} = 4 V_1$$

(b) If both the pressure and the temperature are doubled, the volume remains the same.

$$\frac{P_1 V_1}{T_1} = \frac{P_2 V_2}{T_2}; \qquad P_2 = 2 P_1; \ T_2 = 2 T_1; \ V_2 = ?$$

$$V_2 = \frac{P_1 V_1 T_2}{T_1 P_2} = \frac{P_1 V_1 \ \times \ 2 T_1}{T_1 \ \times \ 2 P_1} = V_1$$

8.66   (a) For one mole of gas, the combined gas law states that P is proportional to T/V, or $P \propto T/V$. Thus, if the temperature doubles and the volume is halved, the new pressure is four times greater than the original pressure:

$$\frac{P_1 V_1}{T_1} = \frac{P_2 V_2}{T_2}; \qquad V_2 = 0.5 V_1; T_2 = 2T_1; P_2 = ?$$

$$P_2 = \frac{P_1 V_1 T_2}{T_1 V_2} = \frac{P_1 V_1 \times 2T_1}{T_1 \times 0.5 V_1} = 4\, P_1$$

(b) If the temperature is halved and the volume is doubled, the new pressure is one-fourth the original pressure:

$$\frac{P_1 V_1}{T_1} = \frac{P_2 V_2}{T_2}; \qquad V_2 = 2V_1; T_2 = 0.5 T_1; P_2 = ?$$

$$P_2 = \frac{P_1 V_1 T_2}{T_1 V_2} = \frac{P_1 V_1 \times 0.5 T_1}{T_1 \times 2V_1} = 0.25 P_1$$

8.67   In this problem, temperature is constant. First, we find the volume of gas at 1.25 atm and 25 °C.
$P_1 V_1 = P_2 V_2$; $P_1 = 1850$ atm; $P_2 = 1.25$ atm; $V_1 = 2.30$ L; $V_2 = ?$

$$V_2 = \frac{P_1 V_1}{P_2} = \frac{1850 \text{ atm} \times 2.30 \text{ L}}{1.25 \text{ atm}} = 3400 \text{ L}$$

If the cylinder holds 3400 L of helium at 1.25 atm and 25 °C, it can be used to fill 3400 L/1.50 L = 2270 balloons.

Avogadro's Law and Standard Molar Volume

8.68   Avogadro's law states that equal volumes of gases at the same temperature and pressure contain equal numbers of molecules. Since the volume of space taken up by gas molecules is so much smaller than the amount of space between molecules, Avogadro's law is true regardless of the chemical identity of the gas.

8.69   The conditions of STP are 760 mmHg (1 atm) pressure and 273 K temperature.

8.70

$$1.0 \text{ L} \times \frac{1 \text{ mol}}{22.4 \text{ L}} \times \frac{6.02 \times 10^{23} \text{ molecules}}{1 \text{ mol}} = 2.7 \times 10^{22} \text{ molecules}$$

$$1.0 \text{ L} \times \frac{1 \text{ mol}}{22.4 \text{ L}} \times \frac{32.0 \text{ g}}{1 \text{ mol}} = 1.4 \text{ g } O_2$$

8.71

$$n_2 = \frac{V_2 \times n_1}{V_1} = 48.6 \text{ L} \times \frac{1.0 \text{ mol}}{22.4 \text{ L}} = 2.17 \text{ mol}$$

8.72

$$16.5 \text{ L} \times \frac{1 \text{ mol}}{22.4 \text{ L}} \times \frac{16.0 \text{ g } CH_4}{1 \text{ mol } CH_4} = 11.8 \text{ g } CH_4$$

8.73

$$1.75 \text{ g HCN} \ \times \ \frac{1 \text{ mol}}{27.0 \text{ g HCN}} \ \times \ \frac{22.4 \text{ L}}{1 \text{ mol HCN}} \ = \ 1.45 \text{ L HCN at STP}$$

8.74

$$V = 4.0 \text{ m} \ \times \ 5.0 \text{ m} \ \times \ 2.5 \text{ m} = 50 \text{ m}^3; \ 50 \text{ m}^3 \ \times \ \frac{10^3 \text{ L}}{1 \text{ m}^3} = 5.0 \times 10^4 \text{ L}$$

$$\frac{1 \text{ mol}}{22.4 \text{ L}} \ \times \ 5.0 \times 10^4 \text{ L} \ = \ 2230 \text{ mol gas}$$

$$\frac{0.21 \text{ mol O}_2}{1 \text{ mol gas}} \ \times \ 2230 \text{ mol gas} \ = \ 470 \text{ mol O}_2$$

$$470 \text{ mol O}_2 \ \times \ \frac{32.0 \text{ g O}_2}{1 \text{ mol O}_2} \ = \ 15{,}000 \text{ g O}_2 \ = \ 15 \text{ kg O}_2$$

8.75

$$\frac{0.79 \text{ mol N}_2}{1 \text{ mol gas}} \ \times \ 2230 \text{ mol gas} \ = \ 1.76 \times 10^3 \text{ mol N}_2$$

$$1760 \text{ mol N}_2 \ \times \ \frac{28.0 \text{ g N}_2}{1 \text{ mol N}_2} \ = \ 49{,}000 \text{ g N}_2 = 49 \text{ kg N}_2$$

The Ideal Gas Law

8.76   The ideal gas law: $PV = nRT$

8.77   The combined gas law can be used to calculate changes in P, V, and T when the amount of gas is fixed. The ideal gas law is valid for any number of moles and uses the ideal gas constant R to calculate any one of the variables P, V, T, and n if the other three are known.

8.78 $PV = nRT$; $n = PV/RT$

For Cl_2: $P = 1.0$ atm; $V = 2.0$ L; $R = 0.082 \dfrac{\text{L} \cdot \text{atm}}{\text{mol} \cdot \text{K}}$; $T = 273$ K

$$n = \frac{1.0 \text{ atm} \ \times \ 2.0 \text{ L}}{0.082 \dfrac{\text{L} \cdot \text{atm}}{\text{mol} \cdot \text{K}} \ \times \ 273 \text{ K}} = 0.089 \text{ mol Cl}_2$$

$$0.089 \text{ mol Cl}_2 \ \times \ \frac{71 \text{ g}}{1 \text{ mol Cl}_2} \ = \ 6.3 \text{ g Cl}_2$$

For CH_4: $P = 1150$ mmHg $= 1.5$ atm; $V = 3.0$ L; $R = 0.082 \dfrac{L \cdot atm}{mol \cdot K}$; $T = 300$ K

$$n = \frac{1.5 \text{ atm } \times 3.0 \text{ L}}{0.082 \dfrac{L \cdot atm}{mol \cdot K} \times 300 \text{ K}} = 0.18 \text{ mol } CH_4$$

$$0.18 \text{ mol } CH_4 \times \frac{16 \text{ g}}{1 \text{ mol } CH_4} = 2.9 \text{ g } CH_4$$

There are more molecules in the CH_4 sample than in the Cl_2 sample. The Cl_2 sample, however, weighs more because the molar mass of Cl_2 is much greater than the molar mass of CH_4.

8.79 As in the preceding problem, use the relationship $n = PV/RT$.

For CO_2: $\quad n = \dfrac{500 \text{ mmHg } \times 2.0 \text{ L } CO_2}{62.4 \dfrac{mmHg \cdot L}{mol \cdot K} \times 300 \text{ K}} = 0.053 \text{ mol } CO_2$

For N_2: $\quad n = \dfrac{760 \text{ mmHg } \times 1.5 \text{ L } N_2}{62.4 \dfrac{mmHg \cdot L}{mol \cdot K} \times 330 \text{ K}} = 0.055 \text{ mol } N_2$

The CO_2 sample has somewhat fewer molecules. Since the molar mass of CO_2 is much greater than the molar mass of N_2, however, the CO_2 sample weighs more (2.3 g) than the N_2 sample (1.5 g).

8.80

$n = 2.3$ mol He; $\quad T = 294$ K; $\quad V = 0.15$ L

$$P = \frac{nRT}{V} = \frac{2.3 \text{ mol He } \times 0.082 \dfrac{L \cdot atm}{mol \cdot K} \times 294 \text{ K}}{0.15 \text{ L}} = 370 \text{ atm}$$

$$370 \text{ atm } \times \frac{14.7 \text{ psi}}{1 \text{ atm}} = 5.40 \times 10^3 \text{ psi}$$

8.81

$n = 3.5$ mol O_2; $\quad P = 1.6$ atm; $\quad V = 27$ L

$$T = \frac{PV}{nR} = \frac{1.6 \text{ atm } \times 27 \text{ L}}{3.5 \text{ mol } \times 0.0821 \dfrac{L \cdot atm}{mol \cdot K}} = 150 \text{ K} = -123 \,°C$$

8.82

$$n = 15.0 \text{ g } CO_2 \times \frac{1 \text{ mol}}{44 \text{ g}} = 0.341 \text{ mol } CO_2; \quad T = 310 \text{ K}; \quad V = 0.30 \text{ L}$$

$$P = \frac{nRT}{V} = \frac{0.341 \text{ mol } CO_2 \times 62.4 \dfrac{mmHg \cdot L}{mol \cdot K} \times 310 \text{ K}}{0.30 \text{ L}} = 2.2 \times 10^4 \text{ mmHg}$$

8.83

$$PV = nRT; \quad n = 20.0 \text{ g } N_2 \text{ x } \frac{1 \text{ mol}}{28.0 \text{ g}} = 0.714 \text{ mol } N_2; \quad V = 4.00 \text{ L}; \quad P = 6.00 \text{ atm}$$

$$T = \frac{PV}{nR} = \frac{6.00 \text{ atm x } 4.00 \text{ L}}{0.714 \text{ mol x } 0.0821 \dfrac{L \cdot atm}{mol \cdot K}} = 409 \text{ K} = 136 \,°C$$

8.84

$$n = 18.0 \text{ g } O_2 \text{ x } \frac{1 \text{ mol}}{32.0 \text{ g}} = 0.563 \text{ mol } O_2; \quad T = 350 \text{ K}; \quad P = 550 \text{ mmHg}$$

$$V = \frac{nRT}{P} = \frac{0.562 \text{ mol } O_2 \text{ x } 62.4 \dfrac{mmHg \cdot L}{mol \cdot K} \text{ x } 350 \text{ K}}{550 \text{ mmHg}} = 22.3 \text{ L } O_2$$

8.85

$$n = \frac{PV}{RT}; \quad P = 2.5 \text{ atm}; \quad V = 0.55 \text{ L}; \quad T = 347 \text{ K}$$

$$= \frac{2.5 \text{ atm x } 0.55 \text{ L}}{0.082 \dfrac{L \cdot atm}{mol \cdot K} \text{ x } 347 \text{ K}} = 0.048 \text{ moles}$$

Dalton's Law and Partial Pressure

8.86   Partial pressure is the pressure contribution of one component of a mixture of gases to the total pressure.

8.87   Dalton's law of partial pressure says that the total pressure exerted by a gas mixture is the sum of the individual pressures of the components in the mixture.

8.88

$$440 \text{ mmHg x } \frac{1 \text{ atm}}{760 \text{ mmHg}} \text{ x } \frac{160 \text{ mmHg}}{1.0 \text{ atm}} = 93 \text{ mmHg}$$

8.89

$$120 \text{ psi x } \frac{760 \text{ mmHg}}{14.7 \text{ psi}} \text{ x } 0.21 = 1300 \text{ mmHg}$$

Liquids

8.90   The vapor pressure of a liquid is the partial pressure of the vapor above the liquid.

8.91   A liquid's heat of vaporization is the amount of heat needed to vaporize one gram of the liquid at its boiling point.

8.92   Increased pressure raises a liquid's boiling point; decreased pressure lowers a liquid's boiling point.

8.93 CH_3Cl has a higher vapor pressure because its intermolecular forces (dipole–dipole) are weaker than the intermolecular forces of CH_3OH (hydrogen bonds).

8.94

(a) $\dfrac{9.72 \text{ kcal}}{1 \text{ mol } H_2O} \times 3.00 \text{ mol } H_2O = 29.2$ kcal of heat required.

(b) $\dfrac{9.72 \text{ kcal}}{1 \text{ mol } H_2O} \times \dfrac{1 \text{ mol } H_2O}{18.0 \text{ g}} \times 320 \text{ g} = 173$ kcal of heat is released.

8.95

$\dfrac{159 \text{ cal}}{1 \text{ g isopropyl alcohol}} \times 190.0 \text{ g isopropyl alcohol} = 3.02 \times 10^4 \text{ cal} = 30.2$ kcal

Solids

8.96 The atoms in a crystalline solid are arranged in a regular, orderly network. The atoms in an amorphous solid have no regular arrangement.

8.97

Type of Solid	Example
Ionic solid	NaCl, $MgBr_2$
Molecular solid	Ice, sucrose
Covalent network solid	Diamond
Metallic solid	Cu, Fe

8.98

$\dfrac{45.9 \text{ cal}}{1 \text{ g}} \times \dfrac{1 \text{ kcal}}{1000 \text{ cal}} \times \dfrac{60.0 \text{ g}}{1 \text{ mol}} \times 1.75 \text{ mol} = 4.82$ kcal

8.99

$\dfrac{630 \text{ cal}}{1 \text{ mol Na}} \times \dfrac{1 \text{ kcal}}{1000 \text{ cal}} \times \dfrac{1 \text{ mol Na}}{23.0 \text{ g Na}} \times 262 \text{ g Na} = 7.18$ kcal

Conceptual Problems

8.100 As the temperature increases, the kinetic energy of gas molecules increases, and the force per unit area that they exert in colliding against the walls of a container increases, thus increasing pressure.

8.101 Two moles of hydrogen react with one mole of oxygen according to the balanced equation. Since equal volumes of gases have equal numbers of moles at STP, 2.5 L of O_2 reacts with 5.0 L of H_2.

8.102 3.0 L of H_2 and 1.5 L of O_2 react completely. At STP, one mole of H_2 occupies 22.4 L. Thus, 3.0 L of hydrogen = 3.0/22.4 or 0.13 mol. This is also the number of moles of H_2O formed.

$n = 0.13 \text{ mol } H_2O; \quad T = 373 \text{ K}; \quad P = 1.0 \text{ atm}$

$V = \dfrac{nRT}{P} = \dfrac{0.13 \text{ mol } H_2O \times 0.0821 \dfrac{L \cdot atm}{mol \cdot K} \times 373 \text{ K}}{1.0 \text{ atm}} = 4.0 \text{ L } H_2O$

8.103 $PV = nRT$; $P = 1.0$ atm; $V = 0.24$ L; $T = 310$ K

$$n = \frac{PV}{RT} = \frac{1.0 \text{ atm } \times 0.24 \text{ L}}{0.0821 \dfrac{\text{L} \cdot \text{atm}}{\text{mol} \cdot \text{K}} \times 310 \text{ K}} = 0.0094 \text{ mol} = \frac{9.4 \text{ mmol CO}_2}{1 \text{ minute}}$$

8.104

$$0.0094 \text{ mol CO}_2 \times \frac{44.0 \text{ g CO}_2}{1 \text{ mol CO}_2} = 0.41 \text{ g CO}_2$$

$$\frac{0.41 \text{ g CO}_2}{1 \text{ min}} \times \frac{60 \text{ min}}{1 \text{ hr}} \times \frac{24 \text{ hr}}{1 \text{ day}} = \frac{590 \text{ g CO}_2}{\text{day}}$$

8.105 $P = 0.975$ atm; $V = 1.6 \times 10^5$ L; $T = 375$ K

$$n = \frac{PV}{RT} = \frac{0.975 \text{ atm } \times 1.6 \times 10^5 \text{ L}}{0.0821 \dfrac{\text{L} \cdot \text{atm}}{\text{mol} \cdot \text{K}} \times 375 \text{ K}} = 5.07 \times 10^3 \text{ mol}$$

$$\text{Density} = \frac{5.07 \times 10^3 \text{ mol}}{1.6 \times 10^5 \text{ L}} \times \frac{29 \text{ g}}{\text{mol}} = \frac{0.92 \text{ g}}{\text{L}}$$

The air in the balloon is less dense than air at STP (density = 1.3 g/L).

8.106

$$n = \frac{745 \text{ mmHg } \times 14.7 \text{ L}}{62.4 \dfrac{\text{mmHg} \cdot \text{L}}{\text{mol} \cdot \text{K}} \times 298 \text{ K}} = 0.589 \text{ mol}; \quad \frac{10.0 \text{ g}}{0.589 \text{ mol}} = 17.0 \text{ g/mol}$$

8.107 Divide the molar mass (molecular weight in grams) by 22.4 L to find the density in grams per liter.

Gas	Molecular Weight	Density (g/L) at STP
(a) CH_4	16.0 g/mol	0.714
(b) CO_2	44.0 g/mol	1.96
(c) O_2	32.0 g/mol	1.43

8.108 $n = 1$ mol; $T = 1$ K ; $P = 1 \times 10^{-14}$ mmHg

$$V = \frac{nRT}{P} = \frac{1 \text{ mol } \times 62.4 \dfrac{\text{mmHg} \cdot \text{L}}{\text{mol} \cdot \text{K}} \times 1 \text{ K}}{1 \times 10^{-14} \text{ mmHg}} = 6 \times 10^{15} \text{ L}$$

$$\text{Density} = \frac{6.022 \times 10^{23} \text{ atoms}}{6 \times 10^{15} \text{ L}} = \frac{1 \times 10^8 \text{ atoms}}{\text{L}}$$

8.109

(a)   (b)

Ethylene glycol   Chloroethane

(c) Ethylene glycol has a higher boiling point than chloroethane because it forms hydrogen bonds.

8.110 (a) $2\ C_8H_{18} + 25\ O_2 \longrightarrow 16\ CO_2 + 18\ H_2O$

(b) $4.6 \times 10^{10}\ \text{L}\ C_8H_{18}\ \times\ \dfrac{0.792\ \text{g}}{1\ \text{mL}}\ \times\ \dfrac{10^3\ \text{mL}}{1\ \text{L}}\ =\ 3.6 \times 10^{13}\ \text{g}\ C_8H_{18}$

$$= 3.6 \times 10^{10}\ \text{kg}\ C_8H_{18}$$

$3.6 \times 10^{13}\text{g}\ C_8H_{18}\ \times\ \dfrac{1\ \text{mol}\ C_8H_{18}}{114\ \text{g}\ C_8H_{18}}\ \times\ \dfrac{16\ \text{mol}\ CO_2}{2\ \text{mol}\ C_8H_{18}}\ \times\ \dfrac{44.0\ \text{g}\ CO_2}{1\ \text{mol}\ CO_2}\ =\ 1.1 \times 10^{14}\ \text{g}\ CO_2$

$$= 1.1 \times 10^{11}\ \text{kg}\ CO_2$$

(c) To find the volume of CO_2, use the ideal gas law at STP; n can be calculated from the mass of CO_2 that was found in part (b).

$$n = 1.1 \times 10^{14}\ \text{g}\ CO_2\ \times\ \dfrac{1\ \text{mol}\ CO_2}{44\ \text{g}\ CO_2}\ =\ 2.5 \times 10^{12}\ \text{mol}\ CO_2$$

$$V = \dfrac{nRT}{P} = \dfrac{2.5 \times 10^{12}\ \text{mol}\ CO_2\ \times\ 0.0821\ \dfrac{\text{L} \cdot \text{atm}}{\text{mol} \cdot \text{K}}\ \times\ 273\ \text{K}}{1.0\ \text{atm}}\ =\ 5.6 \times 10^{13}\ \text{L}\ CO_2$$

Group Problems

8.111 At STP, equal volumes of gases have an equal number of moles. However, O_2 has a greater molecular weight than H_2, and the vessel containing O_2 is heavier.

8.112 Since the pressure at sea level is 1 atm, the external pressure increases by 1 atm for every 10 m of depth.

(a) $\dfrac{1.0\ \text{atm}}{10\ \text{m}}\ \times\ 25\ \text{m}\ =\ 2.5\ \text{atm};\ 2.5\ \text{atm} + 1.0\ \text{atm}\ =\ 3.5\ \text{atm}$

(b) For O_2: $0.20\ \times\ 3.5\ \text{atm}\ =\ 0.7\ \text{atm}$

For N_2: $0.80\ \times\ 3.5\ \text{atm}\ =\ 2.8\ \text{atm}$

8.113 Propane (b.p. –42.2 °C), isobutane (b.p.–16.7 °C), and isopentane (b.p. 27.7 °C) are common propellants. (All of these propellants melt at temperatures below –100 °C). Propane and isobutane are gases at room temperature but are liquids under pressure in an aerosol can. When dispensed, the liquids become gases and propel the active ingredients. Isopentane boils at around room temperature but can be kept as a liquid using a small amount of pressure. It also delivers active ingredients when dispensed.

8.114 (a) A phase diagram of water shows a negative slope for the solid/liquid boundary. This indicates that pressure on ice lowers the melting point. As a person skates on ice, the pressure of the skate blades lowers the melting point, melting some of the ice and forming a layer of liquid water that allows gliding across the ice.
(b) A phase diagram for CO_2 appears in a Chemistry in Action feature in this chapter. At atmospheric pressure, solid CO_2 sublimes to gaseous CO_2. In addition to forming no liquid layer, CO_2 sublimes at –78 °C, a temperature much too cold for ice skating!

8.115 Shifting to biomass-based fuels would decrease dependence on fossil fuels but would have little effect on current atmospheric CO_2 levels. However, biosynthesis of biomass fuels would consume CO_2, which would later be burned and released back into the atmosphere, setting up an equilibrium in which CO_2 levels would remain constant.

Self-Test for Chapter 8

Multiple Choice

1. Which of the following units would you be least likely to use in a chemical laboratory?
   (a) pascal   (b) pounds per square inch   (c) mmHg   (d) atmosphere

2. A fixed amount of a gas has its temperature and volume doubled. What happens to its pressure?
   (a) increases fourfold   (b) doubles   (c) stays the same   (d) is halved

3. In which of the gas laws is the amount of gas not fixed?
   (a) Boyle's law   (b) Charles's law   (c) Gay-Lussac's law   (d) Avogadro's law

4. Which of the following compounds does not exhibit hydrogen bonding?
   (a) CH_3OCH_3 (b) CH_3OH (c) HF   (d) CH_3NH_2

5. Which of the following compounds is the lowest melting?
   (a) NaI   (b) Au   (c) SiO_2 (d) sugar

6. Which term describes the change of state that occurs when a gas changes to a solid?
   (a) fusion   (b) condensation   (c) deposition   (d) sublimation

7. When does a gas obey ideal behavior?
   (a) at low density   (b) at high pressure   (c) at low temperature   (d) in a large container

8. If 22.0 g of CO_2 has a pressure of 1.00 atm at 300 K, what is its volume?
   (a) 22.4 L   (b) 12.3 L   (c) 11.2 L   (d) 9.5 L

9. What volume does the amount of gas in Problem 8 occupy at STP?
    (a) 22.4 L   (b) 12.3 L   (c) 11.2 L   (d) 9.5 L

10. What is the density of CO_2 gas in g/L at STP if one mole of gas has a volume of 22.4 L?
    (a) 44.0 g/L   (b) 11.0 g/L   (c) 6.4 g/L   (d) 1.96 g/L

Sentence Completion

1.   In Boyle's law, the _____ of a gas is inversely proportional to its _____.

2.   A liquid has _____ volume and _____ shape.

3.   _____ law states that the total pressure of a gas mixture is the sum of the individual pressure of the components in the mixture.

4.   A pressure of 760 mmHg and a temperature of 273 K are known as ___ ___ ___ .

5.   _____ law says that equal volumes of gases at the same temperature and pressure contain equal numbers of molecules.

6.   In a closed container, liquid and vapor are at _____.

7.   Units for measuring pressure include _____, _____, _____, _____, and _____.

8.   The _____ __ _____ is the heat necessary to melt one gram of a solid at its melting point.

9.   In Charles's law, _____ and _____ are kept constant.

10. Gas particles move in straight lines, with energy proportional to _____.

11. The transformation of a substance from one phase to another is known as a _____ ___ _____.

12. A liquid that evaporates readily is said to be _____.

True or False

1.   Molecules of ethyl alcohol exhibit hydrogen bonding.

2.   All gases are similar in their physical behavior.

3.   Molecules of CH_3Cl experience both dipole–dipole interactions and London forces.

4.   Doubling the pressure of a gas at constant temperature doubles the volume.

5. $R = 0.082$ L atm/(mol · K) is a value for the gas constant.

6.   Atoms in solids have an orderly arrangement.

7.   Standard temperature and pressure are 760 mmHg and 273 °C.

8.   All substances become solids if the temperature is low enough.

9.   The atmospheric pressure in Death Valley (282 ft below sea level) is lower than the atmospheric pressure at sea level.

10.   The more liquid there is in a closed container, the higher the vapor pressure.

11.   Surface tension is a liquid's resistance to flow.

Match each entry on the left with its partner on the right. Use each answer once.

1. $P_1V_1 = P_2V_2$ (a) Avogadro's law

2.   Dipole–dipole attraction   (b) Gas constant

3. $P_{total} = P_{gas\ 1} + P_{gas\ 2} + ...$ (c) Force per unit area

4.   760 mmHg at 273 K   (d) Charles's law

5.   Hydrogen bonding   (e) Occurs between molecules of CH_3Br

6. $V_1/n_1 = V_2/n_2$ (f) STP

7.   0.007500 mmHg   (g) Boyle's law

8.   London forces   (h) Occurs between molecules of CH_3OH

9.   62.4 mmHg L/mol K   (i) Ideal gas law

10. $V_1/T_1 = V_2/T_2$ (j) Pascal

11.   Pressure   (k) Dalton's law of partial pressure

12. $PV = nRT$ (l) Occurs between molecules of N_2

Chapter Outline

I. Characteristics of solutions (Sections 9.1–9.5).
  A. Mixtures and solutions (Section 9.1).
    1. Mixtures are either heterogeneous or homogeneous.
      a. Heterogeneous mixtures have nonuniform mixing.
      b. Homogeneous mixtures have uniform mixing.
    2. Homogeneous mixtures can be classified by particle size.
      a. In solutions, particles range in size from 0.1 to 2 nm.
      b. In colloids, particles range in size from 2 to 1000 nm.
    3. When a solid is dissolved in a liquid, the liquid is the solvent and the solid is the solute.
  B. The solution process (Sections 9.2).
    1. Solubility depends on the strength of attraction between solute particles and solvent, relative to the attractions in the pure substances.
    2. In predicting solubility, polar solvents dissolve polar substances, and nonpolar solvents dissolve nonpolar substances.
    3. Solvation can be either an exothermic or an endothermic process.
    4. Some ionic compounds attract water to form solid hydrates.
  C. Solubility (Sections 9.3–9.5).
    1. Solubility is a dynamic process (Section 9.3).
      a. When no more of an added solute will dissolve, the solution is said to be saturated.
      b. In a saturated solution, an equilibrium is established between dissolving and crystallizing.
    2. The solubility of a substance is the maximum amount of the substance that will dissolve in a solvent.
    3. Effect of temperature on solubility (Section 9.4).
      a. The effect of temperature on the solubility of a solid solute is unpredictable.
      b. A gas is always less soluble as temperature increases.
      c. A solid that is more soluble at high temperature than at low temperature may form a supersaturated solution.
    4. Effect of pressure on solubility (Section 9.5).
      a. Increased pressure makes gas molecules more soluble in a liquid.
      b. Henry's law: $C = kP_{gas}$.
        i. The solubility of a gas in a liquid is proportional to its partial pressure over the liquid at constant T.
      c. When the partial pressure of a gas changes:

$$\frac{C_1}{P_1} = \frac{C_2}{P_2} = k \text{ (at constant T)}$$

II. Quantitative relationships in solutions (Sections 9.6–9.8).
  A. Concentration (Section 9.6).
    1. Percent concentrations.
      a. (m/m)% concentration $= \dfrac{\text{Mass of solute (g)}}{\text{Mass of solution (g)}} \times 100\%$
      b. (v/v)% concentration $= \dfrac{\text{Volume of solute (mL)}}{\text{Volume of solution (mL)}} \times 100\%$
      c. (m/v)% concentration $= \dfrac{\text{Mass of solute (g)}}{\text{Volume of solution (mL)}} \times 100\%$

2. Parts per million (ppm) $= \dfrac{\text{Mass of solute (g)}}{\text{Mass of solution (g)}} \times 10^6$

$= \dfrac{\text{Volume of solute (mL)}}{\text{Volume of solution (mL)}} \times 10^6$

3. Parts per billion (ppb) $= \dfrac{\text{Mass of solute (g)}}{\text{Mass of solution (g)}} \times 10^9$

$= \dfrac{\text{Volume of solute (mL)}}{\text{Volume of solution (mL)}} \times 10^9$

4. Molarity $= \dfrac{\text{Moles of solute}}{\text{Volume of solution (L)}}$
    a. Molarity can be used as a conversion factor.
B. Dilution (Section 9.7).
    1. During dilution, the number of moles of solute remains constant, while volume changes.
    2. $M_1 \times V_1 = M_2 \times V_2$
    3. V_1/V_2 is known as a dilution factor.
C. Equivalents and milliequivalents (Section 9.8).
    1. Electrolytes.
        a. Substances that dissociate completely are strong electrolytes.
        b. Substances that dissociate partially are weak electrolytes.
        c. Molecular substances that don't produce ions are nonelectrolytes.
    2. The concentration of electrolytes is expressed in equivalents.
        a. 1 gram-equivalent of an ion $= \dfrac{\text{Molar mass of the ion}}{\text{Number of charges on the ion}}$
        b. Milliequivalents are useful when measuring ion concentrations in body fluids.
III. Properties of solutions (Sections 9.9–9.11).
    A. Effects of particles in solution—colligative properties (Section 9.9).
        1. Lowering of vapor pressure.
        2. Boiling point elevation.
        3. Freezing point depression.
        4. These effects do not depend on the identity of the particles.
    B. Osmosis (Section 9.10–9.11).
        1. When two solutions of different concentrations are separated by a semipermeable membrane, water passes through to the more concentrated side. This is known as osmosis.
        2. Osmotic pressure can be applied to establish an equilibrium between the rates of forward and reverse passage of water across the membrane.
        3. The osmotic pressure of a solution depends only on the number of particles in solution.
        4. Osmolarity = Molarity × Number of particles per formula unit.
        5. Two solutions that are isotonic have the same osmolarity.
            a. In cells, a hypotonic solution causes hemolysis.
            b. A hypertonic solution causes crenation.
        6. Dialysis is a process similar to osmosis except that the pores in the membrane allow small solute molecules to pass (Section 9.11).
            a. Hemodialysis is used to cleanse the blood of people whose kidneys malfunction.
            b. Colloidal particles are too large to pass through semipermeable membranes.

Solutions to Chapter 9 Problems

9.1 Orange juice with pulp is heterogeneous, and all of the other mixtures are homogeneous, although hand lotion might be heterogeneous in some cases. Apple juice and tea are solutions because they are nonfilterable and transparent to light. Hand lotion is a colloid.

CIA 9.1 Glauber's salt: $Na_2SO_4 \cdot 10 \, H_2O$

CIA 9.2 Molar mass of Glauber's salt: 322 g/mol. 322 g of Glauber's salt provides 1.00 mol of sodium sulfate.

9.2 Remember the rule "like dissolves like."
(a) CCl_4 and H_2O do not form solutions because CCl_4 is nonpolar and H_2O is polar.
(b) Benzene and $MgSO_4$ don't form solutions because $MgSO_4$ is ionic and benzene is nonpolar.
(c), (d) These two pairs of substances form solutions because they are chemically similar.

9.3 A solution of 12.5 g KBr in 20 mL H_2O is equivalent to a solution of 62.5 g KBr in 100 mL H_2O. According to Figure 9.3, the solution is unsaturated at 60 °C. Cooling the solution to 10 °C makes the solution supersaturated.

9.4

$$\frac{C_1}{P_1} = \frac{C_2}{P_2}; P_1 = 760 \text{ mmHg} ; C_1 = 0.169 \text{ g/100 mL}; P_2 = 2.5 \times 10^4 \text{ mmHg}; C_2 = ?$$

$$C_2 = \frac{C_1 P_2}{P_1} = \frac{\left(\dfrac{0.169 \text{ g}}{100 \text{ mL}}\right) \times 2.5 \times 10^4 \text{ mmHg}}{760 \text{ mmHg}} = 5.6 \text{ g } CO_2/100 \text{ mL}$$

9.5

$$\frac{C_1}{P_1} = \frac{C_2}{P_2}; P_1 = 1.00 \text{ atm}; C_1 = 0.169 \text{ g/100 mL}; P_2 = 4.0 \times 10^{-4} \text{ atm}; C_2 = ?$$

$$C_2 = \frac{C_1 P_2}{P_1} = \frac{\left(\dfrac{0.169 \text{ g}}{100 \text{ mL}}\right) \times 4.0 \times 10^{-4} \text{ atm}}{1.00 \text{ atm}} = 6.8 \times 10^{-5} \text{ g } CO_2/100 \text{ mL}$$

CIA 9.3 At high altitude, P_{O2} is low, and not enough oxygen is available to cause 100% saturation of hemoglobin. In order to deliver enough oxygen to body tissues, the body compensates by manufacturing more hemoglobin, which drives the hemoglobin equilibrium to the right.

CIA 9.4 According to Dalton's Law (Section 8.11), the total pressure exerted by a gas is the sum of the partial pressures of the gases in the mixture. Thus the partial pressure of oxygen at a total pressure of 374 mmHg is 0.18 x 374 mmHg = 67 mmHg. The oxygen-carrying curve shows that hemoglobin is approximately 95% saturated at this partial pressure.

9.6 This is a (mass/mass)% concentration problem in which a mass of 100 g of alloy contains 15.8 g Ni. To find the mass of the alloy that contains 36.5 g Ni:

$$\text{Mass of alloy} = 36.5 \text{ g Ni} \times \frac{100 \text{ g alloy}}{15.8 \text{ g Ni}} = 231 \text{ g alloy}$$

9.7   A 7.5% (v/v) solution contains 7.5 mL of solute per 100 mL of solution.

$$500 \text{ mL solution} \ \times \ \frac{7.5 \text{ mL acetic acid}}{100 \text{ mL solution}} \ = \ 38 \text{ mL acetic acid}$$

To prepare the desired solution, measure 38 mL of acetic acid into a 500.0 mL volumetric flask and add water to the 500.0 mL mark.

9.8   1 dL = 100 mL

$$\frac{8.6 \text{ mg}}{100 \text{ mL}} \ \times \ \frac{1 \text{ g}}{1000 \text{ mg}} \ \times \ 100\% \ = \ 0.0086\% \text{ (m/v) } Ca^{2+}$$

Remember: (m/v)% specifies that the mass of solute be expressed in grams.

9.9   (a) A 16% (m/v) solution contains 16 g of solute per 100 mL of solution.

$$125 \text{ mL} \ \times \ \frac{16 \text{ g glucose}}{100 \text{ mL}} \ = \ 20 \text{ g glucose}$$

(b) A 2.0% (m/v) solution contains 2.0 g of solute per 100 mL of solution.

$$1.2 \text{ g KCl} \ \times \ \frac{100 \text{ mL } H_2O}{2.0 \text{ g KCl}} \ = \ 60 \text{ mL } H_2O$$

9.10

$$\frac{32 \text{ mg NaF}}{20 \text{ kg solution}} \ \times \ \frac{1 \text{ kg}}{10^6 \text{ mg}} \ \times \ 10^6 \ = \ 1.6 \text{ ppm}$$

9.11   For lead:

$$\frac{0.015 \text{ mg Pb}}{1 \text{ kg solution}} \ \times \ \frac{1 \text{ kg}}{10^6 \text{ mg}} \ \times \ 10^6 \ = \ 0.015 \text{ ppm}$$

$$\frac{0.015 \text{ mg Pb}}{1 \text{ kg solution}} \ \times \ \frac{1 \text{ kg}}{10^3 \text{ g}} \ \times \ 100 \text{ g} \ = \ 0.0015 \text{ mg Pb}$$

For copper:

$$\frac{1.3 \text{ mg Cu}}{1 \text{ kg solution}} \ \times \ \frac{1 \text{ kg}}{10^6 \text{ mg}} \ \times \ 10^6 \ = \ 1.3 \text{ ppm}$$

$$\frac{1.3 \text{ mg Cu}}{1 \text{ kg solution}} \ \times \ \frac{1 \text{ kg}}{10^3 \text{ g}} \ \times \ 100 \text{ g} \ = \ 0.13 \text{ mg Cu}$$

9.12

$$\frac{50.0 \text{ g}}{0.160 \text{ L}} \ \times \ \frac{1 \text{ mol}}{337 \text{ g}} \ = 0.927 \text{ M}$$

9.13

$$\text{Moles of solute} = \text{molarity (M)} \times \text{Volume} = \frac{\text{mol}}{\text{L}} \times \text{L}$$

(a) $M = \dfrac{0.35 \text{ mol NaNO}_3}{1 \text{ L}}$; $V = 175 \text{ mL} = 0.175 \text{ L}$

$$\text{Moles} = \frac{0.35 \text{ mol}}{1 \text{ L}} \times 0.175 \text{ L} = 0.061 \text{ mol NaNO}_3$$

(b) $M = \dfrac{1.4 \text{ mol HNO}_3}{1 \text{ L}}$; $V = 480 \text{ mL} = 0.48 \text{ L}$

$$\text{Moles} = \frac{1.4 \text{ mol}}{1 \text{ L}} \times 0.48 \text{ L} = 0.67 \text{ mol HNO}_3$$

9.14 First, find the number of moles of cholesterol in 250 mL of blood:

$$250 \text{ mL} \times \frac{0.0050 \text{ mol cholesterol}}{1000 \text{ mL}} = 0.001\,25 \text{ mol cholesterol}$$

Next, find the molar mass of cholesterol:

$$(27 \times 12.0 \text{ g/mol C}) + (46 \times 1.0 \text{ g/mol H}) + (16.0 \text{ g/mol O}) = 386.0 \text{ g/mol}$$

Now, convert moles into grams:

$$0.001\,25 \text{ mol cholesterol} \times \frac{386.0 \text{ g}}{1 \text{ mol}} = 0.48 \text{ g cholesterol}$$

9.15

(a) $0.065 \text{ L} \times \dfrac{0.12 \text{ mol HCl}}{1 \text{ L}} = 0.0078 \text{ mol HCl}$

(b) $0.0078 \text{ mol HCl} \times \dfrac{1 \text{ mol CaCO}_3}{2 \text{ mol HCl}} \times \dfrac{100.0 \text{ g CaCO}_3}{1 \text{ mol CaCO}_3} = 0.39 \text{ g CaCO}_3$

9.16 $V_2 = 500.0 \text{ mL}$; $M_2 = 1.25 \text{ M}$; $M_1 = 16.0 \text{ M}$; $V_1 = ?$

$$V_1 = V_2 \times \frac{M_2}{M_1} = 500.0 \text{ mL} \times \frac{1.25 \text{ M}}{16.0 \text{ M}} = 39.1 \text{ mL}$$

Ballpark check: Since the final molarity is about 1/12 the initial molarity, the initial volume is about 1/12 the final volume, or about 40 mL.

9.17 $C_1 = 5.0 \text{ ppm}$; $V_1 = 1.5 \text{ L}$; $C_2 = 0.010 \text{ ppm}$; $V_2 = ?$

$$V_2 = V_1 \times \frac{C_2}{C_1} = 1.5 \text{ L} \times \frac{5.0 \text{ ppm}}{0.01 \text{ ppm}} = 750 \text{ L}$$

CIA 9.5 The major electrolytes in sweat are sodium ion (30–40 mEq/L), potassium ion (5–10 mEq/L), small amounts of metals such as magnesium, and chloride to balance charge (35–50 mEq/L).

CIA 9.6 In addition to fluid replacement, sports drinks provide electrolytes to replenish those lost during exercise, they furnish soluble complex carbohydrates for slow-release energy, and they may contain vitamins to protect cells from damage.

CIA 9.7 For Na^+:

$$\frac{20 \text{ mEq Na}^+}{1.0 \text{ L}} \times \frac{23 \text{ mg Na}^+}{1 \text{ mEq}} \times \frac{1 \text{ g}}{1000 \text{ mg}} \times \frac{1.0 \text{ L}}{1000 \text{ mL}} \times 100 \text{ mL} = 0.046 \text{ (m/v)\% Na}^+$$

For K^+:

$$\frac{10 \text{ mEq K}^+}{1.0 \text{ L}} \times \frac{39 \text{ mg K}^+}{1 \text{ mEq}} \times \frac{1 \text{ g}}{1000 \text{ mg}} \times \frac{1.0 \text{ L}}{1000 \text{ mL}} \times 100 \text{ mL} = 0.039 \text{ (m/v)\% K}^+$$

9.18

One equivalent = Molar mass of ion (g) ÷ Number of charges on ion.

Ion	Molar Mass	Charge	Gram-Equivalent	Milligram-Equivalent
(a) K^+	39.1 g	+1	39.1 g	39.1 mg, or 3.91×10^{-2} g
(b) Br^-	79.9 g	−1	79.9 g	79.9 mg, or 7.99×10^{-2} g
(c) Mg^{2+}	24.3 g	+2	12.2 g	12.2 mg, or 1.22×10^{-2} g
(d) SO_4^{2-}	96.0 g	−2	48.0 g	48.0 mg, or 4.80×10^{-2} g
(e) Al^{3+}	27.0 g	+3	9.0 g	9.0 mg, or 9.0×10^{-3} g
(f) PO_4^{3-}	95.0 g	−3	31.7 g	31.7 mg, or 31.7×10^{-2} g

9.19 One gram-equivalent of Mg^{2+} = 12.2 g [Problem 9.18(c)].

$$\frac{\text{g Mg}^{2+}}{1 \text{ L}} = \frac{12 \text{ g Mg}^{2+}}{1 \text{ Eq}} \times \frac{1 \text{ Eq}}{1000 \text{ mEq}} \times \frac{3.0 \text{ mEq}}{1 \text{ L}} = \frac{0.036 \text{ g Mg}^{2+}}{1 \text{ L}}$$

$$\frac{0.036 \text{ g Mg}^{2+}}{\text{L}} \times \frac{1000 \text{ mg}}{1 \text{ g}} \times \frac{1 \text{ L}}{1000 \text{ mL}} \times 250 \text{ mL} = 9.0 \text{ mg Mg}^{2+}$$

9.20 (a) 0.67 mol of $MgCl_2$ in 0.5 kg H_2O yields 2.0 mol ions.
(b) 2.0 mol ions in 0.5 kg H_2O is equivalent to 4.0 mol ions in 1.0 kg H_2O, which raise the boiling point by 4.0 x 0.51 °C = 2.0 °C.

Boiling point = 100.0 °C + 2.0 °C = 102.0 °C

9.21 If HF were a strong electrolyte it would dissociate completely, and a solution of 1.0 mol HF would yield 2.0 mol ions, which would elevate the boiling point of water by 2.0 x 0.51 °C = 1.02 °C. Since the observed boiling point elevation is only 0.5 °C, HF must be a weak electrolyte that is only slightly dissociated.

9.22   (a) The red curve represents the pure solvent, because the solvent boils at a lower temperature than the solution.
(b) A liquid boils when its vapor pressure equals atmospheric pressure. The solvent boils at 62 °C, and the solution boils at 69 °C, an approximately 7° boiling point elevation.
(c) One mole of solute raises the boiling point approximately 3.5 °C, and the observed boiling point elevation is approximately 7 °C. Thus, the concentration of the solute is about 2 M.

9.23   Glucose is not an electrolyte. Thus, 1.0 mol glucose lowers the freezing point of 1.0 kg H_2O by 1.9 °C.
Freezing point = 0.0 °C – 1.9 °C = –1.9 °C

9.24   A freezing point depression of 1.86 °C is produced by 1 mol of ions in 1 kg of water. Thus, a freezing point depression of 2.8 °C must be produced by 2.8 ÷ 1.86 = 1.5 mol of ions. Since 0.5 mol of the ionic substance produces the freezing point depression expected for 1.5 mol ions, the substance gives three ions when it dissolves.

HOC 9.1
(a) Bubbles began to form at around 50 °C; these are due to dissolved gases escaping from solution and were not due to boiling.
(b) Bubbles became larger and reached the surface at 93 °C.
(c) Boiling point of salt water was 97 °C.
There was a lot of uncertainty in the experiment due to the thermometer and the mass of NaCl.

9.25   Osmolarity = molarity x number of particles.
(a) For 0.35 M KBr, osmolarity = 0.35 M x 2 = 0.70 osmolar, since KBr yields two ions (K^+ and Br^-) in solution.
(b) For 0.15 M glucose, osmolarity = 0.15 M x 1 = 0.15 osmolar, since glucose yields only one particle in solution. For K_2SO_4, osmolarity = 0.05 M x 3 = 0.15 osmolar, since K_2SO_4 provides three particles per mole in solution. Total osmolarity = 0.30 osmolar.

9.26   (a) For the oral rehydration solution, osmolarity is equal to the sum of the osmolarities of the individual components. For each of the ionic components, the number of millimoles = the number of mEq, since each ion has one charge. Thus,

$$90 \text{ mM } Na^+ + 20 \text{ mM } K^+ + 110 \text{ mM } Cl^- = 220 \text{ mM ions}$$
$$0.090 \text{ M } Na^+ + 0.020 \text{ M } K^+ + 0.110 \text{ M } Cl^- = 0.220 \text{ M ions}$$

For glucose:

$$\frac{2.0 \text{ g glucose}}{100 \text{ mL}} \times \frac{1000 \text{ mL}}{1 \text{ L}} \times \frac{1 \text{ mol}}{180 \text{ g glucose}} = 0.11 \text{ M glucose}$$

(b) Osmolarity = Molarity x Number of particles. In this problem, all components yield one particle in solution.
Osmolarity = 0.22 M + 0.11 M = 0.33 osmol. This is almost the same osmolarity as found in blood plasma (0.30 osmol).

HOC 9.2
  (a) The lettuce wilted in salt water but stayed crisp in fresh water.
  (b) Water had a net movement out of the lettuce in salt water, causing it to wilt.

CIA 9.8   The hygroscopic material attracts water from the osmotic layer, increasing the osmotic pressure and pushing the medication through the microperforated membrane and into the body.

CIA 9.9   Nylon is a better choice for the hygroscopic material because it is more likely to attract water.

Understanding Key Concepts

9.27

before equilibrium   at equilibrium

The membrane is permeable to the unshaded spheres but impermeable to the shaded spheres. Solvent (unshaded spheres) passes through the membrane until equilibrium is reached.

9.28   The boiling point of water is elevated by 0.5 °C for every mole of dissolved particles. One mol HCl dissolves to form 2 mol particles, which elevate the boiling point of water by 1 °C. Acetic acid exists in solution almost completely as CH_3CO_2H, and 1 mol acetic acid dissolves to form 1 mol particles, which elevate the boiling point of water by only 0.5 °C.

9.29   The same reasoning used in the previous problem applies here. The boiling point of water is elevated by 0.5 °C for every mole of dissolved particles. One mol HBr dissociates to form 2 mol ions, which raise the boiling point of water by 1.0 °C, but 1 mol HF is only slightly dissociated and raises the boiling point by 0.5 °C. Thus, the upper curve (red) represents
1 M HF, and the lower curve (blue) represents 1 M HBr.

9.30

(a)   (b)   (c)

When a beaker of glucose solution (shaded) and a beaker of pure water (unshaded) stand for several days, the liquid levels appear as those pictured in (a). The dissolved glucose particles lower the vapor pressure of water and make evaporation slower.

9.31   Drawing (d) represents the solution that results when 50.0 mL of (a) is withdrawn and diluted by a factor of 4.

Additional Problems

Solutions and Solubility

9.32 In a homogeneous mixture (such as tea), mixing is uniform throughout because the particle size is small. In a heterogeneous mixture (such as chicken soup), mixing is nonuniform because particle size is larger.

9.33 A solution is transparent to light; a colloid usually appears murky or opaque.

9.34 The polarity of water enables it to dissolve many ionic solids.

9.35 Water doesn't dissolve motor oil because oil is nonpolar and water is polar.

9.36 Rubbing alcohol (b) and mouthwash (d) are solutions.

9.37 The pairs in (a), (b), and (d) are miscible.

9.38 (a)

$$C_2 = \frac{C_1 P_2}{P_1}; \ C_1 = 51.8 \text{ g/100 mL}; \ P_1 = 760.0 \text{ mmHg}; \ P_2 = 225.0 \text{ mmHg}$$

$$= \frac{(51.8 \text{ g/100 mL}) \ \text{x} \ 225.0 \text{ mmHg}}{760.0 \text{ mmHg}} = 15.3 \text{ g/100 mL}$$

(b) Subtract the solubility in g/L at 50 °C from the solubility at 25 °C and convert grams to moles.

$$518 \text{ g/L} - 270 \text{ g/L} = 248 \text{ g/L}; \ 248 \text{ g x } \frac{1 \text{ mol}}{17.0 \text{ g}} = 14.6 \text{ mol}$$

9.39 (a)

$$C_2 = \frac{C_1 P_2}{P_1}; \ C_1 = 0.15 \text{ g/100 mL}; \ P_1 = 760.0 \text{ mmHg} = 1.0 \text{ atm}; \ P_2 = 4.5 \text{ atm}$$

$$= \frac{(0.15 \text{ g/100 mL}) \ \text{x} \ 4.5 \text{ atm}}{1.0 \text{ atm}} = 0.68 \text{ g/100 mL}$$

(b) $\frac{3.8 \text{ x } 10^{-4} \text{ atm}}{4.5 \text{ atm}} \text{ x } 100\% = 0.0084\% \ CO_2$ remains.

(c) Initially, both bottles should be equally carbonated, but the warmer bottle will lose its carbonation faster.

Concentration and Dilution of Solutions

9.40 Depending on the solubility of the solute in the solvent, a saturated solution may be either dilute or concentrated. If a solute is only slightly soluble in a solvent, a saturated solution will be dilute. If the solute is very soluble in the solvent, a saturated solution will be concentrated. For a highly soluble solute, even a highly concentrated solution may be unsaturated.

9.41 Mass/volume percent concentration [(m/v)]% is defined as the mass in grams of solute per 100 mL of solution and is used to describe the solution of a solid in a liquid.

9.42 Molarity (M) is defined as the number of moles of solute per liter of solution.

9.43 Volume/volume percent concentration [(v/v)%] is defined as the volume of solute in 100 mL of solution and is used for solutions in which both the solute and the solvent are liquids.

9.44

(a) $\dfrac{21 \text{ mL ethyl alcohol}}{100.0 \text{ mL solution}}$ x 750.0 mL solution = 158 mL ethyl alcohol

(b) 158 mL x $\dfrac{0.789 \text{ g}}{1 \text{ mL}}$ x $\dfrac{1 \text{ mol}}{46.07 \text{ g}}$ x = 2.71 mol ethyl alcohol

$\dfrac{2.71 \text{ mol ethyl alcohol}}{750 \text{ mL}}$ x $\dfrac{1000 \text{ mL}}{1 \text{ L}}$ = 3.61 M

9.45

500.0 mL solution x $\dfrac{0.50 \text{ g B(OH)}_3}{100 \text{ mL solution}}$ = 2.5 g B(OH)$_3$

Dissolve 2.5 g B(OH)$_3$ in water to a final volume of 500.0 mL.

9.46

250 mL solution x $\dfrac{0.10 \text{ mol NaCl}}{1000 \text{ mL solution}}$ x $\dfrac{58.5 \text{ g NaCl}}{1 \text{ mol NaCl}}$ = 1.5 g NaCl

Dissolve 1.5 g NaCl in water to a final volume of 250 mL.

9.47

1.50 L x $\dfrac{1000 \text{ mL}}{1 \text{ L}}$ x $\dfrac{7.50 \text{ g Mg(NO}_3)_2}{100 \text{ mL solution}}$ = 113 g Mg(NO$_3$)$_2$

Dissolve 113 g Mg(NO$_3$)$_2$ in water and dilute to 1.50 L.

9.48

(a) 0.078 mol KCl x $\dfrac{74.6 \text{ g KCl}}{1 \text{ mol KCl}}$ = 5.8 g KCl

$\dfrac{5.8 \text{ g KCl}}{75 \text{ mL}}$ x 100 mL = 7.7 g KCl; $\dfrac{7.7 \text{ g KCl}}{100 \text{ mL}}$ x 100% = 7.7% (m/v) KCl

(b) 0.044 mol sucrose x $\dfrac{342 \text{ g sucrose}}{1 \text{ mol sucrose}}$ = 15 g sucrose

$\dfrac{15 \text{ g sucrose}}{380 \text{ mL}}$ x 100 mL = 3.9 g sucrose; $\dfrac{3.9 \text{ g}}{100 \text{ mL}}$ x 100% = 3.9% (m/v) sucrose

9.49

$$\frac{90 \text{ mg glucose}}{100 \text{ mL}} \times \frac{1 \text{ g}}{1000 \text{ mg}} = \frac{0.090 \text{ g glucose}}{100 \text{ mL}}; \quad 0.090\% \text{ (m/v) glucose}$$

$$\frac{0.090 \text{ g glucose}}{100 \text{ mL}} \times \frac{1000 \text{ mL}}{1 \text{ L}} \times \frac{1 \text{ mol}}{180 \text{ g glucose}} = 0.0050 \text{ M}$$

9.50

(a) $50.0 \text{ mL} \times \dfrac{8.0 \text{ g KCl}}{100 \text{ mL}} = 4.0 \text{ g KCl}; \; 4.0 \text{ g KCl} \times \dfrac{1 \text{ mol KCl}}{74.6 \text{ g KCl}} = 0.054 \text{ mol KCl}$

(b) $200.0 \text{ mL} \times \dfrac{7.5 \text{ g acetic acid}}{100 \text{ mL}} = 15 \text{ g acetic acid}$

$15 \text{ g acetic acid} \times \dfrac{1 \text{ mol acetic acid}}{60 \text{ g acetic acid}} = 0.25 \text{ mol acetic acid}$

9.51

(a) For 0.50 M KCl: $\dfrac{0.50 \text{ mol KCl}}{1 \text{ L}} \times \dfrac{74.6 \text{ g}}{1 \text{ mol}} = \dfrac{37 \text{ g KCl}}{1 \text{ L}} = \dfrac{3.7 \text{ g KCl}}{100 \text{ mL}}$

For 5% (m/v) KCl: $\dfrac{5.0 \text{ g KCl}}{100 \text{ mL}}$

The 5% (m/v) solution is more concentrated.

(b) 2.5% (m/v) $NaHSO_4 = \dfrac{2.5 \text{ g NaHSO}_4}{100 \text{ mL solution}}$

$\dfrac{0.025 \text{ mol NaHSO}_4}{1 \text{ L}} \times \dfrac{120 \text{ g}}{1 \text{ mol}} = \dfrac{3.0 \text{ g NaHSO}_4}{1 \text{ L}} = \dfrac{0.30 \text{ g NaHSO}_4}{100 \text{ mL}}$

The 2.5% (m/v) solution is more concentrated.

9.52

$23 \text{ g KOH} \times \dfrac{100 \text{ mL}}{10.0 \text{ g KOH}} = 230 \text{ mL of a 10.0\% (m/v) solution}$

$23 \text{ g KOH} \times \dfrac{1 \text{ mol}}{56.1 \text{ g KOH}} = 0.41 \text{ mol KOH}$

$0.41 \text{ mol KOH} \times \dfrac{1000 \text{ mL}}{0.25 \text{ mol KOH}} = 1600 \text{ mL of 0.25 M solution}$

9.53

$\dfrac{3 \text{ g}}{100 \text{ mL}} \times \dfrac{1000 \text{ mL}}{1 \text{ L}} \times \dfrac{1 \text{ mol}}{34 \text{ g}} = 0.9 \text{ M}$

9.54

$\dfrac{10 \text{ mg KCN}}{1 \text{ kg body weight}} \times \dfrac{1 \text{ kg}}{10^6 \text{ mg}} \times 10^6 = 10 \text{ ppm}$

9.55

(a) $\dfrac{12.5 \text{ g NaHCO}_3}{0.350 \text{ L}} \times \dfrac{1 \text{ mol}}{84.0 \text{ g}} = 0.425 \text{ M}$

(b) $\dfrac{45.0 \text{ g H}_2\text{SO}_4}{0.300 \text{ L}} \times \dfrac{1 \text{ mol}}{98.1 \text{ g}} = 1.53 \text{ M}$

(c) $\dfrac{30.0 \text{ g NaCl}}{0.500 \text{ L}} \times \dfrac{1 \text{ mol}}{58.5 \text{ g}} = 1.03 \text{ M}$

9.56

(a) $0.200 \text{ L} \times \dfrac{0.30 \text{ mol acetic acid}}{1 \text{ L}} \times \dfrac{60.0 \text{ g acetic acid}}{1 \text{ mol acetic acid}} = 3.6 \text{ g acetic acid}$

(b) $1.50 \text{ L} \times \dfrac{0.25 \text{ mol NaOH}}{1 \text{ L}} \times \dfrac{40.0 \text{ g NaOH}}{1 \text{ mol NaOH}} = 15 \text{ g NaOH}$

(c) $0.750 \text{ L} \times \dfrac{2.5 \text{ mol HNO}_3}{1 \text{ L}} \times \dfrac{63.0 \text{ g HNO}_3}{1 \text{ mol HNO}_3} = 120 \text{ g HNO}_3$

9.57

$0.0040 \text{ mol HCl} \times \dfrac{1000 \text{ mL}}{0.75 \text{ mol}} = 5.3 \text{ mL HCl}$

9.58

$1.5 \text{ mg} \times \dfrac{1 \text{ g}}{1000 \text{ mg}} \times \dfrac{100 \text{ mL}}{0.40 \text{ g}} = 0.38 \text{ mL}$

9.59 First, calculate the number of moles of H_2SO_4 spilled:

$0.450 \text{ L} \times \dfrac{0.50 \text{ mol H}_2\text{SO}_4}{1 \text{ L}} = 0.23 \text{ mol H}_2\text{SO}_4$

According to the equation given in the problem, each mole of H_2SO_4 reacts with two moles of $NaHCO_3$. Thus, the 0.23 mol of H_2SO_4 spilled need to be neutralized with 0.46 mol of $NaHCO_3$.

$0.46 \text{ mol NaHCO}_3 \times \dfrac{84 \text{ g}}{1 \text{ mol}} = 39 \text{ g NaHCO}_3$

9.60

(a) $0.450 \text{ g AgBr} \times \dfrac{1 \text{ mol AgBr}}{187.8 \text{ g}} = 2.40 \times 10^{-3} \text{ mol AgBr}$

According to the equation given, 1 mole of AgBr reacts with 2 moles of $Na_2S_2O_3$. Thus, 2.40×10^{-3} mol AgBr reacts with 4.80×10^{-3} mol $Na_2S_2O_3$.

(b) To calculate the volume of 0.0200 M $Na_2S_2O_3$:

$4.80 \times 10^{-3} \text{ mol} \times \dfrac{1000 \text{ mL}}{0.0200 \text{ mol}} = 240 \text{ mL of } 0.0200 \text{ M Na}_2\text{S}_2\text{O}_3$

9.61

20.0% (v/v) means $\dfrac{20.0 \text{ mL concentrate}}{100.0 \text{ mL juice}}$

Thus, 100.0 mL concentrate x $\dfrac{100.0 \text{ mL juice}}{20.0 \text{ mL concentrate}}$ = 500.0 mL juice

Since the final volume of the diluted juice is 500.0 mL, you would need to add 400.0 mL water to the original 100.0 mL of concentrate.

9.62

$V_2 = V_1 \text{ x } \dfrac{M_1}{M_2}$; $V_1 = 100.0 \text{ mL}$; $M_1 = 0.500 \text{ M}$; $M_2 = 0.150 \text{ M}$

$= 100.0 \text{ mL x } \dfrac{0.500 \text{ M}}{0.150 \text{ M}} = 333 \text{ mL of } 0.150 \text{ M NaOH solution}$

100.0 mL of 0.500 M NaOH is diluted with 233 mL water to give 333 mL of 0.150 M NaOH solution.

9.63

$V_1 = V_2 \text{ x } \dfrac{C_2}{C_1}$; $V_2 = 2.0 \text{ L}$; $C_2 = 75 \text{ ppm}$; $C_1 = 285 \text{ ppm}$

$= 2.0 \text{ L x } \dfrac{75 \text{ ppm}}{285 \text{ ppm}} = 0.53 \text{ L of } 285 \text{ ppm KNO}_3 \text{ solution}$

9.64

$C_2 = C_1 \text{ x } \dfrac{V_1}{V_2}$; $C_1 = 37\% \text{ (m/v)}$; $V_1 = 65 \text{ mL}$; $V_2 = 480 \text{ mL}$

$= 37\% \text{ (m/v) x } \dfrac{65 \text{ mL}}{480 \text{ mL}} = 5.0\% \text{ (m/v) NaCl}$

9.65

$V_2 = V_1 \text{ x } \dfrac{M_1}{M_2}$; $V_1 = 25.0 \text{ mL}$; $M_1 = 12.0 \text{ M}$; $M_2 = 0.500 \text{ M}$

$= 25.0 \text{ mL x } \dfrac{12.0 \text{ M}}{0.500 \text{ M}} = 600 \text{ mL of } 0.500 \text{ M HCl solution}$

9.66

$V_1 = V_2 \text{ x } \dfrac{M_2}{M_1}$; $V_2 = 750.0 \text{ mL}$; $M_1 = 0.100 \text{ M}$; $M_2 = 0.0500 \text{ M}$

$= 750.0 \text{ mL x } \dfrac{0.0500 \text{ M}}{0.100 \text{ M}} = 375 \text{ mL of } 0.100 \text{ M NaHCO}_3$

Electrolytes

9.67 An electrolyte is a substance that conducts electricity when dissolved in water.

9.68 Sodium chloride is an example of a strong electrolyte. Glucose is an example of a nonelectrolyte.

9.69 If the concentration of Ca^{2+} is 3.0 mEq/L, there are 3.0 mmol of charges due to calcium per liter of blood. Since calcium has a charge of +2, there are 1.5 mmol of calcium per liter of blood.

9.70 The total anion concentration in the solution must equal the total cation concentration in order to conserve charge. Thus:

[Cations] = [Anions] = 5.0 mEq/L Na^+ + 12.0 mEq/L Ca^{2+} + 2.0 mEq/L K^+ = 19 mEq/L

9.71

$$10\% \ (m/v) \ = \ \frac{10 \text{ g KCl}}{100 \text{ mL}}; \quad 30 \text{ mL} \ \times \ \frac{10 \text{ g KCl}}{100 \text{ mL}} \ = \ 3.0 \text{ g KCl}$$

The molar mass of KCl is 74.6 g/mol. Thus:

$$3.0 \text{ g KCl} \ \times \ \frac{1 \text{ mol}}{74.6 \text{ g}} \ = \ 0.040 \text{ mol KCl}$$

Since one equivalent equals one mole when an ion has only one charge, there are 0.040 Eq, or 40 mEq, of K^+ in a 30 mL dose.

9.72 One gram-equivalent = molar mass of ion(g) divided by # of charges on ion:

Ion	Molar mass	Charge	Gram-Equivalent	Equivalent	Mass
(a) Ca^{2+}	40.1 g	2+	20.1 g	0.25 Eq	5.0 g
(b) K^+	39.1 g	1+	39.1 g	75 mEq	2.9 g
(c) SO_4^{2-}	96.1 g	2–	48.1 g	199 mEq	9.6 g
(d) PO_4^{3-}	95.0 g	3–	31.7 g	0.65 Eq	21 g

9.73 For Cl^- ion, 100 mEq = 100 mmol = 0.100 mol.

$$1.0 \text{ g Cl}^- \ \times \ \frac{1 \text{ mol}}{35.5 \text{ g}} \ \times \ \frac{1 \text{ L}}{0.100 \text{ mol}} \ \times \ \frac{1000 \text{ mL}}{1 \text{ L}} \ = \ 280 \text{ mL}$$

9.74 For Mg^{2+}, with a molar mass of 24.3 g and a charge of +2, the mass of one equivalent is 24.3 g/2 = 12.2 g, and the mass of 1 milliequivalent is 12.2 mg.

$$\frac{3 \text{ mEq Mg}^{2+}}{1 \text{ L}} \ \times \ \frac{12 \text{ mg Mg}^{2+}}{1 \text{ mEq Mg}^{2+}} \ \times \ \frac{1 \text{ L}}{1000 \text{ mL}} \ \times \ 150.0 \text{ mL} \ = \ 5 \text{ mg Mg}^{2+}$$

Properties of Solutions

9.75 0.20 mol NaOH contains 0.40 mol solute particles, and 0.20 mol $Ba(OH)_2$ contains 0.60 mol solute particles. Since $Ba(OH)_2$ produces more solute particles, it produces greater lowering of the freezing point when dissolved in 2.0 kg of water.

9.76 When 0.300 mol KCl dissolves, it produces 0.600 mol particles; when 0.500 mol glucose dissolves, it produces 0.500 mol particles. The solution with more dissolved particles (0.300 mol KCl) has the higher boiling point.

9.77 Methanol has a molar mass of 32.0 g and provides one mole of solute particles per mole of methanol.

$$10.0\ °C \ \times \ \frac{1\ mol}{1.86\ °C \times 1\ kg} \ \times \ 5.00\ kg \ = \ 26.9\ mol\ methanol$$

$$26.9\ mol \ \times \ \frac{32.0\ g}{1\ mol} \ = \ 861\ g\ of\ methanol\ needed$$

9.78 Cane sugar provides one mole of solute particles per mole. The boiling point elevation:

$$\frac{650\ g\ sugar}{1.5\ kg\ H_2O} \ \times \ \frac{1\ mol}{342\ g} \ \times \ \frac{0.51\ °C \times 1\ kg}{1\ mol} \ = \ 0.65\ °C$$

The boiling point of the resulting solution is $100.0\ °C + 0.65\ °C \ = \ 100.65\ °C$.

Osmosis

9.79 The inside of a red blood cell contains dissolved substances and therefore has a higher osmolarity than pure water. Water thus passes through the cell membrane to dilute the cell contents until pressure builds up and the cell eventually bursts.

9.80 If a 0.15 M NaCl solution is isotonic with blood, then the NaCl solution has the same osmolarity as blood. If distilled water is hypotonic with blood, then the water has a lower osmolarity than blood.

9.81

Solution	Molarity	Number of Particles	Osmolarity
(a) 0.25 M KBr	0.25 M	2	0.50 osmolar
0.20 M Na_2SO_4	0.20 M	3	0.60 osmolar (greater)
(b) 0.30 M NaOH	0.30 M	2	0.60 osmolar
3% (m/v) NaOH	0.75 M	2	1.5 osmolar (greater)

9.82 The solution with the higher osmolarity gives rise to the greater osmotic pressure at equilibrium.

For NaCl: $\dfrac{5.0 \text{ g}}{0.350 \text{ L}}$ x $\dfrac{1 \text{ mol}}{58.5 \text{ g}}$ x 2 particles = 0.49 osmolar

For glucose: $\dfrac{35.0 \text{ g}}{0.400 \text{ L}}$ x $\dfrac{1 \text{ mol}}{180 \text{ g}}$ x 1 particle = 0.49 osmolar

Both solutions give rise to the same osmotic pressure.

9.83 The molar mass of NaCl is 58.5 g, and each mole of NaCl yields two moles of particles.

$\dfrac{270 \text{ g}}{3.8 \text{ L}}$ x $\dfrac{1 \text{ mol}}{58.5 \text{ g}}$ x 2 particles = 2.4 osmolar

9.84 A solution that contains 0.30 osmol KCl is 0.15 M, since KCl produces two particles when dissolved.

$\dfrac{74.5 \text{ g}}{1 \text{ mol}}$ x $\dfrac{0.15 \text{ mol}}{1 \text{ L}}$ x $\dfrac{1 \text{ L}}{1000 \text{ mL}}$ x 175 mL = 2.0 g KCl

Approximately 2.0 g KCl are needed.

Conceptual Problems

9.85 Molar mass of uric acid ($C_5H_4N_4O_3$) = 168 g.

(a) (m/v)%: $\dfrac{0.067 \text{ g}}{1 \text{ L}} = \dfrac{0.0067 \text{ g}}{100 \text{ mL}}$; $\dfrac{0.0067 \text{ g}}{100 \text{ mL}}$ x 100% = 0.0067% (m/v)

(b) ppm: One L of water weighs 1 kg. Thus

$\dfrac{0.067 \text{ g}}{1 \text{ L}} = \dfrac{0.067 \text{ g}}{1 \text{ kg}}$ x $\dfrac{1 \text{ kg}}{1000 \text{ g}}$; $\dfrac{0.067 \text{ g}}{1000 \text{ g}}$ x 10^6 = 67 ppm

(c) $\dfrac{0.067 \text{ g}}{1 \text{ L}}$ x $\dfrac{1 \text{ mol}}{168 \text{ g}}$ = 0.000 40 M = 4.0×10^{-4} M

9.86

5.0% (m/v) = $\dfrac{5.0 \text{ g CaCl}_2}{100 \text{ mL}}$; $\dfrac{5.0 \text{ g CaCl}_2}{100 \text{ mL}}$ x 5.0 mL = 0.25 g $CaCl_2$

0.25 g $CaCl_2$ x $\dfrac{1 \text{ mol}}{111 \text{ g}}$ = 0.0023 mol $CaCl_2$ = 0.0023 mol Ca^{2+}

0.0023 mol Ca^{2+} x $\dfrac{2 \text{ Eq}}{1 \text{ mol}}$ = 0.0046 Eq Ca^{2+} = 4.6 mEq Ca^{2+}

9.87   (a) The volume of concentrated HNO_3 that contains 0.150 mol is:

$$V_2 = V_1 \text{ x } \frac{\text{moles}_2}{\text{moles}_1} = 1.0 \text{ L x } \frac{0.15 \text{ mol}}{16 \text{ mol}} = 0.0094 \text{ L } HNO_3 = 9.4 \text{ mL } HNO_3$$

(b) The volume of a 0.20 M solution that contains 0.15 moles:

$$V_2 = V_1 \text{ x } \frac{\text{moles}_2}{\text{moles}_1} = 1.0 \text{ L x } \frac{0.15 \text{ mol}}{0.20 \text{ mol}} = 0.75 \text{ L}$$

9.88

(a) $13.0 \text{ mL x } \dfrac{0.0100 \text{ mol}}{1000 \text{ mL}} = 1.30 \text{ x } 10^{-4} \text{ mol } I_2$

According to the equation, one mole of I_2 reacts with one mole of $C_6H_8O_6$. Thus, the 25.0 mL sample contains $1.30 \text{ x } 10^{-4}$ mol $C_6H_8O_6$.

(b) $\dfrac{1.30 \text{ x } 10^{-4} \text{ mol}}{25.0 \text{ mL}} \text{ x } \dfrac{1000 \text{ mL}}{1 \text{ L}} = 0.005\ 20 \text{ M} = 5.20 \text{ mM}$

(c) Molar mass of $C_6H_8O_6 = 176$ g.

$$\frac{176 \text{ g}}{1 \text{ mol}} \text{ x } \frac{0.005\ 20 \text{ mol}}{1 \text{ L}} \text{ x } \frac{1000 \text{ mg}}{1 \text{ g}} = \frac{915 \text{ mg ascorbic acid}}{1 \text{ L}}$$

$$60 \text{ mg x } \frac{1 \text{ L}}{915 \text{ mg ascorbic acid}} \text{ x } \frac{1000 \text{ mL}}{1 \text{ L}} = 66 \text{ mL juice}$$

9.89

$$\frac{10 \text{ mg}}{5.0 \text{ L}} \text{ x } \frac{1 \text{ L}}{1000 \text{ mL}} \text{ x } \frac{1 \text{ g}}{1000 \text{ mg}} \text{ x } 100\% = 0.0002\% \text{ (m/v)}$$

9.90   For quantities expressed in parts per million, both the solute and solvent must be expressed in the same units. In this problem, both quantities are expressed in grams. The mass of statin drug is 10 mg = $10 \text{ x } 10^{-3}$ g. The density of blood is 1.05 g/mL.

$$\frac{1.05 \text{ g}}{1 \text{ mL}} \text{ x } \frac{1000 \text{ mL}}{1 \text{ L}} \text{ x } 5.0 \text{ L} = 5.2 \text{ x } 10^3 \text{ g}$$

$$\text{ppm} = \frac{\text{mass of solute (g)}}{\text{mass of solvent (g)}} \text{ x } 10^6 = \frac{10 \text{ x } 10^{-3} \text{ g}}{5.2 \text{ x } 10^3 \text{ g}} \text{ x } 10^6 = 1.9 \text{ ppm}$$

9.91

$$0.080\% \text{ (v/v)} = \frac{0.080 \text{ mL}}{100 \text{ mL}} = \frac{0.80 \text{ mL}}{1 \text{ L}}; \ \frac{0.80 \text{ mL}}{1 \text{ L}} \text{ x } 5.0 \text{ L} = 4.0 \text{ mL alcohol}$$

9.92

(a)

or

(b) $\dfrac{51.8 \text{ g}}{1 \text{ L}} \times \dfrac{1 \text{ mol}}{17.0 \text{ g}} = 3.05 \text{ mol/L}$

9.93   (a) $CoCl_2(s) + 6 H_2O(g) \rightleftharpoons CoCl_2 \cdot 6H_2O(s)$

(b) Molar mass of $CoCl_2 \cdot 6H_2O = 238$ g

$2.50 \text{ g CoCl}_2 \times \dfrac{1 \text{ mol CoCl}_2 \cdot 6 \text{ H}_2\text{O}}{238 \text{ g}} \times \dfrac{6 \text{ mol H}_2\text{O}}{1 \text{ mol CoCl}_2 \cdot 6 \text{ H}_2\text{O}} \times \dfrac{18.0 \text{ g H}_2\text{O}}{1 \text{ mol H}_2\text{O}} = 1.13 \text{ g H}_2\text{O}$

9.94 $BaCl_2(aq) + Na_2SO_4(aq) \longrightarrow BaSO_4(s) + 2 NaCl(aq)$

According to the balanced equation, one mole of $BaCl_2$ reacts with one mole of Na_2SO_4.

moles $Na_2SO_4 = \dfrac{0.200 \text{ mol}}{1 \text{ L}} \times 0.0350 \text{ L} = 0.007\ 00 \text{ mol} = $ # moles $BaCl_2$

$0.007\ 00 \text{ mol BaCl}_2 \times \dfrac{1 \text{ L}}{0.150 \text{ mol}} = 0.0467 \text{ L} = 46.7 \text{ mL BaCl}_2$

$0.007\ 00 \text{ mol} \times \dfrac{233.3 \text{ g}}{1 \text{ mol}} = 1.63 \text{ g BaSO}_4$

9.95   (a) If 36.0% of TCA is dissociated, the solution consists of 0.360 mol TCA anions, 0.360 mol H^+ cations, and 0.640 mol undissociated TCA, for a total of 1.360 mol ions and molecules in 1 kg water.

(b) The freezing point of 1 kg water is depressed by 2.53 °C, resulting in a final temperature of –2.53 °C.

$1.36 \text{ mol} \times \dfrac{1.86 \text{ °C}}{1 \text{ mol}} = 2.53 \text{ °C}$

Group Problems

9.96   (a) N_2 (0.0019 g/100mL), O_2 (0.0043 g/100mL), and CO (0.0028 g/100mL) are all only slightly soluble in water at 25 °C. CO_2 (0.169 g/100mL) is more soluble. CO_2 is the most soluble because of reaction with water to form bicarbonate. The other gases are much less soluble.

(b) For decompression sickness, a hyperbaric chamber reduces the size of N_2 bubbles, preventing gas embolism. For CO poisoning, high O_2 pressure drives the hemoglobin/CO equilibrium to release CO.

9.97 The EPA standard is 15 ppb for lead and 5 ppb for cadmium.
From Section 9.6 we know that 1 ppb = 1 μg solute/1 L solution. Thus, 15 ppb = 15 μg/L.

(a) for lead: $\dfrac{15\ \mu g}{1\ L} \times \dfrac{1\ mg}{10^3\ \mu g} = \dfrac{1.5\ \times\ 10^{-2}\ mg}{1\ L}$; Cd = $\dfrac{5\ \times\ 10^{-3}\ mg}{1\ L}$

for lead: $\dfrac{15\ \mu g}{1\ L} \times \dfrac{1\ \mu mol}{207\ \mu g} = 0.072\ \mu mol$; for Cd: $\dfrac{5\ \mu g}{1\ L} \times \dfrac{1\ \mu mol}{112\ \mu g} = 0.045\ \mu mol$

(b) Cadmium is more toxic because fewer micromoles are needed to reach the maximum level.

(c) $1.0\ \mu g \times \dfrac{1\ L}{15\ \mu g} = 0.067\ L$; $1.0\ \mu g \times \dfrac{1\ L}{5\ \mu g} = 0.2\ L$;

9.98 (a) Ringer's solution is prepared by dissolving 8.6 g NaCl, 0.30 g KCl, and 0.33 g $CaCl_2$ and dilution to 100 mL.

Component	Mass	Molar Mass	Molarity	Osmolarity
(a) NaCl	8.6 g	58.5 g	0.147 M	0.294 osmol
(b) KCl	0.30 g	74.6 g	0.0040 M	0.0080 osmol
(c) $CaCl_2$	0.33 g	111 g	0.0030 M	0.0090 osmol
			Total:	0.31 osmol

(b) Ringer's solution (0.31 osmol) is approximately isotonic with blood plasma (0.30 osmol) to ensure that red blood cells won't be destroyed.

9.99 (a) Most commercial deicing products contain either NaCl or $CaCl_2$, although KCl and $MgCl_2$ may also be used. All of these are ionic compounds. These compounds are used because they lower the melting point of ice to a greater degree than nonionic compounds and because they are relatively inexpensive.

(b) Indicators serve two purposes. First, they show where deicer has been applied. Second, they may show if the salt in an opened bag of deicer has become hydrated by exposure to excess water (see Problem 9.93).

9.100 (a) Scale consists principally of the carbonates of magnesium and calcium, which are formed by the reaction of Ca and Mg ions with dissolved bicarbonate in water.

$$Ca^{2+}(aq) + 2\ HCO_3^-(aq) \rightleftharpoons CaCO_3(s) + H_2O(l) + CO_2(g)$$
$$Mg^{2+}(aq) + 2\ HCO_3^-(aq) \rightleftharpoons MgCO_3(s) + H_2O(l) + CO_2(g)$$

(b) Scale is more of a problem in hot water pipes because both $MgCO_3$ and $CaCO_3$ are less soluble in hot water than in cold water.

9.101 (a) In hemodialysis, blood passes through a dialyzer, which contains a semipermeable membrane through which excess water and small solutes pass. After dialysis, necessary minerals are replaced before blood is returned to the body.

In peritoneal dialysis, dialysate is run into the peritoneal cavity. There, osmosis and diffusion cause waste products to flow through the peritoneal membrane into the dialysate, which is drained and replaced.

(b) Excess water, ions and small molecules such as urea and creatine flow from the blood to the dialysate. Minerals from the dialysate that are needed to replace those that are lost during dialysis flow into the blood.

(c) The replacement level of bicarbonate is slightly higher than in normal blood to act as a buffer and to counteract the higher acidity of blood in patients with renal failure.

Self-Test for Chapter 9

Multiple Choice

1. 50 mL of a 1.0 M NaOH solution is diluted to 1.0 L. What is the dilution factor?
    (a) 1/50   (b) 1/20   (c) 1/10   (d) 1/2

2. Which of the following will not speed up the rate of solution of a solid?
    (a) stirring   (b) heating   (c) grinding the solid into powder
    (d) increasing pressure on the solution

3. To make up 500 mL of a 5% (v/v) solution of methanol (CH_3OH; molar mass = 32 g) in water:
    (a) Dilute 25 mL of CH_3OH with water to a volume of 500 mL.
    (b) Dilute 5 mL of CH_3OH with water to a volume of 500 mL.
    (c) Add 32 g of CH_3OH to 500 g of water.
    (d) Add 25 mL of CH_3OH to 500 mL of water.

4. Which solution has the greatest osmolarity?
    (a) 0.2 M $CaCl_2$ (b) 0.3 M Na_3PO_4 (c) 0.5 M NaCl   (d) 0.8 M glucose

5. Which of the following is a colloid?
    (a) wine   (b) maple syrup   (c) milkshake   (d) salad oil

6. How many mL of a 12.0 M HCl solution are needed to make 1.0 L of 0.10 M HCl?
    (a) 120 mL   (b) 83 mL   (c) 50 mL   (d) 8.3 mL

7. Which of the following properties do not depend on the number of particles in solution?
    (a) boiling point elevation   (b) osmotic pressure
    (c) heat of solution   (d) freezing point depression

8. How many grams of NaOH are needed to make 300 mL of a 0.3 M solution?
    (a) 40 g   (b) 12 g   (c) 3.6 g   (d) 1.0 g

9. How many moles of glucose are present in 250 mL of a 0.25 M solution?
    (a) 0.0625 mol   (b) 0.1 mol   (c) 0.25 mol   (d) 1.0 mol

10. How many mL of a 0.20 M solution of NaF contain 2.1 g of NaF?
    (a) 500 mL   (b) 250 mL   (c) 100 mL   (d) 50 mL

Sentence Completion

1. Two liquids soluble in each other are said to be _____.

2. An ___ ____ is the amount of an ion in grams that contains Avogadro's number of charges.

3. A weight/volume solution can be made up in a piece of glassware called a _____ _____.

4. According to Henry's law, the _____ of a gas varies with its _____ _____.

5. Two solutions that have the same osmolarity are _____.

6. Compounds that attract water from the atmosphere are called _____.

7. The formula used for calculating dilutions is _____.

8. Milk is an example of a _____.

9. _____ and _____ _____ _____ can pass through a dialysis membrane.

10. A solution that has reached its solubility limit is said to be _____.

11. An example of a nonelectrolyte is _____.

12. A solution that is hypotonic with respect to blood has a _____ osmolarity than blood plasma.

True or False

1. A solute is the liquid used to dissolve a substance.

2. It is possible to have a solution of a solid in a solid.

3. In making a volume/volume percent solution, one liquid is added to 100 mL of the other liquid.

4. All ionic compounds are soluble in water.

5. A solution of 0.10 M Na_3PO_4 has a greater osmolarity than a solution of 0.15 M NaCl.

6. Weight/weight percent is a useful way to express concentration.

7. In carrying out a dilution, the number of moles of solute remains constant.

8. The solubility of most substances increases with temperature.

9. A colloid differs from a solution in its ability to transmit light.

10. A blood cell undergoes crenation when placed in distilled water.

11. Particles dissolved in water lower the boiling point of water.

12. The amount of gas dissolved in a liquid increases with increasing pressure.

Match each entry on the left with its partner on the right.

1. Osmotic membrane

2. V_1/V_2

3. Solution

4. Equivalent

5. Suspension

6. Dialysis membrane

7. Osmolarity

8. Hypotonic

9. Colloid

10. Crenation

11. Hydrate

12. Electrolyte

(a) Muddy water

(b) Molarity x number of particles

(c) Conducts electricity in water

(d) Vinegar

(e) Crystalline compound that holds water

(f) Dilution factor

(g) Butter

(h) Permeable only to water

(i) A solution of lower osmolarity than another

(j) Formula weight/number of charges

(k) Permeable to water and small molecules

(l) Happens to cell in hypertonic solution

Chapter 10   Acids and Bases

Chapter Outline

I. Introduction to acids and bases (Section 10.1).
  A. Definition of acids and bases.
    1. Arrhenius definition.
      a. Acids donate H^+ ions in solution.
      b. Bases donate OH^- ions in solution.
      c. Acid + base —> salt + H_2O.
    2. Brønsted–Lowry definition.
      a. A Brønsted–Lowry acid is a proton donor (an H_3O^+ donor).
        i. Some acids can donate more than one proton.
      b. A Brønsted–Lowry base is a proton acceptor.
        i. The base may be negatively charged or neutral.
      c. An acid–base reaction is one in which a proton is transferred.
      d. Products of acid–base reactions are also acids and bases.
        In the reaction $HA + B: —> A^- + BH^+$:
        i. HA and A^- are an acid–conjugate base pair.
        ii. B: and BH^+ are a base–conjugate acid pair.
II. Acid/base strength (Sections 10.2–10.8).
  A. Strong/weak acids and bases (Section 10.2).
    1. Strong acids and bases are 100% dissociated in water.
    2. Weak acids and bases are less than 100% dissociated in solution.
    3. The stronger the acid, the weaker the conjugate base.
      The weaker the acid, the stronger the conjugate base.
    4. Knowing acid strengths allows predicting the direction of an acid–base reaction.
      a. An acid–base proton transfer always favors formation of the weaker acid.
  B. Acid dissociation constants (Section 10.3).
    1. K_a is a measure of the degree to which an acid HA dissociates to H_3O^+ and A^-.
    2. K_a values for weak acids are much less than 1.
    3. Donation of each successive H^+ ion from a polyprotic acid becomes successively more difficult.
    4. Most organic acids have K_a near 10^{-5}.
  C. Water as an acid and a base (Section 10.4).
    1. Water can act both as an acid and a base.
    2. Substances that can act as both acids and bases are amphoteric.
    3. Dissociation of H_2O.
      a. $K_w = [H_3O^+][OH^-] = 1.00 \times 10^{-14}$ at 25 °C.
      b. This relationship is true for any aqueous solution.
      c. Thus, we can calculate $[H_3O^+]$ or $[OH^-]$ for any aqueous solution.
  D. Measuring acidity (Sections 10.5–10.6).
    1. pH.
      a. A pH of < 7 indicates acidity; a pH of > 7 indicates basicity.
      b. pH is the negative logarithm of $[H_3O^+]$.
      c. The pH scale is logarithmic.
      d. pH can be computed with a calculator.
    2. In the laboratory, pH can be measured with indicators or a pH meter.

E. Equivalents of acids and bases (Section 10.7).
   1. Normality.
      a. A gram-equivalent of acid or base $= \dfrac{\text{Molar mass}}{\text{Number of H}^+ \text{ or OH}^- \text{ produced}}$.
      b. One equivalent of acid neutralizes one equivalent of base.
      c. Normality $= \dfrac{\text{Equivalents of acid or base}}{\text{Liters of solution}}$.
      d. Normality = (Molarity of acid or base) × (Number of H$^+$ or OH$^-$ produced).
F Common acid–base reactions (Section 10.8).
   1. Acid + Hydroxide ion —> Water + Salt.
   2. Acid + Carbonate or bicarbonate —> Water + Salt + CO_2.
   3. Acid + Nitrogen–containing compounds —> Ammonium salt.
III. Buffers (Sections 10.9–10.10).
   A. Acidity and basicity of salt solutions (Section 10.9).
      1. The salt of a strong base and a strong acid is neutral in solution.
      2. The salt of a strong base and a weak acid is basic in solution.
      3. The salt of a weak base and a strong acid is acidic in solution.
      4. The pH of the salt of a weak base and a weak acid can be predicted only if K_a values are known.
   B. Buffer solutions (Section 10.10).
      1. A buffer is the solution of a weak acid and its salt (or a weak base and its salt) at similar concentration.
      2. When a small amount of acid or base is added to a buffered solution, pH changes very little.
      3. The effective pH range of a buffer solution is determined by the K_a of the acid or base.
         a. A buffer solution works best when [HA] is close in value to [A$^-$].
         b. A buffer solution works best when [HA] and [A$^-$] are approximately 10 times greater than the amount of acid or base added.
         c. The molar amounts of HA and A$^-$ should be 10 times the molar amounts of acid or base you expect to add.
      4. The Henderson–Hasselbalch equation (pH = pK_a + log ([A$^-$]/[HA]) is useful in buffer calculations. (pK_a = –log K_a)
      5. The carbonate/bicarbonate buffer system is the major regulator of the pH of body fluids.
         a. The phosphate system and proteins are two other buffer systems in the body.
IV. Titration (Section 10.11).
   A. Titration is used to determine the acid or base concentration of a solution.
   B. In titration, a known volume of a solution of unknown acid or base concentration completely reacts with a solution of known concentration.
   C. The volume of solution of known concentration is measured, and the concentration of the unknown is calculated.

Solutions to Chapter 10 Problems

10.1 HCO_2H (a) and H_2S (b) are Brønsted–Lowry acids because they have protons to donate.

10.2 SO_3^{2-} (a) and F^- (c) are Brønsted–Lowry bases because they can be proton acceptors.

10.3   (a)   The conjugate acid of HS^- is H_2S.
          (b)   The conjugate acid of PO_4^{3-} is HPO_4^{2-}.
          (c)   The conjugate base of H_2CO_3 is HCO_3^-.
          (d)   The conjugate base of NH_4^+ is NH_3.

10.4

$$HS^- \;+\; HF \longrightarrow H_2S \;+\; F^-$$
base   acid   acid   base

conjugate acid–base pair

CIA 10.1

(a) A concentration of stomach acid of 0.01 (pH = 2) requires more antacid for neutralization than a concentration of 0.001 (pH = 3).

(b) $NaHCO_3(aq) + HCl(aq) \longrightarrow CO_2(g) + H_2O(l) + NaCl(aq)$

(c) Molarity of acid at pH 1.8 = 0.016 M

$$\frac{0.016 \text{ mol HCl}}{1 \text{ L}} \times 0.0150 \text{ L} = 0.000\,24 \text{ mol HCl}$$

$$0.000\,24 \text{ mol HCl} \times \frac{1 \text{ mol antacid}}{1 \text{ mol HCl}} \times \frac{84.0 \text{ g}}{1 \text{ mol antacid}} = 0.020 \text{ g} = 20 \text{ mg}$$

CIA 10.2

Stomach acid is needed for the digestion of proteins in the diet, for the absorption of micronutrients, and for providing a sterile environment by killing bacteria and yeasts.

CIA 10.3

$$HCl(aq) \;+\; CO_3^{2-}(aq) \rightleftharpoons Cl^-(aq) \;+\; HCO_3^-(aq)$$
acid   base   base   acid

conjugate acid–base pair

10.5   In Table 10.2, the stronger acid is listed *higher* in the table than the weaker acid.

Stronger acids: (a) NH_4^+ (b) H_2SO_4 (c) H_2CO_3

10.6   In Table 10.2, the stronger base is listed *lower* in the table than the weaker base.

Stronger bases: (a) F⁻ (b) OH⁻

10.7

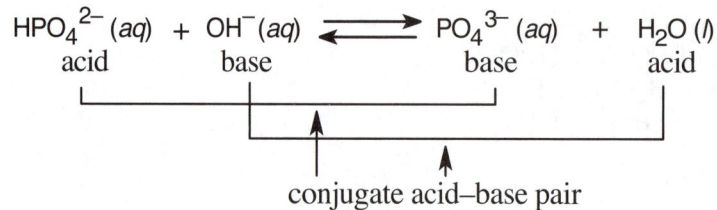

From Table 10.2, we see that OH⁻ is a stronger base than PO_4^{3-} and that HPO_4^{2-} is a stronger acid than H_2O. Thus, the forward direction of the equilibrium is favored.

$HPO_4^{2-}(aq) + OH^-(aq) \rightleftharpoons PO_4^{3-}(aq) + H_2O(l)$
stronger acid   stronger base   weaker base   weaker acid

10.8

$HF(aq) + NH_3(g) \rightleftharpoons F^-(aq) + NH_4^+(aq)$
acid   base   base   acid

conjugate acid–base pair

Since HF is a stronger acid than ammonium ion, the forward reaction is favored

$HF(aq) + NH_3(g) \rightleftharpoons F^-(aq) + NH_4^+(aq)$
stronger acid   stronger base   weaker base   weaker acid

10.9   Remember that an electrostatic potential map shows the most electron-rich area of the molecule as red and the least electron-rich area as blue. In alanine, the $-NH_3^+$ group is the most electron-poor, and thus its hydrogens are the most acidic.

10.10 The value for K_a of benzoic acid is smaller than the value for citrate, indicating that benzoic acid is a weaker acid. Benzoate is thus a stronger base than citrate.

10.11 (a) Ammonia is basic.

$$\left[OH^-\right] = \frac{K_w}{\left[H_3O^+\right]} = \frac{1.0 \times 10^{-14}}{3.1 \times 10^{-12}} = 3.2 \times 10^{-3} \text{ M}$$

(b) Vinegar is acidic. Since $1 \times 10^{-14} = [H_3O^+][OH^-]$,

$$\left[OH^-\right] = \frac{K_w}{\left[H_3O^+\right]} = \frac{1.0 \times 10^{-14}}{4.0 \times 10^{-3}} = 2.5 \times 10^{-12} \text{ M}$$

10.12 (a) pH = $-\log [H^+]$ = $-\log [3.1 \times 10^{-12}]$ = 11.51
(b) pH = $-\log [H^+]$ = $-\log [4.0 \times 10^{-3}]$ = 2.40

10.13 (a) For pH = 13, $[H_3O^+]$ = 1×10^{-13} M; $[OH^-]$ = 0.1 M
(b) For pH = 3, $[H_3O^+]$ = 1×10^{-3} M; $[OH^-]$ = 1×10^{-11} M
(c) For pH = 8, $[H_3O^+]$ = 1×10^{-8} M; $[OH^-]$ = 1×10^{-6} M

The solution of pH = 3 is most acidic, and the solution of pH = 13 is most basic.

10.14 A solution of 0.01 M HNO_2 has a higher pH because HNO_2 is a weaker acid, and its $[H^+]$ is lower.

10.15

Solution	pH	Acidic/Basic	$[H_3O^+]$	$[OH^-]$
(a) Saliva	6.5	acidic	3×10^{-7} M	3×10^{-8} M
(b) Pancreatic juice	7.9	basic	1×10^{-8} M	1×10^{-6} M
(c) Orange juice	3.7	acidic	2×10^{-4} M	5×10^{-11} M
(d) Wine	3.5	acidic	3×10^{-4} M	3×10^{-11} M

Least acidic Pancreatic juice Saliva Orange juice Wine → Most acidic

10.16 Use a calculator to determine pH. Remember that the significant figures in the answer are the digits to the right of the decimal point.

(a) $[H_3O^+]$ = 5.3×10^{-9} mol/L: pH = 8.28
(b) $[H_3O^+]$ = 8.9×10^{-6} mol/L: pH = 5.05

10.17 $[H_3O^+]$ = 2.5×10^{-3}, since HCl is a strong acid. As in the previous problem, use the EE key and the log key on your calculator, and take the negative value of the log.
$[H_3O^+]$ = 2.5×10^{-3}; pH = $-\log (2.5 \times 10^{-3})$ = $-(-2.60)$ = 2.60

CIA 10.4   pH of rain = 5.6; $[H_3O^+]$ = 3×10^{-6} M

CIA 10.5 (a) $[H_3O^+]$ = 0.03 M

(b) $25\ L \times \dfrac{0.03\ mol}{1\ L} \times \dfrac{63\ g}{1\ mol}$ = 47 g, or approximately 50 g HNO_3

10.18, 10.19

Sample	Mass	Molar Mass	# of ions*	Gram Equivalent	# of g-Eq	Normality of 300.0 mL
(a) HNO_3	5.0 g	63.0 g	1	63.0 g	0.079	0.26 N
(b) $Ca(OH)_2$	12.5 g	74.1 g	2	37.0 g	0.338	1.13 N
(c) H_3PO_4	4.5 g	98.0 g	3	32.7 g	0.14	0.47 N

*The number of ions refers to the number of H^+ or OH^- ions produced for each mole of acid or base in solution.

10.20 $3 \text{ HCl}(aq) + \text{Al(OH)}_3(aq) \longrightarrow 3 \text{ H}_2\text{O}(l) + \text{AlCl}_3(aq)$

$2 \text{ HCl}(aq) + \text{Mg(OH)}_2(aq) \longrightarrow 2 \text{ H}_2\text{O}(l) + \text{MgCl}_2(aq)$

10.21 (a) $2 \text{ HCO}_3^-(aq) + \text{H}_2\text{SO}_4(aq) \longrightarrow 2 \text{ H}_2\text{O}(l) + 2 \text{ CO}_2(g) + \text{SO}_4^{2-}(aq)$

(b) $\text{CO}_3^{2-}(aq) + 2 \text{ HNO}_3(aq) \longrightarrow \text{H}_2\text{O}(l) + \text{CO}_2(g) + 2 \text{ NO}_3^-(aq)$

10.22 $2 \text{ NH}_3(aq) + \text{H}_2\text{SO}_4(aq) \longrightarrow (\text{NH}_4)_2\text{SO}_4(aq)$

10.23

HOC 10.1
Vinegar and baking soda react to form bubbles of CO_2; the CO_2 displaces O_2 (necessary for combustion) and extinguishes the flame.

10.24 The salt of a weak acid and a strong base produces a basic solution, the salt of a strong acid and a weak base produces an acidic solution, and the salt of a strong acid and a strong base is neutral.

Salt	Cation is from:	Anion is from:	Solution is:
(a) K_2SO_4	strong base (KOH)	strong acid(H_2SO_4)	neutral
(b) Na_2HPO_4	strong base (NaOH)	weak acid (H_2PO_4^-)	basic
(c) MgF_2	strong base (Mg(OH)_2)	weak acid (HF)	basic
(d) NH_4Br	weak base (NH_3)	strong acid (HBr)	acidic

10.25 Refer to Worked Example 10.16 in the text.
When 0.020 mol of HNO_3 is added, the HF concentration increases from 0.100 M to 0.120 M, and the F^- decreases from 0.120 M to 0.100 M because of the reaction:

$$\text{F}^-(aq) + \text{HNO}_3(aq) \longrightarrow \text{HF}(aq) + \text{NO}_3^-(aq)$$

$$\text{pH} = \text{p}K_a + \log\frac{[\text{A}^-]}{[\text{HA}]} = 3.46 + \log\frac{[0.100]}{[0.120]} = 3.46 - 0.08 = 3.38$$

10.26

$$\text{NH}_4^+(aq) + \text{H}_2\text{O}(aq) \rightleftharpoons \text{H}_3\text{O}^+(aq) + \text{NH}_3(aq)$$

$$\text{pH} = \text{p}K_a + \log\frac{[\text{NH}_3]}{[\text{NH}_4^+]} = 9.25 + \log\frac{[0.080]}{[0.050]} = 9.25 + 0.20 = 9.45$$

10.27

$$pH = pK_a + \log \frac{[HCO_3^-]}{[H_2CO_3]}; \; 7.40 = 6.37 + \log \frac{[HCO_3^-]}{[H_2CO_3]}$$

$$\log \frac{[HCO_3^-]}{[H_2CO_3]} = 1.03; \; \frac{[HCO_3^-]}{[H_2CO_3]} = 10.7$$

The ratio of bicarbonate ion to carbonic acid is 10.7, indicating that more than 90% of dissolved CO_2 exists as bicarbonate at pH = 7.4

10.28 You can see that there are nine HCN molecules and six CN^- molecules in the figure.

$$pH = pK_a + \log \frac{[CN^-]}{[HCN]} = 9.31 + \log \frac{[6]}{[9]} = 9.31 - 0.18 = 9.13$$

10.29 First, write the balanced equation for the neutralization reaction:

$$HCl(aq) + NaOH(aq) \longrightarrow H_2O(l) + NaCl(aq)$$

We see from this equation that one mole of base is needed to neutralize each mole of acid.

$$\text{Moles HCl} = 58.4 \text{ mL} \times \frac{1 \text{ L}}{1000 \text{ mL}} \times \frac{0.250 \text{ mol NaOH}}{1 \text{ L}} \times \frac{1 \text{ mol HCl}}{1 \text{ mol NaOH}}$$

$$= 0.0146 \text{ mol HCl}$$

$$\frac{0.0146 \text{ mol HCl}}{20.0 \text{ mL}} \times \frac{1000 \text{ mL}}{1 \text{ L}} = 0.730 \text{ M HCl}$$

10.30 $2 NaOH(aq) + H_2SO_4(aq) \longrightarrow 2 H_2O(l) + Na_2SO_4(aq)$

Notice that two moles of base are needed to neutralize one mole of acid.

$$\text{Moles NaOH} = 50.0 \text{ mL} \times \frac{1 \text{ L}}{1000 \text{ mL}} \times \frac{0.200 \text{ mol } H_2SO_4}{1 \text{ L}} \times \frac{2 \text{ mol NaOH}}{1 \text{ mol } H_2SO_4}$$

$$= 0.0200 \text{ mol NaOH}$$

$$0.0200 \text{ mol NaOH} \times \frac{1 \text{ L}}{0.150 \text{ mol}} = 0.133 \text{ L NaOH} = 133 \text{ mL NaOH}$$

10.31 $2 \text{ KOH}(aq) + \text{H}_2\text{SO}_4(aq) \longrightarrow 2 \text{ H}_2\text{O}(l) + \text{Na}_2\text{SO}_4(aq)$

(a) Moles H_2SO_4 = 16.1 mL x $\dfrac{1 \text{ L}}{1000 \text{ mL}}$ x $\dfrac{0.150 \text{ mol H}_2\text{SO}_4}{1 \text{ L}}$ = 2.42 x 10^{-3} mol

2.42 x 10^{-3} mol H_2SO_4 x $\dfrac{2 \text{ Eq H}_2\text{SO}_4}{1 \text{ mol H}_2\text{SO}_4}$ = 4.84 x 10^{-3} Eq H_2SO_4

(b) $\dfrac{0.004\ 84 \text{ mol KOH}}{0.0215 \text{ L}}$ = 0.225 M KOH

10.32

Moles NaOH = 9.30 mL x $\dfrac{1 \text{ L}}{1000 \text{ mL}}$ x $\dfrac{0.0012 \text{ mol NaOH}}{1 \text{ L}}$ = 1.1 x 10^{-5} mol NaOH

$\dfrac{1.1 \times 10^{-5} \text{ mol NaOH}}{0.050 \text{ L}}$ = 2.2 x 10^{-4} M KOH = 2.2 x 10^{-4} M H_3O^+; pH = 3.66

CIA 10.6   Bicarbonate reacts with excess H_3O^+ to form H_2CO_3, which goes on to produce H_2O and CO_2 (which is exhaled). This reaction removes H_3O^+ from the blood stream and lowers the serum $[\text{H}^+]$, raising pH.

CIA 10.7   Gastric juice is the most acidic body fluid, and pancreatic juice is the most basic.

Understanding Key Concepts

10.33 (a) The reaction of HF with OH^- is represented by outcome (2). One OH^- reacts with each HF to produce three F^- anions, and nine OH^- anions are left over.
(b) The reaction of H_2SO_3 with OH^- is represented by outcome (3). Two OH^- ions react with each H_2SO_3 molecule to produce three SO_3^{2-} anions, and six OH^- anions are left over.
(c) The reaction of H_3PO_4 with OH^- is represented by outcome (1). Three OH^- anions react with each H_3PO_4 molecule to produce three PO_4^{3-} anions, and three OH^- anions are left over.

10.34 In both molecules, the most acidic hydrogen (deepest blue) is bonded to oxygen. Acetic acid is more acidic because its hydrogen is more positively polarized and less tightly held.

10.35 (a) The acid in the first picture (1) is the weakest because it is the least dissociated.
(b) The acid in the second picture (2) is the strongest acid because all molecules of the acid are dissociated
(c) The acid in the first picture (1) has the smallest value of K_a because it is the weakest acid.

10.36 (a) Picture (3) represents a weak diprotic acid, which has dissociated slightly to form the HA^- anion.
(b) Picture (1) represents an impossible situation. H_2A dissociates stepwise to form HA^-. Only when virtually all H_2A has dissociated to form HA^- is A^{2-} formed by dissociation of HA^-.

10.37 The titration reaction uses up 2/3 of the 1.0 M solution in the buret. Since the solution in the buret is 1.0 M, the solution in the flask must be 0.67 M.

Additional Problems

Acids and Bases

10.38 In water, HBr dissociates almost completely. Water acts as a base to accept a proton, and a solution of H_3O^+ and Br^- results.

10.39 In water, CH_3CO_2H dissociates only to the extent of about 1%. Water acts as a base, yielding a solution of CH_3CO_2H, along with smaller amounts of H_3O^+ and $CH_3CO_2^-$.

10.40 In water, KOH dissociates completely to yield K^+ and OH^- ions.

10.41 In water, NH_3 remains largely unreacted, but a small amount acts as a base and accepts a proton from water to yield NH_4^+ and OH^- ions.

10.42 A monoprotic acid, such as HCl, has only one proton to donate, whereas a diprotic acid, such as H_2SO_4, has two protons to donate.

10.43 H^+ represents a proton, but the species H^+ is too reactive to exist in solution. Instead, the proton reacts with water to form H_3O^+.

10.44 Strong acids: (a) $HClO_4$; (e) HI

10.45 Weak bases: (a) NH_3; (c) HPO_4^{2-}; (e) CN^-

10.46 Brønsted–Lowry acids: (a) HCN; (d) H_2CO_3; (f) $CH_3NH_3^+$
Brønsted–Lowry base: (b) $CH_3CO_2^-$
Neither: (c) $AlCl_3$; (e) Mg^{2+}

10.47 Base and conjugate acid are pairs, as are acid and conjugate base.

(a) $\underset{\text{base}}{CO_3^{2-}(aq)} + \underset{\text{acid}}{HCl(aq)} \longrightarrow \underset{\text{conjugate acid}}{HCO_3^-(aq)} + \underset{\text{conjugate base}}{Cl^-(aq)}$

(b) $\underset{\text{acid}}{H_3PO_4(aq)} + \underset{\text{base}}{NH_3(aq)} \longrightarrow \underset{\text{conjugate base}}{H_2PO_4^-(aq)} + \underset{\text{conjugate acid}}{NH_4^+(aq)}$

(c) $\underset{\text{acid}}{NH_4^+(aq)} + \underset{\text{base}}{CN^-(aq)} \rightleftharpoons \underset{\text{conjugate base}}{NH_3(aq)} + \underset{\text{conjugate acid}}{HCN(aq)}$

(d) $\underset{\text{acid}}{HBr(aq)} + \underset{\text{base}}{OH^-(aq)} \longrightarrow \underset{\text{conjugate acid}}{H_2O(l)} + \underset{\text{conjugate base}}{Br^-(aq)}$

(e) $\underset{\text{acid}}{H_2PO_4^-(aq)} + \underset{\text{base}}{N_2H_4(aq)} \rightleftharpoons \underset{\text{conjugate base}}{HPO_4^{2-}(aq)} + \underset{\text{conjugate acid}}{N_2H_5^+(aq)}$

10.48

	(a)	(b)	(c)	(d)
Base	$ClCH_2CO_2^-$	C_5H_5N	SeO_4^{2-}	$(CH_3)_3N$
Conjugate acid	$ClCH_2CO_2H$	$C_5H_5NH^+$	$HSeO_4^-$	$(CH_3)_3NH^+$

10.49

	(a)	(b)	(c)	(d)
Acid	HCN	$(CH_3)_2NH_2^+$	H_3PO_4	$HSeO_3^-$
Conjugate base	CN^-	$(CH_3)_2NH$	$H_2PO_4^-$	SeO_3^{2-}

Acid and Base Strength: K_a and pH

10.50 For the reaction: $HA(aq) + H_2O(l) \rightleftharpoons H_3O^+(l) + A^-(aq)$

$$K = \frac{[H_3O^+][A^-]}{[HA][H_2O]}$$

Since $[H_2O]$ has no effect on the equilibrium, we define a new constant:

$$K_a = K[H_2O] = \frac{[H_3O^+][A^-]}{[HA]}$$

10.51 K_w is the product of the molar concentrations of H_3O^+ and OH^- in any aqueous solution and is numerically equal to 1.0×10^{-14} at 25 °C.

$$K_a = \frac{[H_3O^+][OH^-]}{[H_2O]}; \quad K_a[H_2O] = K_w = [H_3O^+][OH^-]$$

10.52 The quantity pH is defined as the negative logarithm of the molar H_3O^+ concentration. For example, if $[H_3O^+] = 1.0 \times 10^{-3}$ M, then pH = 3.

10.53 A solution of 0.10 M HCl is 100% dissociated; $[H^+] = [HCl] = 0.10$, and pH = 1. A solution of 0.10 M CH_3COOH is only partially dissociated, and its pH is expected to be higher than 1.0.

10.54 $CH_3COOH(aq) + H_2O(l) \rightleftharpoons CH_3COO^-(g) + H_3O^+(aq)$

$[H_3O^+] = $ antilog $(-2.88) = 1.3 \times 10^{-3}$. 1.3% of acetic acid is dissociated.

10.55 $H_3PO_4(aq) + H_2O(l) \rightleftharpoons H_2PO_4^-(aq) + H_3O^+(aq)$

$$K_a = \frac{[H_2PO_4^-][H_3O^+]}{[H_3PO_4]}$$

$H_2PO_4^-(aq) + H_2O(l) \rightleftharpoons HPO_4^{2-}(aq) + H_3O^+(aq)$

$$K_a = \frac{[HPO_4^{2-}][H_3O^+]}{[H_2PO_4^-]}$$

$HPO_4^{2-}(aq) + H_2O(l) \rightleftharpoons PO_4^{3-}(aq) + H_3O^+(aq)$

$$K_a = \frac{[PO_4^{3-}][H_3O^+]}{[HPO_4^{2-}]}$$

10.56 The most acidic solution (see Table 10.2) has the lowest pH.
Lowest pH   *Highest pH*
$HSO_4^- < HF < HCOOH < H_2CO_3 < NH_4^+$

10.57 Urine (pH = 7.9) is weakly basic, since solutions with pH greater than 7.0 are basic. The $[H_3O^+] = 1 \times 10^{-8}$

10.58 Because the pH of a 0.10 M HCN solution is lower than 7.0, the solution is acidic. $[H_3O^+]$ = antilog $(-5.2) = 6 \times 10^{-6}$. HCN is a weak acid.

10.59 The $[H_3O^+]$ of sweat ranges from 1×10^{-4} to 2×10^{-7} — a range of three orders of magnitude.

10.60 The $[H_3O^+]$ of saliva ranges from 2×10^{-6} to 8×10^{-8}.

10.61 The pH of an 0.02 M solution of a strong monoprotic acid is 1.7. The pH of a 0.02 M solution of a strong base is $14 - 1.7 = 12.3$

10.62

	10.57	**10.58**	**10.59**	**10.60**	**10.61**
pOH	6.1	8.8	7.2–10.0	6.9–8.2	12.3; 1.7

10.63, 10.64

	$[H_3O^+]$	pH	$[OH^-]$	pOH
(a) Egg white	2.5×10^{-8} M	7.60 (iv)	4×10^{-7}	6.40
(b) Apple cider	5.0×10^{-4} M	3.30 (i)	2.0×10^{-11}	10.70
(c) Ammonia	2.3×10^{-12} M	11.64 (iii)	4.3×10^{-3}	2.36
(d) Vinegar	4.0×10^{-3} M	2.40 (ii)	2.5×10^{-12}	11.60

Most acidic: vinegar, apple cider, egg white, ammonia   *Least acidic*

10.65 (a) pH = 4; $[H_3O^+] = 1 \times 10^{-4}$ M $[OH^-] = 1 \times 10^{-10}$ M
(b) pH = 11; $[H_3O^+] = 1 \times 10^{-11}$ M $[OH^-] = 1 \times 10^{-3}$ M
(c) pH = 0; $[H_3O^+] = 1$ M $[OH^-] = 1 \times 10^{-14}$ M
(d) pH = 1.38; $[H_3O^+] = 4.2 \times 10^{-2}$ M $[OH^-] = 2.4 \times 10^{-13}$ M
(e) pH = 7.96; $[H_3O^+] = 1.1 \times 10^{-8}$ M $[OH^-] = 9.1 \times 10^{-7}$ M

10.66 0.12×0.10 M = 0.012 M H_3O^+; $[H_3O^+] = 1.2 \times 10^{-2}$

$$[OH^-] = \frac{K_w}{[H_3O^+]} = \frac{1.0 \times 10^{-14}}{1.2 \times 10^{-2}} = 8.3 \times 10^{-13} \text{ M}$$

pH = 1.92

Reactions of Acids and Bases

10.67 $HCO_3^-(aq) + HCl(aq) \longrightarrow H_2O(l) + CO_2(g) + Cl^-(aq)$

$HCO_3^-(aq) + NaOH(aq) \longrightarrow H_2O(l) + Na^+(aq) + CO_3^{2-}(aq)$

$H_2PO_4^-(aq) + HCl(aq) \longrightarrow H_3PO_4(aq) + Cl^-(aq)$

$H_2PO_4^-(aq) + NaOH(aq) \longrightarrow HPO_4^{2-}(aq) + H_2O(l) + Na^+(aq)$

10.68 The stronger acid and weaker base are conjugate pairs, as are the stronger base and weaker acid. The reaction to form weaker base and weaker acid is favored.

(a) $HCl(aq) + PO_4^{3-}(aq) \rightleftharpoons HPO_4^{2-}(aq) + Cl^-(aq)$
stronger acid   stronger base   weaker acid   weaker base

(b) $CN^-(aq) + HSO_4^-(aq) \rightleftharpoons HCN(aq) + SO_4^{2-}(aq)$
stronger base   stronger acid   weaker acid   weaker base

(c) $HClO_4(aq) + NO_2^-(aq) \rightleftharpoons HNO_2(aq) + ClO_4^-(aq)$
stronger acid   stronger base   weaker acid   weaker base

(d) $HF(aq) + CH_3O^-(aq) \rightleftharpoons CH_3OH(aq) + F^-(aq)$
stronger acid   stronger base   weaker acid   weaker base

10.69 (a) $HCl(aq) + NaHCO_3(s) \longrightarrow H_2O(l) + CO_2(g) + NaCl(aq)$

(b) $H_2SO_4(aq) + 2\,NaHCO_3(s) \longrightarrow 2\,H_2O(l) + 2\,CO_2(g) + Na_2SO_4(aq)$

10.70 (a) $LiOH(aq) + HNO_3(aq) \longrightarrow H_2O(l) + LiNO_3(aq)$

(b) $BaCO_3(aq) + 2\,HI(aq) \longrightarrow H_2O(l) + CO_2(g) + BaI_2(aq)$

(c) $H_3PO_4(aq) + 3\,KOH(aq) \longrightarrow 3\,H_2O(l) + K_3PO_4(aq)$

(d) $Ca(HCO_3)_2(aq) + 2\,HCl(aq) \longrightarrow 2\,H_2O(l) + 2\,CO_2(g) + CaCl_2(aq)$

(e) $Ba(OH)_2(aq) + H_2SO_4(aq) \longrightarrow 2\,H_2O(l) + BaSO_4(s)$

(f) $NH_3(aq) + HCl(aq) \longrightarrow NH_4Cl(aq)$

10.71

$$[H_3O^+] = \frac{K_a\,[HA]}{[A^-]}$$

Salts and Buffers

10.72 Refer to Table 10.4 if necessary.

Salt	Cation is from:	Anion is from:	Solution is:
(a) NH_4Cl	weak base (NH_3)	strong acid (HCl)	acidic
(b) KBr	strong acid (HBr)	strong base (KOH)	neutral
(c) Na_2CO_3	strong base (NaOH)	weak acid (H_2CO_3)	basic
(d) $NaCH_3CO_2$	strong base (NaOH)	weak acid (CH_3CO_2H)	basic

10.73

Buffer	Acid or Base	Salt
(a) NH_3/NH_4Cl	NH_3	NH_4Cl
(c) HCO_3^-/CO_3^{2-}	$NaHCO_3$	Na_2CO_3
(d) $CH_3CO_2H/CH_3CO_2^-$	CH_3CO_2H	$NaCH_3CO_2$

The compounds in (b) don't form a buffer solution.

10.74 A buffer is composed of a weak acid and its conjugate base. Any added H_3O^+ can react with the conjugate base and be neutralized, and any added OH^- can react with the acid. In either case, the ratio of acid to conjugate base changes only slightly, and the pH of a buffered solution remains nearly constant.

10.75 $CH_3CO_2H + CH_3CO_2^-$ Na^+ is a better buffer than $HNO_3 + NaNO_3$ because acetic acid is a weak acid and acetate ion is its conjugate base. Nitric acid is a strong acid, and nitrate ion is nonbasic. Thus, the nitric acid/sodium nitrate mixture is not a buffer and can't control pH.

10.76

(a) $pH = pK_a + \log \dfrac{\left[CH_3CO_2^-\right]}{\left[CH_3CO_2H\right]} = 4.74 + \log \dfrac{[0.100]}{[0.100]} = 4.74$

(b) $CH_3CO_2^-$ $Na^+(aq) + HNO_3(aq) \longrightarrow CH_3CO_2H(aq) + NaNO_3(aq)$.
The added acid is neutralized by sodium acetate.

$CH_3CO_2H(aq) + OH^-(aq) \longrightarrow CH_3CO_2^-(aq) + H_2O(l)$
The added base is neutralized by acetic acid.

10.77 At a specific pH, the most effective buffer is composed of a solution of an acid whose K_a is close to the desired $[H_3O^+]$ and the salt of that acid. Buffer system (b), whose acid (NH_4^+) has a $K_a = 5.6 \times 10^{-10}$, is the best buffer to use.

10.78

$pH = pK_a + \log \dfrac{\left[CN^-\right]}{\left[HCN\right]} = 9.31 + \log \dfrac{[0.150]}{[0.200]} = 9.31 - 0.12 = 9.19$

10.79 (a) Use the formula in the previous problem. Adding 0.020 mol HCl to 1.00 L of the solution in Problem 10.78 changes the value of [HCN] to 0.220 M and the value of [CN$^-$] to 0.130 M.

$pH = pK_a + \log \dfrac{\left[CN^-\right]}{\left[HCN\right]} = 9.31 + \log \dfrac{[0.130]}{[0.220]} = 9.31 - 0.23 = 9.08$

(b) If 0.020 mol NaOH is added to 1.00 L of the HCN/NaCN buffer system, the concentrations change to [HCN] = 0.180 M and [CN$^-$] = 0.170 M.

$pH = pK_a + \log \dfrac{\left[CN^-\right]}{\left[HCN\right]} = 9.31 + \log \dfrac{[0.170]}{[0.180]} = 9.31 - 0.02 = 9.29$

10.80

$pH = pK_a + \log \dfrac{[NH_3]}{[NH_4^+]} = 9.25 + \log \dfrac{[0.10]}{[0.15]} = 9.25 - 0.18 = 9.07$

10.81 A solution with equal amounts of NH_4^+ and NH_3 has a pH of 9.25. If 0.025 moles of NaOH are added to the solution in the previous problem, $[NH_4^+]$ decreases to 0.125 M, $[NH_3]$ increases to 0.125, the base/acid ratio is 1.00, and the solution has pH = 9.25.

Concentrations of Acid and Base Solutions

10.82 An equivalent of an acid or base is its formula weight in grams divided by the number of H_3O^+ or OH^- ions it produces.

10.83 The normality of an acid is equal to its molarity when the acid is monoprotic. For a diprotic or triprotic acid, normality is equivalent to molarity times the number of acidic protons.

10.84, 10.85

Grams needed for 500 mL of 0.15 N solution are found by multiplying the gram equivalent by 0.075 (0.500 L x 0.15 N).

Sample	Molar Mass	# of H^+ or OH^-	Gram Equivalent	Grams Needed for 500 mL of 0.15 N soln.
(a) HNO_3	63.0 g	1	63.0 g	4.73 g
(b) H_3PO_4	98.0 g	3	32.7 g	2.45 g
(c) KOH	56.1 g	1	56.1 g	4.21 g
(d) $Mg(OH)_2$	58.3 g	2	29.2 g	2.19 g

10.86 Since a 0.0050 N solution of any acid has 0.0050 equivalents per liter, 25 mL of a 0.0050 N KOH solution is needed to neutralize 25 mL of a 0.0050 N H_2SO_4. 50 mL of a 0.005 M NaOH solution is needed to neutralize 25 mL of a 0.005 M H_2SO_4 since each H_2SO_4 molecule yields two hydrogens.

10.87

(a) $\dfrac{0.12 \text{ mol } H_2SO_2}{1 \text{ L}} \times \dfrac{2 \text{ Eq}}{1 \text{ mol}} \times 0.0750 \text{ L} = 0.018 \text{ Eq}$

(b) $\dfrac{0.12 \text{ mol } H_3PO_2}{1 \text{ L}} \times \dfrac{3 \text{ Eq}}{1 \text{ mol}} \times 0.0750 \text{ L} = 0.027 \text{ Eq}$

10.88 (a) $0.25 \text{ mol } Mg(OH)_2 \times 2 \text{ Eq/mol} = 0.50 \text{ Eq } Mg(OH)_2$

(b) Molar mass of $Mg(OH)_2 = 58.3$ g;   1 Eq $Mg(OH)_2 = 29.2$ g

$2.5 \text{ g } Mg(OH)_2 \times \dfrac{1 \text{ Eq}}{29.2 \text{ g}} = 0.086 \text{ Eq } Mg(OH)_2$

(c) Molar mass of $CH_3CO_2H = 60.0$ g;   1 Eq $CH_3CO_2H = 60.0$ g

$15 \text{ g } CH_3CO_2H \times \dfrac{1 \text{ Eq}}{60 \text{ g}} = 0.25 \text{ Eq } CH_3CO_2H$

10.89

Molar mass of $C_6H_5O_7H_3 = 192.0$ g; 1 Eq $C_6H_5O_7H_3 = \dfrac{192.0 \text{ g}}{1 \text{ mol}} \times \dfrac{1 \text{ mol}}{3 \text{ Eq}} = 64.0$ g

$\dfrac{64.0 \text{ g}}{1 \text{ Eq}} \times \dfrac{1 \text{ Eq}}{1000 \text{ mEq}} \times 152 \text{ mEq} = 9.73$ g

10.90 Molar mass of $Ca(OH)_2$ = 74.1 g; 1 Eq $Ca(OH)_2$ = 37.1 g

$$\frac{5.0 \text{ g}}{0.500 \text{ L}} \times \frac{1 \text{ mol}}{74.1 \text{ g}} = 0.13 \text{ M}; \frac{0.13 \text{ mol}}{1 \text{ L}} \times \frac{2 \text{ Eq}}{1 \text{ mol}} = \frac{0.26 \text{ Eq}}{1 \text{ L}} = 0.26 \text{ N}$$

10.91 Molar mass of $C_6H_5O_7H_3$ = 192 g; 1 Eq $C_6H_5O_7H_3$ = 64.0 g

$$\frac{25 \text{ g}}{0.800 \text{ L}} \times \frac{1 \text{ mol}}{192 \text{ g}} = 0.16 \text{ M}; \frac{0.16 \text{ mol}}{1 \text{ L}} \times \frac{3 \text{ Eq}}{1 \text{ mol}} = \frac{0.48 \text{ Eq}}{1 \text{ L}} = 0.48 \text{ N}$$

10.92 $HCl(aq) + NaOH(aq) \longrightarrow H_2O(l) + NaCl(aq)$

One mole of HCl is needed to neutralize one mole of NaOH.

$$\text{Moles HCl} = 22.4 \text{ mL} \times \frac{0.12 \text{ mol NaOH}}{1 \text{ L}} \times \frac{1 \text{ L}}{1000 \text{ mL}} \times \frac{1 \text{ mol HCl}}{1 \text{ mol NaOH}}$$

$$= 0.002688 \text{ mol HCl}$$

$$\frac{0.002688 \text{ mol HCl}}{12 \text{ mL}} \times \frac{1000 \text{ mL}}{1 \text{ L}} = 0.22 \text{ M HCl}$$

10.93 $Ba(OH)_2(aq) + 2 HNO_3(aq) \longrightarrow 2 H_2O(l) + Ba(NO_3)_2(aq)$

$$\text{Equivalents Ba(OH)}_2 = 15.0 \text{ mL} \times \frac{0.12 \text{ mol Ba(OH)}_2}{1 \text{ L}} \times \frac{1 \text{ L}}{1000 \text{ mL}} \times \frac{2 \text{ Eq Ba(OH)}_2}{1 \text{ mol Ba(OH)}_2}$$

$$= 0.0036 \text{ Eq Ba(OH)}_2 = 0.0036 \text{ Eq HNO}_3$$

$$0.0036 \text{ Eq HNO}_3 \times \frac{1 \text{ mol HNO}_3}{1 \text{ Eq HNO}_3} \times \frac{1 \text{ L}}{0.085 \text{ mol}} \times \frac{1000 \text{ mL}}{1 \text{ L}} = 42 \text{ mL HNO}_3$$

10.94 $2 NH_3(aq) + H_2SO_4(aq) \longrightarrow 2 H_2O(l) + (NH_4)_2SO_4(aq)$

$$0.0150 \text{ L H}_2SO_4 \times \frac{0.0250 \text{ mol H}_2SO_4}{1 \text{ L H}_2SO_4} \times \frac{2 \text{ mol NH}_3}{1 \text{ mol H}_2SO_4} = 0.000750 \text{ mol KOH}$$

$$\frac{0.000750 \text{ mol NH}_3}{0.0100 \text{ L NH}_3} = 0.0750 \text{ M}$$

10.95 $V_1 \times N_1 = V_2 \times N_2$; V_1 = 35.0 mL; N_1 = 0.100 N; V_2 = 21.5 mL

$$N_2 = \frac{V_1 \times N_1}{V_2} = \frac{35.0 \text{ mL} \times 0.100 \text{ N}}{21.5 \text{ mL}} = 0.163 \text{ N}$$

10.96 The pH of the reactions in Problems 10.92 and 10.93 is near neutral at the equivalence point because the end products are water and a neutral salt.

10.97 The pH of the reaction in Problem 10.94 is acidic at the equivalence point because the product $(NH_4)_2SO_4$ is a salt formed from a strong acid and a weak base.

Conceptual Problems

10.98

(a) The ideal gas law (Chapter 8): $PV = nRT$; $n = PV/RT$

$P = 1.0$ atm; $V = 15$ L HCl; $R = 0.0821 \dfrac{L \cdot atm}{mol \cdot K}$; $T = 25\,°C = 298$ K

$$n = \frac{1 \text{ atm } \times \text{ 15.0 L HCl}}{0.0821 \dfrac{L \cdot atm}{mol \cdot K} \times 298 \text{ K}} = 0.613 \text{ mol HCl}$$

(b) $\dfrac{0.613 \text{ mol HCl}}{0.250 \text{ L}} = 2.45$ M; pH = –0.39

10.99

(a) pH at 0 °C = 7.47   pH at 50 °C = 6.63

(b) $K_w = [H_3O^+][OH^-] = [H_3O^+]^2$ since $[H_3O^+] = [OH^-]$

at 0 °C: $K_w = (3.38 \times 10^{-8})^2 = 1.14 \times 10^{-15}$;

at 50 °C: $K_w = (2.34 \times 10^{-7})^2 = 5.48 \times 10^{-14}$

(c) The reaction is endothermic since heat increases the ratio of products to reactants.

10.100   For simplicity, represent acetylsalicylic acid as HAcsa.

$$HAcsa(aq) + NaOH(aq) \longrightarrow NaAcsa(aq) + H_2O(l)$$

$$13.87 \text{ mL} \times \frac{1 \text{ L}}{1000 \text{ mL}} \times \frac{0.100 \text{ mol}}{1L} \times \frac{40.0 \text{ g}}{1 \text{ mol}} = 0.555 \times 10^{-3} \text{ mol NaOH}$$

Since 1 mol NaOH reacts with 1 mol acetylsalicylic acid:

$$0.555 \times 10^{-3} \text{ mol HAcsa} \times \frac{180.2 \text{ g}}{1 \text{ mol HAcsa}} = 0.1 \text{ g HAcsa}$$

The tablet contains 100 mg acetylsalicylic acid

10.101 $2\,NaOH(aq) + H_2SO_4(aq) \longrightarrow 2\,H_2O(l) + Na_2SO_4(aq)$

$$0.040 \text{ L } H_2SO_4 \times \frac{0.10 \text{ mol } H_2SO_4}{1 \text{ L } H_2SO_4} \times \frac{2 \text{ mol NaOH}}{1 \text{ mol } H_2SO_4} = 0.0080 \text{ mol NaOH}$$

$$0.0080 \text{ mol NaOH} \times \frac{1000 \text{ mL NaOH}}{0.50 \text{ mol NaOH}} = 16 \text{ mL NaOH}$$

10.102

Both 50 mL of a 0.20 N HCl solution and 50 mL of a 0.20 acetic acid solution contain the same number of moles of acid – 0.010 moles. Because HCl is a strong acid, it is almost completely dissociated in water, and the H_3O^+ concentration approaches 0.20 M. Acetic acid, however, is a weak acid that is only slightly dissociated, and the H_3O^+ concentration is much lower – around 0.002 M. Since the HCl solution has a higher H_3O^+ concentration, it has a lower pH.

10.103

(a) $pH = pK_a - \log \dfrac{\left[H_2PO_4^-\right]}{\left[HPO_4^{2-}\right]}$ or $pH = pK_a + \log \dfrac{\left[HPO_4^{2-}\right]}{\left[H_2PO_4^-\right]}$

(b) Using the second of the above expressions:

$$7.40 = 7.21 + \log \frac{\left[HPO_4^{2-}\right]}{\left[H_2PO_4^-\right]} ; 0.19 = \log \frac{\left[HPO_4^{2-}\right]}{\left[H_2PO_4^-\right]} ; \frac{\left[HPO_4^{2-}\right]}{\left[H_2PO_4^-\right]} = 1.55$$

An approximately 3/2 ratio of $HPO_4^{2-}/H_2PO_4^-$ maintains the optimum blood pH.

10.104 $2\ HCl(aq)\ +\ Ca(OH)_2(aq) \longrightarrow 2\ H_2O(l)\ +\ CaCl_2(aq)$

$$0.140\ \text{L HCl}\ \times\ \frac{0.150\ \text{mol HCl}}{1\ \text{L HCl}}\ \times\ \frac{1\ \text{mol Ca(OH)}_2}{2\ \text{mol HCl}} = 0.0105\ \text{mol Ca(OH)}_2$$

$$\frac{0.0105\ \text{mol Ca(OH)}_2}{30.0\ \text{mL Ca(OH)}_2}\ \times\ \frac{1000\text{mL}}{1\ \text{L}} = 0.35\ \text{M Ca(OH)}_2 = 0.70\ \text{N Ca(OH)}_2$$

10.105

Both (a), NaF and HF, and (c), NH_4Cl and NH_3 are effective buffer systems. The NaF/HF system is a solution of a weak acid and the salt of its anion. The NH_4Cl/NH_3 system is a solution of a weak base and the salt of its cation. Neither (b) nor (d) are buffer systems because $HClO_4$ and HBr are both strong acids.

The pH of each buffer solution, when the acid and its conjugate base (or the base and its conjugate acid) are of equal concentration, equals the pK_a of the acid. For (a), pH = 3.46; for (c), pH = 9.25.

10.106

(a) $\underset{\text{acid}}{NH_4^+(aq)}\ +\ \underset{\text{base}}{OH^-(aq)} \longrightarrow \underset{\text{conjugate base}}{NH_3(g)}\ +\ \underset{\text{conjugate acid}}{H_2O(l)}$

(b) $PV = nRT$; $P = 755$ mmHg; $V = 2.86$ L; $T = 333$ K; $R = \dfrac{62.4\ \text{mmHg} \cdot \text{L}}{\text{mol} \cdot \text{K}}$

$$n = \frac{PV}{RT} = \frac{755\ \text{mmHg}\ \times\ 2.86\ \text{L}}{\dfrac{62.4\ \text{mmHg} \cdot \text{L}}{\text{mol} \cdot \text{K}}\ \times\ 333\ \text{K}} = 0.104\ \text{mol NH}_3$$

$$0.104\ \text{mol}\ \times\ \frac{53.5\ \text{g}}{1\ \text{mol}} = 5.56\ \text{g NH}_4\text{Cl}$$

10.107

$$CaO(aq) + SO_2(g) \longrightarrow CaSO_3(s)$$

(a) 1 mol CaO (56 g) reacts with 1 mol SO_2.

(b) 1 kg SO_2 $\times \dfrac{1 \text{ mol } SO_2}{64 \text{ g } SO_2} \times \dfrac{1 \text{ mol } SO_2}{1 \text{ mol } CaO} \times \dfrac{56 \text{ g } CaO}{1 \text{ mol } CaO} = 0.9$ kg CaO

10.108

(a) $Na_2O(aq) + H_2O(l) \longrightarrow 2 \, NaOH(aq)$

(b) 1.55 g Na_2O $\times \dfrac{1 \text{ mol } Na_2O}{62.0 \text{ g } Na_2O} \times \dfrac{2 \text{ mol } NaOH}{1 \text{ mol } Na_2O} = 0.0500$ mol NaOH

$\dfrac{0.0500 \text{ mol } NaOH}{500.0 \text{ mL}} \times \dfrac{1000 \text{ mL}}{1 \text{ L}} = 0.100$ M NaOH; pH $= 13.00$

(c) 0.0500 mol NaOH $\times \dfrac{1 \text{ mol } HCl}{1 \text{ mol } NaOH} \times \dfrac{1 \text{ L}}{0.0100 \text{ mol } HCl} = 5.00$ L HCl

5.00 L HCl $= 5.00 \times 10^3$ mL HCl

Group Problems

10.109

(a) Alka-Seltzer^{TM} contains $NaHCO_3$, aspirin, and citric acid.

(b) Citric acid reacts with sodium bicarbonate to release CO_2 bubbles. Sodium bicarbonate is the antacid.

(c) $C_6H_5O_7H_3(aq) + 3 \, NaHCO_3(aq) \longrightarrow C_6H_5O_7Na_3(aq) + 3 \, H_2O(l) + 3 \, CO_2(g)$

10.110

(a) The active ingredient of smelling salts is ammonium carbonate.

(b) When a packet of smelling salts is activated, solid ammonium carbonate dissolves in water to produce ammonia and ammonium bicarbonate. The ammonia is an irritant that revives an unconscious person.

$(NH_4)_2CO_3(s) \longrightarrow NH_3(g) + NH_4HCO_3{}^-(aq)$

10.111

(a) The designation indicates that the amine has reacted with HCl to form an ammonium salt and is present in that form. Ammonium hydrochlorides are more soluble than amines.

(b) $RNH_2 + HCl \, (aq) \rightleftharpoons RNH_3{}^+Cl^-(aq)$

When the product is dissolved, the solution is acidic, as is shown in the above equilibrium.

Self-Test for Chapter 10

Multiple Choice

1. What volume of 0.10 M H_2SO_4 will neutralize 30 mL of 0.05 M $Ca(OH)_2$?
    (a) 30 mL   (b) 15 mL   (c) 10 mL   (d) 6.0 mL

2. Which of the following salts is acidic in solution?
    (a) NH_4Br (b) NaBr   (c) NH_4CN (d) NaCN

3. KCN is the salt of a:
    (a) strong acid and strong base   (b) strong acid and weak base   (c) weak acid and strong base
    (d) weak acid and weak base

4. A solution has a pH of 9. What is the value of $[OH^-]$?
    (a) 5   (b) 10^{-9} (c) 9   (d) 10^{-5}

5. Which of the following is a diprotic acid?
    (a) CH_3CO_2H (b) $Ba(OH)_2$ (c) H_2SO_3 (d) H_3PO_4

6. Look at Table 10.2 and decide which base is a weaker base than F^-:
    (a) $H_2PO_4^-$ (b) HCO_3^- (c) CN^- (d) NH_3

7. Which of the following substances turns phenolphthalein red?
    (a) urine   (b) milk of magnesia   (c) blood   (d) coffee

8. Which of the following bases is an Arrhenius base?
    (a) NaOH   (b) $Ca_3(PO_4)_2$ (c) NH_3 (d) LiF

9. In the reaction $Mg(OH)_2(aq) + 2\ HBr(aq) \longrightarrow 2\ H_2O(l) + MgBr_2(aq)$, what is the conjugate acid of $Mg(OH)_2$?
    (a) HBr   (b) H_2O (c) $MgBr_2$ (d) can't tell

10. How many water molecules are produced in the balanced neutralization reaction of $H_3PO_4 + Mg(OH)_2$?
    (a) 2   (b) 3   (c) 4   (d) 6

Sentence Completion

1. One _____ of an acid reacts with one _____ of a base.

2. Phenolphthalein turns _____ in basic solution.

3. H_2CO_3 is a _____ acid.

4. The anion of a weak acid is a _____ base.

5. _____ is the splitting apart of an acid into a proton and an anion.

6. An Arrhenius base yields _____ when dissolved in water.

7. _____ is the reaction of an acid with a base.

8. _____ is the measure of a solution's acidity.

9. To completely neutralize 10.0 mL of 1.0 M H_3PO_4, you need _____ mL of 1.0 M NaOH.

10. CH_3CO_2H and $CH_3CO_2^-$ are known as a _____ acid–base pair.

11. The buffer system of blood is the _____ / _____ system.

12. Substances that can act as either acids or bases are _____.

True or False

1. 30.0 mL of 0.10 M H_2SO_4 is neutralized by 30.0 mL of 0.10 M NaOH.

2. A change of one pH unit is a tenfold change in $[H_3O^+]$.

3. H_2SO_4/HSO_4^- is a good buffer system.

4. All bases are negatively charged.

5. Bicarbonate ion neutralizes more acid than carbonate ion.

6. If the pH of a solution is 7.0, $[OH^-] = 10^{-7}$.

7. Water can act as both an acid and a base.

8. According to the Brønsted definition, an acid is a substance that dissolves in water to give H_3O^+ ions.

9. Ammonia reacts with an acid to yield ammonium hydroxide.

10. Whether an acid or a base is strong or weak depends on its percent dissociation in water.

11. The reaction $H_2O + H_2PO_4^- \longrightarrow H_3O^+ + HPO_4^{2-}$ proceeds in the direction written.

12. One equivalent of $Ca(OH)_2$ equals 37 g.

Match each entry on the left with its partner on the right.

1.   Strong acid

2.   1 M HCl

3. $[H_3O^+] = 10^{-8}$

4.   Alkalosis

5. K_w

6.   Weak base

7.   Strong base

8.   Salt

9. $[H_3O^+] = 10^{-6}$

10. Weak acid

11. 1 M H_3PO_4

12. Acidosis

(a) $[H_3O^+]\,[OH^-]$

(b) Cl^-

(c) pH = 6

(d) HCl

(e) 1 N acid

(f) CH_3CO_2H

(g) $[OH^-] = 10^{-6}$

(h) Blood pH lower than 7.35

(i) $CH_3CO_2^-$

(j) Na_2SO_4

(k) 3 N acid

(l) Blood pH higher than 7.35

Chapter 11 Nuclear Chemistry

Chapter Outline

I. Introduction to radioactivity (Sections 11.1–11.3).
- A Nuclear reactions (Section 11.1).
  1. A nuclear reaction occurs when a nuclide changes to a different nuclide.
  2. Nuclear reactions differ from chemical reactions in several ways.
     - a. Nuclear reactions produce different elements.
     - b. Different isotopes have the same behavior in chemical reactions but different behavior in nuclear reactions.
     - c. The rate of a nuclear reaction is unaffected by a change in temperature.
     - d. A nuclear reaction is the same whether an atom is in a compound or is elemental.
     - e. The energy change in a nuclear reaction is immense.
- B. The discovery and nature of radioactivity (Section 11.2).
  1. Radioactivity was discovered by Becquerel and the Curies.
  2. Three types of radiation may be emitted:
     - a. α radiation: He^{2+} nuclei with low penetrating power.
     - b. β radiation: electrons (e^-) with medium penetrating power.
     - c. γ radiation: high-energy light waves with very high penetrating power.
- C. Stable and unstable isotopes (Section 11.3).
  1. All elements have radioactive isotopes.
  2. Most radioisotopes are made in particle accelerators: These are known as artificial radioisotopes.
  3. Radioisotopes have the same chemical properties as their nonradioactive isotopes.

II. Nuclear decay (Sections 11.4–11.6).
- A. Nuclear emissions (Section 11.4).
  1. Alpha emission.
     - a. After α emission, the atomic number of the resulting isotope decreases by 2, and the mass number decreases by 4.
  2. Beta emission.
     - a. In β emission, a neutron decomposes to a proton and an electron.
     - b. The electron is emitted, and the proton is retained.
     - c. The atomic number of the resulting isotope increases by 1, and the mass number is unchanged.
  3. Gamma (γ) emission.
     - a. Gamma (γ) emission usually accompanies α or β emission and does not affect either mass number or atomic number.
  4. Positron emission.
     - a. Positron emission occurs when a proton is converted to a neutron and an ejected positron (positive electron).
     - b. The mass number of the nucleus is unchanged, but the atomic number decreases by 1.
  5. Electron capture (EC).
     - a. EC occurs when a nucleus captures an electron from the surrounding electron cloud, converting a proton into a neutron.
     - b. The mass number of the nucleus is unchanged, but the atomic number decreases by 1.

B. Half-life (Section 11.5).
    1. Half-life is the amount of time it takes for half of a sample of a radioisotope to decay.
    2. Half-life doesn't depend on the amount of sample or on temperature.
    3. Each half-life sees the decay of half of what remains of the sample.
    4. Radiocarbon dating uses the decay of carbon-14 to determine the age of archeological artifacts.
    5. Some heavy radioisotopes undergo a series of disintegrations until a nonradioactive product is reached.
D. Ionizing radiation (Section 11.6).
    1. Ionizing radiation is any high-energy radiation that can create reactive ions when it collides with a chemical compound.
    2. Ionizing radiation converts molecules into highly reactive ions.
    3. Ionizing radiation can injure the body.
        a. Gamma and X radiation are more harmful when radiation comes from outside the body.
        b. Alpha and β radiation are more dangerous when emitted from within the body.
    4. The intensity of radiation decreases with the square of the distance from the source.
III. Detecting and measuring radiation (Section 11.7).
    A. Devices for detecting radiation.
        1. Photographic film badges.
        2. Geiger counter.
        3. Scintillation counter.
    B. Units of radiation.
        1. The *curie* (Ci) measures the number of disintegrations per second.
            a. In the SI system, one *becquerel* is defined as one disintegration per second.
        2. The *roentgen* (R) measures the ionizing intensity of radiation.
        3. The *rad* (*gray* in SI units) measures the energy absorbed per gram of tissue.
        4. The *rem* and the *sievert* (SI) measure tissue damage.
IV. Artificial transmutation (Section 11.8).
    A. Artificial transmutation occurs when nuclei are bombarded with high-energy particles.
    B. After bombardment, a new nucleus is produced.
    C. The transuranium elements were all produced via artificial transmutation.
V. Nuclear fission and nuclear fusion (Section 11.9).
    A. Nuclear fission.
        1. Nuclear fission occurs when a heavy nucleus fragments after bombardment by a small particle, such as a neutron.
        2. The large amounts of energy released are due to mass-to-energy conversions and are measured by $E = mc^2$.
        3. Many different fission products can result from one bombardment.
        4. In some cases, bombardment with one neutron can cause production of more than one neutron, in addition to fragmentation.
            a. This is called a chain reaction.
            b. If the bombarded sample weighs more than a critical mass, a nuclear explosion can result.
            c. A controlled fission reaction can be used to produce energy.
    B. Nuclear fusion.
        1. If two light nuclei are made to collide, a nuclear fusion reaction occurs that produces a combined nucleus plus energy.
        2. Nuclear fusion reactions occur in stars.

Solutions to Chapter 11 Problems

11.1 For α emission, subtract 2 from the atomic number of radon and 4 from the mass number:

$$^{222}_{86}\text{Rn} \rightarrow {}^{4}_{2}\text{He} + {}^{218}_{84} \text{?}$$

Then look in the periodic table for the element with atomic number 84:

$$^{222}_{86}\text{Rn} \rightarrow {}^{4}_{2}\text{He} + {}^{218}_{84}\text{Po}$$

11.2 Add 4 to the mass number of radon-222 to calculate the isotope of radium:

$$^{226}_{88}\text{Ra} \rightarrow {}^{4}_{2}\text{He} + {}^{222}_{86}\text{Rn}$$

11.3

$$^{89}_{38}\text{Sr} \rightarrow {}^{0}_{-1}e + \text{?}$$

The mass number of the product element stays the same, but the atomic number increases by 1, to 39. Looking in the periodic table, we find that ^{89}Y is the element formed:

$$^{89}_{38}\text{Sr} \rightarrow {}^{0}_{-1}e + {}^{89}_{39}\text{Y}$$

11.4

(a) $^{3}_{1}\text{H} \rightarrow {}^{0}_{-1}e + {}^{3}_{2}\text{He}$ (b) $^{210}_{82}\text{Pb} \rightarrow {}^{0}_{-1}e + {}^{210}_{83}\text{Bi}$ (c) $^{20}_{9}\text{F} \rightarrow {}^{0}_{-1}e + {}^{20}_{10}\text{Ne}$

11.5 For positron emission, the atomic number decreases by 1, but the mass number stays the same.

(a) $^{38}_{20}\text{Ca} \rightarrow {}^{0}_{1}e + {}^{38}_{19}\text{K}$ (b) $^{118}_{54}\text{Xe} \rightarrow {}^{0}_{1}e + {}^{118}_{53}\text{I}$ (c) $^{79}_{37}\text{Rb} \rightarrow {}^{0}_{1}e + {}^{79}_{36}\text{Kr}$

11.6 As in positron emission, the atomic number decreases by 1 and the mass number stays unchanged in electron capture.

(a) $^{62}_{30}\text{Zn} + {}^{0}_{-1}e \rightarrow {}^{62}_{29}\text{Cu}$ (b) $^{110}_{50}\text{Sn} + {}^{0}_{-1}e \rightarrow {}^{110}_{49}\text{In}$ (c) $^{81}_{36}\text{Kr} + {}^{0}_{-1}e \rightarrow {}^{81}_{35}\text{Br}$

11.7 Table 11.2 shows the changes brought about by various radioactive decay patterns. Only β emission produces a change of +1 to the atomic number and a decrease in the number of neutrons. The original element was indium ($Z = 49$), and tin ($Z = 50$) is the decay product.

$$^{120}_{49}\text{In} \rightarrow {}^{0}_{-1}e + {}^{120}_{50}\text{Sn}$$

CIA 11.1

(1) *In vivo procedures* take place inside the body and assess the functioning of organs or body systems. Determination of whole-blood volume using chromium-51 is an example. (2)*Therapeutic procedures* are used to kill diseased tissue. For example, irradiation of a tumor with cobalt-60 is a treatment for cancer.

(3) In *boron neutron capture therapy (BNCT),* boron-containing drugs are administered to a cancer patient and are concentrated in the tumor, which is then irradiated with a neutron beam. The boron captures a neutron and undergoes transmutation to produce a lithium nucleus and an alpha particle, which kills the tumor.

CIA 11.2

The total sample dose is 2.0 mL x 1.25 μCi $=$ 2.5 μCi. The calculated blood volume:

$$2.5 \ \mu\text{Ci} \ \times \ \frac{1.0 \ \text{mL}}{2.6 \times 10^{-4} \ \mu\text{Ci}} \ \times \ \frac{1 \ \text{L}}{1000 \ \text{mL}} = 9.6 \ \text{L}$$

11.8 $17{,}000 \div 5730 = 2.97$, or approximately 3 half-lives.

The percentage of $^{14}_{6}\text{C}$ is $(1/2) \times (1/2) \times (1/2) \times 100\% = 12.5\%$ of the original sample.

11.9 To determine blood volume, use the formula:

$$R_{\text{sample}} = R_{\text{tracer}}\left(\frac{V_{\text{sample}}}{V_{\text{system}} + V_{\text{tracer}}}\right)$$

Where $R_{\text{sample}} = 0.10\%$, $R_{\text{tracer}} = 100\%$, $V_{\text{sample}} = 5.00$ mL, and $V_{\text{tracer}} = 1.00$ mL As an approximation, ignore the volume of the tracer, and rearrange the formula:

$$V_{\text{system}} = (R_{\text{tracer}} \times V_{\text{sample}}) / R_{\text{sample}} = (100\% \times 5.00 \ \text{mL}) / 0.10\% = 5000 \ \text{mL} = 5.0 \ \text{L}$$

It is apparent that the volume of the tracer is very small, relative to the blood volume of the patient (V_{system}).

11.10 (a) According to Figure 11.2, all four isotopes have greater mass than their nonradioactive counterparts.

(b)

$$^{3}_{1}\text{H} \ \rightarrow \ ^{0}_{-1}e + ^{3}_{2}\text{He} \qquad\qquad ^{14}_{6}\text{C} \ \rightarrow \ ^{0}_{-1}e + ^{14}_{7}\text{N}$$

$$^{24}_{11}\text{Na} \ \rightarrow \ ^{0}_{-1}e + ^{24}_{12}\text{Mg} \qquad\qquad ^{32}_{15}\text{P} \ \rightarrow \ ^{0}_{-1}e + ^{32}_{16}\text{S}$$

The isotopes resulting from decay have a smaller neutron/proton ratio.

11.11 Find the point on the graph that represents 50% of the sample remaining. Locate the time that corresponds to this point; this is the half-life–3 days.

11.12

$$\frac{I_1}{I_2} = \frac{d_2^2}{d_1^2}: I_1 = 250 \text{ units}; I_2 = 25 \text{ units}; d_1^2 = (4.0 \text{ m})^2 = 16 \text{ m}^2; d_2^2 = ?$$

$$d_2^2 = \frac{d_1^2 \times I_1}{I_2} = \frac{16 \text{ m}^2 \times 250 \text{ units}}{25 \text{ units}} = 160 \text{ m}^2$$

$$d_2 = 13 \text{ m}$$

CIA 11.3

The purpose of food irradiation is to kill harmful microorganisms by exposing food to ionizing radiation, which destroys the genetic material of the microorganisms.

CIA 11.4   Gamma rays from cobalt-60 are used to treat food.

CIA 11.5

1 Gray = 100 rad.

$$0.20 \text{ kGy} \times \frac{1000 \text{ Gy}}{1 \text{ kGy}} \times \frac{100 \text{ rad}}{1 \text{ Gy}} = 2.0 \times 10^4 \text{ rad}$$

To convert the dose to rem, use the formula rem = rad x RBE, where RBE = relative biological effectiveness factor.

The dose is 2.0×10^4 rem if the radiation is primarily γ rays (RBE = 1). Since α radiation (RBE = 20) is 20 times more penetrating than γ radiation, the dose is 20 x 2.0×10^4 rad = 4×10^5 rem for α radiation.

11.13

$$\frac{44 \ \mu Ci}{1 \text{ mL}} \times 3.98 \text{ mL} = 175 \ \mu Ci$$

11.14

$$0.02 \text{ mSv} \times \frac{1 \text{ Sv}}{1000 \text{mSv}} \times \frac{100 \text{ rem}}{1 \text{ Sv}} = 2.0 \times 10^{-3} \text{ rem}$$

Each X ray exposes a patient to 2.0×10^{-3} rems of radiation.

$$25 \text{ rem} \times \frac{1 \text{ Xray}}{2.0 \times 10^{-3} \text{ rem}} = 1.2 \times 10^3 \text{ Xrays}$$

A patient can receive more than 1000 chest X rays before biological effects might be observed.

CIA 11.6

The body-imaging techniques CT and PET are noninvasive and yield a large array of images that can be processed by computer to produce a three-dimensional image of an organ.

CIA 11.7

MRI imaging does not involve radiation and produces images that have greater contrast than those from CT and PET scans.

CIA 11.8

Try to estimate the value of n in the expression $0.1\% = 100\% \times (0.5)^n = 100\% \times 1/2^n$. This expression shows that the amount of radiation decreases to 1/1000 of its initial value. A table of powers of two shows that after 10 half-lives ($n = 10$) the intensity of radiation will have dropped to approximately 1/1000 of its initial activity. If a half-life is 6 hours, it will take 60 hours for the activity of the radiation to reach this level.

11.15

$$^{241}_{95}\text{Am} \rightarrow {}^{4}_{2}\text{He} + {}^{237}_{93}\text{Np}$$

11.16

$$^{241}_{95}\text{Am} + {}^{4}_{2}\text{He} \rightarrow 2\,{}^{1}_{0}n + {}^{243}_{97}\text{Bk}$$

11.17

$$^{40}_{18}\text{Ar} + {}^{1}_{1}\text{H} \rightarrow {}^{1}_{0}n + {}^{40}_{19}\text{K}$$

11.18

$$^{235}_{92}\text{U} + {}^{1}_{0}n \rightarrow {}^{137}_{52}\text{Te} + 2\,{}^{1}_{0}n + {}^{97}_{40}\text{Zr}$$

11.19

$$^{238}_{92}\text{U} + {}^{1}_{0}n \rightarrow {}^{239}_{92}\text{U} \rightarrow {}^{0}_{-1}e + {}^{239}_{93}\text{Np} \rightarrow {}^{0}_{-1}e + {}^{239}_{94}\text{Pu}$$

11.20

$$^{2}_{1}\text{H} + {}^{2}_{1}\text{H} \rightarrow {}^{1}_{0}n + {}^{3}_{2}\text{He}$$

HOC 11.1

(a) Both land-based and naval nuclear power sources use nuclear power to generate heat. The naval power sources use the heat to turn turbines that propel the boat and that generate electricity. The land-based generators also produce electricity from heat that turns steam turbines. The land-based reactors are larger than the naval reactors.

(b) The key component of a nuclear reactor is the containment vessel that contains the reactor (which contains uranium-based fuel rods and control rods), pressurized steam, water pumps, and a steam generator.

Understanding Key Concepts

11.21 After one half-life, the sample would consist of eight ${}^{28}_{13}\text{Al}$ atoms and eight ${}^{28}_{12}\text{Mg}$ atoms.

After two half-lives, the remaining eight ${}^{28}_{12}\text{Mg}$ atoms would have decayed to four ${}^{28}_{13}\text{Al}$ atoms, producing the outcome shown in the picture.

11.22

$$^{28}_{12}\text{Mg} \rightarrow {}^{0}_{-1}e + {}^{28}_{13}\text{Al}$$

11.23

Aluminum-28

Magnesum-28

11.24 The illustrated isotope, with six protons and eight neutrons, is $^{14}_{6}C$. Based on Figure 11.2, one would expect this isotope to be radioactive.

11.25 The shorter arrows represent β emission because they show the decomposition of a neutron with an accompanying one-unit increase in atomic number.
The longer arrows represent α emission because they indicate a decay that results in the loss of two protons and two neutrons (α particle).

11.26

$$^{241}_{94}Pu \ \rightarrow \ ^{241}_{95}Am \ \rightarrow \ ^{237}_{93}Np \ \rightarrow \ ^{233}_{91}Pa \ \rightarrow \ ^{233}_{92}U$$

11.27 The isotope can decay by either positron emission or electron capture since the atomic number decreases by 1 and the mass number remains the same.
$$^{148}_{69}Tm \ \rightarrow \ ^{0}_{1}e \ + \ ^{148}_{68}Er \quad \text{or} \quad ^{148}_{69}Tm \ + \ ^{0}_{-1}e \ \rightarrow \ ^{148}_{68}Er$$

11.28 The half-life is approximately 3.5 years.

11.29 The curve does not represent nuclear decay. According to the curve, half of the sample has decayed after 5 days. Half of the remaining sample should have decayed after 10 days, but on this curve the second half-life is shown at 13 days. A decay curve should approach zero after six half-lives, but this curve shows 10% activity remaining at that time.

Additional Problems

Radioactivity

11.30 A substance is said to be radioactive if it emits radiation by decay of an unstable nucleus.

11.31

Radiation	Particle Produced	Atomic Number	Mass Number	Penetrating Power
α emission	$^{4}_{2}He$	decreases by 2	decreases by 4	low
β emission	$^{0}_{-1}e$	increases by 1	unchanged	medium
γ emission	none	unchanged	unchanged	high
positron emission	$^{0}_{1}e$	decreases by 1	unchanged	medium
electron capture	$^{0}_{-1}e$ (captured, not emitted)	decreases by 1	unchanged	high

11.32

	Nuclear Reaction	*Chemical Reaction*
	Involves change in an atom's nucleus	Involves change in an atom's outer-shell electrons
	Usually produces a different element	Never produces a different element
	Different isotopes have different behavior in a nuclear reaction	Different isotopes have the same behavior in a chemical reaction
	Rate is unaffected by a change in temperature or by a catalyst	Rate changes with a change in temperature or by using a catalyst
	Rate is the same, whether the atom is in a compound or uncombined	
	Energy change is millions of times larger for a nuclear reaction	

11.33 When ionizing radiation strikes a molecule, a high-energy reactive ion with an odd number of electrons is produced. This reactive ion immediately reacts with other chemical compounds, creating other reactive fragments that can cause further reactions.

11.34 Ionizing radiation causes cell damage by breaking bonds in DNA. The resulting damage may lead to mutation, cancer, or cell death. Ionizing radiation may also damage other biomolecules.

11.35 Background radiation may arise from naturally occurring radioactive isotopes or from cosmic rays.

11.36 A neutron in the nucleus decays to a proton and an electron, which is emitted as a β particle.

11.37 An α particle is a helium nucleus, $_2^4\text{He}$; a helium atom is a helium nucleus plus two electrons.

Nuclear Decay and Transmutation

11.38 A nuclear equation is balanced if the number of nucleons is the same on both sides and if the sums of the charges on the nuclei and on any other elementary particles are the same on both sides.

11.39 A transuranium element is an element with an atomic number greater than 92. It is produced by bombardment of an element of slightly smaller mass with high-energy particles. All transuranium elements are unstable.

11.40 For an atom emitting an α particle, the atomic number decreases by 2, and the mass number decreases by 4.
For an atom emitting a β particle, the atomic number increases by 1, and the mass number is unchanged.

11.41 For an atom emitting a γ ray, both atomic number and mass number are unchanged.
For an atom emitting a positron, the atomic number decreases by 1, and the mass number is unchanged.

11.42 In nuclear fission, bombardment of a nucleus causes fragmentation in many different ways to yield a large number of smaller fragments. Normal radioactive decay of a nucleus produces an atom with a similar mass and yields a predictable product.

11.43 Bombardment of a $^{235}_{92}U$ nucleus with one neutron produces three neutrons, which, on collision with three more $^{235}_{92}U$ nuclei yield nine neutrons, and so on. Because the reaction is self-sustaining, it is known as a chain reaction.

11.44

(a) $^{35}_{16}S \rightarrow ^{0}_{-1}e + ^{35}_{17}Cl$ (b) $^{24}_{10}Ne \rightarrow ^{0}_{-1}e + ^{24}_{11}Na$

(c) $^{90}_{38}Sr \rightarrow ^{0}_{-1}e + ^{90}_{39}Y$

11.45

(a) $^{190}_{78}Pt \rightarrow ^{4}_{2}He + ^{186}_{76}Os$ (b) $^{208}_{87}Fr \rightarrow ^{4}_{2}He + ^{204}_{85}At$

(c) $^{245}_{96}Cm \rightarrow ^{4}_{2}He + ^{241}_{94}Pu$

11.46

(a) $^{109}_{47}Ag + ^{4}_{2}He \rightarrow ^{113}_{49}In$ (b) $^{10}_{5}B + ^{4}_{2}He \rightarrow ^{13}_{7}N + ^{1}_{0}n$

11.47

(a) $^{26}_{11}Na \rightarrow ^{0}_{-1}e + ^{26}_{12}Mg$ (b) $^{212}_{83}Bi \rightarrow ^{4}_{2}He + ^{208}_{81}Tl$

11.48

(a) $^{235}_{92}U + ^{1}_{0}n \rightarrow ^{160}_{62}Sm + ^{72}_{30}Zn + 4\,^{1}_{0}n$

(b) $^{235}_{92}U + ^{1}_{0}n \rightarrow ^{87}_{35}Br + ^{146}_{57}La + 3\,^{1}_{0}n$

11.49

(a) $^{126}_{50}Sn \rightarrow ^{0}_{-1}e + ^{126}_{51}Sb$ β decay   (b) $^{210}_{88}Ra \rightarrow ^{4}_{2}He + ^{206}_{86}Rn$ α decay

(c) $^{76}_{36}Kr + ^{0}_{-1}e \rightarrow ^{76}_{35}Br$ electron capture

11.50

$^{198}_{80}Hg + ^{1}_{0}n \rightarrow ^{198}_{79}Au + ^{1}_{1}H$

A proton is produced in addition to gold-198.

11.51

$^{59}_{27}Co + ^{1}_{0}n \rightarrow ^{60}_{27}Co$

$^{60}_{27}Co \rightarrow ^{0}_{-1}e + ^{60}_{28}Ni$

11.52 The four α decays result in a loss of 16 nucleons from the mass number and a reduction of 8 in atomic number. The β decay results in a gain of 1 in atomic number and no change in mass number. The parent isotope, which has seven more protons and a mass 16 units greater than bismuth-212, is thorium-228.

11.53

$$^{209}_{83}\text{Bi} + {}^{58}_{26}\text{Fe} \rightarrow {}^{266}_{109}\text{Mt} + {}^{1}_{0}n \quad \text{A neutron is also produced.}$$

Half-Life

11.54 If strontium-90 has a half-life of 28.8 years, half of a given quantity of strontium-90 will have decayed after 28.8 years.

11.55 Use Figure 11.4, which shows a radioactive decay curve. After 1.5 half-lives, about 35% of a radioactive sample remains. After 3.2 half-lives, 10% remains.

11.56 (a) An 80% decrease in activity takes place in 2.3 half-lives (Figure 11.4). For selenium-75, this is equivalent to 2.3 x 120 days = 280 days.

(b) $365 \div 120 = 3.04$, or approximately 3 half-lives.

$$0.050 \text{ g } \times (0.5)^3 = 0.050 \times 0.125 = 0.0063 \text{ g}$$

Approximately 0.0063 grams of $^{75}_{34}\text{Se}$ will remain after a year.

11.57 (a) A 75% loss of activity takes place in 2 half-lives (240 days).

(b) After 7 half-lives, 99% of any sample will have decayed. Since the half-life of $^{75}_{34}\text{Se}$ is approximately 0.33 year, 99% of the sample will have disintegrated in 7 x 0.33 = 2.3 years.

11.58 If the half-life of mercury-97 is 64.1 hours:
7 days = 168 hours; 168/64.1 = 2.6, or approximately 3 half-lives
30 days = 720 hours; 720/64.1 = approximately 11 half-lives
$5.0 \text{ ng } \times (0.5)^3$ = approx. 1 ng of mercury-197 remains after 7 days.
$5.0 \text{ ng } \times (0.5)^{11}$ = approx. 2×10^{-3} ng of mercury-197 remains after 30 days.

11.59

(a) $^{198}_{79}\text{Au} \rightarrow {}^{0}_{-1}e + {}^{198}_{80}\text{Hg}$

(b) $\dfrac{3.75 \text{ mCi}}{30.0 \text{ mCi}} = 0.125 = 3.00 \text{ half-lives}; \quad 3.00 \times 2.695 \text{ days} = 8.09 \text{ days}$

(c) $\dfrac{1.0 \text{ mCi}}{1 \text{ kg}} \times 70.0 \text{ kg} = 70 \text{ mCi}$

Measuring Radioactivity

11.60 The inside walls of a Geiger counter tube are negatively charged, and a wire in the center is positively charged. Radiation ionizes argon gas inside the tube, which creates a conducting path for current between the wall and the wire. The current is detected, and the Geiger counter makes a clicking sound.

11.61 A film badge is protected from light exposure, but exposure to other radiation causes the film to get cloudy. Photographic development of the film and comparison with a standard allows the amount of exposure to be calculated.

11.62 In a scintillation counter, a phosphor emits a flash of light when struck by radiation. The number of flashes are counted and converted to an electrical signal.

11.63 Rems indicate the amount of tissue damage from any type of radiation, and allow comparisons between equivalent doses of different types of radiation to be made.

11.64 According to Table 11.6, any amount of radiation above 25 rems produces detectable effects in humans.

11.65 (1) curie   (c) Number of disintegrations per second
       (2) rem   (b) Amount of tissue damage
       (3) rad   (d) Amount of radiation per gram of tissue
       (4) roentgen   (a) Ionizing intensity of radiation

11.66

$$28 \text{ mCi} \times \frac{1 \text{ mL}}{15 \text{ mCi}} = 1.9 \text{ mL}$$

11.67

(a) $68 \text{ kg} \times \dfrac{180 \ \mu\text{Ci}}{1 \text{ kg}} \times \dfrac{1 \text{ mCi}}{10^3 \ \mu\text{Ci}} = 12 \text{ mCi}$

(b) $12 \text{ mCi} \times \dfrac{1 \text{ mL}}{6.5 \text{ mCi}} = 1.8 \text{ mL}$

11.68

(a) $\dfrac{I_1}{I_2} = \dfrac{d_2^2}{d_1^2}$; $I_1 = 300$ rem; $d_1^2 = (2.0 \text{ m})^2 = 4.0 \text{ m}^2$; $d_2^2 = (16.0 \text{ m})^2 = 256 \text{ m}^2$; $I_2 = ?$

$I_2 = \dfrac{d_1^2 \times I_1}{d_2^2} = \dfrac{4.0 \text{ m}^2 \times 300 \text{ rem}}{256 \text{ m}^2} = 4.7 \text{ rem}$

(b) $\dfrac{I_1}{I_2} = \dfrac{d_2^2}{d_1^2}$; $I_1 = 300$ rem; $d_1^2 = (2.0 \text{ m})^2 = 4.0 \text{ m}^2$; $d_2^2 = (25.0 \text{ m})^2 = 625 \text{ m}^2$; $I_2 = ?$

$I_2 = \dfrac{d_1^2 \times I_1}{d_2^2} = \dfrac{4.0 \text{ m}^2 \times 300 \text{ rem}}{625 \text{ m}^2} = 1.9 \text{ rem}$

11.69

$\dfrac{I_1}{I_2} = \dfrac{d_2^2}{d_1^2}$; $I_1 = 650$ rem; $I_2 = 25$ rem; $d_1^2 = 1.0 \text{ m}^2$; $d_2^2 = ?$

$d_2^2 = \dfrac{d_1^2 \times I_1}{I_2} = \dfrac{1.0 \text{ m}^2 \times 650.0 \text{ rem}}{25 \text{ rem}} = 26 \text{ m}^2$

$d_2 = 5.1 \text{ m}$

Conceptual Problems

11.70 The region with no filter monitors α radiation, the region with a plastic filter monitors β radiation, and the region with a tinfoil filter monitors γ radiation. This type of badge filters the types of radiation by penetrating power.

11.71 It takes 3 half-lives for radiation to decay to one-eighth of its original value.
3 x 5730 years = 17,200 years old.

11.72 Nuclear reactions occur spontaneously, and no substance can "react with" radioactive emissions to neutralize them.

11.73 A film badge is more useful for measuring radiation exposure over a period of time. Scintillation counters and Geiger counters are more useful for measuring a current source of radiation.

11.74

$$\frac{I_1}{I_2} = \frac{d_2^2}{d_1^2}; I_1 = 28 \text{ cpm}; d_1^2 = (10 \text{ m})^2 = 100 \text{ m}^2; d_2^2 = (5.0 \text{ m})^2 = 25 \text{ m}^2; I_2 = ?$$

$$I_2 = \frac{d_1^2 \times I_1}{d_2^2} = \frac{100 \text{ m}^2 \times 28 \text{ cpm}}{25 \text{ m}^2} = 110 \text{ cpm}$$

11.75 If an isotope has more protons than neutrons, either electron capture or positron emission is the most likely decay route because they both reduce the number of protons and increases the number of neutrons. If an isotope has more neutrons than protons, β emission is the most likely decay route because it both decreases the number of neutrons and increases the number of protons.

11.76

(a) $^{99}_{42}\text{Mo} \rightarrow \,^{0}_{-1}e + \,^{99}_{43}\text{Tc}$ ^{99}Mo decays to Tc-99m by β emission.

(b) $^{98}_{42}\text{Mo} + \,^{1}_{0}n \rightarrow \,^{99}_{42}\text{Mo}$

11.77 24 hours $\div$ 6.01 hours = approx. 4 half-lives. 15 μCi x $(0.5)^4$ = 0.94 μCi

11.78

(a) $^{238}_{94}\text{Pu} \rightarrow \,^{4}_{2}\text{He} + \,^{234}_{92}\text{U}$
(b) The metal case serves as a shield to protect the wearer from α radiation.

11.79

(a) $^{24}_{11}\text{Na} \rightarrow \,^{0}_{-1}e + \,^{24}_{12}\text{Mg}$

(b) 50 hr $\div$ 15 hr = 3.3 half-lives. $(0.5)^3$ = 0.125. Somewhat less than 12.5% of the original sodium-24 remains after 50 hrs.

11.80 Embryos and fetuses are particularly susceptible to the effects of radiation because their rate of cell division is high.

11.81 (a) Iodine-131 decays by β-emission to produce Xe-131. Gamma radiation is also produced.

$$^{131}_{53}\text{I} \rightarrow {}^{0}_{-1}e + {}^{131}_{54}\text{Xe}$$

(b) Use Figure 11.4. A period of 4 weeks is about 3 1/2 half-lives. After this time, about 8% of I-131 remains. After 8 weeks, less than 1% remains.

11.82 Nuclear fission does not require high temperatures or holding materials long enough for nuclei to react. Drawbacks of fission include disposal of radioactive waste and issues of safety.

11.83

$$^{197}_{79}\text{Au} + 7\,{}^{1}_{0}n \rightarrow 3\,{}^{0}_{-1}e + {}^{204}_{82}\text{Pb}$$

Gold-197 is bombarded by 7 neutrons and yields lead-204 plus 3 β particles

11.84

(a) $^{253}_{99}\text{Es} + {}^{4}_{2}\text{He} \rightarrow {}^{256}_{101}\text{Md} + {}^{1}_{0}n$ (b) $^{250}_{98}\text{Cf} + {}^{11}_{5}\text{B} \rightarrow {}^{257}_{103}\text{Lr} + 4\,{}^{1}_{0}n$

11.85

$$^{10}_{5}\text{B} + {}^{1}_{0}n \rightarrow {}^{7}_{3}\text{Li} + {}^{4}_{2}\text{He}$$

11.86

$$^{232}_{90}\text{Th} \rightarrow {}^{208}_{82}\text{Pb} + 6\,{}^{4}_{2}\text{He} + 4\,{}^{0}_{-1}e$$

The 24-amu loss in mass in going from $^{232}_{90}\text{Th}$ to $^{208}_{82}\text{Pb}$ is due to the emission of six α particles, which also account for a reduction in atomic number from 90 to 78. The emission of four β particles increases the atomic number from 78 to 82 and yields $^{208}_{82}\text{Pb}$.

11.87 Alpha particles are used for the bombardment.

$$^{238}_{92}\text{U} + 3\,{}^{4}_{2}\text{He} \rightarrow {}^{246}_{98}\text{Cf} + 4\,{}^{1}_{0}n$$

11.88

$$^{48}_{20}\text{Ca} + {}^{249}_{97}\text{Bk} \rightarrow {}^{294}_{117}\text{Uus} + 3\,{}^{1}_{0}n$$
$$^{48}_{20}\text{Ca} + {}^{249}_{97}\text{Bk} \rightarrow {}^{293}_{117}\text{Uus} + 4\,{}^{1}_{0}n$$

Group Problems

11.89 This question illustrates the relative dangers of internal radiation and external radiation. Alpha and beta radiation, because of their short penetrating distance, are dangerous inside the body. You can put the alpha cookie in your pocket, but don't try this with the beta cookie, which might damage sensitive genetic material; hold it your hand (wearing gloves). The gamma cookie, with a large penetrating distance, can probably be eaten because the radiation will pass through the body before it can cause damage. Throw away the neutron cookie.

11.90 BNCT is a therapy for destroying cancerous tumors that involves irradiating tissue that has taken up boron-containing compounds. Irradiation causes radioactive decay of boron molecules that sends high-energy radiation to destroy the tumor without damaging tissue that hasn't taken up the boron-containing compound.
(a) The two most commonly used boron-containing compounds, either boronophenylalanine and sodium borocaptate, are introduced by intravenous infusion.
(b) Neutrons are produced by either cyclotrons or by other types of particle accelerators.
(c) $^{10}_{5}B + ^{1}_{0}n \rightarrow [^{11}_{5}B] \rightarrow ^{4}_{2}He + ^{7}_{3}Li$
The high-energy products are α particles and γ rays.

11.91 (a) The primary nuclear fuel was U-235. The major radioactive materials initially released were I-131 and Xe-131, but these have short half-lives. Later, Cs-137 and Cs-134 became the major components of fallout.
(b) Although estimates vary, at least one billion curies of radiation were released into the environment, much of it initially as inert gases and I-131. Cs-13, with a half-life of 30 years, later became the principal component of fallout, as other components with shorter half-lives decayed.
(c) People near, but not in, Chernobyl and who were evacuated had a minimum exposure of 350 Sv. The current estimate of background radiation due to Chernobyl is 0.002 mSv. If yearly worldwide exposure to background radiation is 3 mSv, the percentage due to Chernobyl:
$$\frac{0.002 \text{ mSv}}{3 \text{ mSv}} \times 100\% = 0.07\%$$

Self-Test for Chapter 11

Multiple Choice

1. Bombardment of a uranium-235 atom with a neutron produces three neutrons, which can go on to bombard three more U-235 atoms. How many neutrons are produced after the fourth cycle?
(a) 12   (b) 36   (c) 81   (d) 108

2. The product of the α emission of $^{146}_{62}Sm$ is:
(a) $^{146}_{63}Eu$ (b) $^{144}_{61}Pm$ (c) $^{144}_{60}Nd$ (d) $^{142}_{60}Nd$

3. A half-inch-thick piece of plastic blocks 50% of radiation. How much radiation passes through a one-inch-thick piece of plastic?
(a) 50%   (b) 25%   (c) 12%   (d) 0%

4. The *curie* measures:
(a) amount of radioactivity   (b) ionizing intensity of radiation
(c) amount of radiation absorbed   (d) tissue damage

5. A sample of a radionuclide has a half-life of 40 days. How much radioactivity remains after 160 days?
(a) 25%   (b) 12.5%   (c) 6.25%   (d) 0%

6. How many disintegrations per second does one microcurie of a sample emit?
(a) 3.7×10^{10} (b) 3.7×10^{7} (c) 3.7×10^{6} (d) 3.7×10^{4}

7. If a person standing 100 m from a radiation source approaches to within 10 m of the source, how much does the intensity of the radiation change?
    (a) increases by a hundredfold   (b) increases tenfold
    (c) stays the same   (d) decreases tenfold

8. Artificial transmutation was probably used to produce which one of the following radionuclides?
    (a) $^{238}_{92}U$ (b) $^{210}_{82}Pb$ (c) $^{60}_{27}Co$ (d) $^{239}_{94}Pu$

9. Which of the following radioisotopes is not medically useful?
    (a) $^{131}_{53}I$ (b) $^{14}_{6}C$ (c) $^{24}_{11}Na$ (d) $^{60}_{27}Co$

10. In an experiment by the Curies, aluminum-27 was bombarded with α particles. If each aluminum-27 atom captured one α particle and emitted one neutron, what other atom was produced?
    (a) $^{29}_{13}Al$ (b) $^{28}_{14}Si$ (c) $^{30}_{15}P$ (d) $^{31}_{15}P$

Sentence Completion

1. The change of one element to another is called _____.

2. _____ _____ causes the most tissue damage.

3. A chemical compound tagged with a radioactive atom can be used as a _____.

4. All people are exposed to _____ radiation.

5. A nuclear equation is balanced when the number of _____ is the same on both sides of the equation.

6. The age of an organic sample can be determined by measuring the amount of _____ it contains.

7. The _____ is a unit for measuring the number of radioactive disintegrations per second.

8. In _____ _____, an atom is split apart by neutron bombardment to give small fragments.

9. People who work around radiation sources wear _____ _____ for detection of radiation.

10. PET and CT are techniques for _____ _____.

11. Very light elements release large amounts of energy when they undergo _____ _____.

12. The _____ _____ is the size of a radioactive sample that is needed for a nuclear reaction to be self-sustaining.

True or False

1. If a sample has a half-life of 12 days, it will have decayed completely in 24 days.

2.   X rays are considered to be ionizing radiation.

3.   The ratio of carbon-14 to carbon-12 in the atmosphere is constant.

4.   Pierre and Marie Curie discovered radioactivity.

5.   A β particle travels at nearly the speed of light.

6.   An α particle travels at nearly the speed of light.

7.   The product of β emission has an atomic number 1 amu smaller than the starting material.

8.   A nuclear power plant can undergo a nuclear explosion in an accident.

9.   The rad is the unit most commonly used in medicine to measure radiation dosage.

10.   Radiation causes injury by ionizing molecules.

11.   Most of the known radioisotopes are naturally occurring.

12.   Many isotopes undergo nuclear fission.

Match each entry on the left with its partner on the right.

1.   Rutherford   (a) α particle

2. $^{245}_{96}\text{Cm}$ (b) Used in cancer therapy

3. $t_{1/2}$ (c) Discovered radioactivity

4. $^{0}_{-1}e$ (d) Neutron

5.   Becquerel   (e) Used for archeological dating

6. $^{90}_{38}\text{Sr}$ (f) Discovered α and β particles

7. $^{1}_{0}n$ (g) Natural radioactive element that undergoes fission

8.   P. and M. Curie   (h) Discovered radium

9. $^{60}_{27}\text{Co}$ (i) Transuranium element

10. $^{4}_{2}\text{He}$ (j) Component of radioactive waste

11. $^{14}_{6}\text{C}$ (k) β particle

12. $^{235}_{92}\text{U}$ (l) Half-life

Chapter Outline

I. Introduction to organic chemistry (Sections 12.1–12.5).
  A. The nature of organic molecules (Section 12.1).
    1. Carbon always forms four bonds.
    2. Almost all of the bonds in organic molecules are covalent.
      a. These bonds are polar covalent bonds when carbon forms a bond to an element on the far right of the periodic table.
      b. Carbon can form multiple bonds by sharing more than two electrons.
    3. Organic molecules have specific three-dimensional shapes.
    4. In addition to hydrogen, oxygen and nitrogen are often present in organic molecules.
    5. Organic molecules are low-melting and low-boiling, relative to inorganic compounds.
    6. Organic molecules are usually insoluble in water and do not conduct electricity.
  B. Functional groups (Section 12.2).
    1. A functional group is an atom or group of atoms that has a characteristic reactivity.
    2. The chemistry of organic molecules is determined by functional groups.
    3. Functional groups fall into four categories.
      a. Hydrocarbons.
      b. Compounds with C–X single bonds, where X is electronegative.
      c. Compounds with C=O double bonds.
      d. Compounds that have functional groups that contain sulfur.
    4. Many molecules have more than one functional group.
  C. Structures of organic molecules: alkane isomers (Sections 12.3–12.4).
    1. When hydrocarbons contain more than three carbons, there is more than one way to arrange the carbon atoms (Section 12.3).
      a. Some alkanes are straight-chain alkanes.
      b. Other alkanes are branched.
    2. Compounds with the same formula but different orders of connecting the carbon atoms are called constitutional isomers.
      a. Constitutional isomers have different structures and different properties.
      b. Isomers having different functional groups are functional group isomers.
    3. Drawing structures (Section 12.4).
      a. Condensed structures are simpler to draw than structural formulas.
        i.   In condensed structures, C—H and C—C single bonds are implied rather than drawn.
        ii. Vertical bonds are often shown for clarity.
        iii. Occasionally, parentheses are used to show a row of $-CH_2-$ groups.
      b. In line structures, C and H do not appear.
        i   Each C–C bond is represented as a line.
        ii. The beginning or end of a line represents a carbon atom.
        iii. Any atom other than C or H must be shown.
  D. The shapes of organic molecules (Section 12.5).
    1. The groups around a C—C bond are free to assume an infinite number of conformations.
    2. Molecules adopt the least crowded conformation.
    3. Two structures with identical connections between atoms are identical, no matter how they are drawn.

II. Alkanes (Sections 12.6–12.8).
   A. Naming alkanes (Section 12.6).
      1. Straight-chain alkanes.
         a. Count the carbons.
         b. Find the root name.
         c. Add -*ane* to the root name.
      2. Alkyl groups.
         a. Remove a hydrogen from an alkane and replace -*ane* with -*yl*.
         b. More than one alkyl group can be formed from some alkanes.
            i. Isopropyl and propyl are the two groups that can be formed from propane.
            ii. Four groups can be formed from butane.
            iii. Alkyl groups can be substituents on a straight chain.
      3. Substitution patterns.
         a. Primary carbon: $R–CH_3$.
         b. Secondary carbon: $R–CH_2–R'$.
         c. Tertiary carbon:

$$R–\underset{\underset{R}{|}}{C}H–R$$

         d. Quaternary carbon:

      4. Naming branched-chain alkanes.
         a. Name the main chain by finding the longest continuous chain of carbons and naming it.
         b. Number the carbons in the main chain, starting from the end nearer the first branch point.
         c. Identify and number each branching substituent according to its point of attachment to the main chain.
         d. Write the name as a single word.
   B. Properties of alkanes (Section 12.7).
      1. The boiling points and melting points of alkanes increase with the number of carbons.
      2. Alkanes are insoluble in water but are soluble in nonpolar organic solvents.
      3. Alkanes are colorless and odorless.
      4. Alkanes are flammable.
      5. Alkanes with four or fewer carbons are gases, those from C_5 to C_{15} are liquids, and those with more than 15 carbons are solids.
   C. Reactivity of alkanes (Section 12.8).
      1. Alkanes are very unreactive.
      2. Reactions that alkanes undergo:
         a. Combustion: alkane + O_2 —> CO_2 + H_2O.
         b. Halogenation: alkane + X_2 —> halogenated alkane + HCl.
III. Cycloalkanes (Sections 12.9–12.10).
   A. Properties of cycloalkanes (Section 12.9).
      1. Cycloalkanes contain rings of carbon atoms.
      2. Cyclopropane and cyclobutane are more reactive than other cycloalkanes.
      3. Cyclohexane exists in a chairlike conformation.
      4. Cycloalkanes have properties similar to those of acyclic alkanes.
      5. Free rotation is not possible around the carbon–carbon bonds of a cycloalkane ring.
   B. Forming a ring requires forming a new C–C bond and removal of two –Hs.
   C. Drawing and naming cycloalkanes (Section 12.10).
      1. In drawing cycloalkanes, the ring is represented as a polygon.

2. Naming cycloalkanes.
   a. Use the cycloalkane as the parent.
   b. Number the substituents.
      i. Start numbering at the group that has alphabetical priority.
      ii. Proceed around the ring to give to the second group the lowest possible number.

Solutions to Chapter 12 Problems

12.1

(a)

Propylene glycol

(b)

Glutaric acid

(c)

Lactic acid

(d)

Phenylalanine

12.2

(a)

an aldehyde

(b)

a ketone

(c)

a carboxylic acid

HOC 12.1

(a)

Acetic acid

Acetic acid is found in products containing vinegar: salad dressing, mustard, taco sauce, barbecue sauce.

(b)

Citric acid

a = amide
b = amine
c = aromatic ring
d = carboxylic acid

Folic acid (vitamin B_9, vitamin M, vitamin B_C)

Dextrose (glucose)

a = amine
b = aromatic ring
c = alcohol
d = sulfide

Thiamine (vitamin B_1)   Both rings are aromatic.

Citric acid is used as a preservative in jam, soup, and soda.
Folic acid is an essential vitamin and is found in leafy green vegetables, bread, and flour.
Dextrose is used for energy and is found in candy, milk, and nutrition bars.
Thiamine, which prevents Beriberi, is found in bread, cereal, and flour.

(c)

a = amide
b = carboxylic acid
c = sulfide
d = aromatic ring
e = amine
f = double bond

Bacitracin

a = alcohol
b = amine
c = ether

Neosporin

$(R = C_8H_{17})$

Polymyxin B   a = amide
                 b = amine
                 c = alcohol
                 d = aromatic ring

12.3

(a)

   C_7H_{16} Heptane

(b)

   C_9H_{20} Nonane

12.4

12.5

(a)   (b)   (c)

$CH_3CH_2CH_2CH_2CH_3$ $CH_3\overset{\displaystyle CH_3}{\underset{|}{C}}HCH_2CH_3$ $CH_3\overset{\displaystyle CH_3}{\underset{\displaystyle CH_3}{C}}CH_3$

Pentane   2-Methylbutane

                                                 2,2-Dimethylpropane

12.6 *Step 1:* Find the longest straight carbon chain and draw it as a zigzag line.
Step 2: Add groups to the appropriate carbons, remembering that atoms other than carbon and hydrogen must be indicated.

Condensed *Structure*	*Straight-Chain* *Backbone*	*Line-Bond* *Structure*

(a)

CH$_3$CHCHCH$_2$CH$_2$CH$_3$

CH$_2$OH

OH

(b)

CH$_3$ CH$_3$

CH$_3$CHCHCH$_2$CHCH$_3$

CH$_2$CH$_3$

(c)

Br   OH

CH$_3$CHCHCH$_2$CH$_2$CHCH$_3$

CH$_3$

Br   OH

12.7 Reverse the steps in the previous problem.
Step 1: Draw a chain of carbons that is the same length as the longest chain in the line structure.
Step 2: Add condensed groups to the condensed backbone.

Line *Structure*	*Condensed* *Backbone*	*Condensed* *Structure*

(a)

Cl

C—C—C—CH$_3$

C—C—CH$_3$

H$_3$C   Cl

CH$_3$CH$_2$C—CHCH$_2$CH$_3$

CH$_2$CHCH$_3$

CH$_3$

(b)

CH$_3$C—C—CCH$_3$

C
CH$_3$

H$_3$C CH$_3$ H$_3$C CH$_3$

CH$_3$C—CH—CCH$_3$

CH$_2$CH$_3$

12.8

(a)

CH₃CH₂CHO

(b)

CH₃COCH₃

(c)

CH₃CH₂CO₂H

12.9 All three structures have the same molecular formula (C_7H_{16}). Structures (a) and (c) represent the same structures.

12.10 First, check to see if both compounds have the same molecular formula. Then, check to see if each compound has the same number of carbons in its principal carbon chain. Once you have done this, check to see if the functional groups are bonded to the same carbon in each structure.
(a) same compounds (b) different compounds (c) same compounds

12.11

p = primary
s = secondary
t = tertiary
q = quaternary

12.12

(a)

3-methyl group

3-Methylheptane

longest chain—a heptane

(b)

longest chain—a heptane

4-isopropyl group

4-Isopropylheptane

12.13 To answer this problem, draw the straight-chain hydrocarbon corresponding to the parent name, and replace —H's with the groups indicated.

(a)

$$\underset{p}{CH_3}\underset{s}{CH_2}\underset{s}{CH_2}\underset{t}{\overset{\overset{\displaystyle p\,CH_3}{|}}{CH}}\underset{s}{CH_2}\underset{p}{CH_3}$$

3-Methylhexane

(b)

$$\underset{p}{CH_3}\underset{s}{CH_2}\underset{s}{CH_2}\underset{s}{CH_2}\underset{t}{\overset{\overset{\displaystyle p\,CH_3}{\overset{t}{|}}}{CH}}\underset{t}{\overset{\overset{\displaystyle }{}}{CH}}\underset{s}{CH_2}\underset{p}{CH_3}$$
$$p\,CH_3$$

3,4-Dimethyloctane

(c)

$$\underset{p}{CH_3}\overset{\overset{\displaystyle p\,CH_3}{|}}{CH}\underset{s}{CH_2}\overset{\overset{\displaystyle p\,CH_3}{|}}{\underset{q}{C}}CH_3$$
$$p\,CH_3$$

2,2,4-Trimethylpentane

Where p = primary, s = secondary, t = tertiary, and q = quaternary

12.14 There are many answers to this question. For example,

(a)

$$CH_3CH_2\overset{\overset{\displaystyle CH_3}{|}}{\underset{t}{CH}}CH_3$$

2-Methylbutane

(b)

$$CH_3\overset{\overset{\displaystyle H_3C}{|}}{\underset{t}{CH}}\overset{\overset{\displaystyle CH_3}{|}}{\underset{q}{C}}CH_3$$
$$CH_3$$

2,2,3-Trimethylbutane

12.15

(a)

H₃C CH₃ H H ... (structure)

2,2-Dimethylpentane

(b)

H H H₃C CH₃ H H ... (structure)

2,3,3-Trimethylpentane

CIA 12.1
    An anabolic pathway creates new molecules.

CIA 12.2
    A cytotoxic compound is poisonous to cells.

CIA 12.3
    The cofactors are vitamins B-6, B-12, and folic acid.

MR 12.1

$$CH_3-CH_2-\overset{..}{\underset{..}{O}}-H \ + \ H^+ \ \rightleftharpoons \ CH_3-CH_2-\overset{H}{\underset{+}{\overset{|}{\underset{..}{O}}}}-H$$

MR 12.2

$$CH_3-\overset{..}{\underset{..}{S}}-H \ + \ ^-\!\!:\!\overset{..}{\underset{..}{O}}H \ \rightleftharpoons \ CH_3-\overset{..}{\underset{..}{S}}:^- \ + \ H\!:\!\overset{..}{\underset{..}{O}}H$$

$$CH_3-\overset{..}{\underset{..}{S}}:^- \ + \ CH_3-\overset{..}{\underset{..}{I}}: \ \longrightarrow \ CH_3-\overset{..}{\underset{..}{S}}-CH_3 \ + \ :\overset{..}{\underset{..}{I}}:^-$$

12.16

$$CH_4 \ + \ 2\,O_2 \ \longrightarrow \ CO_2 \ + \ 2\,H_2O$$

12.17

$$\underset{CH_3CHCH_2CHCH_3}{\overset{\overset{CH_3}{|} \quad \overset{CH_3}{|}}{}} \ + \ Cl_2 \ \longrightarrow \ \underset{CH_3CHCH_2CHCH_2Cl}{\overset{\overset{CH_3}{|} \quad \overset{CH_3}{|}}{}} \ + \ \underset{\underset{Cl}{|}}{\overset{\overset{CH_3}{|} \quad \overset{CH_3}{|}}{CH_3CHCH_2CCH_3}}$$

$$+ \ \underset{\underset{Cl}{|}}{\overset{\overset{CH_3}{|} \quad \overset{CH_3}{|}}{CH_3CHCHCHCH_3}}$$

Since 2,4-dimethylpentane has three different kinds of hydrogens, three different monosubstitution products can be formed.

12.18

(a)

1-Ethyl-4-methylcyclohexane

The parent ring is a cyclohexane. The two substituents are an ethyl group and a methyl group. The ethyl group receives the smaller number because it has alphabetical priority.

(b)

1-Ethyl-3-isopropylcyclopentane

The parent ring is a cyclopentane. The two substituents are an ethyl group and an isopropyl group. The ethyl group receives the smaller number because it has alphabetical priority.

12.19

(a)

1,1-Diethylcyclohexane

(b)

1,3,5-Trimethylcycloheptane

12.20

(a)

1,2,4-Trimethylcyclohexane

(b)

Cyclopentylcyclohexane

(c)

1,3 Diethyl-2-methylcyclopentane

(a) The substituents should have the lowest possible numbers.
(b) The smaller ring is a substituent of the larger ring.
(c) When two substituents are the same, use the prefix *di–*.

12.21

Propylcyclohexane

CIA 12.4

(a) Synthetic rubber is more flexible and behaves better in extreme temperatures than natural rubber.
(b) Butadiene is a readily available product of petroleum cracking.

CIA 12.5

(a) Paraffin waxes are used on fruits and vegetables to keep them from bruising.
(b) Ethylene and propylene are the starting materials for industrial chemicals and plastics. Plastic bottles, toys, and many other items are manufactured from polyethylene and polypropylene.

Understanding Key Concepts

12.22 In a line drawing, a carbon is assumed to be at the intersection of lines.

(a)

(b)

12.23

(a)

C—C double bond

ketone

—O ← ether

(b)

amine

H₂N

carboxylic acid

C—C double bond

12.24

(a)

CH₃
|
CH₃CH₂CHCHCH₃
|
CH₃

2,3-Dimethylpentane

(b)

CH₃
|
CH₃CHCH₂CH₂CHCH₃
|
CH₃

2,5-Dimethylhexane

12.25

(a)

CH₃

CH₃

1,1-Dimethylcyclopentane

(b)

CH₃

CH

CH₃

Isopropylcyclobutane

12.26 In one of the isomers (b), the methyl groups are on the same side of the ring, and in the other isomer (a), they are on opposite sides.

(a)

H   CH₃

H₃C—   —H

opposite sides

(b)

H   H

H₃C—   —CH₃

same side

Additional Problems

Organic Molecules and Functional Groups

12.27 If you look at the periodic table you will see that carbon belongs to group 4A, whose elements can form four bonds. Because carbon is in the middle of the periodic table, these bonds are covalent, and because it is in period 2, these bonds are strong. Consequently, carbon is unique in that it can form four strong bonds to other elements and to other carbon atoms, making possible a great many different compounds.

12.28 Functional groups are groups of atoms within a molecule that have a characteristic chemical behavior. They are important because most of the chemistry of organic compounds is determined by functional groups. The reactivity of a functional group is similar in all compounds in which it occurs.

12.29 Most organic compounds do not dissolve in water because they are nonpolar. They do not conduct electricity because they are covalent, not ionic.

12.30 A polar covalent bond is a covalent bond in which electrons are shared unequally, being more attracted to one atom than the other. For example, the electrons in the C—Br bond of bromomethane are attracted more strongly to the electronegative bromine than to carbon.

12.31 There are many possible answers to this question. For example,

12.32

(a)

(b)

12.33

(a)

Donepezil

a = ether
b = ketone
c = aromatic ring
d = amine

(b)

Aripiprazole

a = halide (aromatic)
b = amine
c = ether
d = amide
e = aromatic ring

12.34 There are many possible answers to each of these questions. For example,

(a)

$$CH_3CH_2CH_2CH_2CH$$

aldehyde

(b)

$$CH_3CH_2CH_2C-OCH_2CH_3$$

ester

(c)

$$HS-CH_2CH_2C-NH_2$$

amide
thiol

12.35 Other structures are possible.

(a)

$$CH_3CH_2CH_2C-NH_2$$

amide

(b)

aldehyde

(c)

aromatic ether

Alkanes and Isomers

12.36 For two compounds to be isomers, they must have the same molecular formula but different structures.

12.37 Compounds with the formulas C_5H_{10} and C_4H_{10} are not isomers because they do not have the same molecular formulas.

12.38 (a), (b) A primary carbon is bonded to one other carbon; a secondary carbon is bonded to two other carbons; a tertiary carbon is bonded to three other carbons; and a quaternary carbon is bonded to four other carbons.
(c) Aripiprazole has 10 secondary carbons (starred).

Aripiprazole

12.39 A compound can't have a quintary carbon because carbon forms only four bonds, not five.

12.40

(a)

2,3-Dimethylbutane

(b)

1,2-Dimethyl-   1,3-Dimethyl-   1,4-Dimethyl-
cyclohexane   cyclohexane   cyclohexane

12.41

(a)

2,2-Dimethylpropane

(b)

1,2,4-Trimethylcyclohexane

12.42

(a)

(b)

12.43

(a)

(b)

(c)   (d)

(e)

12.44

(a)

CH₃CH₂CH₂CH₂CH₂CH₂OH   CH₃CH₂CH₂CH₂CHCH₃ (OH)   CH₃CH₂CH₂CHCH₂CH₃ (OH)

(b)

CH₃CH₂CH₂CH₂CH₂CH₂CH₂NH₂   CH₃CH₂CH₂CH₂CH₂CHCH₃ (NH₂)

CH₃CH₂CH₂CH₂CHCH₂CH₃ (NH₂)   CH₃CH₂CH₂CHCH₂CH₂CH₃ (NH₂)

12.45

(a)

CH₃CH₂CH₂CHCH₂Br (CH₃)   CH₃CH₂CH₂CCH₃ (CH₃)(Br)   CH₃CH₂CHCHCH₃ (CH₃)(Br)

CH₃CHCH₂CHCH₃ (Br)(CH₃)   BrCH₂CH₂CH₂CHCH₃ (CH₃)

(b)

$$CH_3CH_2\overset{\overset{\displaystyle CH_3}{|}}{C}HCH_2CH_2Cl \qquad CH_3CH_2\overset{\overset{\displaystyle CH_3}{|}}{C}H\underset{\underset{\displaystyle Cl}{|}}{C}HCH_3 \qquad CH_3CH_2\overset{\overset{\displaystyle CH_3}{|}}{\underset{\underset{\displaystyle Cl}{|}}{C}}CH_2CH_3 \qquad CH_3CH_2\overset{\overset{\displaystyle CH_2Cl}{|}}{C}HCH_2CH_3$$

(c)

$$CH_3CH_2CH_2CH_2\overset{\overset{\displaystyle CH_3}{|}}{C}HCH_2OH \qquad CH_3CH_2CH_2CH_2\overset{\overset{\displaystyle CH_3}{|}}{\underset{\underset{\displaystyle OH}{|}}{C}}CH_3 \qquad CH_3CH_2CH_2\overset{\overset{\displaystyle CH_3}{|}}{C}H\underset{\underset{\displaystyle OH}{|}}{C}HCH_3$$

$$CH_3CH_2\overset{\overset{\displaystyle CH_3}{|}}{C}HCH_2\underset{\underset{\displaystyle OH}{|}}{C}HCH_3 \qquad CH_3\underset{\underset{\displaystyle OH}{|}}{C}HCH_2CH_2\overset{\overset{\displaystyle CH_3}{|}}{C}HCH_3 \qquad HOCH_2CH_2CH_2CH_2\overset{\overset{\displaystyle CH_3}{|}}{C}HCH_3$$

12.46

(a)

$$CH_3CH_2CH_3 \quad \text{and} \quad CH_3CH_2 \quad \text{are identical.}$$
$$\qquad\qquad\qquad\qquad\quad |$$
$$\qquad\qquad\qquad\qquad CH_3$$

(b)

$$CH_3\text{—}\overset{|}{\underset{|}{N}}\text{—}CH_3 \quad \text{and} \quad CH_3CH_2\text{—}\overset{|}{\underset{|}{N}}\text{—}H \quad \text{are isomers.}$$
$$\quad\quad\; H \qquad\qquad\qquad\qquad\qquad H$$

(c)

$$CH_3CH_2CH_2\text{—}O\text{—}CH_3 \quad \text{and} \quad CH_3CH_2CH_2\overset{\overset{\displaystyle O}{\parallel}}{C}\text{—}CH_3 \quad \text{are unrelated.}$$

(d)

$$CH_3\overset{\overset{\displaystyle O}{\parallel}}{C}\text{—}CH_2CH_2\overset{\overset{\displaystyle CH_3}{|}}{C}HCH_3 \quad \text{and} \quad CH_3CH_2\overset{\overset{\displaystyle O}{\parallel}}{C}\text{—}CH_2CH_2CH_2CH_3 \quad \text{are isomers.}$$

(e)

$$CH_3CH=CHCH_2CH_2\text{—}O\text{—}H \quad \text{and} \quad CH_3CH_2\overset{\overset{\displaystyle O}{\parallel}}{\underset{\underset{\displaystyle CH_3}{|}}{C}}H\text{—}C\text{—}H \quad \text{are isomers.}$$

12.47

(a)

H H H H
| | | |
H—C—C—C—C—H
| | | |
H H H H

H
|
H—C—H
|
H H H
| | |
H—C—C—C—H
| | |
H H H

H H H
| | |
H—C—C—C—H
| | |
H H H
|
H—C—H
|
H

The first two structures are identical; the third is an isomer.

(b)

CH₃
|
CH₃CHCHCH₃
|
Br

CH₃
|
CH₃CHCHCH₃
|
Br

CH₃
|
CH₂CHCH₂CH₃
|
Br

The first two structures are identical; the third is an isomer.

(c)

The first two structures are isomers, and the third structure is unrelated to them.
(d) The two compounds are stereoisomers of carvone. Although they have the same two-
dimensional structure, they are not identical. As you can see, the structures are mirror
images. (Note: bonds pointing out of the page are wedges, and bonds pointing behind the
page are dashed.)

Spearment   Caraway

12.48 All three structures have a carbon atom with five bonds.

(a)

$CH_3{=}CHCH_2CH_2OH$

↑
5 bonds

(b)

$$\overset{O}{\overset{\|}{}}$$
$CH_3CH_2CH{=}CCH_3$

↗
5 bonds

(c)

CH₃
|
$CH_2CH_2CH_2C{\equiv}CCH_3$

↑   ↗
3 bonds   5 bonds

12.49

a. The carbon lettered "a" has five bonds.
b. The chlorine–carbon bond should be a single bond.

Alkane Nomenclature

12.50

(a)

$$CH_2CH_3$$
$$CH_3CH_2CH_2CH_2CHCHCH_2CH_3$$
$$CH_3$$

4-Ethyl-3-methyloctane

(b)

$$CH_3CHCH_3$$
$$CH_3CH_2CH_2CHCH_2CHCH_3$$
$$CH_2CH_3$$

5-Isopropyl-3-methyloctane

(c)

$$CH_3 \qquad CH_3$$
$$CH_3CCH_2CH_2CH_2CHCH_3$$
$$CH_3$$

2,2,6-Trimethylheptane

(d)

$$CH_2CH_2CH_2CH_3$$
$$CH_3CH_2CH_2CCH_3$$
$$CH_3CHCH_3$$

4-Isopropyl-4-methyloctane

(e)

$$CH_3 \ CH_3$$
$$CH_3CCH_2CCH_3$$
$$CH_3 \ CH_3$$

2,2,4,4-Tetramethylpentane

(f)

$$CH_3CH_2 \ CH_3$$
$$CH_3CH_2CCH_2CH$$
$$CH_3CH_2 \ CH_3$$

4,4-Diethyl-2-methylhexane

(g)

$$CH_3$$
$$CH_3(CH_2)_7C-CH_3$$
$$CH_3$$

2,2-Dimethyldecane

12.51

$$CH_3CH_2CH_2CH_2CH_2CH_3$$
Hexane

$$CH_3$$
$$CH_3CH_2CH_2CHCH_3$$
2-Methylpentane

$$CH_3$$
$$CH_3CH_2CHCH_2CH_3$$
3-Methylpentane

$$CH_3$$
$$CH_3CH_2CCH_3$$
$$CH_3$$
2,2-Dimethylbutane

$$CH_3$$
$$CH_3CHCHCH_3$$
$$CH_3$$
2,3-Dimethylbutane

12.52

(a)

$$C(CH_3)_3$$
$$CH_3CH_2CH_2CHCH_2CHCH_3$$
$$CH_3$$

4-*tert*-Butyl-2-methylheptane

(b)

$$CH_3 \quad CH_3$$
$$CH_3CHCH_2CHCH_3$$

2,4-Dimethylpentane

(c)

$$H_3C \quad CH_2CH_3$$
$$CH_3CH_2CHCCH_2CH_2CH_2CH_3$$
$$CH_2CH_3$$

4,4-Diethyl-3-methyloctane

(d)

3-Ethyl-1-isopropyl-5-methylcycloheptane

(e)

1,1,3-Trimethylcyclopentane

12.53

(a)

1,1-Dimethylclopropane

(b)

1,3-Dimethylcyclopentane

(c)

Ethylcyclohexane

(d)

Cycloheptane

(e)

1-Methyl-3-propylcyclohexane

(f)

1-Ethyl-4-isopropylcyclooctane

12.54

(a)

H₃C—⬦—CH₂CH₃

1-Ethyl-3-methylcyclobutane

(b)

H₃C, CH₃
H₃C⟋⟍CH₃

1,1,3,3-Tetramethylcyclopentane

(c)

CH₂CH₂CH₃

CH₂CH₃

1-Ethyl 3-propylcyclohexane

(d)

H₃C
H₃C—
—CH₂CH₂CH₂CH₃
H₃C
H₃C

4-Butyl-1,1,2,2-tetramethylcyclopentane

12.55

(a)

Isopropylcyclooctane

(b)

CH₂CH₃

CH₂CH₃

1,2-Diethylcyclobutane

(c)

H₃C—⬡—CH₃

1,4-Dimethylcyclohexane

12.56   *Structure*   *Error*

(a)   CH₃
           |
    CH₃CCH₂CH₂CH₃   The prefix *di-* must appear when
           | two substituents are the same.
           CH₃

    2,2-Dimethylpentane

(b)

        CH₃   CH₃
         | | You must choose the longest
   CH₃CHCH₂CHCH₃   carbon chain as the parent name.

    2,4-Dimethylpentane

(c)

        CH₃
         | This compound is an alkyl-
   CH₃CHCH₂—⬦   substituted cycloalkane.
                            Choose the cycloalkane
    Isobutylcyclobutane   as the parent name.

12.57

	Structure	*Error*

(a)

CH₂CH₃
|
CH₃CH₂CHCH₃

3-Methylpentane

The longest carbon chain is a pentane and should be used as the root name.

(b)

CH₃CHCH₃
|
CH₃CH₂CH₂CCH₃
|
CH₃

2,3,3-Trimethylhexane

The longest carbon chain is a hexane and should be used as the root name.

(c)

CH₂CH₃

CH₃

CH₃

2-Ethyl-1,1-dimethylcyclopentane

The substituents should be given the lowest possible numbers. The prefix *di-* should be used when two substituents are the same.

(d)

CH₃ CH₂CH₃
| |
CH₃CCH₂CCH₂CH₃
| |
CH₃ CH₃

4-Ethyl-2,2,4-trimethylhexane

Numbering must start from the end nearer the first substituent.

(e)

CH₃

CH₃CH₂

CH₃

4-Ethyl-1,2-dimethylcyclohexane

Substituents must be cited in alphabetical order (prefixes are not used for alphabetizing).

(f)

CH₂CH₃
|
CH₃CHCH₂CHCH₃
|
CH₂CH₃

3,5-Dimethylheptane

The longest carbon chain is a heptane and should be used as the root name.

(g)

CH₃CH₂ CH₃ CH₃
| | |
CH₃CH₂CH₂C — C — CCH₂CH₂CH₂CH₃
| | |
CH₃CH₂ CH₃ CH₃

4,4-Diethyl-5,5,6,6-tetramethyldecane

The prefixes *tetra-* and *di-* must be included in the name. Substituents must be cited in alphabetical order. Numbering must start from the end nearer to the first substituent.

12.58

$$CH_3CH_2CH_2CH_2CH_2CH_2CH_3$$
Heptane

CH₃
|
$$CH_3CH_2CH_2CH_2CHCH_3$$
2-Methylhexane

CH₃
|
$$CH_3CH_2CH_2CHCH_2CH_3$$
3-Methylhexane

CH₃
|
$$CH_3CH_2CH_2CCH_3$$
|
CH₃
2,2-Dimethylpentane

H₃C CH₃
| |
$$CH_3CH_2CHCHCH_3$$
2,3-Dimethylpentane

CH₃ CH₃
| |
$$CH_3CHCH_2CHCH_3$$
2,4-Dimethylpentane

CH₃
|
$$CH_3CH_2CCH_2CH_3$$
|
CH₃
3,3-Dimethylpentane

CH₂CH₃
|
$$CH_3CH_2CHCH_2CH_3$$
3-Ethylpentane

CH₃ CH₃
| |
$$CH_3CH-CCH_3$$
|
CH₃
2,2,3-Trimethylbutane

12.59

Cyclopentane   Methylcyclobutane   Ethylcyclopropane

1,1-Dimethylcyclopropane   1,2-Dimethylcyclopropane

Reactions of Alkanes

12.60

$$CH_3CH_2CH_3 + 5\ O_2 \longrightarrow 3\ CO_2 + 4\ H_2O$$

12.61

$$2\ C_8H_{18} + 25\ O_2 \longrightarrow 16\ CO_2 + 18\ H_2O$$

12.62

CH₃
|
$$CH_3CH_2CHCH_3 + Cl_2 \xrightarrow{light}$$

CH₃
|
$$CH_3CH_2CHCH_2Cl$$ +

CH₃
|
$$CH_3CH_2CCH_3$$
|
Cl

+

Cl CH₃
| |
$$CH_3CHCHCH_3$$ +

CH₃
|
$$ClCH_2CH_2CHCH_3$$

12.63

$$CH_3CHCH_3 + Br_2 \xrightarrow{\text{light}} CH_3CHCHBr_2 + CH_3CCH_2Br + CH_3CHCH_2Br$$

Conceptual Problems

12.64

(a)

a = alcohol
b = ketone
c = carbon–carbon
    double bond

Testosterone

(b)

a = amine
b = amide
c = alcohol
d = carboxylic acid
e = sulfide
f = carbon–carbon
    double bond

Thienamycin

(c)

a = amine
b = amide
c = aromatic ring

Lisdexamfetamine

Lisdexamfetamine increases neurotransmitter levels in the brain. It is used for both ADHD and/or performance enhancement.

12.65

Pregalbin

Carbon (a) is primary, carbons (c) and (d) are secondary, and carbon (b) is tertiary.

12.66 Both carbon (b) and the starred carbon are tertiary.

12.67 Because carbon forms only four bonds, the largest number of hydrogens that can be bonded to three carbons is eight.

C_3H_8

12.68 Since "like dissolves like," lipstick, which is composed primarily of hydrocarbons, is more soluble in the hydrocarbon petroleum jelly than in water.

12.69

(a)

monobromination products

(b)

dibromination products

12.70 Pentane has a higher boiling point than neopentane because London forces, which must be overcome in boiling, are greater for linear than for more spherical or nonlinear compounds with the same number of carbons.

12.71 Other structures are also possible.

(a)   (b)   (c)   (d)

$CH_3CH_2CH_2\overset{\overset{O}{\|}}{C}-OH$ $CH_3CH_2CH=CHCH_2I$

Group Problems

12.72 Compounds (a) and (d) are the same. Compound (b) is a pentane, and compound (c) has nine carbons.

12.73 In Problem 12.3, we drew heptane, the straight-chain isomer, and in Problem 12.4, we drew the two hexane isomers. Using line-bond structures for the other isomers:

12.74 Among Vaseline's uses: wound ointment, prevention of moisture loss after skin surgery, skin lotion, anti-chapping agent, anti-chafing agent, hair pomade, wood finishing, lubricating metal parts, anti-corrosion coating, moisturizing paws of animals.

12.75

(a)
CH_3CH_2CH with CH_3 and CH_3

12 hydrogens

(b)
$CH_3CH=C$ with CH_3 and CH_3

10 hydrogens

(c)
$HC\equiv CCH$ with CH_3 and CH_3

8 hydrogens

12.76 It is not possible to interconvert the two isomers without breaking bonds.

Self-Test for Chapter 12

Multiple Choice

1. Which of the following functional groups doesn't contain a carbon–oxygen double bond?
   (a) ether (b) aldehyde (c) ketone (d) ester

2. In which of the following alkanes are carbons not tetrahedral?
   (a) ethane (b) propane (c) cyclopropane (d) cyclohexane

3. How many products can result from chlorination of ethane? (Include products with more than one chlorine.)
   (a) 2 (b) 4 (c) 6 (d) 9

4. How many branched-chain isomers of C_6H_{14} are there?
   (a) 2 (b) 3 (c) 4 (d) 5

5. Which of the following condensed structures doesn't represent a cycloalkane?
   (a) C_3H_6 (b) C_4H_{10} (c) C_5H_{10} (d) C_6H_{12}

6.

$$CH_3CH_2\overset{\overset{\displaystyle CH_3}{\displaystyle |}}{C}CH_2\overset{\overset{\displaystyle CH_3}{\displaystyle |}}{C}HCH_3$$
$$\underset{\displaystyle |}{\overset{}{}}CH_2CH_3$$

The correct name for the above structure is:
(a) 3-ethyl-3,5-dimethylhexane   (b) 3-methyl-3-*sec*-butylpentane
(c) 2,4-dimethyl-4-ethylhexane   (d) 4-ethyl-2,4-dimethylhexane

7. How many secondary carbons are in the structure shown in Problem 6?
    (a) 1   (b) 2   (c) 3   (d) 4

8. Which of the following structures is incorrectly drawn?

(a) $CH_3CH_2\overset{\overset{\displaystyle CH_2}{\displaystyle |}}{}\qquad\overset{\overset{\displaystyle CH_3}{\displaystyle |}}{}$
$CH_3CH_2\overset{}{C}CH_2CH_2\overset{}{C}CH_3$
$\qquad\underset{\displaystyle CH_3}{\overset{\displaystyle |}{}}$

(b) $\qquad\overset{\overset{\displaystyle CH_3}{\displaystyle |}}{}\qquad\overset{\overset{\displaystyle CH_2CH_2CH_3}{\displaystyle |}}{}$
$CH_3\overset{}{C}CH_2CH_2\overset{}{C}HCH_3$
$\qquad\underset{\displaystyle CH_3}{\overset{\displaystyle |}{}}$

(c) $\qquad\overset{\overset{\displaystyle CH_3}{\displaystyle |}}{}\quad\overset{\overset{\displaystyle CH_3}{\displaystyle |}}{}$
$CH_3CH_2\overset{}{C}HCH_2\overset{}{C}HCH_3$

(d) $\qquad\overset{\overset{\displaystyle CH_3}{\displaystyle |}}{}\quad\overset{\overset{\displaystyle CH_3}{\displaystyle |}}{}$
$CH_3CH_2CH_2\overset{}{C}H\overset{}{C}HCH_3$
$\qquad\underset{\displaystyle CH_3}{\overset{\displaystyle |}{}}$

9. How many moles of O_2 are necessary for the complete combustion of one mole of C_6H_{12}?
    (a) 4   (b) 6   (c) 9   (d) 10

10. Which alkyl group has a quaternary carbon when part of a hydrocarbon?
    (a) *tert*-butyl   (b) isobutyl   (c) *sec*-butyl   (d) butyl

Sentence Completion

1. C_{20} to C_{36} alkanes are known as _____.

2. Alkynes are compounds that contain _____ bonds.

3. Alcohols and ethers are functional groups that contain _____.

4. Mixtures of hydrocarbons can be separated by _____.

5. A _____ _____ is a shorthand way of drawing a chemical structure.

6. A _____carbon is bonded to four other carbons.

7. Organic compounds generally have _____ melting points than inorganic compounds.

8. Cycloalkanes are compounds that contain carbon atoms joined in a _____.

9. Organic molecules are named by the _____ system.

10. Compounds with the same formula but different structures are called _____.

11. _____ is the reaction of an alkane with oxygen.

12. A straight-chain alkane is _____ boiling than a branched-chain alkane.

True or False

1. Cyclohexane and hexane are isomers.

2. A molecule with the formula C_5H_{12} can have the root name pentane, butane, or propane.

3. A cyclohexane ring is flat.

4. The correct name of the following alkane is 1,3-dimethylpentane.

$$\begin{array}{cc} CH_3 & CH_3 \\ | & | \\ CH_3CH_2CHCH_2CH_2 & \end{array}$$

5. The compound 2,3-dimethylbutane has only primary and tertiary carbons.

6. Acyclic alkanes and cycloalkanes have similar chemical reactivity.

7. The C—Cl bond in CH_3Cl is ionic.

8. The compound 1,4-dimethylcyclohexane is correctly named.

9. Alkanes with one to four carbons exist as gases at room temperature.

10. Compounds with many functional groups are more reactive than compounds with few functional groups.

11. 2-Methylpentane and 3-methylpentane have nearly identical boiling points.

12. At room temperature, an alkane exists in a single conformation.

Match each entry on the left with its partner on the right.

1. $RCH=O$ (a) Contains polar covalent bonds

2. C_5H_{12} (b) Butyl group

3. $CH_3CH(CH_3)_2$ (c) Ketone

4. $CH_3CH_2CH_2CH_3$ (d) Natural gas

5. $CH_3CH_2CH_2-$ (e) Formula of methylcyclobutane

6. $R_2C=O$ (f) Branched-chain alkane

7. CH_2Cl_2 (g) An alkene

8. $C_{30}H_{62}$ (h) Propyl group

9. C_5H_{10} (i) Aldehyde

10. $CH_3CH_2CH_2CH_2-$ (j) Formula of 2-methylbutane

11. $CH_3CH=CH_2$ (k) A solid

12. CH_4 (l) Straight-chain alkane

Chapter 13 Alkenes, Alkynes, and Aromatic Compounds

Chapter Outline

I. Alkenes and alkynes (Sections 13.1–13.7).
  A. Introduction (Section 13.1).
    1. Alkenes are compounds that have double bonds (—C=C—).
    2. Alkynes are compounds that have triple bonds (—C≡C—).
  B. Naming alkenes and alkynes (Section 13.2).
    1. Find the longest chain containing the double or triple bond and name the parent compound by adding -*ene* or -*yne* to the root.
    2. Number the carbons in the main chain, beginning at the end nearer the multiple bond.
      a. If the bond is an equal distance from each end, begin numbering at the end nearer the first branch point.
      b. Cyclic alkenes are named cycloalkenes, and the double-bond carbons are carbons 1 and 2.
    3. Write out the full name.
      a. Number the substituents, and list them alphabetically.
      b. Indicate the position of the multiple bond by giving it the number of the first carbon in the bond.
      c. If more than one double bond is present, use the endings -*diene*, -*triene*, -*tetraene*.
  C. Cis–trans isomerism in alkenes (Section 13.3).
    1. Atoms attached to a double bond lie in a plane.
    2. If each end of the bond is bonded to two different groups, isomerism can result.
      a. When the larger groups are on the same side of the bond, the isomer is cis.
      b. When the larger groups are on opposite sides of the bond, the isomer is trans.
  D. Properties of alkenes and alkynes (Section 13.4).
    1. Nonpolar, insoluble in water, flammable.
    2. Alkane-like in physical properties.
    3. Alkenes show cis–trans isomerism.
    4. Both alkenes and alkynes react at the multiple bond to form addition products.
  E. Organic reactions (Sections 13.5–13.7).
    1. Types of organic reactions (Section 13.5).
      a. Addition reactions occur when two reactants add to form a single product.
      b. Elimination reactions occur when a reactant splits into two products.
      c. Substitution reactions occur when two reactants exchange parts.
      d. Rearrangement reactions occur when a single reactant undergoes a rearrangement of bonds to yield an isomeric product.
    2. Addition reactions of alkenes and alkynes (Section 13.6).
      a. Alkene + H_2 —> alkane (hydrogenation).
        i. Alkyne + 2 H_2 —> alkane.
      b. Alkene + X_2 —> dihaloalkane (X = halogen) (halogenation).
      c. Alkene + HX —> alkyl halide.
        i. Direction of addition: X is bonded to the more substituted carbon (Markovnikov's rule).
      d. Alkene + H_2O —> alcohol (hydration).
        i. Addition also follows Markovnikov's rule.
    3. Mechanism of addition reactions, using HBr addition as an example.
      a. H^+ adds to the alkene to form a carbocation.
      b. Br^- adds to the carbocation to form the product.

4. Alkene polymers (Section 13.7).
    a. Simple alkenes of the type H$_2$C=CHZ (monomer) can add to each other in long chains.
    b. The reaction is started by an initiator, which adds to an alkene to form a reactive intermediate.
    c. The reactive intermediate starts the growth of the chain.
    d. The product is a chain-growth polymer.
    e. Varying –Z varies the properties of the polymer.
    f. Properties of polymers also depend on chain length and degree of branching.
II. Aromatic compounds (Sections 13.8–13.10).
    A. Structure of benzene (Section 13.8).
        1. Benzene is a flat, six-membered ring containing three double bonds.
            a. All carbons in benzene are equivalent.
            b. The structure of benzene is an average of the two conventional structures.
            c. This phenomenon is known as resonance.
        2. Despite having double bonds, benzene is less reactive than alkenes.
        3. Aromatic compounds are nonpolar, water-insoluble, volatile, and flammable.
        4. Some aromatic compounds are toxic and carcinogenic.
    B. Naming aromatic compounds (Section 13.9).
        1. Use "benzene" as the parent name.
        2. For monosubstituted benzenes, the name of the substituent is followed by -*benzene* (no number needed).
        3. For disubstituted benzenes:
            a. Ortho-disubstituted benzenes have substituents in the 1,2 positions of the ring.
            b. Meta-disubstituted benzenes have substituents in the 1,3 positions of the ring.
            c. Para-disubstituted benzenes have substituents in the 1,4 positions of the ring.
        4. Many aromatic compounds have trivial names.
    C. Reactions of aromatic compounds (Section 13.10).
        Aromatic compounds undergo substitution reactions.

        a. Aromatic ring + HNO$_3$ $\xrightarrow{\text{H}_2\text{SO}_4}$ Nitroaromatic ring

        b. Aromatic ring + X$_2$ $\xrightarrow{\text{FeX}_3}$ Haloaromatic ring

        c. Aromatic ring + SO$_3$ $\xrightarrow{\text{H}_2\text{SO}_4}$ Aromatic sulfonic acid

Solutions to Chapter 13 Problems

13.1

(a)   2-methyl → CH$_3$
        CH$_3$CH$_2$CH$_2$CH=CHCHCH$_3$
         7   6   5   4   3 2 1

        2-Methyl-3-heptene

(b)   6   5   4   3   2   1
        H$_2$C=CHCH$_2$CH$_2$C=CH$_2$
                          |
                  2-methyl → CH$_3$

        2-Methyl-1,5-hexadiene

(c)   3-Methyl-3-hexene

(d)   3-Ethyl-6-methyl-4-octyne

13.2

(a)

$$CH_3CH_2CH_2CH_2CHCH=CH_2$$
with CH_3 attached

3-Methyl-1-heptene

(b)

$$H_3C-\overset{\overset{\displaystyle CH_3}{|}}{\underset{\underset{\displaystyle CH_3}{|}}{C}}-C\equiv C-CH_3$$

$$-C\equiv C-$$

4,4-Dimethyl-2-pentyne

(c)

$$CH_3CH_2CH=CHCHCH_3$$
with CH_3 attached

2-Methyl-3-hexene

(d)

with CH_3

H_3C CH_3

1,3,3-Trimethylcyclohexene

13.3

(a)

2,3-Dimethyl-1-pentene

(b)

2,3-Dimethyl-2-hexene

13.4   (a) Attached to C2: $-CH_3$, $-CH_3$
    Attached to C3: $-H$, $-CH_2CH_3$
2,3-Dimethyl-2-pentene does not exhibit cis–trans isomerism because the groups bonded to C2 are the same.

$$\underset{CH_3CH_2}{\overset{H_3C}{\diagdown}}C=C\underset{CH_3}{\overset{CH_3}{\diagup}}$$

2,3-Dimethyl-2-pentene

(b) Attached to C2: $-CH_3$, $-CH_3$
    Attached to C3: $-H$, $-CH_2CH_2CH_3$
As in (a), since the two groups attached to C2 are identical, 2-methyl-2-hexene does not exist as cis–trans isomers:

$$\underset{CH_3CH_2CH_2}{\overset{H}{\diagdown}}C=C\underset{CH_3}{\overset{CH_3}{\diagup}}$$

2-Methyl-2-hexene

(c) Attached to C2: —H, —CH$_3$
    Attached to C3: —H, —CH$_2$CH$_2$CH$_3$

Since each carbon has two different groups attached to it, 2-hexene exists as cis–trans isomers:

trans-2-Hexene   *cis*-2-Hexene

HOC 13.1

(a) The structure in which three groups on one carbon are eclipsed by three groups on an adjacent carbon is highest in energy.

(b) The cis isomer experiences a similar type of eclipsing interaction as takes place in alkanes.

(c) Only one isomer is possible.

(d) Molecules with all trans bonds are less kinked and can form crystals easier.

13.5

cis-3,4-Dimethyl-3-hexene   *trans*-3,4-Dimethyl-3-hexane

13.6

(a)   (b)

cis-4-Methyl-2-hexene   *trans*-5,6-Dimethyl-3-heptene

13.7

(a)   CH$_3$Br + NaOH ⟶ CH$_3$OH + NaBr

This reaction is a substitution because two reagents exchange parts to give two different products.

(b)   H$_2$C=CH$_2$ + HCl ⟶ CH$_3$CH$_2$Cl

This reaction is an addition because two reactants combine to give one product.

(c) CH_3CH_2Br $\longrightarrow$ $H_2C{=}CH_2$ + HBr

In this elimination reaction, two products are formed from one reactant.

13.8 (a) This reaction is an addition because water adds to the alkene double bond.
(b) This reaction is an elimination.

CIA 13.1
(a) The reaction is a rearrangement.
There are 28 hydrogens present in 11-*cis*-retinal.
One aldehyde and 5 carbon–carbon double bonds are present in 11-*cis*-retinal.
(b) The rod cells are responsible for vision in dim light, and the cone cells are responsible for vision in bright light and for color vision.
(c) Tetrabromofluorescein absorbs in the yellow range.

13.9

(a)

(b)

(c)

(d)

13.10

(a)

(b) $CH_3CH_2CH_2CH{=}CH_2$ + Cl_2 $\longrightarrow$ $CH_3CH_2CH_2\underset{\underset{Cl}{|}}{CH}{-}CH_2Cl$

(c)

CH₃CH₂CH=CCH₂CHCH₃ + Cl₂ ⟶ CH₃CH₂CH—CCH₂CHCH₃

(d)

+ Br₂ ⟶

13.11

CH₃CH=CCH₃ $\xrightarrow{\text{HCl}}$ CH₃CH₂CCH₃ + CH₃CHCHCH₃

major Cl   minor Cl

In the major product, hydrogen adds to the alkene carbon that already has a hydrogen bonded to it.

13.12

(a)

+ HCl ⟶

(b)

+ HBr ⟶

(c)

$\xrightarrow{\text{HCl}}$

13.13

(a)

CH₃CH₂C=CHCH₃ + HCl ⟶ CH₃CH₂CCH₂CH₃

3-Ethyl-2-pentene   3-Chloro-3-ethylpentane

3-Chloro-3-ethylpentane is the major product because hydrogen adds to the double bond carbon that has the most hydrogens already bonded to it.
(b) This is not the major product. In the major product, hydrogen adds to the carbon that already has a hydrogen bonded to it.

$\xrightarrow{\text{HBr}}$

13.14

MR 13.1

2-Methylpropene   carbocation   2-Chloro-2-
                                   intermediate   methylpropane

MR 13.2

4-Methyl-1-pentyne   actual product

Addition of HBr to 4-methyl-1-pentyne might form either of the two bromoalkenes pictured above. The second product is actually formed because its intermediate carbocation is the one predicted by Markovnikov's rule.

MR 13.3

Addition of H$^+$ produces a carbocation that can be represented as two resonance forms. Br$^-$ adds to produce the observed alkene (see MR 13.2).

13.15

(a)

major   minor

(b)

major   minor

(c)

The products are formed in approximately equal amounts.

13.16

$$CH_3CH_2\overset{\overset{\displaystyle CH_3}{|}}{C}=CHCH_3$$
3-Methyl-2-pentene

or

$$CH_3CH_2\overset{\overset{\displaystyle CH_2}{||}}{C}CH_2CH_3$$
2-Ethyl-1-butene

$+ \quad H_2O \quad \xrightarrow{H_2SO_4} \quad CH_3CH_2\overset{\overset{\displaystyle CH_3}{|}}{\underset{\underset{\displaystyle OH}{|}}{C}}CH_2CH_3$

13.17 To draw a vinyl polymer: (1) Orient the monomer unit so that the CH₂= end is on the left. (2) Draw the functional group(s) at the other end of the molecule pointing up (or up and down). (3) Break the double bond, and draw a single bond extending from each carbon. (4) Connect a monomer unit to each of the new bonds.

13.18 Use the procedure described in the previous problem.

13.19

(a)

HO

—CH₂CH₃

m-Ethylphenol

(b)

Cl

—CH₃

m-Chlorotoluene

(c)

1-Ethyl-3-isopropylbenzene

13.20

(a)

NO₂

Cl

m-Chloronitrobenzene

(b)

H₃C   NO₂

o-Nitrotoluene

(c)

H₃C—   —NH₂

p-Methylaniline

(d)

O₂N—   —OH

p-Nitrophenol

13.21

(a)

OH

CHCH₃
|
CH₃

o-Isopropylphenol

(b)

Br

NH₂

p-Bromoaniline

13.22 For reasons not explained here, not all of these products are formed in equal amounts.

(a)

(b)

(c)

13.23

o-Bromophenol   m-Bromophenol   p-Bromophenol

CIA 13.4

Rasagiline is nontoxic to humans and is neuroprotective.

CIA 13.5

The antibody directs the chemotherapy drug to tumor tissue without affecting surrounding tissue.

CIA 13.6

Many tumors develop drug resistance and many therapies are nonselective and toxic.

Understanding Key Concepts

13.24

(a)

$$CH_3CH_2CHCH_2CH=CCH_3$$

with CH_3 substituent, **2,5-Dimethyl-2-heptene**

HBr →

$$CH_3CH_2CHCH_2CH_2CCH_3$$ with CH_3 and Br substituents

H_2O / acid catalyst →

$$CH_3CH_2CHCH_2CH_2CCH_3$$ with CH_3 and OH substituents

(b)

3.3-Dimethylcyclopentene

HBr → (two bromo products)

H_2O / acid catalyst → (two hydroxy products)

13.25

(a)

$$HC\equiv CCH_2CCH_2CH_3$$ with two CH_3 substituents

4,4-Dimethyl-1-hexyne

(b)

$$CH_3CHCH_2C\equiv CCH_2CHCH_3$$ with CH_3 substituents

2,7-Dimethyl-4-octyne

13.26

(a)

m-Ethylphenol

(b)

o-Bromobenzoic acid

13.27

(a)

(b)

13.28

(a)

4,4-Dimethyl-1-hexyne   3,3-Dimethylhexane

(b)

2,7-Dimethyl-4-octyne   2,7-Dimethyloctane

13.29

Additional Problems

Naming Alkenes, Alkynes, and Aromatic Compounds

13.30 (a) Alkenes, alkynes, and aromatic compounds are said to be unsaturated because they have carbon–carbon multiple bonds to which hydrogen can be added.
(b) $CH_3CH_2CH_2CH_3$ is saturated, and $CH_3CH_2CH=CH_2$ is unsaturated.

13.31 (a) The term *aromatic* in chemistry refers to compounds that have a six-membered ring containing three double bonds. The association with aroma is of historical origin but no longer has any meaning.
(b) Resonance describes a phenomenon in which the true structure of a compound is an average of two or more possible structures.

13.32

Family:	Alkene	Alkyne	Aromatic Compound
Name Ending:	*-ene*	*-yne*	*-benzene*

13.33 (a) The name of a 1,3-disubstituted benzene includes the prefix *meta- (m-)*.
(b) The name of a 1,4-disubstituted benzene includes the prefix *para- (p-)*.

13.34 There are many possible answers to this question. For example,

(a)

(b)

$CH_3CH_2CH_2C{\equiv}CH$ $CH_3CH_2C{\equiv}CCH_3$ $CH_3CHC{\equiv}CH$ with CH_3 substituent

(c)   (d)

13.35 Other structures are possible.

(a)

$CH_3CH_2CH=C$ with CH_3 and CH_3 substituents

(b)

13.36

(a)

CH₃CH=CHCH₂CH₃

2-Pentene

(b)

2,5-Dimethyl-3-hexyne

(c)

3,4-Diethyl-3-hexene

(d)

$$CH_3C=C-C=CHCH_3$$
$$\quad | \quad \quad |$$
$$CH_3 \quad CH_3$$

2,4-Dimethyl-2,4-hexadiene

(e)

3,6-Dimethylcyclohexene

(f)

4-Ethyl-1,2-dimethylcyclopentene

13.37

(a)

o-Chloroisopropylbenzene

(b)

Br—⟨⟩—NO₂

p-Bromonitrobenzene

(c)

O₂N

m-Nitrophenol

13.38

(a)

CH₃CH₂   H
    \\   /
     C=C
    /   \\
   H   CH₃

trans-2-Pentene

(b)

CH₃CH₂   CH₃
    \\   /
     C=C
    /   \\
  H₃C   CH₂CH₃

trans-3,4-Dimethyl-3-hexene

(c)

      CH₃
       |
H₂C=CHC=CH₂

2-Methyl-1,3-butadiene

(d)

CH₃CH₂CH₂   H
     \\   /
      C=C
     /   \\
    H   CH₂CH₃

trans-3-Heptene

(e)

O₂N—⟨⟩—CH₃

p-Nitrotoluene

(f)

o-Chlorophenol

(g)

1,2-Dimethylcyclobutene

(h)

        CH₃   CH₂CH₃
         | |
CH₃CH₂CH₂CHCH=CHCCH₂CH₃
                    |
                   CH₂CH₃

3,3-Diethyl-6-methyl-4-nonene

13.39

(a) Aniline

(b) Phenol

(c) *o*-Xylene

(d) 2,4,6-Trinitrobenzene

(e) *p*-Chlorobenzoic acid

(f) *m*-Nitroaniline

(g) *o*-Chlorobenzaldehyde

(h) Anisole

13.40

CH₃CH₂CH₂CH₂C≡CH

1-Hexyne

CH₃CH₂CH₂C≡CCH₃

2-Hexyne

CH₃CH₂C≡CCH₂CH₃

3-Hexyne

CH₃CH₂CHC≡CH
3-Methyl-1-pentyne

CH₃CHCH₂C≡CH
4-Methyl-1-pentyne

CH₃CHC≡CCH₃
4-Methyl-2-pentyne

CH₃CC≡CH
3,3-Dimethyl-1-butyne

13.41

o-Methylphenol
(*o*-Cresol)

m-Methylphenol
(*m*-Cresol)

p-Methylphenol
(*p*-Cresol)

13.42

![Reaction of ethylbenzene with HNO3/H2SO4 producing o-, m-, and p-nitroethylbenzene]

HNO₃
————→
H₂SO₄

NO₂

+

NO₂

+

NO₂

Ethylbenzene   *o*-Nitroethylbenzene   *m*-Nitroethylbenzene

p-Nitroethylbenzene

13.43

$$CH_3$$
$$H_2C=CCH_2CH_2CH_3$$
2-Methyl-1-pentene

$$CH_3$$
$$CH_3C=CHCH_2CH_3$$
2-Methyl-2-pentene

$$CH_3$$
$$CH_3CHCH=CHCH_3$$
4-Methyl-2-pentene
(cis and trans)

$$CH_3$$
$$CH_3CHCH_2CH=CH_2$$
4-Methyl-1-pentene

Alkene Cis–Trans Isomers

13.44 For an alkene to show cis–trans isomerism, each carbon of the double bond must be bonded to two different groups.

13.45 Alkynes do not show cis–trans isomerism because a triple bond is linear and only one group is bonded to each alkyne carbon.

13.46

(a)

2-Methyl-2-octene

(b)

3-Heptene
trans and cis

(c)

3,4-Dimethyl-3-hexene
trans and cis

13.47 4-Methyl-2-pentene is the only compound in Problem 13.43 that can show cis–trans isomerism because it is an unsymmetrically substituted alkene.

13.48

(a)

cis-3-Heptene

(b)

cis-4-Methyl-2-pentene

(c)

trans-2,5-Dimethyl-3-hexene

13.49

(a)

(b)

13.50 (a) These compounds are identical:

and

(b) These compounds are also identical:

and

13.51

(a)

(b)

Kinds of Reactions

13.52 In a substitution reaction, two reactants exchange parts to give two new products. In an addition reaction, two reactants add together to form a single product, with no atoms left over.

13.53

$$CH_3CH=CHCH_3 \ + \ H_2 \ \xrightarrow[\text{catalyst}]{\text{Pd}} \ CH_3CH_2CH_2CH_3$$

13.54 The conversion of 2-methyl-2-pentene to 1-hexene would be a rearrangement because it is the conversion of a single reactant into a single product by the reorganization of bonds.

13.55 The conversion of bromocyclohexane to cyclohexene would be an elimination reaction because a single reactant splits to give two products (cyclohexene and HBr).

13.56

(a)

This reaction is a substitution because two reactants exchange parts to give two new products.

(b)

This is a rearrangement.

13.57

(a)

substitution reaction

(b)

addition reaction

Reactions of Alkenes and Alkynes

13.58

(a)

$$CH_3CH_2CH=CHCH_3 \xrightarrow[Pd]{H_2} CH_3CH_2CH_2CH_2CH_3$$

(b)

$$CH_3CH_2CH=CHCH_3 \xrightarrow{Br_2} CH_3CH_2\underset{\underset{Br}{|}}{\overset{\overset{Br}{|}}{C}}HCHCH_3$$

(c)

$$CH_3CH_2CH=CHCH_3 \xrightarrow{HCl} CH_3CH_2\underset{\underset{Cl}{|}}{C}HCH_2CH_3 + CH_3CH_2CH_2\underset{\underset{Cl}{|}}{C}HCH_3$$

(d)

$$CH_3CH_2CH=CHCH_3 \xrightarrow[H_2SO_4]{H_2O} CH_3CH_2\underset{\underset{OH}{|}}{C}HCH_2CH_3 + CH_3CH_2CH_2\underset{\underset{OH}{|}}{C}HCH_3$$

13.59 In (c) and (d), the major products are shown.

13.60

(a)

(b)

$$CH_3CH=CH_2 \xrightarrow[Pd\ catalyst]{H_2} CH_3CH_2CH_3$$

(c)

$$CH_3CH=CHCH_3$$

or

$$H_2C=CHCH_2CH_3 \xrightarrow{HBr} CH_3\underset{\underset{Br}{|}}{C}HCH_2CH_3$$

(d)

$$\text{cyclopentene} \xrightarrow[\text{H}_2\text{SO}_4]{\text{H}_2\text{O}} \text{cyclopentanol (OH)}$$

(e)

$$\text{methylenecyclohexane} (=\text{CH}_2) \xrightarrow{\text{Cl}_2} \text{1-chloro-1-(chloromethyl)cyclohexane (Cl, CH}_2\text{Cl)}$$

13.61

$$\text{CH}_3\text{CH}_2\text{C}\equiv\text{CCH}_3 \xrightarrow{\text{2 Br}_2} \text{CH}_3\text{CH}_2\overset{\overset{\displaystyle Br}{|}}{\underset{\underset{\displaystyle Br}{|}}{C}}-\overset{\overset{\displaystyle Br}{|}}{\underset{\underset{\displaystyle Br}{|}}{C}}\text{CH}_3$$

2-Pentyne   2,2,3,3-Tetrabromopentane

13.62

$$\underset{\text{CH}_3\text{CHCH}_2\text{C}\equiv\text{CH}}{\overset{\overset{\displaystyle \text{CH}_3}{|}}{}} \xrightarrow{\text{HBr}} \underset{\text{CH}_3\text{CHCH}_2\text{C}=\text{CH}_2}{\overset{\overset{\displaystyle \text{CH}_3}{|}\,\,\,\overset{\displaystyle Br}{|}}{}} + \underset{\text{CH}_3\text{CHCH}_2\text{CH}=\text{CHBr}}{\overset{\overset{\displaystyle \text{CH}_3}{|}}{}}$$

Addition of HBr to 4-methyl-1-pentyne might form either of the two bromoalkenes pictured above.

13.63

$$\text{polymerization} \longrightarrow \text{PVP}$$

$$+(\text{CH}_2-\text{CH}-\text{CH}_2-\text{CH}-\text{CH}_2-\text{CH})+$$

13.64

$$\text{H}_2\text{C}=\text{CCl}_2 \xrightarrow{\text{polymerization}} +\left(\overset{\overset{\displaystyle H}{|}}{\underset{\underset{\displaystyle H}{|}}{C}}-\overset{\overset{\displaystyle Cl}{|}}{\underset{\underset{\displaystyle Cl}{|}}{C}}-\overset{\overset{\displaystyle H}{|}}{\underset{\underset{\displaystyle H}{|}}{C}}-\overset{\overset{\displaystyle Cl}{|}}{\underset{\underset{\displaystyle Cl}{|}}{C}}-\overset{\overset{\displaystyle H}{|}}{\underset{\underset{\displaystyle H}{|}}{C}}-\overset{\overset{\displaystyle Cl}{|}}{\underset{\underset{\displaystyle Cl}{|}}{C}}\right)+$$

Reactions of Aromatic Compounds

13.65 Under conditions in which an alkene would react with all three reagents, chlorobenzene reacts only with Br_2 (a) and HNO_3 (c) to give the products shown below.

(a)

o-Bromochloro-benzene m-Bromochloro-benzene p-Bromo-chlorobenzene

(c)

o-Nitrochloro-benzene m-Nitrochloro-benzene p-Nitro-chlorobenzene

13.66

13.67

Toluene   +   3 H_2 $\xrightarrow{\text{catalyst}}$ Methylcyclohexane

13.68

Conceptual Problems

13.69

Salicylic acid

13.70

Structure	*Error*
(a) CH₃ on chain, CH₃CH=CHCH₂CHCH₃ 5-Methyl-2-hexene	Numbering should start from the end nearer the double bond.
(b) CH₃ on chain, CH₃CH₂CH₂CHC≡CCH₃ 4-Methyl-2-heptyne	The methyl group cannot be bonded to carbon 1.
(c) CH₃CHCH₃ / CH₃C=CH₂ 2,3-Dimethyl-1-butene	The longest chain containing the double bond should be used as the base name.
(d) O₂N— benzene ring —NO₂, NO₂ 1,2,4-Trinitrobenzene	Substituents should receive the lowest possible numbers.

(a)

$$CH_3CH=CHCH_2CHCH_3 \quad (CH_3)$$

5-Methyl-2-hexene

(b)

$$CH_3CH_2CH_2CHC\equiv CCH_3 \quad (CH_3)$$

4-Methyl-2-heptyne

(c)

$$CH_3CHCH_3$$
$$CH_3C=CH_2$$

2,3-Dimethyl-1-butene

(d)

$$O_2N- \quad -NO_2$$
$$NO_2$$

1,2,4-Trinitrobenzene

Structure   *Error*

(e)

The double bond receives the lowest number. (For cyclic alkenes, the double bond receives no number but is understood to be between carbons 1 and 2.)

3,4-Dimethylcyclohexene

(f)

$$CH_3$$
$$|$$
$$H_2C{=}CHC{=}CHCH_3$$

The double bonds should receive the lowest possible numbers.

3-Methyl-1,3-pentadiene

None of the compounds are symmetrically substituted.

13.71 Compounds in parts (a) and (f) can show cis–trans isomerism.

(a)

(f)

13.72 None of the reagents that add to alkenes react with alkanes such as cyclohexane. Thus, you can mix a reagent such as Br_2 with a sample from each bottle. The mixture of cyclohexene and Br_2 turns from reddish-brown to colorless, but the mixture of cyclohexane and Br_2 remains colored.

13.73 The same reasoning used in Problem 13.72 applies here. Benzene is much less reactive than alkenes. Thus, treatment of a sample from each bottle with an alkene addition reagent such as Br_2 gives a rapid reaction with cyclohexene but no reaction with benzene.

13.74

Cl—◯—Cl   *p*-Dichlorobenzene

13.75

Menthene

13.76

Cinnamaldehyde
The other addition product does not form.

13.77

(a)

$$CH_3CH_2CH=CHCHCH_3 \xrightarrow[\text{catalyst}]{\overset{H_2}{\text{Pd}}} CH_3CH_2CH_2CH_2CHCH_3$$

(b)

(c)

(d)

(e)

$$CH_3C{\equiv}CCH_2CH_3 \xrightarrow[\text{catalyst}]{\overset{H_2}{\text{Pd}}} CH_3CH_2CH_2CH_2CH_3$$

13.78

$$CH_3CH_2CH\!=\!CHCH_3 \xrightarrow{\ HBr\ } CH_3CH_2\overset{\overset{\displaystyle Br}{|}}{C}H-\overset{\overset{\displaystyle H}{|}}{C}HCH_3 \ + \ CH_3CH_2\overset{\overset{\displaystyle H}{|}}{C}H-\overset{\overset{\displaystyle Br}{|}}{C}HCH_3$$
2-Pentene

The products are formed in equal amounts because each end of the double bond has the same number of hydrogens and the two intermediate carbocations are formed in approximately equal amounts.

13.79

(a)

Ocimene   (3,7-Dimethyl-1,3,6-octatriene)

(b)

Ocimene $\xrightarrow{\ xs\ HBr\ }$ $CH_3\overset{\overset{\displaystyle CH_3}{|}}{\underset{\underset{\displaystyle Br}{|}}{C}}CH_2CH_2CH_2\overset{\overset{\displaystyle CH_3}{|}}{\underset{\underset{\displaystyle Br}{|}}{C}}-\overset{}{\underset{\underset{\displaystyle Br}{|}}{C}}HCH_3$ major product

13.80

3,4,4-Trimethyl-2-pentene

or

2-Ethyl-3,3-dimethyl-1-butene

$\xrightarrow[\substack{H_2SO_4\\ catalyst}]{H_2O}$ $CH_3CH_2-\overset{\overset{\displaystyle HO}{|}}{\underset{\underset{\displaystyle H_3C}{|}}{C}}-\overset{\overset{\displaystyle CH_3}{|}}{\underset{\underset{\displaystyle CH_3}{|}}{C}}-CH_3$

13.81 Compound (b) does not exhibit cis–trans isomerism.

(a)

(c)

Group Problems

13.82 A trans double bond is too strained to exist in a small ring like cyclohexene, but a large ring is more flexible and can include a trans double bond:

The double bond must be cis in this six-membered ring.

The double bond can be trans in this 10-membered ring.

13.83

polymerization

13.84

2-Methyl-1-butene   3-Methyl-1-butene   2-Methyl-2-butene

Self-Test for Chapter 13

Multiple Choice

1. Which of the following polymers is a hydrocarbon?
   (a) Teflon   (b) poly(vinyl chloride)   (c) Lucite   (d) polystyrene

2.
   $$CH_3CH=CHCHCH_3$$
   with CH_2CH_3 substituent

   The name of this compound is:
   (a) 3-methyl-4-hexene   (b) 4-methyl-2-hexene   (c) 2-ethyl-3-pentene   (d) 3-ethyl-2-pentene

3. Which of the following statements about the reaction of benzene with Br_2 is true?
   (a) The reaction is catalyzed by iron.   (b) The reaction is an addition reaction.
   (c) Two different products are formed.
   (d) When bromine is added to benzene, the bromine color disappears.

4. Reaction of a hydrocarbon with 2 H_2 gives a product with the formula C_6H_{12}. Which of the following is the original hydrocarbon?
   (a) benzene   (b) 3-methyl-1,3-pentadiene   (c) 1,4-cyclohexadiene   (d) 2-hexyne

5. Which of the following molecules shows cis–trans isomerism?
   (a) 2-hexene   (b) 2-hexyne   (c) 2-methyl-2-pentene   (d) 2-ethyl-1-hexene

6. Which of the following statements about carbocations is true?
   (a) A carbocation has seven electrons in its outer shell.
   (b) Two different carbocations can be formed in the reaction of HBr with 3-hexene.
   (c) A carbocation is an intermediate in the reaction of HBr with an alkene.
   (d) Carbocations are stable ions.

7. How many isomers with the formula C_5H_8 can be drawn? (Don't include isomers with rings or cis–trans isomers.)
   (a) 5   (b) 7   (c) 9   (d) 11

8. Which of the following is probably not a monomer in a polymerization reaction?
   (a) $H_2C=CHCN$ (b) $H_2C=CHCH_3$ (c) $CH_3CH=C(CH_3)_2$ (d) $H_2C=CCl_2$

9. If toluene reacted with one mole of Cl_2, and each of the products of the reaction reacted with one mole of Cl_2, how many different products could be formed? (Not all of these can be formed in the laboratory.)
   (a) 3   (b) 4   (c) 5   (d) 6

10. Which of the following alkenes is not a starting material for 3-bromo-3-methylhexane?
    (a) 2-ethyl-1-pentene   (b) 3-methyl-2-hexene   (c) 3-methyl-3-hexene   (d) 4-methyl-2-hexene

Sentence Completion

1. The name of $CH_3CH_2CH_2CH_2C{\equiv}CCH_3$ is _____.

2. Addition of hydrogen to an alkene or alkyne is known as _____.

3. A benzene ring with substituents at the 1 and 4 positions is said to be _____ disubstituted.

4. In the addition of HX to an alkene, the H becomes attached to the carbon that already has ____ H's, and the X becomes attached to the carbon that has _____ H's.

5. Ethanol can be produced by the _____ of ethylene.

6. In the nitration of benzene, H_2SO_4 is used as a _____.

7. The intermediate formed during the addition of HBr to an alkene is called a _____.

8. Compounds with double or triple bonds are said to be _____.

9. _____ is another name for methylbenzene.

10. The reaction of Br_2 with benzene is an example of a _____ reaction.

11. _____ _____ _____ are aromatic compounds consisting of two or more rings joined together by a common bond.

12. Simple alkenes are made by _____ _____ of natural gas and petroleum.

True or False

1.   Cis–trans isomers have identical physical properties.

2.   Addition of one equivalent of H_2 to an alkyne yields an alkene.

3.   Aromatic compounds are less reactive than alkenes.

4.   The major product of addition of HBr to 1-butene is 1-bromobutane.

5.   The reaction of HBr with an alkene is known as halogenation.

6.   A compound with the formula C_4H_6 can be either an alkyne or an alkene.

7.   1-Methylcyclohexene can exhibit cis–trans isomerism.

8.   Mixing an alkene with water causes the hydration of the double bond.

9.   Addition of HBr to an alkene is a two-step reaction.

10.   Benzene and 1-butene react with Br_2 under the same reaction conditions.

11.   Only aromatic compounds with more than one ring are carcinogenic.

12.   Two different compounds result from the reaction of HBr with 2-pentene.

Match each entry on the left with its partner on the right.

1. HNO_3, H_2SO_4 (a) Alkene without cis–trans isomers

2. H_2O, H_2SO_4 (b) Used to hydrogenate an alkene

3. $CH_3C{\equiv}CH$ (c) Polycyclic aromatic hydrocarbon

4.   Hydroxybenzene   (d) Addition reaction

5. H_2, Pd   (e) Used to nitrate aromatic compounds

6. $(CH_3)_2CH^+$ (f) Substitution reaction

7.   Propene + Br_2 (g) Exhibits cis–trans isomerism

8.   Naphthalene   (h) Teflon monomer

9.   2-Methylpropene   (i) Alkyne

10.   Benzene + Br_2 (j) Phenol

11.   2-Butene   (k) Used to hydrate an alkene

12. $CF_2{=}CF_2$ (l) Carbocation

Chapter Outline

I. Compounds containing oxygen (Sections 14.1–14.8).
    A. Introduction (Section 14.1).
        1. In all of these compounds, oxygen forms a single bond to either carbon or hydrogen.
        2. Alcohols (R—OH), phenols (Ar—OH), and ethers (R—O—R) are examples of these compounds.
            a. Many alcohols resemble water in their physical properties.
            b. Alcohols (and phenols) can form hydrogen bonds, which raise their boiling points.
    B. Alcohols (Sections 14.2–14.4).
        1. Naming alcohols (Section 14.2).
            a. Find the longest chain containing the hydroxyl group, and name the chain by replacing the -e ending of the corresponding alkane with -ol.
            b. Number the carbon atoms in the main chain, beginning at the end nearer the –OH group.
            c. Write the name.
                i.   Place the number of the —OH group immediately before the parent compound name.
                ii. Number all other substituents according to their positions and list them alphabetically.
            d. Diols are often known as glycols when the two –OH groups are adjacent.
            e. Alcohols can be classified according to the number of substituents bonded to the —OH carbon.
                i.   If there is one substituent (RCH_2OH), the alcohol is primary.
                ii. If there are two substituents (R_2CHOH), the alcohol is secondary.
                iii. If there are three substituents (R_3COH), the alcohol is tertiary.
        2. Properties of alcohols (Section 14.3).
            a. Hydrogen bonding makes alcohols higher boiling than other compounds of similar mass.
            b. Hydrogen bonding also makes smaller alcohols miscible with water, as well as with organic solvents.
                i.   Alcohols of higher mass are not soluble in water.
                ii. Diols (glycols) are even more soluble than alcohols of similar mass.
            c. Alcohols are weak acids.
        3. Reactions of alcohols (Section 14.4).
            a. Dehydration: alcohol $\xrightarrow{\text{H}_2\text{SO}_4}$ alkene + H_2O
                If more than one product is possible, the product with the more substituted double bond will be favored.
            b. Oxidation:
                i.   Primary alcohol $\xrightarrow{[O]}$ aldehyde $\xrightarrow{[O]}$ carboxylic acid
                ii. Secondary alcohol $\xrightarrow{[O]}$ ketone
                iii. Tertiary alcohol $\xrightarrow{[O]}$ no reaction

C. Phenols (Sections 14.5–14.6).
  1. Properties of phenols (Section 14.5).
    a. Phenols are named by replacing *-benzene* with *-phenol*.
    b. The properties of phenols are influenced by hydrogen bonding.
      i. Many phenols are water-soluble.
      ii. Phenols are higher boiling than alkylbenzenes.
    c. Phenols are commonly used as disinfectants.
  2. Acidity of alcohols and phenols (Section 14.6).
    a. Alcohols and phenols are weak acids.
      i. Methanol and ethanol are as acidic as water.
      ii. Dissociation of alcohols in water, or reaction of alcohols with sodium metal, produces an alkoxide anion (^-OR).
    b. Phenols are more acidic than water.
      i. K_a of phenol is 1.0×10^{-10}.
      ii. Phenols are soluble in dilute aqueous NaOH.
D. Ethers (Section 14.7).
  1. Names and properties of ethers.
    a. Ethers have two organic groups bonded to the same oxygen.
    b. Ethers are named by identifying the two groups and adding the word *ether*.
      i. Some cyclic ethers have common names (ethylene oxide, 1,4-dioxane).
      ii. An $-OR$ group is an alkoxy group.
    c. Properties of ethers.
      i. Ethers are low-boiling and don't form hydrogen bonds.
      ii. Ethers are unreactive.
      iii. Ethers are good solvents for organic compounds.
      iv. Ethers are volatile and flammable.
      v. Ethers are only slightly soluble in water.
      vi. Ethers form explosive peroxides on standing.
  2. Some common ethers.
    a. Diethyl ether is commonly used as a solvent but was formerly an anesthetic.
    b. Ether groups occur in many essential oils.
II. Thiols and disulfides (Section 14.8).
  A. Thiols.
    1. Thiols are the sulfur analogs of alcohols ($R-SH$).
    2. Naming thiols is similar to naming alcohols, except that the suffix *-thiol* is added to the parent name.
    3. Thiols are important biologically because they occur in proteins.
    4. Thiols stink!
  B. Disulfides.
    1. Reaction of a thiol with a mild oxidizing agent produces a disulfide ($RS-SR$).
    2. The reverse reaction occurs when a disulfide is treated with a reducing agent.
III. Halogen-containing compounds (Section 14.9).
  A. Haloalkanes are named by considering the halogen as a substituent on a parent alkane.
  B. A few haloalkanes are named by naming the alkyl group and then naming the halide.
  C. Haloalkanes do not occur frequently in nature but are widely used in industry and agriculture.

IV. Stereochemistry and Chirality (Section 14.10).
  A. Stereochemistry.
    1. Stereochemistry is the study of molecules that have the same overall connectivity but differ in the arrangement of these atoms in space.
    2. Stereoisomers such as cis–trans isomers have different configurations.
  B. Chirality.
    1. Objects and molecules may have chirality.
    2. Chirality is the property of handedness.
    3. A molecule is chiral if it contains a tetrahedral carbon bonded to four different groups.
      a. A carbon that is tetrahedral and bonded to four different groups is a chiral carbon.
    4. The two mirror images of a molecule with a chiral carbon are enantiomers, or optical isomers.
      a. Optical isomers are identical in all properties except for their effect on polarized light and their interactions with other chiral compounds.
      b. Pairs of enantiomers may differ in their activity as drugs.

Solutions to Chapter 14 Problems

14.1

(a)

$CH_3CH_2CHCH_3$
     |
     OH

an alcohol

(b)

an alcohol

(c)

a phenol

(d)

an alcohol

(e)

an ether

(f)

$CH_3CHOCH_2CH_3$
    |
    CH_3

an ether

14.2 Even though ethers can't form hydrogen bonds with each other, they can form hydrogen bonds with water: The ether oxygen lone-pair electrons can hydrogen-bond with the hydrogens of water.

14.3

(a)

$$CH_3CH_2CH_2\overset{\overset{\displaystyle CH_3}{|}}{C}HCH_2CH_2OH$$

3-Methyl-1-hexanol
primary alcohol

(b)

1-Methyl-3-propylcyclopentanol
tertiary alcohol

(c)

$$CH_3CH_2CH_2\overset{}{\underset{\underset{\displaystyle OH}{|}}{C}}H-\overset{\overset{\displaystyle CH_3}{|}}{\underset{\underset{\displaystyle CH_3}{|}}{C}}CH_3$$

2,2-Dimethyl-3-hexanol
secondary alcohol

(d)

$$CH_3CH_2CH_2CH_2\overset{\overset{\displaystyle OH}{|}}{C}HCH_2CH_3$$

3-Heptanol
secondary alcohol

(e)

2,3-Diethylcyclohexanol
secondary alcohol

14.4

(a)

$$H_3C-\overset{\overset{\displaystyle CH_3}{|}}{\underset{\underset{\displaystyle CH_3}{|}}{C}}-OH$$

2-Methyl-2-propanol
(*tert*-Butyl alcohol)
tertiary alcohol

(b)

3-Methyl-2-pentanol
secondary alcohol

(c)

$$CH_3CH_2\overset{\overset{\displaystyle CH_2OH}{|}}{C}HCH_2CH_2\overset{\overset{}{}}{\underset{\underset{\displaystyle Cl}{|}}{C}}HCH_3$$

5-Chloro-2-ethyl-1-hexanol
primary alcohol

(d)

1,2-Cyclopentanediol
secondary alcohol

14.5 See Problems 14.3 and 14.4.

14.6

Highest boiling $HOCH_2\overset{\overset{}{}}{\underset{\underset{\displaystyle OH}{|}}{C}}HCH_2OH,$ $CH_3\overset{\overset{}{}}{\underset{\underset{\displaystyle OH}{|}}{C}}HCH_2OH,$ $CH_3CH_2CH_2OH,$ $CH_3CH_2CH_3$ *Lowest boiling*

14.7

(i) (a)

(b)

(c)

(ii) Hydrophilic parts are circled; the rest of the molecule is hydrophobic.
(iii) Compound (c) is most soluble (two hydrophilic groups), and (a) is least soluble (large hydrophobic part).

MR 14.1

$$CH_3CH_2CH_2OH \xrightleftharpoons[]{H_2SO_4} CH_3CH=CH_2 + H_2O$$

The conversion of an alcohol, such as propanol, to the related alkene is an equilibrium process in which sulfuric acid serves as a catalyst for both the forward and the reverse reaction. In alkene formation, sulfuric acid is the source of H^+, which protonates –OH, creating $-OH_2^+$, which leaves the alcohol, forming a carbocation (see Problem MR 14.2). In alcohol formation, H^+ protonates the double bond, forming a carbocation, which adds water to form a protonated alcohol that loses $-H^+$. The concentration of water determines the direction of reaction.

MR 14.2

MR 14.3

Since all of the reactions are equilibrium processes, the double bond of 4-methyl-2-pentene can react with H_2O to form 2-methyl-3-pentanol, which can be dehydrated to yield 2-methyl-2-pentene, the product with the more substituted double bond.

4-Methyl-2-pentanol   4-Methyl-2-pentene   2-Methyl-3-pentanol   2-Methyl-2-pentene

14.8

(a)

$$CH_3CH{-}CH_2 \xrightarrow{\;H_2SO_4\;} CH_3CH{=}CH_2 \;+\; H_2O$$

(the circled H and OH are on adjacent carbons)

(b)

$$\xrightarrow{\;H_2SO_4\;}$$ (cyclohexene) $+\; H_2O$

(c)

$$CH_2{-}CH{-}CHCHCH_3 \xrightarrow{\;H_2SO_4\;} H_2C{=}CHCH_2\underset{CH_3}{CH}CH_3 \;+\; CH_3CH{=}CH\underset{CH_3}{CH}CH_3 \;+\; H_2O$$

minor   major

14.9

(a)

$$CH_3\underset{OH}{CH}\underset{CH_3}{CH}CH_3 \;or\; CH_3CH_2\underset{CH_3}{\underset{OH}{C}}CH_3 \xrightarrow{\;H_2SO_4\;} CH_3CH{=}\underset{CH_3}{C}CH_3 \;+\; H_2O$$

(b)

$$\xrightarrow{\;H_2SO_4\;}$$ (product) $+\; H_2O$

(c)

$$\xrightarrow{\;H_2SO_4\;}$$ (stilbene)

14.10

$$\xrightarrow{\;H_2SO_4\;}$$ (major) ${=}CH{-}CH_3 \;+$ (minor) ${-}CH{=}CH_2 \;+\; H_2O$

major   minor

$$\xrightarrow{\;H_2SO_4\;}$$ (major) $+$ (minor) $+\; H_2O$

major   minor

The major product has three alkyl groups bonded to it, and the minor product has only one.

14.11

(a)

(b)

(c)

14.12

(a)

(b)

(c)

14.13

(a)

(b)

14.14

(a)

2,4-Dinitrophenol

(b)

m-Ethylphenol

14.15

(a)

2,4-Dibromophenol

(b)

3-Iodo-2-methylphenol

CIA 14.1

Diethyl ether was used as the first general anesthetic.

CIA 14.2

The solubility of inhaled anesthetics in the bloodstream correlates with their potency.

CIA 14.3

Minimum alveolar concentration (MAC) is defined as the concentration of anesthetic in inhaled air that results in anesthesia in 50% of patients.

14.16

(a)

CH₃OCH₂CHCH₃
         |
         OCH₃

1,2-Dimethoxypropane

(b)

p-Methoxynitrobenzene

(c)

tert-Butyl methyl ether

14.17

(a)

$$2\ CH_3CH_2CH_2SH \xrightarrow{[O]} CH_3CH_2CH_2S-SCH_2CH_2CH_3 + H_2O$$

(b)

$$2\ CH_3\underset{CH_3}{CH}CH_2CH_2SH \xrightarrow{[O]} CH_3\underset{CH_3}{CH}CH_2CH_2S-SCH_2CH_2\underset{CH_3}{CH}CH_3 + H_2O$$

14.18

(a)

Cl / CH₂CH₃ on cyclopentane

1-Chloro-1-ethylcyclopentane

(b)

$$CH_3CH_2CHCH_2CHCH_2CH_3$$

with CH₃ and Br substituents

3-Bromo-5-methylheptane

HOC 14.1

(a) Molecules of CH_2BrCl can be superimposed and have a plane of symmetry.

(b) Molecules of $CHIBrCl$ can't be superimposed and do not have a plane of symmetry.

(c) Both sets have two chiral carbons. In Set 1, the two molecules are superimposable; in Set 2, they are not superimposable.

(d) The stereoisomers are not mirror images.

14.19

$$H_3C-\underset{\underset{H}{|}}{\overset{\overset{NH_2}{|}}{C}}-CH_3$$

2-Aminopropane

$$CH_3CH_2-\underset{\underset{H}{|}}{\overset{\overset{NH_2}{|}}{\overset{*}{C}}}-CH_3$$

2-Aminobutane

2-Aminopropane is achiral because no carbon has four different groups bonded to it. 2-Aminobutane is chiral because four different groups are bonded to carbon 2 ($-H$, $-CH_3$, $-NH_2$ and $-CH_2CH_3$).

14.20

(a)

$$CH_3CH_2\underset{\underset{H}{|}}{\overset{\overset{Cl}{|}}{C}}CH_2CH_3$$

achiral

(b)

$$CH_3CH_2CH_2\underset{\underset{H}{|}}{\overset{\overset{Cl}{|}}{\overset{*}{C}}}CH_3$$

chiral

(c)

$$CH_3CH\underset{\underset{\,}{|}}{\overset{\overset{CH_3}{|}}{\,}}CH_2\underset{\underset{H}{|}}{\overset{\overset{CH_3}{|}}{\overset{*}{C}}}CH_2CH_3$$

chiral

The starred carbon atoms in (b) and (c) each have four different groups bonded to them.

CIA 14.4

Ethanol is a depressant. Its effects on the central nervous system are like those of anesthetics and other central nervous system depressants.

CIA 14.5

At a blood alcohol concentration of 80–300 mg/dL, speech becomes slurred. At a blood alcohol concentration above 600 mg/dL, death may result.

CIA 14.6

A syndrome is a group of symptoms that are linked to one another and to a specific disease.

CIA 14.7

CNS disorders may include delayed growth, facial abnormalities, delayed development of motor skills, hyperactivity, attention deficit disorder, and mental retardation.

Understanding Key Concepts

14.21

(a)

5-Methyl-3-hexanol

(b)

m-Methoxytoluene

(c)

3-Methylcyclohexanol

14.22

14.23

14.24

14.25

(a)

(b)

(c)

Additional Problems

Alcohols, Ethers, and Phenols

14.26 *Alcohols* have an —OH group bonded to an alkane-like carbon atom.
Ethers have an oxygen atom bonded to two carbon atoms.
Phenols have an —OH group bonded to a carbon of an aromatic ring.

an alcohol   an ether   a phenol

14.27 A primary alcohol has one carbon substituent bonded to the OH-bearing carbon; a secondary alcohol has two carbon substituents bonded to the OH-bearing carbon; a tertiary alcohol has three carbon substituents bonded to the OH-bearing carbon.

primary alcohol   secondary alcohol   tertiary alcohol

14.28 Alcohols contain —OH groups, which can form hydrogen bonds to each other. Since extra energy (heat) must be supplied to break these hydrogen bonds, alcohols have higher boiling points than ethers, which can't form hydrogen bonds.

14.29 Phenol is a stronger acid than ethanol.

14.30

a = alcohol
b = ether
c = ketone
d = carbon–carbon
      double bond

Taxane nucleus

All alcohols are secondary except for two tertiary alcohols (starred).

14.31

14.32, 14.36

(a)

2-Methyl-2-propanol
tertiary alcohol

(b)

2-Methyl-1-propanol
primary alcohol

(c)

1,2,4-Butanetriol
∗ primary alcohol
secondary alcohol

(d)

2-Methyl-2-phenyl-
1-propanol
primary alcohol

(e)

3-Methylcyclohexanol
secondary alcohol

(f)

3-Ethyl-3-methyl-2-hexanol
secondary alcohol

14.33

(a)

3,4,5-Trimethylphenol

(b)

$CH_3CH-O-CH_2CH_3$
with CH_3 above

Ethyl isopropyl ether

(c)

2,4,6-Trinitrophenol

(d)

Cyclobutyl cyclopentyl
ether

(e)

o-Butylphenol

(f)

$CH_3CH_2CH_2OCH_2CH_2CH_3$

Dipropyl ether

14.34

(a)

$$CH_3CH_2CH_2CHCH_2CCH_3$$

with CH_3 groups above and OH below

2,4-Dimethyl-2-heptanol

(b)

OH, CH_2CH_3, CH_2CH_3 on cyclohexane

2,2-Diethylcyclohexanol

(c)

$$CH_3CH_2CCH_2CH_2CH_2CH_2OH$$

with CH_2CH_3 above and CH_3 below

5-Ethyl-5-methyl-1-heptanol

(d)

$$CH_3CH_2CHCH_2CHCH_3$$

with CH_3CH_2 and OH

4-Ethyl-2-hexanol

(e)

OCH_3, OH on cyclooctane

3-Methoxycyclooctanol

(f)

$$CH_3CHCH_2CH_2CCH_2CH_2OH$$

with OH, CH_3, CH_3

3,3-Dimethyl-1,6-heptanediol

14.35

(a)

CH_3OCHCH_3 with CH_3

Isopropyl methyl ether

(b)

OH, OH on benzene

o-Dihydroxybenzene (catechol)

(c)

$$O-C-CH_3$$

with CH_3, CH_3 on phenyl

Phenyl *tert*-butyl ether

(d)

OH, I on benzene

m-Iodophenol

(e)

$$CH_3CHCHCHCH_3$$

with CH_3O, OCH_3, CH_3

2,4-Dimethoxy-3-methylpentane

(f)

$$CH_3CHCHCH=CH_2$$

with CH_3, OCH_3

3-Methoxy-4-methyl-1-pentene

14.36 (a) See Problem 14.32.

(b) Compounds (a), (b), and (c) are water-soluble. The hydroxyl groups (hydrophilic) are responsible for solubility. As the rest of the molecule (hydrophobic) becomes larger, solubility decreases.

14.37 See Problem 14.30.

14.38

Compound	Boiling Point	Reason
1-Hexanol	Highest	Forms hydrogen bonds
Dipropyl ether	Middle	Polar, but does not form hydrogen bonds
Hexane	Lowest	Nonpolar

14.39 Glucose, with five hydroxyl groups, can form more hydrogen bonds with water and thus is more soluble than 1-hexanol, which has only one hydroxyl group.

Reactions of Alcohols

14.40 A ketone is formed on oxidation of a secondary alcohol:

A secondary alcohol   A ketone

14.41 In order to undergo oxidation, an alcohol must have a –H bonded to the carbon that is bonded to the –OH group. Tertiary alcohols are not oxidized because they have no –H bonded to the OH-bearing carbon, and thus can't form a C=O double bond.

No hydrogen bonded to this carbon

A tertiary alcohol

14.42 A carboxylic acid is formed when a primary alcohol is oxidized with an excess of oxidizing agent.

a primary alcohol   an aldehyde   a carboxylic acid

14.43 An alkoxide anion is formed on reaction of an alcohol with Na metal.

$$2 \text{ H—OR} + 2 \text{ Na} \longrightarrow 2 \text{ Na}^{+-}\text{OR} + \text{H}_2$$

alcohol   sodium
                          alkoxide

14.44 Phenols are more acidic than alcohols and are converted into their sodium salts on reaction with NaOH. Thus, the phenol dissolves in aqueous NaOH but the alcohol does not.

14.45

(a)

(b)

(c)

14.46

(a)

(b)

(c)

or

The product shown is the major product because three groups are attached to the double bond.

(d)

(e)

$$HOCH_2CH_2CH_2CH_2CH_2OH \xrightarrow{\text{xs } H_2SO_4} H_2C=CHCH_2CH=CH_2 + 2\ H_2O$$

This is the only route that gives the desired product cleanly.

14.47

(a)

major   minor

(b)

minor   major

(c)

(d)

(e)

14.48

(a)

(b)

(c)

(d)

(e)

(f)

14.49

(a)

(b)

(c)

Thiols and Disulfides

14.50 The most noticeable characteristic of thiols is their nasty odor.

14.51 A thiol is the sulfur analog of an alcohol.

$$R-S-H \qquad\qquad R-O-H$$
$$\text{a thiol} \qquad\qquad \text{an alcohol}$$

14.52

14.53

14.54 Propanol is high-boiling because its –OH groups can form hydrogen bonds. The –SH groups of ethanethiol don't form hydrogen bonds, and thus ethanethiol has a lower boiling point, as does chloroethane, which also doesn't form hydrogen bonds.

14.55 The explanation in the previous problem applies here. Propanol is very soluble in water because it can hydrogen-bond with water. Chloroethane and ethanethiol don't form hydrogen bonds with water and are thus only slightly soluble in water.

Stereochemistry and Chirality

14.56 (a) A *chiral* object is one that has handedness.
(b) An *achiral* object has no handedness and possesses a plane of symmetry such that one half of the object is an exact mirror image of the other half.
(c) A *chiral carbon* is a carbon having four different substituents.
(d) An *enantiomer* is one of a pair of compounds that have a mirror image relationship.

14.57 chiral: (c) your right hand   achiral: (a) a fork; (b) this book; (c) index card

14.58 Chiral carbons are starred.

14.59 Chiral carbons are starred. If a compound is not chiral, a plane of symmetry is indicated.
chiral: (b), (c)   achiral: (a), (d)

Conceptual Problems

14.60 Straight-chain alcohol isomers of $C_5H_{12}O$:

Straight-chain ether isomers of $C_4H_{10}O$:

14.61

Thyronine

$$\xrightarrow{4\ I_2}$$

Thyroxine

Thyroxine results from the aromatic substitution reaction of thyronine and I_2.

14.62 Alcohols become less soluble with increasing hydrocarbon chain length because the effect of the lengthening water-insoluble hydrocarbon chain becomes more important than the effect of the polar hydroxyl group.

14.63

14.64 An antiseptic kills microorganisms on living tissue; a disinfectant is toxic and is used only to kill microorganisms on surfaces.

14.65 Alcohols (a), (b), and (e) in Problem 14.47 are achiral. Alcohols (c) and (d) are chiral because each has at least one carbon bonded to four different groups.

14.66 Alcohols (b), (c), (e), and (f) in Problem 14.48 are chiral because each has at least one carbon bonded to four different groups.

14.67

(a)

$$\overset{OH}{\underset{|}{CH_3CHCH_3}}$$

2-Propanol

(rubbing alcohol)

(b)

$$CH_3OH$$

Methanol

(wood alcohol)

(c)

$$CH_3CH_2OH$$

Ethanol

(grain alcohol)

(d)

$$HOCH_2CH_2OH$$

Ethylene glycol

(antifreeze)

$$\overset{OH}{\underset{|}{CH_3CHCH_2OH}}$$

Propylene glycol

(antifreeze)

14.68

(a)

Br— ⟨benzene ring⟩ —Br

p-Dibromobenzene

(b)

BrCH=CCH₂CH₃
        |
        Br

1,2-Dibromo-1-butene

(c)

OCH₃

⟨benzene ring⟩

CH₂CH₂CH₃

m-Propylanisole
or
m-Methoxypropylbenzene

(d)

⟨cyclopentane ring with two Br⟩
Br
Br

1,1-Dibromocyclopentane

(e)

    OH   OH
    | |
CH₃CCH₂CCH₃
    | |
    CH₃ CH₃

2,4-Dimethyl-2,4-pentanediol

(f)

        HO OH   OH
        | | |
CH₃CH₂CHCHCH₂CHCH₃
            |
            CH₃

4-Methyl-2,4,5-heptanetriol

(g)

        Br   CH₃
        | |
CH₃C≡CCHCH₂CCH₃
              |
              CH₃

4-Bromo-6,6-dimethyl-2-heptyne

(h)

⟨cyclobutane ring⟩
Cl
I

1-Chloro-2-iodocyclobutane

14.69 Major products are shown.

(a)

    CH₃
    |
CH₃C=CHCH₃ + HBr ⟶

    CH₃
    |
CH₃CCH₂CH₃
    |
    Br

(b)

  [O] ⟶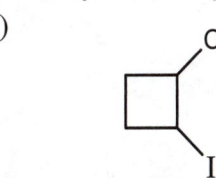

H₃C OH
 | |
CH₃CH₂CH₂C—CHCH₃
     |
     H₃C

H₃C O
 | ‖
CH₃CH₂CH₂C—CCH₃
     |
     H₃C

(c)

H₃C OH
 | |
CH₃CH₂CH₂C—CHCH₃   H₂SO₄ ⟶
     |
     H₃C

H₃C
 |
CH₃CH₂CH₂C—CH=CH₂
     |
     H₃C

(d)

(e)

(f)

(g)

14.70

3,7-Dimethyl-2,6-octadien-1-ol   Citral

14.71

Ethanol   Acetic acid

14.72

$$CH_3CH_2OH + 3 O_2 \longrightarrow 2 CO_2 + 3 H_2O$$

Group Problems

14.73

(a)

(b) Chiral carbons are starred. Notice that each chiral alcohol has two chirality centers.

14.74

The major product has a tetrasubstituted double bond, and the two minor products have trisubstituted double bonds. Since dehydration can occur in three different directions, it is reasonable for there to be three different products and for one of them to be more stable than the others.

14.75

Three products can be formed from dehydration of the alcohol shown. The first product (major) and the third product (minor) each have a double-bond carbon bonded to identical groups and do not show cis–trans isomerism. In the second product, each double-bond carbon is bonded to two different groups; those alkene products show cis–trans isomerism. In one compound, the two larger groups are cis, and in the other compound, the larger groups are trans.

Self-Test for Chapter 14

Multiple Choice

1. Which of the following alcohols yields an aldehyde on oxidation?
   (a) 2-hexanol   (b) 2-methyl-2-hexanol   (c) 2-methyl-1-hexanol   (d) *p*-chlorophenol

2. Which of the following compounds is least acidic?
   (a) *p*-chlorophenol   (b) 2-butanol   (c) HCl   (d) acetic acid

3. How many alcohols of the formula $C_4H_{10}O$ are primary alcohols?
   (a) 1   (b) 2   (c) 3   (d) 4

4. Which of the following is known as wood alcohol?
   (a) CH_3OH (b) CH_3CH_2OH (c) $CH_3CH_2CH_2OH$ (d) $HOCH_2CH_2OH$

5.   HO $CH_2CH_2CH_3$
            | |
         $CH_3CHCH_2CHCH_2CH_3$

   What is the name of the compound shown?
   (a) 4-propyl-2-hexanol   (b) 4-ethyl-6-heptanol   (c) 4-ethylheptanol   (d) 4-ethyl-2-heptanol

6. Which of the following is not a property of alcohols?
   (a) weakly acidic   (b) flammable   (c) react with acids   (d) low-boiling

7. Which of the following is not a product of dehydration of 2,3-dimethyl-3-pentanol?
   (a) 2,3-dimethyl-1-pentene   (b) 2,3-dimethyl-2-pentene   (c) 3,4-dimethyl-2-pentene
   (d) 2-ethyl-3-methyl-1-butene

8. Which of the following is a cyclic ether?
   (a) enflurane   (b) ethylene oxide   (c) methoxybenzene (anisole)   (d) methyl cyclohexyl ether

9. An alcohol of the formula $C_5H_{12}O$ forms a carboxylic acid on oxidation yet can't be dehydrated by acid. What is its name?
   (a) 1-pentanol   (b) 2-methyl-2-butanol   (c) 2,2-dimethyl-1-propanol   (d) 2-methyl-1-butanol

10. Which reagent does not react with alcohols?
    (a) Na   (b) H_2SO_4 (c) $KMnO_4$ (d) NaOH

Sentence Completion

1. The common name for methanol is ___ ____.

2. Dialcohols are often called _____.

3. Aromatic compounds called _____ react with NaOH to give salts.

4. The dehydration of an alcohol yields an _____.

5.   Compounds called _____ are noted for their foul odors.

6.   _____ can be used to oxidize an alcohol.

7.   Ethers are _____ boiling than alcohols of similar molecular weight.

8.   _____ is a phenol used as a food additive.

9.   The major product of alcohol dehydration has the _____ number of alkyl groups attached to the double-bond carbons.

10.   A _____ _____ is a reactive intermediate containing an unpaired electron.

11.   On prolonged contact with air, an ether forms a _____.

12.   _____ is another name for a chlorofluorocarbon.

True or False

1.   Phenols can form hydrogen bonds.

2.   An aldehyde can be formed by oxidation of a secondary alcohol.

3.   Two different alkenes result from dehydration of 2-pentanol.

4.   Another name for phenol is carbonic acid.

5.   Phenols are more acidic than alcohols.

6.   Another name for a thiol is a mercaptan.

7.   Phenol, isopropyl alcohol, and chloroethane are all antiseptics.

8.   Oxidation is the removal of two hydrogen ions.

9.   Several halogenated alkanes are used as anesthetics.

10.   Under careful reaction conditions, a primary alcohol is oxidized to an aldehyde.

11.   Isopropyl alcohol is more toxic than methyl alcohol.

12.   Hydrogen bonding accounts for both boiling-point elevation and water solubility.

Match each entry on the left with its partner on the right (use each answer once).

1. CCl_3F (a) Peroxide

2. $(CH_3)_2CHS-SCH(CH_3)_2$ (b) Thiol

3. $CH_3CH(OH)CH_2CH_2CH_3$ (c) Phenol

4. $CH_3CH_2OCH_3$ (d) Alcohol that is not water-soluble

5. CH_3OH (e) Tertiary alcohol

6. CH_3SH (f) Forms two alkenes on dehydration

7. $(CH_3)_3C-OH$ (g) Disulfide

8. $CH_3CH_2OOCH_3$ (h) Ether

9. $(CH_3)_3CCOCH_3$ (i) Glycol

10. $CH_3(CH_2)_{10}CH_2OH$ (j) Chlorofluorocarbon

11. $CH_3CH(OH)CH_2OH$ (k) Oxidation product of 3,3-dimethyl-2-butanol

12. C_6H_5OH (l) Formed from CO and H_2

Chapter Outline

I. Characteristics of aldehydes and ketones (Sections 15.1–15.4).
  A. The carbonyl group (Section 15.1).
    1. Carbonyl groups are polarized, with carbon having a partial positive charge and oxygen having a partial negative charge.
    2. Carbonyl groups are planar, with a 120° bond angle.
    3. Carbonyl compounds can be divided into two groups:
      a. In aldehydes and ketones, the carbonyl carbon is bonded to atoms that don't attract electrons strongly.
      b. In carboxylic acids, esters, anhydrides, and amides, the carbonyl carbon is bonded to nitrogen or oxygen.
  B. Naming simple aldehydes and ketones (Section 15.2).
    1. Aldehydes are named by replacing the suffix of the parent compound with -al.
      a. For substituted aldehydes, the carbon chain is numbered with the aldehyde carbon as carbon 1.
      b. Aldehydes with common names are formaldehyde, acetaldehyde, and benzaldehyde.
    2. Ketones are named by replacing the suffix of the parent compound with -one.
      a. Numbering starts at the end nearer the carbonyl carbon.
      b. Ketones with common names include acetone and acetophenone.
      c. Sometimes ketones are named by naming the two alkyl groups, followed by the word *ketone*.
  C. Properties of aldehydes and ketones (Section 15.3).
    1. Aldehydes and ketones are moderately polar.
    2. Aldehydes and ketones don't hydrogen-bond with each other and have lower boiling points than alcohols.
    3. Aldehydes and ketones are soluble in organic solvents.
    4. Simple aldehydes and ketones are water-soluble because they can hydrogen-bond with water.
    5. Aldehydes and ketones have distinctive odors.
    6. Aldehydes and ketones are liquids.
  D. Common aldehydes and ketones (Section 15.4).
    1. Many aromas and flavors are due to aldehydes and ketones.
    2. Aldehydes and ketones are used as starting materials in industry and pharmaceutical manufacture.
    3. Formaldehyde, acetaldehyde, acetone, and benzaldehyde are among the most common aldehydes and ketones.
II. Reactions of aldehydes and ketones (Sections 15.5–15.7).
  A. Oxidation (Section 15.5).
    1. Aldehydes can be oxidized to carboxylic acids.
      a. Tollens' reagent converts aldehydes to carboxylic acids and leaves a shiny silver residue on the flask.
      b. Benedict's reagent leaves a brick-red precipitate.
    2. Ketones don't react with most oxidizing agents.

B. Reduction (Section 15.6).
   1. Reducing agents reduce aldehydes to primary alcohols and ketones to secondary alcohols in a two-step reaction.
      a. Hydride ion adds to the carbonyl carbon.
      b. Hydrogen ion adds to the carbonyl oxygen.
   2. In the body, the coenzyme NADH is the reducing agent.
C. Formation of hemiacetals and acetals (Section 15.7).
   1. Hemiacetals and hemiketals.
      a. An alcohol can add to an aldehyde or ketone to form a hemiacetal or hemiketal.
         i. The negatively polarized alcohol oxygen adds to the positively polarized carbonyl carbon.
         ii. The reaction is reversible: Hemiacetals and hemiketals easily revert to an aldehyde or ketone.
      b. Although hemiacetals and hemiketals are often too unstable to isolate, the hemiacetals of sugars are stable.
   2. Acetals and ketals.
      a. Hemiacetals and hemiketals can react with a second molecule of an alcohol to form an acetal or ketal.
      b. Acetals and ketals are stable and can be isolated.
   3. Hydrolysis.
      a. Aqueous acid reacts with acetals and ketals to regenerate the original aldehyde or ketone, plus alcohol.

Solutions to Chapter 15 Problems

15.1

(a)

(b)

Prostaglandin E_1

Testosterone

(c)

Vanillin

(d)

$C_4H_9COCH_3$
ketone

(e)

C_4H_9CHO
aldehyde

15.2

(d)

H–C–C–C–C–C–C–H (structural formula with H atoms and O)

(e)

H–C–C–C–C–C–H (structural formula with H atoms and O)

15.3

(a)

CH₃CH₂CH₂CH₂CH₂CH₂CH₂CH
(with O double bond)

Octanal

(b)

Methyl phenyl ketone

(c)

CH₃CH₂CHCH₂CH₂CH (with CH₃ branch and O)

4-Methylhexanal

(d)

H₃C–C–C–CH₃ (with H₃C groups and O)

tert-Butyl methyl ketone

CIA 15.1

(a)

CH₃CH₂CH₂CH₂C–CN (with OH and H)

(b)

CH₃CH₂CCH₂CH₃ (with HO and CN)

Break the carbonyl C=O bond, add –H to the –O–, and add –CN to the carbon.

CIA 15.2

The cyanohydrin that decomposes to form HCN is nontoxic and is stable inside the millipede's body.

15.4

(a)

CH₃CH₂CH₂CH₂CH (with O)

Pentanal   H

(b)

CH₃CH₂CCH₂CH₃ (with O)

3-Pentanone
(Diethyl ketone)

(c)

CH₃CH₂CHCH₂CH₂CH (with CH₃ branch and O)

4-Methylhexanal   H

(d)

CH₃CH₂CH₂CCH₂CH₂CH₃

4-Heptanone
(Dipropyl ketone)

15.5

(a)

Benzyl methyl ketone

(b)

Ethyl isopropyl ketone

(c)

tert-Butyl methylcyclohexyl ketone

15.6

CH₃CHCH₂CCH₂CH₃
|
CH₃

C₇H₁₄O

5-Methyl-3-hexanone
a ketone

CH₃CHCH₂CH₂CH
|
CH₃

C₆H₁₂O

4-Methylpentanal
an aldehyde

15.7

(a)

O
‖
CH₃CCH₂CH₃

polar, soluble in water

(b)

O
‖
CH₃CH₂C—H

polar, soluble in water

(c)

CH₃CH₂CH₂CH₂CH₃

nonpolar, insoluble
in water

(d)

H₃C

O

polar, insoluble in water

15.8   Alcohols are higher boiling than aldehydes and ketones of similar molecular weight because hydrogen bonding takes place between alcohol molecules, raising their boiling points. Aldehydes and ketones, however, are higher boiling than alkanes of similar molecular weight because aldehydes and ketones are polar and show dipole–dipole interactions, which elevate their boiling points relative to the boiling points of alkanes.

15.9

(a)

(b)

(c)

aldehyde

(d)

$H_2NCH_2CH_2COCH_3$

amine   ketone

15.10

(a)   (b)   (c)   (d)

(a) This compound, an aldehyde, gives a positive reaction with both Tollens' reagent and Benedict's reagent.
(b) This compound, a ketone, gives a negative reaction with both Tollens' reagent and Benedict's reagent.
(c) This compound, an aldehyde, gives a positive reaction with both Tollens' reagent and Benedict's reagent.
(d) Even though this compound is a ketone, it gives a positive test with Benedict's reagent because it is an alpha hydroxy ketone. It gives a negative reaction with Tollens' reagent.

15.11 First, draw the aldehyde or ketone so that the C=O group is vertical. Next, break the double bond, and add one hydrogen to the carbon and one to the oxygen. Finally, redraw the product.

(a)

(b)

(c)

15.12 If the target alcohol is primary, start with an aldehyde. If the target alcohol is secondary, start with a ketone.

(a)

(b)

(c)

15.13 To identify a hemiacetal, look for a carbon with single bonds to two different oxygens. If one group bonded to the carbon is —OH and the other is —OR (R = organic group), the compound is a hemiacetal. If that carbon is also bonded to two alkyl groups, the compound is a ketal.

(a)

This compound is a hemiacetal.

(b)

OH   OH
 | |
CH₃CH—CHCH₃

This compound is not a hemiacetal.

(c)

O⟍ ⟍OH

OCH₃

This compound is not a hemiacetal.

(d)

O⟍ *⟍OH

OH

This compound is a hemiacetal
(the hemiacetal carbon is starred).

15.14 First, locate the carbon bearing both –OR groups (starred below). Then check to see if one of the other two groups bonded to that carbon is a hydrogen. If there is a hydrogen, the compound was formed from an aldehyde and is an acetal. If neither of the other groups is hydrogen, the compound was formed from a ketone and is a ketal.

(a)   (b)   (c)

Compound:

OCH₃
|
*C—CH₂CH₃
|
OCH₂CH₃

CH₃ at top; O—*—O ring

cyclopentane with *C(CH₃)(O-isopropyl)(O)

	(a)	(b)	(c)
Groups bonded to starred carbon:	2 –OR, 2 –R	2 –OR, – R, –H	2 –OR, 2 – R
Acetal or ketal?	ketal	acetal	ketal

Aldehyde/ketone:

O
‖
—CCH₂CH₃ (phenyl)

O
‖
CH₃C—H

(methyl isopropyl ketone structure)

15.15

(a)

(isobutyraldehyde) H + CH₃CH₂OH ⇌ (acid catalyst) → product
HO   OCH₂CH₃

(b)

CH₃ (cyclohexanone ring) =O + CH₃OH ⇌ (acid catalyst) → CH₃ (ring) OCH₃, OH

15.16

(a)

(b)

15.17 Follow the instructions for Worked Example 15.5.

(a)   (b)   (c)   (d)

hemiacetal   acetal   ketal   hemiketal

15.18 Use the instructions from previous problems. If the anomeric carbon also has –H bonded to it, the sugar is a hemiacetal; otherwise, it is a hemiketal.

Tagatose
hemiketal

Idose
hemiacetal

15.19 Find the two —OR groups that form the acetal or ketal. Remove them and replace them with =O to form the original aldehyde or ketone. Add –H to the –OR groups to form the original alcohol.

(a)

(b)

acetal bonds

CH₃CH₂CH₂O—CH—OCH₂CH₂CH₃ $\xrightarrow{H_3O^+}$ CH₃CH₂CH + 2 CH₃CH₂CH₂OH
　　　　　　　|
　　　　　　CH₂CH₃　　　　　　　　　　　　Propanal　　　　1-Propanol

(c)

acetal bonds

CH₃CH₂CH₂O—CH₂—OCH₂CH₂CH₃ $\xrightarrow{H_3O^+}$ H—C—H + 2 CH₃CH₂CH₂OH
　　　　　　　　　　　　　　　　　　　Formaldehyde　　　1-Propanol

HOC 15.1

(a)　　　　　　　　　　　　　　　　(b)

α-Glucose

β-Glucose

(a) α-Glucose has five chiral carbons (starred). An arrow points to the hemiacetal carbon.
(b) β-Glucose also has five chiral carbons (starred). An arrow points to the hemiacetal carbon.
(c) The two glucose anomers are not mirror images.

(d)　　　　　　　　　　(e)

α-Galactose　　　　　α-Fructose　　　　β-Fructose

(d) α-Galactose has five chiral carbons (starred). It differs from α-glucose in the configuration at the indicated carbon (#).
(e) α-Fructose has four chiral carbons (starred).

MR 15.1

Cl₃C—C—OH　Chloral hydrate
　　　|
　　OH (top), H (bottom)

MR 15.2

MR 15.3

CIA 15.3

The five classes of chemotherapy drugs: (1) alkylating agents; (2) antimetabolites; (3) antitumor antibiotics; (4) topoisomerase inhibitors; (5) mitotic inhibitors.

CIA 15.4

Microtubules are long, cable-like proteins that are assembled in a cell when they are needed to move organelles and when it is necessary to move and separate chromosomes when a cell undergoes division.

CIA 15.5

$$\frac{8\ \mu g}{1\ kg} \times \frac{1\ kg}{2.205\ lb} \times 200\ lb = 70\ \mu g$$

Understanding Key Concepts

15.20 (a) The hydride ion is added to the carbonyl carbon. The carbon end of the polar C=O bond has a partial positive charge, and thus reaction with negatively charged reagents occurs at carbon.
(b) The top arrow (reaction to the right) represents reduction, and the bottom arrow (reaction to the left) represents oxidation.

15.21 Aldehydes can be oxidized to carboxylic acids, but ketones cannot. Two tests that can distinguish aldehydes from ketones take advantage of selective aldehyde oxidation. Tollens' reagent produces a deposit of silver when it is mixed with an aldehyde, but no reaction takes place with a ketone. Benedict's reagent yields a red copper oxide precipitate when it reacts with an aldehyde, but reaction occurs with a ketone only if the ketone has an adjacent hydroxyl group.

15.22 In a solution of an aldehyde in water, hydrogen bonds form between the oxygen atoms of the aldehyde and the hydrogen atoms of water. Molecules of aldehyde don't form hydrogen bonds with each other, and the hydrogen atoms bonded to the aldehyde carbon don't take part in hydrogen bonding.

15.23 (a) Under acidic conditions, an alcohol adds to the carbonyl group of an aldehyde to form a hemiacetal. Although some hemiacetals are stable, most hemiacetals are either reconverted to aldehydes or further react with a second molecule of the alcohol to form acetals. The yield of acetal can be improved if the water by-product is removed from the reaction mixture, driving the equilibrium to the right.
(b)

15.24 In solution, glucose exists primarily in the cyclic hemiacetal form shown below because it is more stable in this form.

15.25 In addition to the two oxygens, a ketal carbon is bonded to two carbons. The acetal carbon of an aldehyde is bonded to a carbon and a hydrogen.

ketone   aldehyde

Additional Problems

Aldehydes and Ketones

15.26

15.27 There are many possible answers to this question.

The C–O bond angle is close to 120°.

15.28

(a)

CH₃CH₂CCH₃

ketone

(b)

NH₂

ketone   amide

(c)

CH₃CH₂—O—CH₂—CH

ether   aldehyde

(d)

CH₃CH₂CHCH₂OH
        |
       OCH₃

no carbonyl group

(e)

ester (lactone)

(f)

CH₃CCH₂CH₂OH

ketone

Structure (c) is an aldehyde, and structures (a), (b), and (f) are ketones.

15.29

(a)

aldehyde

(b)

no carbonyl group

(c)

amide

(d)

no carbonyl group

(e)

ketone

Structure (a) is an aldehyde, and structure (e) is a ketone.

15.30

(a)

CH₃CH₂CHCHCH
        |
       CH₃

3-Methylpentanal

(b)

ClCH₂CH₂CHCH
            |
           OH

4-Chloro-2-hydroxybutanal

(c)

H₃C—

—CH

p-Methylbenzaldehyde

(d)

CH₂CH₃

2-Ethylcycloheptanone

(e)

CCH₃

Cyclopropyl methyl ketone

(f)

CCH₃

Acetophenone

15.31

(a)

CH₃CH₂CH₂C̈CH₂C̈—CH
with CH₃, CH₃, O above and OH, CH₃ below

4-Hydroxy-2,2,4-
trimethylheptanal

(b)

CH₃CH₂CHCH₂CH—CH
with CH₂CH₃, O above and CH₃CHCH₃ below

4-Ethyl-2-isopropylhexanal

(c)

Br— benzene ring —CH with =O

p-Bromobenzaldehyde

(d)

2,4-Dihydroxyclohexanone

(e)

CH₃CH₂CCH₂CCl₃ with =O

1,1,1-Trichloro-3-pentanone

(f)

CH₃CH₂CH₂CCHCH₃ with =O above and CH₃ below

2-Methyl-3-hexanone

15.32

(a)

CH₃CH₂CCHO with CH₃ above and CH₃ below

2,2-Dimethylbutanal

(b)

CH₃CH₂CH₂CCH₃ with CHO above and OH below

2-Hydroxy-2-methylpentanal

(c)

CHO structure

3-Methylbutanal

(d)

4-Methyl-3-hexanone

(e)

3-Hydroxy-2-methylcyclohexanone

15.33

(a)

CHO on benzene ring

Benzaldehyde

(b)

CH₃CH₂CCH(CH₃)₂ with =O

2-Methyl-3-pentanone

(c)

CH₃CH—CHCH₂CH with CH₃CH₂, O above and CH₃ below

3-Ethyl-4-methylpentanal

(d)

Dicyclopentylketone

(e)

3-Hydroxy-4-heptanone

15.34 (a) The name 1-pentanone is incorrect because a ketone group can't occur at the end of a carbon chain. Correct name: pentanal.
(b) The methyl group should have the lowest possible number. Correct name: 2-methyl-3-pentanone.
(c) Numbering must start at the end of the carbon chain that is closer to the ketone functional group. Correct name: 2-butanone.

15.35 (a) The name cyclohexanal is incorrect because an aldehyde carbon can't be part of a ring.
(b) The name 2-butanal is incorrect because an aldehyde group must be at the end of a chain, not in the middle.
(c) The name 1-methyl-1-pentanone is incorrect because a ketone group can't occur at the end of a carbon chain.

Reactions of Aldehydes and Ketones

15.36 A *hemiacetal* is produced when an aldehyde reacts with an alcohol in a 1:1 ratio. No second product is formed.

15.37 An *ketal* is produced when a ketone reacts with an alcohol in a 1:2 ratio in the presence of an acid catalyst: Water is a second product.

15.38 (a) Cyclopentanone reacts with neither Tollens' reagent nor Benedict's reagent.
(b), (c) Both compounds react with Tollens' reagent and Benedict's reagent because both are aldehydes.

15.39

(a)

(b)

(c)

15.40

15.41

15.42

(c)

$$CH_3CH_2CH_2\overset{\overset{\displaystyle O-CH_2CH_3}{|}}{CH}-OCH_3 + H_2O \underset{}{\overset{H_3O^+}{\rightleftharpoons}} CH_3CH_2CH_2\overset{\overset{\displaystyle O}{\|}}{CH} + CH_3CH_2OH + CH_3OH$$

(d)

$+ H_2O \underset{}{\overset{H_3O^+}{\rightleftharpoons}}$ $C=O + HOCH_2CH_2OH$

15.43

(a)

$$CH_3\overset{\overset{\displaystyle O}{\|}}{C}CH_3 + CH_3CH_2OH \underset{catalyst}{\overset{acid}{\rightleftharpoons}} CH_3\overset{\overset{\displaystyle OH}{|}}{\underset{\underset{\displaystyle CH_3}{|}}{C}}-OCH_2CH_3$$

hemiketal

(b)

$$CH_3(CH_2)_4\overset{\overset{\displaystyle O}{\|}}{CH} + CH_3CH_2\overset{\overset{\displaystyle OH}{|}}{CH}CH_3 \underset{catalyst}{\overset{acid}{\rightleftharpoons}} CH_3(CH_2)_4\overset{\overset{\displaystyle OH}{|}}{\underset{\underset{\displaystyle H}{|}}{C}}-O\overset{\overset{\displaystyle CH_3}{|}}{CH}CH_2CH_3$$

hemiacetal

(c)

$-OCH_3 + H_2O \underset{}{\overset{H_3O^+}{\rightleftharpoons}} HOCH_2CH_2CH_2CH_2\overset{\overset{\displaystyle O}{\|}}{C}-H + CH_3OH$

(d)

$+ H_2O \underset{}{\overset{H_3O^+}{\rightleftharpoons}}$ $+ 2\ CH_3OH$

15.44

$\underset{catalyst}{\overset{acid}{\rightleftharpoons}}$

5-Hydroxyhexanal

15.45

Glucosamine

15.46

15.47

hemiketal

15.48

Aldosterone

The hemiacetal group is derived from an aldehyde because the hemiacetal carbon has a hydrogen atom bonded to it.

15.49

Carvone

Conceptual Problems

15.50

CH₃O— ⟨benzene ring⟩ —CH=O   *p*-Methoxybenzaldehyde

15.51 The given alcohol, 2,2-dimethyl-1-propanol, can't be formed from reduction of an aldehyde or ketone because it is a tertiary alcohol. Only primary and secondary alcohols result from reduction of aldehydes and ketones, respectively.

15.52 The portion of the odor due to aldehyde is less stable because aldehydes are easily oxidized to carboxylic acids.

15.53 Formaldehyde in the air can cause bronchial pneumonia and dermatitis. Ingestion of formaldehyde can cause kidney damage, coma, and death.

15.54

(a)

$$CH_3CH_2CCH(CH_3)_2$$
2-Methyl-3-pentanone

(b)

$$CH_3CH_2CH_2CH=CH_2$$
1-Pentene

(c)

m-Bromotoluene

(d)

$$(CH_3)_3CCH_2CCH_2CH_3$$
5,5-Dimethyl-3-hexanone

15.55

(a)

o-Chloroethylbenzene

(b)

3-Hydroxycyclohexanone

(c)

trans-2-Methyl-3-hexene

(d)

$$(CH_3)_2CH-C-H$$
2-Methylpropanal

15.56

(a)

2,4-Dinitroacetophenone

(b)

2,4-Dihydroxycyclopentanone

(c)

2-Methoxy-2-methylpropane

(d)

2,3,4-Trimethyl-3-pentanol

15.57

(a)

2,3-Dimethylpentanal

(b)

1,3-Dibromopropanone

(c)

4-Hydroxy-4-methyl-
2-hexanone

15.58

(a)

(b)

(c)

(d)

15.59

(a)

(b)

(c)

15.60 Treat samples of the two compounds with Tollens' reagent. The aldehyde reacts to yield a silver mirror, but the alcohol doesn't react.

15.61 1-Butanol (boiling point [bp] 117 °C) is the higher boiling of the two compounds because it forms stronger hydrogen bonds. Butanal (bp 75 °C) is polar but doesn't form hydrogen bonds.

15.62 2-Heptanone is less soluble in water than 2-butanone because 2-heptanone has a longer hydrocarbon chain.

Group Problems

15.63

2-Octanone   3-Octanone   4-Octanone

15.64 Using α-glucose as an example:

(a)   (b)

α-Glucose

15.65 Using α-fructose as an example:

α-Fructose

15.66

Diynemicin A

a = alcohol (phenol)
b = ketone
c = carboxylic acid
d = ether
e = amine
f = alkene
g = alkyne
h = aromatic ring

Self-Test for Chapter 15

Multiple Choice

1. Which of the following alcohols can't be produced by reduction of an aldehyde or ketone?
   (a) 2-pentanol   (b) cyclohexanol   (c) 2-methyl-2-pentanol   (d) 2-methyl-1-butanol

2. The term LD_{50} refers to:
   (a) 50% of the amount of a substance that will kill a population
   (b) the amount of a substance that will kill 50% of a population
   (c) the name of a substance, 50% of which will kill a population
   (d) none of the above

3. 3-Pentanone is unlikely to undergo which of the following reactions?
   (a) oxidation   (b) reduction   (c) acetal formation   (d) hemiacetal formation

4. Vanillin contains all of the following functional groups except:
   (a) aldehyde   (b) ether   (c) aromatic ring   (d) ketone

5. How many ketone isomers of the formula $C_6H_{12}O$ are there?
   (a) 4   (b) 5   (c) 6   (d) 7

6. Which of the following aldehydes is least soluble in water?
   (a) propanal   (b) benzaldehyde   (c) formaldehyde   (d) acetaldehyde

7. Which of the following compounds undergoes reaction with Tollens' reagent?
   (a) 3-methyl-3-pentanol   (b) glucose   (c) acetone   (d) cyclohexanone dimethyl acetal

8. The main difference between aldehydes and ketones is:
   (a) reactivity   (b) solubility   (c) flammability   (d) polarity

9. Formaldehyde is used for all of the following except:
   (a) polymers   (b) disinfectant   (c) preservative   (d) food additive

10. Which of the following statements about 3-heptanone is true?
    (a) It undergoes reaction with Benedict's reagent.   (b) It forms a cyclic ketal with 1,3-propanediol.   (c) It is soluble in water.   (d) It is a solid at room temperature.

Sentence Completion

1. Aldehydes, esters, and ketones are all _____ compounds.

2. The shiny product of oxidation of an aldehyde by Tollens' reagent is _____.

3. _____ is a reaction that converts an acetal to an aldehyde or a ketone.

4. A carbonyl group is polarized, with a partial _____ charge on carbon and a partial _____ charge on oxygen.

5. _____ is a ketone that is widely used as a solvent.

6. Chronic exposure to the aldehyde _____ can cause symptoms resembling those of alcoholism.

7. Bonds to the carbonyl carbon of esters and amides are _____ polar than the bonds to the carbonyl carbon of ketones and aldehydes.

8. The conversion of an aldehyde or ketone into an alcohol is said to be a _____ reaction.

9. Formaldehyde is widely used as a _____.

10. The initial product of reaction between an aldehyde and an alcohol is called a _____.

11. _____ reagent is used to detect sugar in urine.

12. _____ can be detected on the breath during starvation.

True or False

1.   Glucose contains an acetal link.

2.   A bombardier beetle protects itself by secreting benzoquinone.

3.   Ketones are oxidized by Tollens' reagent to carboxylic acids.

4.   The reduction of aldehydes and ketones is carried out by using NaOH.

5.   An acetal is an alternative name for an ester.

6.   Only aldehydes form acetals.

7.   An aldehyde group always occurs at the end of a carbon chain.

8.   In the chemical industry, formaldehyde is used to synthesize polymers.

9. NAD^+ reduces ketones and aldehydes to alcohols.

10.   Acetaldehyde results from the oxidation of ethanol.

11.   Both aldehydes and ketones can be reduced to alcohols.

12.   Aldehydes and ketones are not capable of forming hydrogen bonds with each other.

Match each entry on the left with its partner on the right (use each answer once).

1. $AgNO_3$, NH_3, H_2O (a) Reagent used to form an acetal

2. $CH_3CH_2COCH_3$ (b) Reduces carbonyl groups to alcohols

3. CH_3CH_2COOH (c) An amide

4. CH_3OH, H^+ catalyst   (d) Product of an aldehyde plus Tollens' reagent

5. CH_3CH_2OH (e) Tollens' reagent

6. $CH_3CH_2COOCH_3$ (f) Colored product of Benedict's test

7. $CH_3CH(OH)CH_3$ (g) Formaldehyde

8.   NADH   (h) Product of aldehyde reduction

9. CH_3CHO (i) Yields secondary alcohol when reduced

10. Cu_2O (j) Ester

11.   HCHO   (k) Can be oxidized to a ketone

12. CH_3CONH_2 (l) Product of primary alcohol oxidation

Chapter Outline

I. General characteristics of amines (Sections 16.1–16.4).
  A. Classification of amines (Section 16.1).
    1. Amines are classified by the degree of substitution at nitrogen.
      a. Primary amines have the structure RNH_2.
      b. Secondary amines have the structure R_2NH.
      c. Tertiary amines have the structure R_3N.
      d. Quaternary ammonium salts have the structure $R_4N^+ X^-$.
    2. The groups bonded to nitrogen may be alkyl or aryl groups and may also contain functional groups.
  B. Naming amines (Section 16.2).
    1. Primary amines are named by identifying the alkyl group attached to nitrogen and adding the suffix -*amine*.
    2. Secondary and tertiary amines.
      a. If the groups are identical, the prefix *di-* or *tri-* is added to the name that would be given if the amine were primary.
      b. If the groups are different, the compound is named as an *N*-substituted derivative of a primary amine.
    3. The simplest aromatic amine is aniline.
    4. When the amino group is a substituent, the prefix *amino-* is used.
  C. Properties of amines (Section 16.3).
    1. Amines are weak bases.
    2. Primary and secondary amines form hydrogen bonds to each other.
    3. Amines can form hydrogen bonds with water.
    4. Simple amines are water-soluble.
    5. Amines have unpleasant odors.
    6. Amines are an important class of biomolecules.
  D. Heterocyclic amines (Section 16.4).
    1. In some nitrogen-containing compounds, nitrogen is part of a ring.
    2. Heterocyclic compounds may be aromatic or nonaromatic.
II. Reactions of amines (Sections 16.5–16.6).
  A. Acid–base reactions of amines (Section 16.5).
    1. Aqueous amines are weak bases that can accept a proton to form ammonium ions.
      a. Nonaromatic amines are stronger bases than aromatic amines.
    2. Amines can react with acids to form ammonium salts.
      a. Ammonium cations of alkylamines are named by replacing the ending -*amine* by the ending -*ammonium*.
      b. For cations of heterocyclic amines, -*e* is replaced by -*ium*.
    3. Ammonium salts react with hydroxide ion to yield an amine plus water.
    4. In the body, many amines are present as ammonium ions.
  B. Amine salts (Section 16.6).
    1. Amine salts are composed of an ammonium ion (the cation) and an anion.
    2. Most amine salts are much more soluble in water than the amines from which they were derived.
    3. Sometimes amine salts are written as amine·HX.
    4. An amine can be regenerated from an amine salt by reaction with base.
    5. Amine salts can be formed by reaction of an amine with an alkyl halide.

6. If four alkyl groups are bonded to nitrogen, the compound is a quaternary ammonium salt.
   a. Quaternary ammonium salts are usually neither acidic or basic.
III. Alkaloids: amines in plants (Section 16.7).
   1. Many alkaloids are toxic and physiologically active.
   2. Familiar alkaloids are coniine, atropine, solanine, and morphine.

Solutions to Chapter 16 Problems

16.1

(a)

$CH_3(CH_2)_4CH_2NH_2$

primary amine

(b)

$CH_3CH_2CH_2NHCH(CH_3)_2$

secondary amine

(c)

$CH_3\underset{\underset{CH_3}{|}}{\overset{\overset{CH_3}{|}}{C}}-NH_2$

primary amine

(d)

secondary amine

(e)

tertiary amine

16.2

(a)

$(CH_3CH_2CH_2)_2NH$

Dipropylamine

(b)

N-Ethyl-N-methylcyclopentylamine

(c)

N-Isopropylaniline

16.3, 16.4

(a)

$CH_3CH_2CH_2CH_2CH_2CH_2CH_2CH_2NH_2$

Octylamine
primary amine

(b)

$CH_3CH_2CH_2CH_2CH_2\underset{\underset{CH_3}{|}}{NH}$

N-Methylpentylamine
secondary amine

(c)

N-Ethylaniline
secondary amine

(d)

$\underset{CH_2CH_2CHCH_3}{\overset{NH_2\quad OH}{|\quad\quad|}}$

4-Amino-2-butanol
primary amine

16.5

$$H_3C - \overset{\overset{\displaystyle CH_3}{|}}{\underset{\underset{\displaystyle CH_3}{|}}{N^+}} - CH_3$$

Tetramethylammonium ion

The atoms of tetramethylammonium ion form 16 bonds (12 C—H bonds and 4 C—N bonds) that contain 32 valence electrons. Four neutral carbon atoms, 12 hydrogen atoms, and one nitrogen atom have a total of 33 valence electrons. Since the ion has one electron fewer than the neutral atoms, the ion has a +1 charge.

16.6

$CH_3CH_2CH_2CH_2NHCH_2CH_3$

N-Ethylbutylamine

N-Ethylbutylamine is a secondary amine because two alkyl groups are bonded to the amine nitrogen.

16.7

Lowest Boiling —————————————⟶ *Highest Boiling*

$$\underset{(a)}{\overset{\overset{\displaystyle CH_3}{|}}{CH_3NCH_2CH_3}} \quad < \quad \underset{(c)}{CH_3CH_2CH_2CH_2NH_2} \quad < \quad \underset{(b)}{CH_3CH_2CH_2CH_2OH}$$

Alcohol (b) is highest boiling because —OH bonds of alcohols are more polar and form stronger hydrogen bonds than —NH bonds of amines. Amine (a) is lowest boiling because tertiary amines such as (a) don't form hydrogen bonds to each other.

16.8 In Section 8.2, we learned that a hydrogen bond forms between an electron lone pair of an electronegative atom and a hydrogen bonded to a second electronegative atom. To draw the structures: (1) draw the amine, including the lone pair electrons; (2) form hydrogen bonds between: (i) the amine lone pair and a hydrogen atom of water; (ii) the amine hydrogens bonded to —N and an electron lone pair of water. In (b), only hydrogen bonds of type (i) can form.

HOC 16.1(a)

Amoxicillin

a = amine
b = ketone
c = sulfide
d = carboxylic acid
e = alcohol (phenol)
f = amide
g = aromatic ring

Amoxicillin is used for bacterial infections such as sinus infections, bladder infections, and pneumonia.

Doxycycline

a = amine
b = ketone
c = alcohol
d = amide
e = C–C double bond
f = aromatic ring

Doxycycline is used for bacterial and protozoal infections such as Lyme disease, Rocky Mountain spotted fever, cholera, sinusitis, and malaria.

Ciprofloxacin

a = amine
b = ketone
c = halide
d = carboxylic acid
e = C–C double bond
f = aromatic ring

Ciprofloxacin is used to treat bacterial infections of bones and joints, respiratory and urinary tract infections.

Metronidazole

a = amine
b = alcohol
c = nitro group
d = double bond

Metronidazole is used for bacterial infections such as giardia, vaginosis, rosacea, and lung abscesses.

(b)

Drug	Generic Name	Used for	Amines?	Other Groups
Crestor	rosuvastatin	high cholesterol	tertiary (3)	sulfonamide, $-CO_2H$, $-OH$, aromatic, halide
Synthroid	levothyroxine	thyroid hormone	primary	$-CO_2H$, ether, aromatic, $-I$
Nexium	esomeprizole	stomach acid reduction	secondary, tertiary (2)	aromatic rings, ethers, sulfonate
Ventalin	salbutamol	bronchodilator	secondary	aromatic ring, alcohol
Advair	fluticasone + salmeterol	asthma	secondary	alcohol, ester, aromatic ring, sulfide, ether, halide
Lantus	synthetic insulin	blood sugar regulator	primary, secondary	amide, $-CO_2H$, aromatic ring, $-OH$ and others
Vyvanse	lisdexamfetamine	ADHD	primary	$-CO_2H$, aromatic
Lyrica	pregabalin	epilepsy fibromyalgia	primary	$-CO_2H$
Spiriva	tiotropium bromide	COPD	quaternary	heterocyclic rings, $-OH$, ester, ether
Diovan	valsertan	high blood pressure	secondary, tertiary (3)	$-CO_2H$, amide, aromatic, hetero-cyclic ring

(c) Since amines are found in all of these medications, amines must be an important part of their biological activity.

16.9

(a)

CH₃NH₂	CH₃CH₂NH₂	CH₃NHCH₃	CH₃NCH₃
Methylamine	Ethylamine	Dimethylamine	$\overset{\displaystyle |}{CH_3}$
			Trimethylamine

Amines that are gases at room temperature.

(b)

(c)

Pyridine

a heterocyclic amine

Aniline

a compound with an amine group
on an aromatic ring

16.10

(a)

(b)

Pyrimidine

$C_4H_4N_2$

Purine

$C_5H_4N_4$

16.11 In a heterocyclic compound containing nitrogen, nitrogen must form part of the ring. Thus, compounds (a) and (d) are heterocyclic nitrogen compounds.

(a)   (b)   (c)

heterocyclic nitrogen
compound

(d)

heterocyclic nitrogen
compound

16.12

(a)

base

+ H_2O ⇌

acid

acid

+ OH^-

base

(b)

16.13–16.14

(a)

N-Methylisopropylamine N-Methylisopropylammonium bromide

(b)

Aniline   Anilinium chloride

(c)

Piperidine   Piperidinium chloride

(d)

tert-Butylammonium ion   *tert*-Butylamine

16.15 Review Section 16.5 to choose the stronger base.
     Basicity: nonaromatic amine > ammonia > aromatic amine

     (a) Ethylamine is a stronger base than ammonia.
     (b) Triethylamine (nonaromatic amine) is a stronger base than pyridine (aromatic amine).

16.16

(a)

Epinephrine
amine

ammonium ion

(b)

Amphetamine
amine

ammonium ion

16.17–16.18

(a)

$$CH_3CH_2CH_2CH_2\overset{CH_2CH_3}{\underset{CH_2CH_3}{\overset{|}{\underset{|}{N}}}}H^+ Br^-$$

Butyldiethylammonium bromide
or N,N–Diethylbutylammonium bromide
salt of a tertiary amine

(b)

$$(CH_3CH_2CH_2CH_2)_4N^+ OH^-$$

Tetrabutylammonium hydroxide

salt of a quaternary amine

(c)

$$CH_3CH_2CH_2NH_3^+I^-$$

Propylammonium iodide
salt of a primary amine

(d)

$$CH_3\overset{CH_3}{\underset{CH_3}{\overset{|}{\underset{|}{C}}}}HNH_2^+ Cl^-$$

Isopropylmethylammonium chloride
salt of a secondary amine

16.19

$$CH_3CH_2CH_2CH_2NH_3^+ Cl^-(aq) + NaOH(aq) \longrightarrow CH_3CH_2CH_2CH_2NH_2 + H_2O(l) + NaCl(aq)$$

16.20 Benadryl has the general antihistamine structure illustrated below. Attached to the central skeleton (–Z–CH$_2$CH$_2$N–) of Benadryl are two methyl groups bonded to nitrogen (R = –CH$_3$) and two phenyl groups (R' = R" = C$_6$H$_5$–) bonded to Z (Z = –CHO–).

Benadryl   general antihistamine structure

16.21

Benzylammonium chloride

16.22

(a)

$$(CH_3CH_2)_3NH^+ \; Br^- + LiOH \longrightarrow (CH_3CH_2)_3N + H_2O + LiBr$$

(b)

(c)

CIA 16.1

(a) An anxiolytic is a medication used to treat anxiety.
(b) Benefits of diazepines: can be used "as needed"; good for short-term use.
Side effects: tolerance; addiction; cognitive impairment.
(c) Benefits of SSRIs: relieve mood swings; safe; nonaddictive.
Side effects: take weeks to become effective; weight gain, insomnia, headaches.

Understanding Key Concepts

16.23 (a)

s = secondary
q = quaternary
ha = heterocyclic aromatic amine

(b) The secondary amine group can provide a hydrogen bond and accept a hydrogen bond. The heterocyclic amine can accept hydrogen bonds from primary and secondary amine groups but not from other tertiary amine groups. The quaternary group doesn't participate in hydrogen bonding.

16.24

Lysine NH_2

(a) Both amine groups (as well as the carboxylic acid group) can participate in hydrogen bonding.
(b) Lysine is likely to be water-soluble because it can form hydrogen bonds with water.

16.25

(a)

(b)

(c)

16.26

An O—H bond of water is broken, and a N—H bond is formed in the reaction to form OH⁻, an ammonium ion and an amine. The electrons remain with their original atoms.

16.27

Most basic ⟶ *Least basic*

$(CH_3)_2NH$ > NH_3 > ⟨benzene ring⟩—NH_2

16.28

(a)

⟨pyridinium⟩ N⁺—H + OH⁻ ⟶ ⟨pyridine⟩ N: + H_2O

(b)

⟨isopropyl⟩—$\overset{..}{N}H_2$ + H_2O ⇌ ⟨isopropyl⟩—NH_3^+ + OH⁻

(c)

$(CH_3CH_2)_3\overset{..}{N}$ + HBr ⟶ $(CH_3CH_2)_3NH^+$ Br⁻

(d)

⟨pyrrole⟩ N—H + HCl ⟶ ⟨pyrrolium⟩ N⁺ H, H Cl⁻

Additional Problems

Amines and Ammonium Salts

16.29

(a) ⟨cyclohexyl⟩—$NHCH_3$

N-Methylcyclohexylamine

(b) $CH_3CH_2CH_2\overset{\overset{\text{H}}{|}}{N}CH_2CH_2CH_3$

Dipropylamine

(c) $CH_3CH_2CH_2CH_2CH_2NH_2$

Pentylamine

16.30

(a)

$$CH_3CH_2CH_2CH_2CH_2\overset{\overset{\displaystyle H}{|}}{N}CH_3$$

N-Methylpentylamine

(b)

⬜—$\overset{\overset{\displaystyle H}{|}}{N}CH_2CH_3$

N-Ethylcyclobutylamine

(c)

$CH_3CH_2CH_2$—⬡—NH_2

p-Propylaniline

16.31

(a)

N,N-Dimethyl-*sec*-butylamine
tertiary amine

(b)

⬡—$NHCH_3$

N-Methylaniline
secondary amine

16.32

(a)

N-Ethyclopentylamine
secondary amine

(b)

⬡—NH_2

Cycloheptylamine
primary amine

16.33 Water is a weaker base. In the equilibrium shown below, it is observed that products are favored over reactants.

$$NH_3 + H_2O \rightleftharpoons NH_4^+ + OH^-$$

If water were a stronger base, the following equilibrium (not observed) would occur:

$$NH_3 + H_2O \rightleftharpoons NH_2^- + H_3O^+$$

16.34 Diethylamine is a stronger base than diethyl ether because amines are more basic than ethers.

16.35

(a)

$$CH_3CH_2CH_2-\overset{\overset{\displaystyle +}{N}H_2}{\underset{\displaystyle CH_3}{|}} \; Br^-$$

N-Methylpropylammonium bromide
(salt of a secondary amine)

(b)

$$\overset{\overset{\displaystyle CH_3}{|}}{\underset{\overset{\displaystyle |}{CH_3}}{\overset{+}{N}H}} \; Cl^-$$

N,N-Dimethylisopropylammonium chloride
(salt of a tertiary amine)

(c)

$$\overset{\displaystyle CH_3CH_2CH_2}{\underset{\displaystyle CH_3CH_2CH_2CH_2NH_2^+ \; Br^-}{|}}$$

N-Propylbutylammonium bromide
(salt of a secondary amine)

(d)

⬜—$NH_3^+ \; Br^-$

Cyclobutylammonium bromide
(salt of a primary amine)

16.36

(a)

$$CH_3CH_2\overset{|}{C}HCH_3$$
$$\overset{|}{C}H_3\overset{+}{N}H_2\ NO_3^-$$

N-Methyl-2-butylammonium nitrate
(salt of a secondary amine)

(b)

Pyridinium chloride
(salt of a tertiary amine)

(c)

$$CH_3\overset{|}{C}HCH_3$$
$$CH_3CH_2CH_2CH_2CH_2CH_2\overset{|}{\overset{+}{N}}H^+\ Cl^-$$
$$\overset{|}{C}H_2CH_2CH_2CH_3$$

N-Butyl-*N*-isopropylhexylammonium chloride
(salt of a tertiary amine)

16.37

Lidocaine

16.38

Cocaine

16.39

(a)

(b)

$$CH_3CH_2CH_2\overset{\overset{+}{\overset{|}{H}}}{N}CH_2CH_2CH_3$$
$$\underset{H\ \ X^-}{}$$

(c)

$$CH_3CH_2CH_2CH_2CH_2\overset{+}{N}H_3$$
$$X^-$$

16.40

(a)

$$CH_3CH_2CH_2CH_2CH_2\overset{\overset{+}{\overset{|}{H}}}{N}CH_3$$
$$\underset{H\ \ X^-}{}$$

(b)

(c)

$$CH_3CH_2CH_2-\text{(aromatic ring)}-\overset{+}{N}H_3$$
$$X^-$$

Reactions of Amines

16.41

(a)

(b)

(c)

16.42

(a)

(b)

(c)

16.43 The ammonium salt pictured does not react with acids or bases because it is neither acidic nor basic.

16.44 Choline doesn't react with HCl because its nitrogen atom isn't basic.

Conceptual Problems

16.45 (a) Six secondary amines of the formula $C_5H_{13}N$ can be drawn.

$$\underset{\text{H}}{CH_3CH_2CH_2\overset{\displaystyle |}{N}CH_2CH_3}$$

$$CH_3\underset{\underset{\displaystyle CH_3}{\displaystyle |}}{\overset{\overset{\displaystyle H}{\displaystyle |}}{C}H}NCH_2CH_3$$

$$\underset{\text{H}}{CH_3CH_2CH_2CH_2\overset{\displaystyle |}{N}CH_3}$$

$$CH_3\underset{\underset{\displaystyle \text{}}{\overset{\overset{\displaystyle CH_3}{\displaystyle |}}{C}H}}CH_2\overset{\overset{\displaystyle H}{\displaystyle |}}{N}CH_3$$

$$CH_3CH_2\underset{\underset{\displaystyle CH_3}{\displaystyle |}}{\overset{\overset{\displaystyle H}{\displaystyle |}}{C}H}NCH_3$$

$$CH_3\underset{\underset{\displaystyle CH_3}{\displaystyle |}}{\overset{\overset{\displaystyle CH_3}{\displaystyle |}}{C}}—\overset{\overset{\displaystyle H}{\displaystyle |}}{N}CH_3$$

(b) Many tertiary amines of the formula $C_6H_{13}N$ can be drawn. Here are several of them:

$$H_2C=CH—\underset{\underset{\displaystyle CH_2CH_3}{\displaystyle |}}{N}—CH_2CH_3$$

(c) Other structures are possible.

16.46

PABA

16.47

Sodium sulfanilamide

(a) Both PABA and sulfanilamide have *p*-amino-substituted aromatic rings.
(b) The sodium salt form of the drug is used because it is more soluble.

16.48

amide

Acyclovir — related to purine

amine

H₂N

amine

CH₂OCH₂CH₂OH ← alcohol

ether

16.49 Trimethylamine is a stronger base than ammonia. Thus, the reaction of trimethylamine with ammonia proceeds in the following direction:

$$NH_4^+ + :N(CH_3)_3 \rightleftharpoons :NH_3 + H-N^+(CH_3)_3$$

stronger acid   stronger base

16.50

Amines	Alcohols
(a) Foul smelling	Pleasant smelling
(b) Basic	Not basic
(c) Lower boiling, due to weaker hydrogen bonds, but higher boiling than ethers or alkyl halides	Higher boiling, due to strong hydrogen bonds

16.51 Alkaloids are often bitter or poisonous.

16.52

(a) CH₃CHCH₂CH₂CH=CHCH₃ with CH₃ branch

6-Methyl-2-heptene

(b) HO—C₆H₄—CHCH₃ with CH₃

p-Isopropylphenol

(c) (CH₃CH₂CH₂CH₂)₂NH

Dibutylamine

16.53

(a) CH₃CH₂CCH₂CH=CCH₃ (with CH₃ and CH₂CH₃) + HCl → CH₃CH₂CCH₂CH₂CCH₃ (with CH₃, Cl, CH₂CH₃)

(b) CH₃CH₂CHCH(CH₃)₂ (with OH) + H₂SO₄ → CH₃CH₂CH=C(CH₃)₂ + CH₃CH=CHCH(CH₃)₂

(c)

$$2 \ CH_3CH_2SH \xrightarrow{[O]} CH_3CH_2S{-}SCH_2CH_3$$

(d)

(e) $(CH_3)_3N + H_2O \rightleftharpoons (CH_3)_3NH^+ + OH^-$

(f) $(CH_3)_3N + HCl \longrightarrow (CH_3)_3NH^+ \ Cl^-$

(g) $(CH_3)_3NH^+ + OH^- \longrightarrow (CH_3)_3N + H_2O$

16.54 Molecules of hexylamine can form hydrogen bonds to each other. Heat must be supplied to break these hydrogen bonds, and thus the boiling point of hexylamine is higher than that of triethylamine, a tertiary amine whose molecules don't hydrogen-bond to each other.

16.55

Baeocystin   Indole

Baeocystin is related to indole.

16.56 Cyclohexylamine is not a heterocycle because its nitrogen atom is not part of a ring.

16.57 Pyridine is soluble in water because the ring nitrogen can form hydrogen bonds with water. No hydrogen bonding is possible for benzene.

16.58 (a) *N*-Ethylcyclohexylamine   (b) Anilinium bromide   (c) *N*-Methylethylamine

Group Problems

16.59 (a) Acetic acid has the highest boiling point of the compounds listed because it is more polar and undergoes more extensive hydrogen bonding than the other compounds listed. (b),(c),(d) Butane is the lowest boiling, least soluble in water, and least reactive of the compounds listed because it is an alkane.

16.60 Decylamine is much less soluble in water than ethylamine because it resembles a hydrocarbon more than it resembles an amine.

16.61 The amine group is responsible for the odor of fish. The citric acid in lemon juice reacts with amines to form ammonium salts, which are odorless and are no longer volatile.

Self-Test for Chapter 16

Multiple Choice

1. Which of the following amines is heterocyclic?
   (a) aniline   (b) cyclohexylamine   (c) pyridine   (d) amphetamine

2. Which of the following amines is a secondary amine?
   (a) *N*-methylaniline   (b) histamine   (c) triethylamine   (d) codeine

3. The name of the compound shown below is:
   (a) ethylcyclopentylamine   (b) cyclopentylethylamine   (c) *N*-cyclopentylethylamine
   (d) *N*-ethylcyclopentylamine

4. An amine and its ammonium salt differ in all respects except:
   (a) charge   (b) solubility   (c) carbon skeleton   (d) basicity

5. Which of the following amines has the highest boiling point?
   (a) triethylamine   (b) hexylamine   (c) tetramethylammonium chloride   (d) dipropylamine

6. Which of the following is an alkaloid?
   (a) atropine   (b) amphetamine   (c) pyrimidine   (d) aniline

7. How many isomers of the formula $C_4H_{11}N$ are secondary amines?
   (a) 2   (b) 3   (c) 4   (d) 5

8. All of the following amines have a heterocyclic ring except:
   (a) nicotine   (b) caffeine   (c) morphine   (d) aniline

9. Propylamine and trimethylamine differ in all respects except:
   (a) formula weight   (b) boiling point   (c) solubility in water   (d) melting point

10. Which of the following is a quaternary ammonium salt?
    (a) anilinium chloride   (b) triethylpropylammonium chloride   (c) pyridinium chloride
    (d) trimethylammonium chloride

Sentence Completion

1. When the $-NH_2$ group is a substituent, _____ is used as a prefix.

2. Tetramethylammonium bromide is a _____ amine salt.

3. _____ contains nitrogen heterocyclic rings derived from either purine or pyrimidine.

4. _____ is the science devoted to poisons.

5. A _____ has a nitrogen atom contained in a ring.

6. _____ are a class of amines derived from plants.

7. Simple amines are water-soluble because of _____ _____.

8. Most amines can be made water-soluble by conversion to _____ salts.

9. Nitric oxide is a molecule that is a ____ _____ because it has an unpaired electron.

10. Ammonia is less basic than _____ _____.

True or False

1. Amine groups in biomolecules are usually protonated in body fluids.

2. Ammonium salts are water-soluble.

3. Tertiary amines are higher boiling than primary or secondary amines.

4. Morphine and codeine are alkaloids.

5. Quaternary ammonium salts are acidic.

6. Caffeine is a heterocyclic amine.

7. Heroin is a naturally occurring alkaloid.

8. The physical properties of low-molecular-weight amines and ammonia are similar.

9. NO is responsible for elevating blood pressure.

10. The correct name for $CH_3CH_2CH_2CH_2NHCH_3$ is methylbutylamine.

Match each item in the left column with its partner on the right (use each answer once).

1. Trimethylamine   (a) Amino acid

2. Atropine   (b) Vitamin

3. Piperidine   (c) Primary aromatic amine

4. Tetramethylammonium chloride   (d) Tertiary amine

5. Alanine   (e) Aromatic heterocycle

6. Dimethylamine   (f) Neurotransmitter

7. Pyridoxine   (g) Primary amine

8. Cadaverine   (h) Nonaromatic heterocycle

9. Serotonin   (i) Quaternary ammonium salt

10. Methylamine   (j) Secondary amine

11. Pyridine   (k) Diamine

12. Aniline   (l) Alkaloid

Chapter 17   Carboxylic Acids and
Their Derivatives

Chapter Outline

I. Properties and names of carboxylic acids and derivatives (Section 17.1).
  A. General characteristics.
    1. Groups bonded to the carbonyl carbon are polar.
      a. In carboxylic acids, —OH is bonded to the carbonyl carbon.
      b. In amides, —NH$_2$, —NHR, or —NR$_2$ is bonded to the carbonyl carbon.
      c. In esters, —OR is bonded to the carbonyl carbon.
    2. All of these functional groups undergo carbonyl group substitution reactions.
  B. Carboxylic acids.
    1. Properties.
      a. All have high boiling points.
      b. Carboxylic acids can form hydrogen bonds with each other.
      c. Smaller carboxylic acids are water-soluble.
      d. Volatile carboxylic acids have sharp odors.
    2. Naming carboxylic acids.
      a. For simple carboxylic acids, the *-e* of the corresponding alkane is replaced by *-oic acid*.
      b. Many simple carboxylic acids have common names.
      c. In some cases, the carbon next to the —COOH group is referred to as the α carbon.
      d. The group remaining when a carboxylic acid loses —OH is an acyl group.
      e. For dicarboxylic acids, *-dioic acid* replaces the *-e* of the corresponding alkane.
      f. Unsaturated carboxylic acids are named by using *-enoic acid*.
  C. Esters.
    1. Properties.
      a. Esters are lower boiling than carboxylic acids.
      b. Esters have pleasant odors.
      c. Simple esters are liquids.
    2. Naming esters.
    Ester names have two parts.
      a. The —OR fragment has the name of the alkyl group R.
      b. The remaining portion is the name of the parent carboxylic acid, with *-ate* replacing *-oic acid*.
  D. Amides.
    1. Properties.
      a. Amides are higher boiling than carboxylic acids; most are solids.
      b. Amides are not basic.
    2. Naming amides.
      a. An amide has the name of the parent acid, with *-amide* replacing *-oic acid*.
      b. If nitrogen is substituted, the alkyl substituents are specified, preceded by *N-* (to indicate that the substituents are bonded to nitrogen).

II. Carboxylic acids (Sections 17.2–17.3).
  A. Acidity of carboxylic acids (Section 17.2).
    1. Carboxylic acids are weak acids, with $pK_a \sim 10^{-5}$.
    2. Carboxylate anions are given the name of the parent carboxylic acid, with *-ate* replacing *-oic acid*.
    3. Carboxylic acids react with base to give carboxylic acid salts.
      a. The amount of salt or acid present in solution depends on pH.
      b. Carboxylic acid salts are much more soluble in water than the parent acids.
  B. Reactions of carboxylic acids (Section 17.3).
    1. Ester formation.

      a. carboxylic acid + alcohol $\underset{\text{catalyst}}{\overset{H^+}{\rightleftharpoons}}$ ester + H_2O

      b. Esterification reactions are reversible.
    2. Amide formation.
      a. Carboxylic acid + ammonia (or an amine) —> amide + H_2O
III. Esters, amides, and carboxylic acid anhydrides (Sections 17.4–17.5).
  A. Hydrolysis of esters and amides (Section 17.4).
    1. Esters.

      a. Ester + H_2O $\xrightarrow{\text{acid catalyst}}$ carboxylic acid + alcohol

      b. Ester + H_2O $\xrightarrow{\text{base}}$ carboxylate anion + alcohol
    2. Amides.

      Amide + H_2O $\xrightarrow{\text{acid or base}}$ carboxylic acid + amine (or ammonia)
  B. Polyesters and polyamides (Section 17.5).
    1. Nylons are polyamides formed in the reaction of a diacid with a diamine.
    2. Polyesters result from the reaction of a diacid with a dialcohol.
IV. Phosphoric acid derivatives (Section 17.6).
  A. Structure of phosphate esters.
    1. Phosphoric acid has three ionizable hydrogens and can form three different anions.
    2. Phosphates can react with alcohols at one, two, or three oxygens.
    3. Mono- and diphosphate esters are present in the body as anions.
  B. Phosphate esters in the body.
    1. In the body, phosphate esters exist as anions.
    2. In the body, phosphate esters can form anhydrides.
      a. Diphosphates and triphosphates are the anhydrides that occur in living things.
      b. Triphosphate esters are used for energy storage in the body.

Solutions to Chapter 17 Problems

17.1

(a)

$$CH_3\overset{\overset{O}{\|}}{C}NH_2$$

amide

(b)

$$CH_3OCH_3$$

none
(ether)

(c)

$$CH_3COOH$$

carboxylic acid

(d)

$$CH_3COOCH_2CH_3$$

ester

(e)

$$CH_3COCH_3$$

none
(ketone)

(f)

$$CH_3CH_2CONHCH_3$$

amide

(g)

$$CH_3CH_2NH_2$$

none
(amine)

(h)

$$CH_3CH_2\overset{\overset{O}{\|}}{C}NH_2$$

amide

17.2

(a)

$$\underset{6}{CH_3}\underset{5}{CH_2}\underset{4}{CH_2}\underset{3}{\overset{\overset{OH}{|}}{CH}}\underset{|2}{CH_2}\underset{1}{\overset{\overset{O}{\|}}{C}OH}$$
$$\underset{}{CH_2CH_3}$$

2-Ethyl-3-hydroxyhexanoic acid

(b)

m-Nitrobenzoic acid

17.3

Succinic acid ($C_4H_6O_4$)

17.4

$$H_2C{=}CH{-}\overset{\overset{O}{\|}}{C}{-}OH \;+\; Br_2 \longrightarrow BrCH_2\overset{\overset{O}{\|}}{C}H\overset{}{C}{-}OH$$
$$\underset{}{\overset{|}{Br}}$$

Acrylic acid
(Propenoic acid)

2,3-Dibromopropanoic acid

17.5

(a)

from benzoic acid

$$\text{C}_6\text{H}_5{-}\overset{\overset{O}{\|}}{C}OCH_2CH_2CH_2CH_2CH_2CH_3$$

from hexanol

Hexyl benzoate

(b)

from formic acid

$$H\overset{\overset{O}{\|}}{C}OCH_3$$

from methanol

Methyl formate

(c)

from acrylic acid

$$CH_2{=}CH\overset{\overset{O}{\|}}{C}OCH_2CH_3$$

from ethanol

Ethyl acrylate

17.6 (a) CH_3OCH_3

Low boiling.
Polar, but does not
form hydrogen bonds.

(b) CH_3COOH

Highest boiling.
Forms strongest
hydrogen bonds.

(c) $CH_3CH_2CH_3$

Lowest boiling.
Nonpolar–Doesn't
form hydrogen bonds.

17.7 *More Soluble*   *Less Soluble*   *Reason*

(a) $CH_3CH_2CH_2COOH$ $C_8H_{17}COOH$ The large hydrocarbon part of $C_8H_{17}COOH$ makes it insoluble in water.

(b) $(CH_3)_2CHCOOH$ $CH_3CH_2COOCH(CH_3)_2$ Carboxylic acids form stronger hydrogen bonds with water.

HOC 17.1

(a) The two penicillins have a close resemblance. Each has two amide groups, one of which is a cyclic lactam, and they each have a carboxylic acid group, a sulfide group in a five-membered ring, and an aromatic ring. Penicillin V also has an ether oxygen. Both penicillins have many uses: Penicillin G is used to treat syphilis, meningitis, pneumonia and gangrene; Penicillin V is used to treat streptococcal infections, tonsillitis, anthrax, and Lyme disease.

Penicillin G

Penicillin V

a = amide
b = carboxylic acid
c = sulfide
d = aromatic ring
e = ether

(b) No carboxylic acid, no amide: Nexium, Ventolin, Advair, Spiriva.
Carboxylic acid, no amide: Synthroid, Lyrica.
Amide, no carboxylic acid: Vyvanse (secondary).
Both: Crestor (tertiary sulfonamide), Diovan (tertiary), Lantus (all types).

(c)

Paclitaxel
(esters are starred)

17.8

(a)

CH₃CH₂CH₂CO—⟨cyclopentyl⟩

(b)

CH₃CH₂CH₂CNHCH(CH₃)₂

(c)

CH₃CH₂CH₂CN(CH₂CH₃)₂

(d) Cyclopentyl butyrate   *N*-Isopropylbutyramide   *N,N*-Diethylbutyramide

17.9

(a)   (b)

N-Methyl-2-bromobutanamide   *N*-Ethyl-*N*-methylbenzamide
(*N*-Methyl-2-bromobutyramide)

(c)

3-Hydroxy-2-methylpropanamide

17.10

(a)

$$CH_3CHCH_2CH_2CNH_2$$

with CH_3 above and O (double bond) above the C

4-Methylpentanamide

(b)

$$CH_3CH_2C-NCH_2CH_3$$

with O (double bond) above C and CH_3 above N

N-Ethyl-N-methylpropanamide

17.11

17.12

(a)

CH_3COOCH_3

ester (ii)

(b)

RCONHR

amide (i)

(c)

C_6H_5COOH

carboxylic acid (iii)

(d)

$$CH_3CH_2CN(CH_3)_2$$

with O (double bond) above C

amide (i)

(e)

$CH_3CH_2CH_2CONH_2$

amide (i)

(f)

$$HOOCCH_2CHCH_3$$

with CH_3 above

carboxylic acid (iii)

17.13

(a)

amide ($C_6H_5CONH_2$)

(b)

carboxylic acid (CH_3CH_2COOH)

(c)

ester ($CH_3COOCH_2CH_3$)

CIA 17.1

Salicylic acid   Salsalate

CIA 17.2   NSAID = <u>n</u>on-<u>s</u>teroidal <u>a</u>nti-<u>i</u>nflammatory <u>d</u>rug

CIA 17.3

Aspirin   Acetaminophen

Benzocaine   Lidocaine

17.14

(a)

(b)

17.15

Potassium salicylate   Disodium oxalate

17.16 When potassium acetate and disodium glutarate are dissolved in water, acetate ion, glutarate ion, potassium ion, and sodium ion are present; the first two are organic ions.

acetate ion   glutarate ion   potassium ion   sodium ion

17.17 Redraw the structures so that the alcohol hydroxyl group and the —OH group of the carboxylic acid are facing each other. Remove water, and draw the ester bond.

Octyl acetate

17.18 Use the same technique described in the previous problem.

2-Methyl-1-propyl formate

17.19

(a)

(b)

17.20 Use the technique shown in Problem 17.17 to draw the amide product.

(a)

(b)

17.21

p-Ethoxyaniline   Acetic acid   Phenacetin   + H₂O

17.22 Moisture in the air causes hydrolysis of the ester bond of aspirin, yielding acetic acid.

17.23

(a)

Benzoic acid   2-Propanol

(b)

bond broken   Phenol   Cyclopentane-
carboxylic acid

(c)

bond broken   Propanoic acid   Ethanol

17.24

(a)

2-Butenoic acid   Methylamine

(b)

p-Nitrobenzoic acid

+   H—NCH₃
            |
           CH₃
      Dimethylamine

17.25

Nomex

17.26 As in Problems 17.17–17.20, draw the reacting partners so that the atoms involved in the reaction face each other. Remove one molecule of water for each bond formed.

(a)

(b)

17.27

Isopropyl phosphate (unionized)

Isopropyl phosphate (ionized)

17.28

(a)

amide   Acetic acid   Ammonia

(b)

$CH_3CH_2OPO_3^{2-} + H_2O \longrightarrow CH_3CH_2OH + HOPO_3^{2-}$

phosphate monoester   Ethanol   Hydrogen phosphate anion

(c)

carboxylic acid ester   Propanoic acid   Methanol

CIA 17.4

Promazine   Promazine hydrochloride

The nitrogen that is not part of a ring is protonated because it is more basic.

CIA 17.5

Naproxen is converted to its sodium salt because it is more readily absorbed in that form.

17.29

Understanding Key Concepts

17.30 (a) At physiological pH (7.4), pyruvic acid exists as the pyruvate anion, and lactic acid exists as the lactate anion.

(b)

Pyruvic acid   Lactic acid

(c) Hydrogen bonds between the carboxylate oxygen and water are shown. In addition, the carbonyl oxygens can also hydrogen-bond with water. For lactate, hydrogen bonding takes place between the hydroxyl hydrogen and water (shown), as well as between molecules of lactate. You would expect pyruvate and lactate to have comparable solubility in water.

17.31

(a) Under conditions that favor amide hydrolysis (H_2O, plus either acid or base), the acetyl group can be removed from *N*-acetylglucosamine.

(b)

17.32

(a) Glycerate and phosphate are connected by a phosphate monoester linkage.
(b)

1,3-Bisphosphoglycerate

17.33

Oxalate   Malonate   Succinate   Glutarate

Adipate   Pimelate

17.34

(a)

$$2 \; HOC-CH-CH_2-CHOH \longrightarrow HOC-CH-CH_2-CHO-C-CH-CH_2-CHOH \; + \; H_2O$$

(with O (double bond), NH$_2$, CH$_3$ substituents)

(b)

$$2 \; HOC-CH-CH_2-CHOH \longrightarrow HOC-C-N-C-CH-CH_2-CHOH \; + \; H_2O$$

CH$_2$CH(CH$_3$)OH

(c)

$$HOC-CH-CH_2-CH-OH \longrightarrow$$

2-Amino-4-hydroxypentanoic acid
(2-Amino-4-hydroxyvaleric acid)

lactone

17.35

(a)(i)   (ii)   (iii)

Formic acid   Methyl formate   Formamide

(b)   *Lowest boiling*   – – – – – – >   *Highest boiling*
       Methyl formate,   Formic acid,   Formamide

Methyl formate is lowest boiling because it doesn't form hydrogen bonds. Formamide is higher boiling than formic acid because formic acid forms dimers (pictured below). These dimers do not form the extensively hydrogen-bonded array of molecules that occurs in formamide, and they lower the boiling point of formic acid, relative to formamide (see Section 17.1).

17.36 The hydrolyzed bonds are indicated on the structure.

The charge of the products depends on whether the hydrolysis was carried out under acidic or basic conditions.

17.37

(a)

N-Ethyl acetamide

(b)

Methyl 2-chlorocyclopentanecarboxylate

(c)

Ethyl pentanoate

(d)

N,N-Diethylformamide

Additional Problems

Carboxylic Acids

17.38

17.39

(a) in water

(b) at pH = 12

(c) at pH = 2

17.40

$CH_3CH_2CH_2COOH$
Butanoic acid

CH_3
|
$CH_3CHCOOH$
2-Methylpropionic acid

17.41

H_3C O
| ||
$CH_3CH_2CHC—OH$
2-Methylbutanoic acid

CH_3 O
| ||
$CH_3CHCH_2C—OH$
3-Methylbutanoic acid

17.42

(a)

OH O
| ||
$CH_3CHCHCH_2COH$
|
CH_3

3-Hydroxy-4-methylpentanoic acid

(b)

$CH_3(CH_2)_7COOH$

Nonanoic acid

(c)

—COOH

Cyclohexanecarboxylic acid

(d)

$H_2N—$ $\overset{O}{\overset{||}{C}}—OH$

p-Aminobenzoic acid

17.43

(a)

$$BrCH_2CH_2\underset{\underset{CH_3}{|}}{CH}\overset{\overset{O}{\|}}{C}OH$$

4-Bromo-2-methylbutanoic acid

(b)

![structure of o-methylbenzoic acid with CH₃ and COOH on benzene ring]

o-Methylbenzoic acid

(c)

$(CH_3CH_2)_3CCOOH$

2,2-Diethylbutanoic acid

(d)

$CH_3(CH_2)_5COOH$

Heptanoic acid

17.44

(a)

$$CH_3CH_2\underset{\underset{CH_2CH_3}{|}}{CH}CH_2\overset{\overset{O}{\|}}{C}O^-\ K^+$$

Potassium 3-ethylpentanoate

(b)

![benzene ring with CO⁻ NH₄⁺ substituent]

Ammonium benzoate

(c)

$$\left[CH_3CH_2\overset{\overset{O}{\|}}{C}O^- \right]_2 Ca^{2+}$$

Calcium propanoate

17.45

(a)

$$CH_3\overset{\overset{O}{\|}}{C}O^-\ NH_4^+$$

(b)

$$^-O\overset{\overset{O}{\|}}{C}\underset{\underset{CH_2CH_3}{|}}{CH}(CH_2)_2\overset{\overset{O}{\|}}{C}O^-\ 2\,Na^+$$

(c)

$$\underset{O=\overset{|}{C}-O^-}{\overset{O=\overset{|}{C}-O^-}{}}\ Ca^{2+}$$

Systematic name:

Ammonium ethanoate   Disodium 2-ethylpentanedioate   Calcium ethanedioate

Common name:

Ammonium acetate   Disodium 2-ethylglutarate   Calcium oxalate

17.46

(a)

$$CH_3CH_2\underset{\underset{CH_3}{|}}{CH}\underset{\underset{CH_3}{|}}{\overset{\overset{CH_3}{|}}{CH}}CH_2\overset{\overset{O}{\|}}{C}OH$$

3,4-Dimethylhexanoic acid

(b)

![benzene ring with CH₂-COOH substituent]

Phenylacetic acid

(c)

O$_2$N

O$_2$N

—COH (with =O)

3,4-Dinitrobenzoic acid

(d)

CH$_3$CH$_2$CH$_2$CO$^-$ $^+$NH(CH$_2$CH$_3$)$_3$ (with =O)

Triethylammonium butanoate

17.47

(a)

CH$_3$CHC—COH (with F substituents, F, F, and C=O)

2,2,3-Trifluorobutanoic acid

(b)

CH$_3$CHCH$_2$COH (with OH and C=O)

3-Hydroxybutanoic acid

(c)

—CHCCH$_2$COH (with CH$_3$, H$_3$C, CH$_3$, and C=O, phenyl ring)

3,3-Dimethyl-4-phenyl-
pentanoic acid

17.48

HOCCH$_2$CH—COH (with O, OH, O)   Malic acid

17.49

C=C with H, COOH, HOOC, H   Fumaric acid

17.50

C=C with H, COO$^-$ NH$_4$$^+$, NH$_4$$^+$ $^-$OOC, H   Diammonium fumarate

17.51

[CH$_3$CO$^-$ (with =O)]$_3$ Al^{3+} Aluminum acetate

Esters and Amides

17.52 There are many answers to this question. Here are some possibilities:

(a)

$$
\begin{array}{c}
O \\
\parallel \\
CH_3CH_2CH_2CH_2CNH_2
\end{array}
$$

Pentanamide

$$
\begin{array}{c}
O \\
\parallel \\
CH_3CH_2CNHCH_2CH_3
\end{array}
$$

N-Ethylpropanamide

$$
\begin{array}{c}
O \\
\parallel \\
HCNCH_2CH_3 \\
| \\
CH_2CH_3
\end{array}
$$

N,N-Diethylformamide

(b)

$$
\begin{array}{c}
O \\
\parallel \\
CH_3CH_2CH_2CH_2COCH_3
\end{array}
$$

Methyl pentanoate

$$
\begin{array}{c}
O \\
\parallel \\
CH_3CH_2COCH_2CH_2CH_3
\end{array}
$$

Propyl propanoate

$$
\begin{array}{c}
O \\
\parallel \\
HCOCH_2CH_2CH_2CH_2CH_3
\end{array}
$$

Pentyl formate

17.53 There are many possible answers.

(a)

$$
\begin{array}{c}
CH_3 \quad O \\
| \quad \parallel \\
CH_3CHCHCNH_2 \\
| \\
CH_3
\end{array}
$$

2,3-Dimethylbutanamide

$$
\begin{array}{c}
O \\
\parallel \\
CH_3CHCNHCH_2CH_3 \\
| \\
CH_3
\end{array}
$$

N-Ethyl-2-methylpropanamide

$$
\begin{array}{c}
O \\
\parallel \\
CH_3CNCH_2CH_3 \\
| \\
CH_2CH_3
\end{array}
$$

N,N-Diethylacetamide

(b)

$$
\begin{array}{c}
O \\
\parallel \\
CH_3COCH_2CH_2CH_3
\end{array}
$$

Propyl acetate

$$
\begin{array}{c}
O \\
\parallel \\
CH_3COCHCH_3 \\
| \\
CH_3
\end{array}
$$

Isopropyl acetate

$$
\begin{array}{c}
O \\
\parallel \\
CH_3CHCOCH_3 \\
| \\
CH_3
\end{array}
$$

Methyl 2-methylpropanoate

17.54

(a)

$$
\begin{array}{c}
O \quad\quad CH_3 \\
\parallel \quad\quad | \\
CH_3COCH_2CH_2CHCH_3
\end{array}
$$

3-Methylbutyl acetate
(3-Methylbutyl ethanoate)

(b)

$$
\begin{array}{c}
CH_3 \quad\quad O \\
| \quad\quad \parallel \\
CH_3CHCH_2CH_2COCH_3
\end{array}
$$

Methyl 4-methylpentanoate

(c)

$$
\begin{array}{c}
O \\
\parallel \\
CH_3CO-
\end{array}
$$

Cyclohexyl acetate
(Cyclohexyl ethanoate)

(d)

Phenyl o-hydroxybenzoate

17.55

(a)

Cyclopentyl cyclohex-
anecarboxylate

(b)

CH₃CHCOCH₂CH₃
　　｜
　　OH

Ethyl 2-hydroxypropanoate

(c)

Propyl benzoate

(d)

　　　　　　CH₃　O
　　　　　　｜　‖
CH₃CH₂CH₂CCH₂COCH₂CH₂CH₂CH₃
　　　　　　｜
　　　　　　CH₃

Butyl 3,3-dimethylhexanoate

(e)

　　　　　O
　　　　　‖
(CH₃)₂CHCOC(CH₃)₃

1,1-Dimethylethyl 2-methylpropanoate
(*tert*-Butyl 2-methylpropanoate)

17.56

(a)

(b)

(c)

(d)

17.57

(a)

Cyclopentyl cyclohexanecarboxylate

(b)

$$\text{CH}_3\text{CHC}-(\text{OH} + \text{H})-\text{OCH}_2\text{CH}_3 \underset{\text{catalyst}}{\overset{\text{acid}}{\rightleftharpoons}} \text{CH}_3\text{CHCOCH}_2\text{CH}_3 + \text{H}_2\text{O}$$

with OH substituents below the CH and product CH.

Ethyl 2-hydroxypropanoate

(c)

$$\cdots \text{C}-(\text{OH} + \text{H})-\text{OCH}_2\text{CH}_2\text{CH}_3 \underset{\text{catalyst}}{\overset{\text{acid}}{\rightleftharpoons}} \cdots -\text{COCH}_2\text{CH}_2\text{CH}_3 + \text{H}_2\text{O}$$

Propyl benzoate

(d)

$$\text{CH}_3\text{CH}_2\text{CH}_2\overset{\text{CH}_3}{\underset{\text{CH}_3}{\text{C}}}\text{CH}_2\text{C}-(\text{OH} + \text{H})-\text{OCH}_2\text{CH}_2\text{CH}_2\text{CH}_3 \underset{\text{catalyst}}{\overset{\text{acid}}{\rightleftharpoons}}$$

$$\text{CH}_3\text{CH}_2\text{CH}_2\overset{\text{CH}_3}{\underset{\text{CH}_3}{\text{C}}}\text{CH}_2\text{COCH}_2\text{CH}_2\text{CH}_2\text{CH}_3$$

Butyl 3,3-dimethylhexanoate

(e)

$$(\text{CH}_3)_2\text{CHC}-(\text{OH} + \text{H})-\text{OC}(\text{CH}_3)_3 \underset{\text{catalyst}}{\overset{\text{acid}}{\rightleftharpoons}} (\text{CH}_3)_2\text{CHCOC}(\text{CH}_3)_3 + \text{H}_2\text{O}$$

tert-Butyl 2-methylpropanoate

17.58

(a)

$$\text{CH}_3\text{CH}_2\overset{\text{O}}{\underset{\text{CH}_2\text{CH}_3}{\text{CHCNH}_2}}$$

2-Ethylbutanamide

(b)

N-Phenylbenzamide

(c)

N-Ethyl-*N*-methylbenzamide

(d)

$$\text{CH}_3\text{CH}_2\text{CH}_2\overset{\text{Br}}{\text{CH}}\overset{\text{O}}{\underset{\text{Br}}{\text{CHCNH}_2}}$$

2,3-Dibromohexanamide

17.59

(a)

$$CH_3CH_2CHCH_2CNH_2$$

with CH₃ and C=O groups

3-Methylpentanamide

(b)

$$CH_3CNH-C_6H_5$$

N-Phenylacetamide

(c)

$$HCN(CH_3)_2$$

N,N-Dimethylformamide

(d)

$$CH_3CH_2CNHCHCH_3$$

with O and CH₃ groups

N-1-Methylethylpropanamide
(N-Isopropylpropanamide)

17.60

(a)

$$CH_3CH_2CHC(-OH + H-)NH_2 \longrightarrow CH_3CH_2CHC-NH_2 + H_2O$$

with CH₂CH₃ group

2-Ethylbutanamide

(b)

$$C_6H_5-C(-OH + H-)NH-C_6H_5 \longrightarrow C_6H_5-CNH-C_6H_5$$

N-Phenylbenzamide + H₂O

(c)

$$C_6H_5-C(-OH + H-)NCH_2CH_3 \longrightarrow C_6H_5-CNCH_2CH_3 + H_2O$$

with CH₃ group

N-Ethyl-N-methylbenzamide

(d)

$$CH_3CH_2CH_2CHCHC(-OH + H-)NH_2 \longrightarrow CH_3CH_2CH_2CHCHCNH_2 + H_2O$$

with Br and Br groups

2,3-Dibromohexanamide

17.61

(a)

CH$_3$CH$_2$CHCH$_2$CNH$_2$ + H$_2$O $\xrightarrow[\text{base}]{\text{acid or}}$ CH$_3$CH$_2$CHCH$_2$COH + NH$_3$

3-Methylpentanamide   3-Methylpentanoic acid Ammonia

(b)

CH$_3$CNH— + H$_2$O $\xrightarrow[\text{base}]{\text{acid or}}$ CH$_3$COH + H$_2$N—

N-Phenylacetamide   Acetic acid   Aniline

(c)

HCN(CH$_3$)$_2$ + H$_2$O $\xrightarrow[\text{base}]{\text{acid or}}$ HCOH + HN(CH$_3$)$_2$

N,N-Dimethylformamide   Formic acid   Dimethylamine

(d)

CH$_3$CH$_2$CNHCHCH$_3$ + H$_2$O $\xrightarrow[\text{base}]{\text{acid or}}$ CH$_3$CH$_2$COH + H$_2$NCHCH$_3$

N-Isopropylpropanamide   Propanoic acid   Isopropylamine

Reactions of Carboxylic Acids and Their Derivatives

17.62

17.63

17.64

17.65

(a)

(b)

17.66

* = C–C double bond

17.67

Polyesters and Polyamides

17.68

17.69

$$3\ H_2NCH_2CH_2NH_2\ +\ 3\ HO-\overset{\displaystyle O}{\overset{\|}{C}}-\overset{\displaystyle O}{\overset{\|}{C}}-OH\ \xrightarrow[-\ H_2O]{heat}$$

$$NH_2CH_2CH_2NH-\overset{\displaystyle O}{\overset{\|}{C}}-\overset{\displaystyle O}{\overset{\|}{C}}-NHCH_2CH_2NH-\overset{\displaystyle O}{\overset{\|}{C}}-\overset{\displaystyle O}{\overset{\|}{C}}-NHCH_2CH_2NH-\overset{\displaystyle O}{\overset{\|}{C}}-\overset{\displaystyle O}{\overset{\|}{C}}-OH$$

Phosphate Esters and Anhydrides

17.70

$$\begin{matrix} CH_2OH \\ | \\ C=O \\ | \\ CH_2-O-\overset{\overset{\displaystyle O}{\|}}{\underset{\underset{\displaystyle O^-}{|}}{P}}-O^- \end{matrix}\ +\ H_2O\ \longrightarrow\ \begin{matrix} CH_2OH \\ | \\ C=O \\ | \\ CH_2OH \end{matrix}\ +\ HO-\overset{\overset{\displaystyle O}{\|}}{\underset{\underset{\displaystyle O^-}{|}}{P}}-O^-$$

The products of hydrolysis are dihydroxyacetone and hydrogen phosphate anion.

17.71

$$HO-\overset{\overset{\displaystyle O}{\|}}{\underset{\underset{\displaystyle OH}{|}}{P}}-O-\overset{\overset{\displaystyle O}{\|}}{\underset{\underset{\displaystyle OH}{|}}{P}}-O-CH_2-\!\!\bigcirc\!\!\ \xrightarrow{H_3O^+}\ 2\ HO-\overset{\overset{\displaystyle O}{\|}}{\underset{\underset{\displaystyle OH}{|}}{P}}-OH\ +\ HOCH_2-\!\!\bigcirc$$

anhydride   ester
  bond   bond

17.72

$$HO-\overset{\overset{\displaystyle O}{\|}}{\underset{\underset{\displaystyle OH}{|}}{P}}-O-\overset{\overset{\displaystyle O}{\|}}{C}-CH_3$$

Acetyl phosphate

17.73

$$CH_3CH_2OH\ +\ HO-\overset{\overset{\displaystyle O}{\|}}{\underset{\underset{\displaystyle OH}{|}}{P}}-O-\overset{\overset{\displaystyle O}{\|}}{C}-CH_3\ \longrightarrow\ CH_3CH_2O-\overset{\overset{\displaystyle O}{\|}}{\underset{\underset{\displaystyle OH}{|}}{P}}-OH\ +\ HO-\overset{\overset{\displaystyle O}{\|}}{C}-CH_3$$

Ethanol   Acetyl phosphate   Ethyl phosphate   Acetic acid

17.74

A cyclic phosphate diester

A cyclic phosphate diester is formed when one phosphate forms ester bonds with two hydroxyl groups in the same molecule.

17.75

a phosphate diester

A phosphate diester is formed when one phosphate group reacts with two alcohol hydroxyl groups.

Pyrophosphoric acid   a diphosphate

A diphosphate is formed when a phosphoric acid anhydride reacts with an alcohol.

Conceptual Problems

17.76

N,N-Dimethylformamide   *N*-Methylacetamide   Propanamide
         A   **B**   **C**

A is the lowest boiling isomer. Since no hydrogens are bonded to nitrogen, no hydrogen bonding between molecules of **A** can take place. Hydrogen bonding can occur between molecules of **B**, and thus **B** is higher boiling than **A**. **C** is the highest boiling isomer.

17.77

Salol (Phenyl salicylate)

17.78

CH₃CH₂CNH₂   CH₃COCH₃
Propanamide   Methyl acetate

Propanamide and methyl acetate are both soluble in water because both are polar molecules and can hydrogen-bond with water. Propanamide is higher boiling because molecules of propanamide can form hydrogen bonds with each other.

17.79 (1) Measure the pH of solutions of the two compounds. Benzoic acid is acidic; benzaldehyde is nonacidic.
(2) Benzaldehyde reacts with Tollens' reagent to form a silver mirror; benzoic acid doesn't react with Tollens' reagent.

17.80

Glycerol   Glyceryl tristearate

17.81

(a)

2-Chloro-3,4-dimethyl-3-hexene

(b)

CH₃CH₂CNCH₃

N-Methyl-*N*-phenylpropanamide

(c)

(CH$_3$CH$_2$)$_3$CCO—

Phenyl 2,2-diethylbutanoate

(d)

—CNHCH$_2$CH$_3$

NO$_2$

N-Ethyl-*o*-nitrobenzamide

Group Problems

17.82 (a) Although the identity of the odor of Juicy Fruit Gum is a trade secret, people think that the ester responsible for the odor is isoamyl acetate.

Isoamyl acetate + H$_2$O ← 3-Methyl-1-butanol + Acetic acid

(b) Peach aroma is due to γ-decalactone, the product formed when 4-hydroxydecanoic acid cyclizes to form a lactone.

γ-Decalactone + H$_2$O ← 4-Hydroxydecanoic acid

(c) Several compounds may cause the aroma of apple; among them are methyl butyrate, pentyl pentanoate, and hexyl acetate. Using the first ester as an example:

CH$_3$CH$_2$CH$_2$C—OCH$_3$ ← CH$_3$CH$_2$CH$_2$C—OH + HOCH$_3$

Methyl butyrate + H$_2$O   Butyric acid   Methanol
(Methyl butanoate)   (Butanoic acid)

(d) Methyl propanoate is responsible for the odor of rum.

CH$_3$CH$_2$C—OCH$_3$ ← CH$_3$CH$_2$C—OH + HOCH$_3$

Methyl propanoate + H$_2$O   Propanoic acid   Methanol

17.83

Pentanoic acid

2-Methylbutanoic acid

3-Methylbutanoic acid

2,2-Dimethylpropanoic acid

17.84

Penicillin G

a = amide
b = carboxylic acid
c = sulfide
d = aromatic ring
e = ether
f = alcohol (phenol)
g = ester
h = amine

Penicillin V

Amoxicillin

Cephalosporin C

In addition to the beta-lactam ring, all four antibiotics pictured have an additional amide group, a carboxylic acid group, and a sulfide.

Self-Test for Chapter 17

Multiple Choice

1. Which of the following amides does not form hydrogen bonds?
    (a) benzamide   (b) formamide   (c) *N*-methylformamide   (d) *N,N*-dimethylformamide

2. Which of the following statements about phosphoric acid anhydrides is untrue?
    (a) Energy is released when triphosphate anhydrides react with water.
    (b) Both mono- and dianhydrides occur in the body.
    (c) Hydrolysis of phosphate ester anhydrides yields caustic products.
    (d) Two molecules of water are necessary to completely hydrolyze a triphosphate anhydride.

3. Which class of compounds is lowest boiling?
    (a) carboxylic acid esters   (b) amides   (c) carboxylic acids   (d) carboxylic acid salts

4. The common name of an *-enoic* acid is:
    (a) succinic acid   (b) benzoic acid   (c) oxalic acid   (d) acrylic acid

5. Amide bonds occur in all of the following except:
    (a) nylon   (b) Dacron   (c) urea   (d) proteins

6. In the reaction $CH_3COOH + CH_3OH \rightleftharpoons CH_3COOCH_3 + H_2O$, the yield of CH_3COOCH_3 can be improved by:
   (a) distilling off the CH_3COOCH_3 (b) removing the water
   (c) using more CH_3COOH and CH_3OH (d) all of the above

7. Reaction of methanol with oxalic acid in the presence of an acid catalyst forms:
   (a) methyl oxalate   (b) dimethyl oxalate   (c) methyl acetate   (d) reaction doesn't occur

8. Polypropylene and nylon resemble each other in which respect?
   (a) They are both polymers.   (b) They are both hydrocarbons.   (c) Concentrated acid will dissolve them.   (d) An initiator is needed for polymerization to occur.

9. Which of the following compounds is an α-amino acid?

   (a)   (b)   (c)   (d)

   $H_2NCH_2CH_2COH$ $H_2NCCH_2CH_3$ $H_2NCH_2COCH_3$ H_2NCH_2COH

10. How many amides of the formula C_3H_7NO can be drawn?
    (a) 3   (b) 4   (c) 5   (d) 6

Sentence Completion

1. The most common general reaction of carboxylic acids, esters, and amides is called a _____ ___ _____ reaction.

2. Kevlar is a type of _____.

3. A carboxylic acid can be dissolved in water by converting it into its _____ ____.

4. The reaction of a carboxylic acid with an alcohol in the presence of an acid catalyst is called an _____ reaction.

5. In carboxylic acids, esters, and amides, the carbonyl-group carbon is bonded to an atom that strongly _____ electrons.

6. Carboxylic acids are named by using the name ending ___ ____.

7. $CH_3CH_2COOCH_2CH_2CH_3$ is named _____ _____.

8. Many flavors and fragrances are due to _____.

9. _____ is the transfer of a $-PO_3^{2-}$ group from one molecule to another.

10. Long-chain carboxylic acids can be found in nature as components of _____.

11. The reaction of an ester with aqueous NaOH to yield a salt and an alcohol is called a _____ reaction.

12. Malonic acid is also known as _____ acid.

True or False

1. A carboxylic acid salt is more soluble in water than a carboxylic acid.

2. A *saponification* reaction is the acid-catalyzed hydrolysis of an ester.

3. Amides and amines are both basic.

4. A carboxylic acid is higher boiling than an ester of the same molecular weight.

5. Basic hydrolysis of ethyl acetate yields acetic acid and ethanol.

6. The compound $HCON(CH_3)_2$ is named dimethylformamide.

7. Nitrate esters are important in living systems.

8. Acetic acid is about 1% dissociated in a 1 M aqueous solution.

9. Carboxylic acids are either solids or liquids at room temperature.

10. Esterification can be brought about by treating a carboxylic acid with an alcohol in the presence of NaOH.

11. Pyrophosphoric acid is a phosphate ester.

12. Propanamide is higher boiling than methyl acetate.

Match each entry on the left with its partner on the right.

1. $CH_3COOH + CH_3OH + HCl$ (a) Amide hydrolysis

2. $HOOCCH_2COOH$ (b) Ester hydrolysis

3. $(CH_3COO^-)_2Mg^{2+}$ (c) Thioester

4. $CH_3COOCH_3 + NaOH, H_2O$ (d) *-enoic* acid

5. HCOOH   (e) Phosphate ester

6. $CH_3CH=CHCOOH$ (f) Dioic acid

7. $CH_3CONH_2 + NaOH, H_2O$ (g) Amide

8. CH_3OPO_3H (h) Polyester

9. Nylon   (i) Formic acid

10. Dacron   (j) Ester formation

11. Acetyl-SCoA   (k) Acid salt

12. Lidocaine   (l) Polyamide

Chapter Outline

I.   Introduction to biochemistry and proteins (Sections 18.1–18.2).
   A. Biochemistry (Section 18.1).
      1. Biochemistry is the study of the structure and function of biomolecules.
      2. The principal classes of biomolecules are proteins, carbohydrates, lipids, and nucleic acids.
      3. Biomolecules range in size from small organic molecules to polymers with millions of subunits.
      4. The three-dimensional shape of biomolecules is critical to their ability to function.
   B. Proteins—an overview (Section 18.2).
      1. Proteins are polymers of amino acids.
      2. Proteins have numerous functions in living things.
         a. Among these functions are structure, support, storage, transport, and control of biochemical reactions.
II. Amino acids (Sections 18.3–18.4).
   A. Structure of amino acids (Section 18.3).
      1. Twenty amino acids commonly occur in nature.
         a. Each amino acid consists of a carboxylic acid, an amine, and a side chain, all bonded to the same carbon.
         b. All 20 are α-amino acids.
         c. Nineteen of the twenty amino acids are primary amines.
      2. In this book, each amino acid is represented by a three-letter code.
      3. Amino acids are classified as neutral, acidic, or basic.
         a. Of the common amino acids, 15 are neutral, 3 are basic, and 2 are acidic.
      4. Amino acid side chains are classified as hydrophilic (polar) or hydrophobic (nonpolar).
         a. Hydrophobic side chains cluster in the middle of a protein.
         b. Hydrophilic side chains are positioned on the surface of the protein.
         c. Interactions between amino acid side chains are determined by noncovalent forces.
      5. All amino acids (except for glycine) are chiral.
   B. Acid–base properties of amino acids (Section 18.4).
      1. Amino acids exist as dipolar zwitterions.
         a. Amino acids have properties similar to salts: They are high-melting and soluble in water.
         b. Amino acids can react as acids or bases.
      2. The pH at which the number of positive charges on an amino acid equals the number of negative charges is called the isoelectric point (pI).
         a. Neutral amino acids have pIs in the range 5.0–6.5.
         b. Acidic amino acids are negatively charged at physiological pH.
III. Peptides (Section 18.5).
   A. The bonds that form between amino acids are amide bonds (peptide bonds).
      1. Two amino acids bond together to form a dipeptide.
      2. Three amino acids bond together to form a tripeptide.
      3. A chain with 10–100 amino acids is a polypeptide.
      4. The atoms that form a peptide bond have a planar relationship.

B. Peptide structure.
  1. The atoms that form the peptide bond lie in a plane.
  2. The amino-terminal end of the peptide is written on the left, and the carboxyl-terminal end is on the right.
  3. A peptide is named by listing the residues in order, from left to right.

IV. Proteins (Sections 18.6–18.9).
  A. Primary protein structure (Section 18.6).
    1. The repeating chain of peptide bonds and α carbons is known as the protein backbone.
    2. Protein function is dependent on the exact sequence of amino acids in protein primary structure.
    3. Primary structure is responsible for the shape of a protein molecule.
      a. In sickle-cell anemia, a change in only one amino acid is responsible for the disease.
  B. Secondary protein structure (Section 18.7).
    1. Secondary structure refers to the ordered relationships of backbone atoms resulting from hydrogen bonding.
    2. There are two types of secondary structures.
      a. The α-helix is a right-handed coil formed by hydrogen bonds between an N—H group and a carbonyl group four amino acids away.
      b. The β-sheet occurs when polypeptide chains line up next to each other, and hydrogen bonds form between adjacent chains.
    3. Fibrous and globular proteins show examples of secondary structure.
      a. Some fibrous proteins are composed almost entirely of α-helices.
      b. Globular proteins have smaller regions of α-helix and β-sheet secondary structure.
    4. The parts of the polypeptide chain not involved in a secondary structural element are called random coil segments.
  C. Tertiary protein structure (Section 18.8).
    1. Tertiary structure.
      a. Tertiary structure describes the overall shape of proteins.
      b. The interactions responsible for tertiary structure may be between amino acids many units apart.
      c. Interactions give each protein its unique three-dimensional shape.
      d. Noncovalent interactions.
        i. Hydrogen bonds can occur between atoms on the protein backbone.
        ii. Hydrogen bonds can occur between side chains of amino acids or between side-chain amino acids and the polypeptide backbone.
        iii. Ionic interactions can occur between ionized acidic and basic amino acids; these are called salt bridges.
        iv. Hydrophobic interactions occur between nonpolar side chains.
      e. Covalent interactions. The thiol groups of cysteine residues form disulfide bonds.
    2. Examples.
      a. Insulin.
        i. Insulin was the first polypeptide whose sequence was determined.
        ii. Insulin consists of two polypeptide chains linked by disulfide bonds.
      b. Ribonuclease—a globular protein.
        i. Hydrophobic side-chains congregate in the middle, and hydrophilic side chains are distributed on the outside.
      c. Myoglobin—a globular protein.
        i. Myoglobin has eight α-helix segments.
        ii. A heme group is embedded within the polypeptide chain.

D. Quaternary structure (Section 18.9).
  1. Quaternary structure describes the noncovalent interactions between the polypeptide chains of proteins that consist of more than one chain.
  2. Examples:
    a. Hemoglobin consists of four polypeptide chains held together by hydrophobic interactions.
    b. Collagen is formed by several tropocollagen strands overlapping lengthwise.
  3. Classification of proteins.
    a. Fibrous proteins are insoluble and are composed of fibers and sheets.
    b. Globular proteins are water-soluble and have chains folded into compact shapes.
    c. Simple proteins contain only amino acids.
    d. Conjugated proteins contain one or more non-amino acid units.
V. Chemical properties of proteins (Section 18.10).
  A. Hydrolysis.
    1. Polypeptide amide bonds can be cleaved by acid or enzymes to yield amino acids.
  B. Denaturation.
    1. Denaturation disturbs a protein's shape without disrupting its primary structure.
    2. Denaturation changes the properties of proteins.
    3. Denaturation can be caused by various agents.
      a. Heat disturbs side-chain interactions.
      b. Mechanical agitation.
      c. Detergents disrupt the interactions of hydrophobic side chains.
      d. Organic compounds disrupt hydrogen bonds.
      e. pH changes disrupt salt bridges.
      f. Inorganic salts disrupt salt bridges.
  C. Occasionally, denatured proteins can undergo renaturation.

Solutions to Chapter 18 Problems

18.1   Alcohol dehydrogenase is an enzyme.

18.2   Cortisol and oxytocin are hormones.

18.3

Alanine has a neutral, nonpolar side chain.

18.4

18.5

(a)

H₂N—CH—C—OH with =O on C, branch CH—OH, CH₃

α-amino acid
(Threonine)

(b)

H₂N—C—CH₂CH₂CH₃ with =O on C

not an α-amino acid
(amide)

(c)

CH₃CH₂CHCH₂—NH₂ with OH branch

not an α-amino acid
(amino alcohol)

(d)

HO—C—CHCH₂CH(CH₃)₂ with =O on C, branch NH₂

α-amino acid
(Leucine)

18.6 (a) Amino acids containing an aromatic ring: phenylalanine, tyrosine, tryptophan

Phenylalanine   Tyrosine   Tryptophan

(b) Amino acids containing sulfur: cysteine, methionine

Cysteine   Methionine

(c) Amino acids that are alcohols: serine, threonine, tyrosine (a phenol)

Serine   Threonine   Tyrosine

(d) Amino acids with alkyl group side chains: alanine, valine, leucine, isoleucine

Alanine   Valine   Leucine   Isoleucine

18.7

Serine   Valine

(a) The serine side chain has a polar hydroxyl group; the valine side chain has a nonpolar isopropyl group.
(b) Valine has a hydrophobic side chain, and serine has a hydrophilic side chain.

18.8 Aspartic acid has a hydrophilic, polar side chain.

18.9 Tryptophan has a hydrophobic, nonpolar side chain.

18.10, 18.11
Serine is chiral because four different groups are bonded to a central carbon atom.

Serine

18.12 Threonine and isoleucine each have two chiral carbon atoms (starred in the following structures).

Threonine   Isoleucine

CIA 18.1

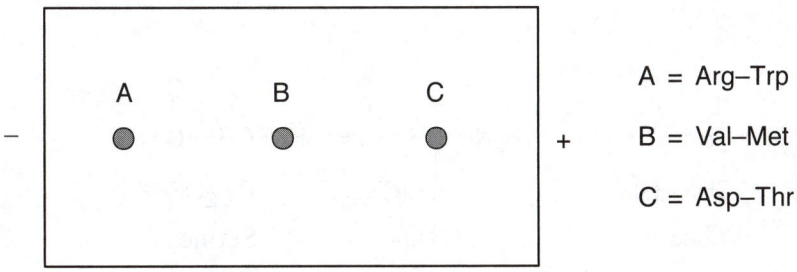

C = collagen
BI = bovine insulin
HH = human hemoglobin

At pH = 6.6, collagen (isoelectric point = 6.6) has as many positive charges as negative charges and does not migrate. Bovine insulin (pI = 5.4) is negatively charged and migrates to the positively charged electrode. Human hemoglobin (pI = 7.1) is positively charged and migrates to the negative electrode.

CIA 18.2

Remember these facts to answer this question:
(1) In a peptide, the acidic and basic amino acid side chains determine the charge of the peptide.
(2) An amino acid that is acidic is negatively charged at a neutral pH, and a basic amino acid is positively charged at a neutral pH.
(3) A positively charged molecule migrates to the cathode (–), and a negatively charged molecule migrates to the anode (+).

A = Arg–Trp

B = Val–Met

C = Asp–Thr

18.13

at low pH   Glutamic acid   at high pH

At low pH, both of the carboxylic acid groups and the amino group are protonated. At high pH, the carboxylic acid groups exist as carboxylate anions and the amino group is not protonated. The second form represents the zwitterion since it has an equal number of positive and negative charges.

18.14 In the zwitterionic form of an amino acid, the $-NH_3^+$ group is an acid (because it can donate H^+), and the $-COO^-$ group is a base (because it can accept H^+).

HOC 18.1

Alanine   Glycine

Alanylglycine   peptide bond $+ H_2O$

Amino groups are circled.

Carboxyl groups are bracketed.

Chiral carbons are starred.

Water is eliminated.

18.15

Serine   Valine   Valine   Serine

18.16 (a) Gly–Ser–Tyr   Tyr–Ser–Gly   Ser–Tyr–Gly
Gly–Tyr–Ser   Tyr–Gly–Ser   Ser–Gly–Tyr

(b)

Gly–Ser–Tyr   Gly–Tyr–Ser

18.17   Ile–Arg–Val   Arg–Ile–Val   Val–Arg–Ile
Ile–Val–Arg   Arg–Val–Ile   Val–Ile–Arg

18.18

(a)

Leu–Asp

Leucine (Leu)   Aspartate (Asp)
neutral   polar, acidic

(b)

Tyr–Ser–Lys

Serine (Ser)   Lysine (Lys)
basic

Tyrosine (Tyr)
All residues in (b) are polar

18.19 Vasopressin has eight peptide bonds.

CIA 18.3

People need a daily source of protein in the diet because the human body does not store the protein or the amino acids needed for its continuous synthesis of proteins. The body does store fats and carbohydrates.

CIA 18.4

An incomplete protein lacks one or more of the nine essential amino acids.

CIA 18.5

Casein and egg white protein are well-balanced proteins because they must provide complete nutrition to newborn and embryonic organisms. It is thus not surprising that they are also well-balanced proteins for human growth and development.

CIA 18.6

Symptoms of sickle-cell anemia include inflammation, pain, restricted blood flow that may damage organs, and a shortened life span.

CIA 18.7

A person with sickle-cell anemia carries two copies of the defective hemoglobin gene; a person with sickle-cell trait carries only one copy.

18.20 (a) A planar unit consists of a carbonyl group, the adjacent –NH group, and the two α-carbons.
(b) These atoms come from two different amino acids. The units are planar because of electron delocalization over the amide bond from the carbonyl group, resulting in partial double-bond character for the amide bond and restricting rotation around that bond.

18.21 There are n! different arrangements of amino acids in a peptide containing n different amino acids. For four different amino acids, $n = 4$ x 3 x 2 x $1 = 24$. Thus, there are 24 different ways that four amino acids can be arranged in a peptide.

18.22 A protein can't function unless its amino acids are in the exact order needed.

18.23 In an α-helix, a carbonyl oxygen forms a hydrogen bond with an amide hydrogen 12 atoms away. Thus, 11 backbone atoms lie between the atoms that take part in an α-helix hydrogen bond.

18.24 (a) Hydrogen bonding is responsible for sheet formation.
(b) Hydrogen bonding takes place between an amide hydrogen and an amide carbonyl oxygen on an adjacent chain.

18.25 (a) fibrous (b) fibrous

18.26 The skin is composed of fibrous protein that is insoluble in water.

18.27

| | *Side-Chain Functional Groups* | *Hydrogen Bonding?* |

(a)

Phe, Thr   —CH$_2$—⟨benzene ring⟩   —CHCH$_3$ no
                                          |
                                          OH

(b)

Asn, Ser   —CH$_2$CNH$_2$ —CH$_2$OH   yes
                    ||
                    O

—CH$_2$O:
         |
         H

:O:
 || ⋮
—CH$_2$C—NH$_2$

(c)

Thr, Tyr   —CHCH$_3$ —CH$_2$—⟨benzene ring⟩—OH   yes
                |
                OH

—CHCH$_3$
        |
       :OH
        ⋮
⟨benzene ring⟩—O:
               H

(d)

Gly, Trp   —H   —CH$_2$—⟨indole ring, N–H⟩   no

Amino acids that have amine, amide, alcohol, or phenol groups are capable of hydrogen bonding to other amino acids having these groups. Only in (b) and (c) do both amino acids have groups that can participate in hydrogen bonding. Other combinations of hydrogen bonds can be drawn in (b) and (c).

18.28 (a) Glutamine and serine interact through *hydrogen bonds* between the terminal —NH$_2$ group of glutamine and the —OH of serine.
(b) Isoleucine and proline are pulled together through *hydrophobic interactions* between the alkyl side chain of isoleucine and the ring portion of proline.
(c) A *salt bridge* occurs between the negatively charged aspartate carboxyl group and the terminal amine group of lysine.
(d) The alkyl side chain of alanine and the aromatic ring of phenylalanine are held together by *hydrophobic interactions*.

18.29   (a) Hydrogen bonds from side chains: tyrosine, asparagine, serine.
(b) Hydrophobic group interactions: alanine, isoleucine, valine, leucine.

18.30 Use Table 18.5 to answer this question.
(a) lipoprotein   (b) metalloprotein   (c) phosphoprotein
(d) glycoprotein   (e) hemoprotein   (f) nucleoprotein

18.31 (a) Alpha-keratin consists of many different amino acids but has a relatively large percentage of cysteine residues, forming disulfide bonds. In all collagen molecules, every third residue is a glycine, and many proline and hydroxyproline residues are also found.
(b) In α-keratin, pairs of α-helixes twist together into small fibrils that are twisted into larger bundles. In tropocollagen, three coiled chains wrap around each other to form a triple helix.

18.32 Trypsin cleaves a peptide on the carboxyl side of arginine and lysine. Treatment of this peptide with trypsin yields three fragments.

$$Ala–Phe–Lys \}\ Cys–Gly–Asp–Arg \}\ Leu–Leu–Phe–Gly–Ala$$

$$\downarrow \text{ Trypsin}$$

$$Ala–Phe–Lys \ + \ Cys–Gly–Asp–Arg \ + \ Leu–Leu–Phe–Gly–Ala$$

18.33 Acid hydrolysis would yield 12 individual amino acids because all peptide bonds would be cleaved.

HOC 18.2
1. Clear egg white becomes opaque and white.
2. Coagulation occurs as clumps form.
3. Strings form in the egg whites.
4. The egg whites were whipped (mechanical agitation).

CIA 18.8
Osteogenesis imperfecta is a dominant genetic defect in which improperly synthesized collagen leads to frequent bone fractures.

CIA 18.9
In order to support bones, collagen needs to be strong.

Understanding Key Concepts

18.34 Peptides that contain neither acidic nor basic amino acids have a net charge of +1 at pH = 2 because both the N-terminal amino group and the C-terminal carboxylic acid group are protonated. At pH = 14, these peptides have a net charge of –1, because the N–terminal group is uncharged and the C-terminal group is a carboxylate anion.

(a) Val

(b) Arg

(c) Tyr–Ser

(d) Glu–Asp

at pH = 2

at pH = 14

(e) Gln–Ala–Asn

at pH = 2

at pH = 14

(f) Met–Trp–Cys

at pH = 2

at pH = 14

18.35 In order for two amino acids to interact, they must both be of the same type (both polar or both nonpolar, for example).
(a) The pairs Pro...Phe and Ala...Gly show hydrophobic interactions because they are both nonpolar.
(b) Glu...Lys and Asp...His show ionic interactions because they are both charged.
(c) Asp...Ser form hydrogen bonds.

18.36

Asp–Gly–Phe–Leu–Glu–Ala

Hydrogen bonds in an α-helix are indicated by dashed lines.

18.37

	Fibrous Protein	*Globular Protein*
Biological function:	Structural proteins	Enzymes, hormones, transport
Water solubility:	Water-insoluble	Usually water-soluble
Amino acid composition:	Contains many Gly and Pro residues	Varies; hydrophilic groups on the outside account for solubility
Secondary structure:	Large regions of α-helix or β-sheet	Smaller regions of α-helix or β-sheet
Tertiary Structure:	Few interactions between side chains on the same backbone	Complex: determined by hydrophobic and hydrophilic groups on side chains, as well as by disulfide bonds
Examples:	Collagen (connective tissue) α-Keratin (hair) Fibroin (silk)	Ribonuclease (enzyme) Hemoglobin (oxygen transport) Insulin (hormone)

18.38 (a) Leu, Phe, Ala, or any amino acid with a nonpolar side chain might be found in the part of the protein that lies within the cell membrane.
(b), (c) Asp, Lys, Thr, or any amino acid with a hydrophilic side chain might be found in the part of the protein that either lies outside the cell membrane or inside the cell membrane.

18.39

The upper chiral carbon is responsible for the D,L configuration.

Additional Problems

Proteins and Their Functions

18.40 | *Type of Protein* | *Function* | *Example* |
|---|---|---|
| Enzymes | Catalyze biochemical reactions | Ribonuclease |
| Hormones | Regulate body functions | Insulin |
| Storage proteins | Store essential substances | Myoglobin |
| Transport proteins | Transport substances through body fluids | Serum albumin |
| Structural proteins | Provide shape and support | Collagen |
| Protective proteins | Defend the body against foreign matter | Immunoglobulins |
| Contractile proteins | Do mechanical work | Myosin and actin |

18.41 (a) Human growth hormone regulates growth.
(b) Myosin is a contractile protein that makes muscles function.
(c) A protease is an enzyme that catalyzes the hydrolysis of peptide bonds.
(d) Myoglobin stores oxygen in muscles.

Amino Acids

18.42

(a)
$$H_3\overset{+}{N}-\underset{\underset{CH(CH_3)_2}{|}}{CH}-\overset{\overset{O}{\|}}{C}-O^-$$
Valine (Val)

(b)
$$H_3\overset{+}{N}-\underset{\underset{CH_2OH}{|}}{CH}-\overset{\overset{O}{\|}}{C}-O^-$$
Serine (Ser)

(c)
$$H_3\overset{+}{N}-\underset{\underset{CH_2CH_2\underset{\underset{O^-}{|}}{C}=O}{|}}{CH}-\overset{\overset{O}{\|}}{C}-O^-$$
Glutamic acid (Glu)

18.43

(a)
$$H_3\overset{+}{N}-\underset{\underset{\underset{CH_3}{|}}{CHCH_2CH_3}}{CH}-\overset{\overset{O}{\|}}{C}-O^-$$
Isoleucine (Ile)

(b)
$$H_3\overset{+}{N}-\underset{\underset{\underset{CH_3}{|}}{CHOH}}{CH}-\overset{\overset{O}{\|}}{C}-O^-$$
Threonine (Thr)

(c)
$$H_3\overset{+}{N}-\underset{\underset{CH_2CH_2\overset{\overset{O}{\|}}{C}-NH_2}{|}}{CH}-\overset{\overset{O}{\|}}{C}-O^-$$
Glutamine (Gln)

18.44

(a)
$$H_3\overset{+}{N}-\underset{\underset{\underset{SH}{|}}{CH_2}}{CH}-\overset{\overset{O}{\|}}{C}-O^-$$
Cysteine (Cys)

(b)
$$H_3\overset{+}{N}-\underset{\underset{\underset{OH}{|}}{CH_2}}{CH}-\overset{\overset{O}{\|}}{C}-O^-$$
Tyrosine (Tyr)

18.45

(a)
Valine (Val)

(b)
Threonine (Thr)

18.46 A chiral object is one that has handedness. Examples include a glove and scissors.

18.47 An achiral object has no handedness and possesses a plane of symmetry such that one half of the object is an exact mirror image of the other half. Examples include a spoon and a pencil.

18.48

Leucine

18.49

Isoleucine

18.50 The side chain of phenylalanine is hydrophobic because it is nonpolar.

18.51 The side chain of histidine is hydrophilic because it can form hydrogen bonds with water.

Acid–Base Properties of Amino Acids

18.52 At a neutral pH (pH = 5–8), an amino acid with a pI less than 5 is negatively charged, and an amino acid with a pI greater than 8 is positively charged. Otherwise, an amino acid has no net charge at neutral pH and is neutral. Asparagine (pI = 5.4) and proline (pI = 6.3) are neutral. Lysine (pI = 9.7) is positively charged.

18.53 (a) Aspartic acid (pI = 3.0) is negatively charged. (b) Histidine (pI = 7.8) is positively charged. (c) Valine (pI = 6.0) is neutral.

18.54

This structure represents aspartic acid at pH = 3, its isoelectric point. (low pH)

This structure represents aspartic acid at pH = 13. (high pH)

This structure represents aspartic acid at pH = 1. (low pH)

At neutral pH, a structure with two carboxylates and a protonated amino group predominates.

18.55 The zwitterion is represented by (a), the form present at its isoelectric point (pH = 3).

18.56

(a)

H₃N⁺—CH—C—O⁻ (C=O double bond)
          |
       (CH₂)₄
          |
        ⁺NH₃

at pH = 7
(neutral pH)

(b)

H₃N⁺—CH—C—OH (C=O double bond)
          |
       (CH₂)₄
          |
        ⁺NH₃

at pH = 3
(low pH)

(c)

H₃N⁺—CH—C—O⁻ (C=O double bond)
          |
       (CH₂)₄
          |
        NH₂

at pH = 9.7 (pI)
(high pH)

18.57 The zwitterion is represented by (c), the form present at its isoelectric point (pH = 9.7).

18.58 At its isoelectric point, a protein has the same number of positive and negative charges and is thus electrically neutral. On either side of the isoelectric point, the protein is charged and is more soluble in a polar solvent such as water.

18.59 To increase solubility, buffer the water solution (Section 10.10) to a pH at which the amino acid does not exist in its zwitterionic form.

Peptides

18.60 Val–Met–Leu   Met–Val–Leu   Leu–Met–Val
Val–Leu–Met   Met–Leu–Val   Leu–Val–Met

18.61

H₃N⁺—CH—C—NH—CH—C—O⁻ (two C=O double bonds)
          | |
        CH₂   CH₂
          | |
         CH   COO⁻
       /   \
    H₃C   CH₃

Leu – Asp

H₃N⁺—CH—C—NH—CH—C—O⁻ (two C=O double bonds)
          | |
        CH₂   CH₂
          | |
        COO⁻   CH
                       /   \
                    H₃C   CH₃

Asp –Leu

18.62

N-terminal
amino acid

Tyr – Gly – Gly – Phe – Met

C-terminal
amino acid

18.63

N-terminal
amino acid

Tyr – Gly – Gly – Phe – Leu

C-terminal
amino acid

18.64

N-terminal
amino acid

C-terminal
amino acid

Valine	Glycine	Serine	Alanine	Aspartate
Val	Gly	Ser	Ala	Asp

18.65

N-terminal
amino acid

C-terminal
amino acid

Cysteine	Lysine	Glutamate	Proline
Cys	Lys	Glu	Pro

Protein Structure

18.66 Primary structure refers to the sequence of connection of amino acids in a protein.

18.67 The backbone consists of repeating units of the sequence ($-\alpha$-carbon$-$NH$-$carbonyl carbon$-$).

18.68 (a) Bradykinin:

Arg$-$$-Pro-$$-$$-Pro-$$-$$-$$-Gly-$$-$$-$$-Phe-$

(b) The rigid proline rings introduce kinks and bends in bradykinin and prevent hydrogen bonds from forming (because proline has no hydrogens to form hydrogen bonds).

18.69 Red blood cells of people with sickle-cell anemia contain hemoglobin that, instead of retaining its globular form while transporting and releasing oxygen, forms fibrous chains after releasing oxygen. These fibrous chains deform red blood cells.

Secondary Protein Structure

18.70 Hydrogen bonding, which is responsible for protein secondary structure, takes place between a carbonyl oxygen on the backbone of an amino acid chain and an amide hydrogen on the backbone of the same chain or on a neighboring chain.

18.71 Hydrogen bonding is noncovalent.

18.72 Hydrogen bonds in an alpha helix form between carbonyl oxygens and amide hydrogens thirteen atoms away on the same polypeptide chain.

18.73 A beta-pleated sheet forms between carbonyl oxygens and amide hydrogens on different polypeptide chains.

18.74 Keratin, a protein consisting of primarily alpha helices, is a fibrous protein.

18.75 Fibroin, a protein consisting primarily of beta-pleated sheets, is also a fibrous protein.

Tertiary Protein Structure

18.76 (a) Two cysteines form a disulfide bond.
   (b) Alanine and leucine experience hydrophobic interactions.
   (c) Aspartic acid and asparagine form hydrogen bonds.
   (d) Serine and lysine form hydrogen bonds.

18.77 The bond in (a) is covalent; all other bonds are noncovalent.

18.78 The side chains of proteins orient themselves to optimize interactions with both each other and with the aqueous environment that surrounds them in a way that produces the lowest energy state.

18.79 Proteins called "chaperones" guide the folding of polypeptide chains.

18.80 A simple protein is composed only of amino acids. A conjugated protein consists of a simple protein associated with one or more nonprotein molecules.

18.81 (a) Metalloproteins contain metal ions and protein.
(b) Hemoproteins contain heme and protein.
(c) Lipoproteins contain lipids and protein.
(d) Nucleoproteins contain ribonucleic acid and protein.

18.82 The disulfide bonds between two cysteine residues help to stabilize a protein's tertiary structure by drawing parts of the polypeptide chain together.

18.83 In the presence of an oxidizing agent, thiol groups of two different cysteine residues can come together to form a disulfide bond. Disulfide bonds can connect two different peptide chains or can connect two distant parts of the same chain, introducing a loop in the chain.

18.84 (a) *Primary structure* refers to the sequence of connection of amino acids in a protein.
(b) *Secondary structure* refers to the orientation of segments of the protein chain into a regular pattern, such as an α-helix or a β-sheet, by hydrogen bonding between backbone atoms.
(c) *Tertiary structure* refers to the coiling and folding of the entire protein chain into a three-dimensional shape as a result of interactions between amino acid side chains.
(d) *Quaternary structure* refers to the aggregation of several protein chains to form a larger structure.

18.85 (a) Primary structure   (b) Secondary structure   (c) Tertiary structure

18.86 (a) *Hydrophobic interactions* occur between hydrocarbon side chains of amino acids. In a protein, these side chains cluster in the center of the molecule to exclude water, and are responsible for the nearly spherical tertiary shape of globular proteins. Alanine and isoleucine take part in hydrophobic interactions.
(b) *Salt bridges* occur between negatively charged and positively charged amino acid side chain groups. They can stabilize the tertiary structure of a protein by connecting two distant parts of a polypeptide chain or by pulling the protein backbone together in the middle of the chain. They can also stabilize quaternary structure by bringing together two polypeptide chains. Lysine and aspartate can form salt bridges.

18.87 (a) Side chains that can form *hydrogen bonds* are located on the outside of a protein, where they can stabilize tertiary structure by forming hydrogen bonds with water. Hydrogen bonds stabilize quaternary protein structure by holding together several protein strands. Threonine and glutamine can form hydrogen bonds.
(b) *Disulfide bonds* connect distant parts of a polypeptide chain and introduce loops into the chain. Two cysteine residues form disulfide bridges.

18.88 A minimum of two polypeptide chains are necessary for quaternary structure to exist.

18.89 Hemoglobin has four polypeptide chains that contribute to its quaternary structure.

18.90 A conjugated protein consists of a simple protein plus associated non-amino acid units.

18.91 Among the possible nonprotein units of conjugated proteins are carbohydrates, lipids, metal ions, phosphate groups, heme, and RNA. Lipoproteins are an example of conjugated proteins.

Chemical Properties of Proteins

18.92 When a protein is denatured, its three-dimensional structure is disrupted, and its ability to catalyze reactions is impaired. Primary structure is not affected by denaturation.

18.93 (a) *Heat* denatures proteins by disrupting side-chain interactions.
(b) *Addition* of a *strong acid* disrupts side chain interactions and interferes with salt bridges.
(c) *Organic solvents* disrupt the association of hydrophobic side chains.

18.94 Protein digestion is the hydrolysis of peptide bonds to form amino acids. Protein denaturation is the disruption of secondary, tertiary, or quaternary structure without disrupting peptide bonds.

18.95 Denaturation disrupts a protein's shape without disrupting its primary structure; hydrolysis breaks bonds and thus destroys a protein's primary structure.

18.96 Pineapple that is canned has been heated to inactivate enzymes. Thus, no enzymatic hydrolysis occurs when canned pineapple is added to gelatin.

18.97 (a) charcoal grilled steak–heat   (b) pickled pigs' feet–inorganic salts, pH change
(c) meringue–mechanical agitation   (d) steak tartare–none
(e) salt pork–inorganic salts

Conceptual Problems

18.98 Hydrophobic interactions: (e) tryptophan, (f) alanine, (g) leucine, (h) methionine
Hydrogen bonding: (a) tyrosine, (c) asparagine, (d) lysine
Salt bridges: (d) lysine
Covalent bonding: (b) cysteine

18.99

```
        Asn—Cys—Pro—Leu—Gln
       /       \
    Gln         S
     |          |            Oxytocin
    Ile         S
       \       /
        Tyr—Cys
```

18.100

The sulfur in methionine is a sulfide ($-SCH_3$) sulfur instead of a thiol ($-SH$) sulfur and can't form disulfide bridges.

18.101

Leucine, alanine, glycine, and valine all have nonpolar side chains and are likely to be found in the interior of proteins, where they can avoid water.

18.102

Polar, hydrophilic amino acids that can form hydrogen bonds with water are found on the outside of globular proteins. Examples: (b) Glu, (e) Ser.
Nonpolar amino acids that have small side chains are found in fibrous proteins. Examples: (a) Ala, (f) Val. Phe and Leu are nonpolar but are too bulky to be often found in fibrous proteins.

18.103

Glycine is found in bends because of its small size. Proline, because of its rigid structure, is often the initiator of bends.

18.104

If an aspartate residue were substituted for a glutamate residue, a health problem might not occur because both amino acids have similar physical properties and chemical behavior.

18.105

(a) The blood samples will be analyzed by electrophoresis.
(b) The sick child's blood sample will show the electrophoresis pattern of sickle-cell anemia. The other children may show either of the other two patterns, which will indicate if they have sickle-cell trait or no sickle-cell gene.

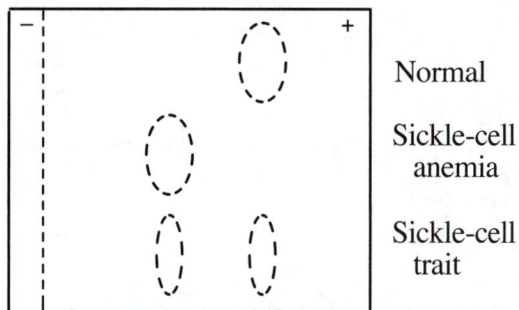

(c) People with sickle-cell anemia carry two copies of the defective gene and suffer from anemia, inflammation, and pain. People with sickle-cell trait have only one copy of the gene and suffer from these symptoms only under conditions of extreme stress.

18.106

If a diabetic took insulin orally, digestive enzymes would catalyze its hydrolysis, and the individual amino acids would be absorbed as food.

18.107

When aspartame is hydrolyzed by digestive enzymes, it produces phenylalanine, which is harmful to people who have PKU.

18.108

A combination of grains, legumes, and nuts in each meal provides all of the essential amino acids.

Group Problems

18.109

A peptide rich in Lys and Asp residues is more soluble in water than a peptide rich in Ala and Leu residues. The side chains of Lys and Asp are polar and can form hydrogen bonds with water, unlike the nonpolar, hydrophobic side chains of Ala and Leu.

18.110

Amino acids with polar side chains (aspartate and histidine, for example) are likely to be found on the outside of a soluble protein, where they can form hydrogen bonds with water and with each other. Amino acids with nonpolar side chains (alanine and valine, for example) are likely to be found on the inside of globular proteins, where they can escape from water.

18.111

Leucine and phenylalanine, amino acids with nonpolar side chains, are found on the inside of a soluble protein, where they can avoid water. Glutamate and glutamine (amino acids with polar side chains) are found on the outside of a globular protein, where they can form hydrogen bonds with water or can form salt bridges.

18.112

Amino acids that can form hydrogen bonds contain the following functional groups: amine, alcohol, amide, and carboxylic acid (carboxylate). These amino acids are arginine, aspartic acid, asparagine, glutamic acid, glutamine, histidine, lysine, serine, threonine, and tyrosine. Shown below is hydrogen bonding between asparagine and serine.

Hydrogen bonds between the amino acids and water:

Self-Test for Chapter 18

Multiple Choice

1. Which level of protein structure doesn't involve hydrogen bonds?
   (a) primary   (b) secondary   (c) tertiary   (d) quaternary

2. Which of the following amino acids has a hydrophobic side chain?
   (a) asparagine   (b) threonine   (c) histidine   (d) phenylalanine

3. The isoelectric point of an amino acid is 7.6. Which of the following statements about it is untrue?
   (a) At pH = 7.6, the number of positive charges equals the number of negative charges.
   (b) It is a basic amino acid.
   (c) At pH = 5.5, both the carboxylic acid group and the amine group are protonated.
   (d) It forms salt bridges when it is part of a polypeptide chain.

4. Which of the following organic molecules is chiral?
   (a) 1-pentanol   (b) 2-pentanol   (c) 3-pentanol   (d) 2-methyl-2-butanol

5. How many tetrapeptides can be formed from alanine, lysine, and two valines?
   (a) 6   (b) 12   (c) 18   (d) 24

6. Which of the following is an amino acid with a nitrogen heterocyclic ring?
   (a) histidine   (b) threonine   (c) tyrosine   (d) cysteine

7. Addition of heavy metal ions to a protein disrupts which structural element?
   (a) hydrogen bonds   (b) primary structure   (c) hydrophobic interactions   (d) disulfide bonds

8. A table in which proteins are described as collagens and immunoglobulins classifies them by:
   (a) size   (b) shape   (c) function   (d) the type of nonprotein associated with them

9. In a nucleoprotein, the protein is conjugated with:
   (a) carbohydrate   (b) metal ions   (c) RNA   (d) lipids

10. Substitution of one amino acid for another in a polypeptide chain disrupts:
    (a) secondary structure   (b) tertiary structure   (c) isoelectric point   (d) all or none of the above, depending on the amino acids involved

Sentence Completion

1. The isoelectric point of a neutral amino acid is near pH = _____.

2. Two molecules that differ only in the arrangement of groups around a chiral carbon atom are called _____.

3. Amino acids exist as dipolar ions called _____.

4. The repeating chain of amide bonds in a peptide is called the _____.

5. A disulfide bond between two cysteines in the same chain produces a _____ in the peptide.

6. A protein bonded to a carbohydrate is called a _____.

7. _____ structure refers to how the entire protein is folded and coiled into a specific three-dimensional shape.

8. In the secondary structure called a _____ _____, polypeptide chains line up in a parallel arrangement held together by hydrogen bonds.

9. The tripeptide Ser-Gln-Lys contains side chains that are _____.

10. The amino acids methionine and cysteine are the only two that contain the element _____.

11. A chain with fewer than 50 amino acids is called a _____.

12. _____ interactions pull nonpolar side chains together to exclude water.

True or False

1. All amino acids have at least one chiral carbon atom.

2. Tyrosine and valine can react to form a dipeptide.

3. Protein denaturation disrupts the primary structure of a protein.

4. Lysylalanine is identical to alanyllysine.

5. Some amino acids have an isoelectric point near 10.

6. Simple proteins are more common than conjugated proteins.

7. β-Sheets and α-helices occur mostly in fibrous proteins.

8. Amino acids are quite water-soluble.

9. Proteins, as well as amino acids, have isoelectric points.

10. Both fibrous and globular proteins are water-soluble.

11. Proteins may be classified by biological function.

12. 2-Butanol has a chiral carbon atom.

Match each entry on the left with its partner on the right.

1. Isoleucine

2. Insulin

3. Glycylglycine

4. Glycine

5. β-Sheet

6. Aspartic acid

7. Collagen

8. Cysteine

9. Albumin

10. Arginine

11. Proline

12. Tryptophan

(a) Achiral amino acid

(b) Secondary structure

(c) Fibrous protein

(d) Amino acid that is a secondary amine

(e) Amino acid with hydrocarbon side chain

(f) Globular protein

(g) Peptide hormone

(h) Basic amino acid

(i) Aromatic amino acid

(j) Dipeptide

(k) Acidic amino acid

(l) Can form disulfide bridges

Chapter 19   Enzymes and Vitamins

Chapter Outline

I. Introduction to enzymes (Sections 19.1–19.3).
  A. Catalysis by enzymes (Section 19.1).
    1. Enzymes are globular proteins that catalyze biochemical reactions.
      a. The active site is the region of the enzyme where catalysis takes place.
      b. The reactant in an enzyme-catalyzed reaction is called the substrate.
      c. Some enzymes are specific for one particular substrate, whereas others catalyze reactions involving a number of substrates.
      d. Most enzymes are specific with respect to stereochemistry.
    2. Enzymes affect only the rate of reaction and not the position of equilibrium.
    3. The catalytic efficiency of an enzyme is measured by its turnover number.
  B. Enzyme cofactors (Section 19.2).
    1. Many enzymes include protein portions and nonprotein portions that are called cofactors.
    2. The cofactors are either metal ions or small organic molecules, called coenzymes.
    3. Enzymes can obtain from cofactors chemically reactive groups not available as side chains.
    4. Some cofactors are covalently bonded; others serve as cosubstrates.
  C. Enzyme classification (Section 19.3).
    1. Naming enzymes.
      a. The first part of the name identifies the substrate.
      b. The second part of the name identifies the enzyme subclass.
      c A few enzymes have older, common names.
    2. Classes of enzymes.
      a. Oxidoreductases catalyze oxidation–reduction reactions.
      b. Transferases catalyze the transfer of a specific group from one substrate to another.
      c. Hydrolases catalyze the hydrolysis of substrates.
      d. Isomerases catalyze the isomerization of substrates.
      e. Lyases catalyze the addition or elimination of small molecules.
      f. Ligases catalyze the bonding of two substrate molecules.
II. Enzyme function (Sections 19.4–19.5).
  A. How enzymes work (Section 19.4).
    1. Models for representing enzyme action.
      a. The shape of the active site is complementary to the shape of the substrate.
        i. This description of enzyme–substrate is called the lock-and-key model.
      b. A more modern interpretation of enzyme–substrate interaction is the induced-fit model.
        i. In this model, an enzyme is flexible enough to change shape to fit the spatial requirement of the substrate.
        ii. The enzyme and substrate induce each other to change shape.
    2. Enzyme-catalyzed reactions are often described in the following way:
      a. $E + S \rightarrow ES \rightarrow EP \rightarrow E + P$
    3. Mechanism of enzyme catalysis.
      a. An enzyme–substrate complex is formed when the substrate migrates to the active site, and atoms that will take part in the reaction are held in place (proximity effect).

      b. Reactants are positioned at the exact distance necessary for reaction to occur (orientation effect).

      c. The active site provides the kinds of functional groups necessary for catalysis (catalytic effect).

      d. The activation energy for reaction is lowered by introducing strain into the bonds of the substrate (energy effect).

B. Factors affecting enzyme activity (Section 19.5).

  1. Effect of concentration on enzyme activity.

      a. When substrate concentration varies, and enzyme concentration remains constant:

         i.   Additional substrate increases the rate until all active sites are occupied.

         ii. At this point, reaction rate becomes constant.

         iii. The enzyme is said to be saturated.

      b. Under most conditions, reaction rate is controlled by enzyme efficiency.

         i.   The upper limit to reaction rate is about 10^8 collisions per mole per second.

      c. When substrate concentration is high, and enzyme concentration varies:

         i.   Additional enzyme increases the reaction rate.

         ii. The increase in reaction rate is directly proportional to enzyme concentration.

  2. Effect of temperature.

      a. All enzymes have an optimum temperature for greatest catalytic efficiency. In the human body, this temperature is 37 °C.

      b. At temperatures above 50–60 °C, enzymes denature.

  3. Effect of pH.

      a. Enzymes also have an optimum pH.

      b. If pH is too high or too low, enzymes lose their ability to catalyze reactions.

III. Enzyme regulation (Sections 19.6–19.8).

A. General strategies of control (Section 19.6).

  1. Activation is any process that increases the action of an enzyme.

  2. Inhibition is any process that slows or stops the action of an enzyme.

B. Enzyme inhibition .

  1. Reversible uncompetitive inhibition.

      a. An inhibitor binds to the enzyme at a site other than the active site and changes the shape of the enzyme.

      b. The enzyme is a less effective catalyst, and reaction rate is slowed.

      c. The inhibitor does not compete with the substrate.

  2. Reversible competitive inhibition.

      a. In competitive inhibition, a substrate that resembles the normal substrate occupies the enzyme's active site.

      b. Competitive inhibition is reversible.

         i.   The degree of inhibition can be regulated by controlling the amounts of substrate and inhibitor.

         ii. Increasing the concentration of substrate displaces inhibitor from the active site.

  3. Irreversible inhibition.

      a. The inhibitor bonds covalently to the active site, and the enzyme is irreversibly inhibited.

C. Allosteric control and feedback inhibition (Section 19.7).

  1. Allosteric control.

      a. Most enzymes are regulated by allosteric control.

      b. Allosteric enzymes have more than one polypeptide chain.

      c. Allosteric enzymes have catalytic sites and regulatory sites.

      d. Binding of a positive regulator increases reaction rate.

      e. Binding of a negative regulator decreases reaction rate.

2. Feedback inhibition.
  a. The product of a series of reactions is an inhibitor for the first reaction.
  b. This strategy keeps excessive amounts of products from accumulating.
D. Covalent modification and genetic control (Section 19.8).
  1. Zymogens.
    a. Some enzymes (zymogens) are synthesized in a form that differs from their catalytic form.
    b. Zymogens are activated by enzymatic addition or removal of a part of the zymogen.
    c. Enzymes that might injure the body if constantly present in their active form are synthesized as zymogens.
  2. Phosphorylation/dephosphorylation.
    a. Phosphate groups can be added to serine and threonine side chains to activate certain enzymes.
  3. Genetic control.
    a. Synthesis of some enzymes is regulated by genes.
    b. Genetic control is used for enzymes needed only at certain stages of development.
IV. Vitamins, antioxidants, and minerals (Section 19.9).
  A. Vitamins are small organic molecules that must be supplied in the diet.
  B. Vitamins are classified by solubility.
    1. Water-soluble vitamins are either coenzymes or precursors to coenzymes.
      a. Water-soluble vitamins include vitamin C and the B vitamins.
    2. Fat-soluble vitamins are stored in the body's fatty tissues.
      a. Vitamin A is important for vision.
      b. Vitamin D regulates calcium absorption.
      c. Vitamin E is an antioxidant.
      d. Vitamin K is involved in blood clotting.
  C. Antioxidants.
    1. Antioxidants include vitamins C and E, β-carotene, and selenium.
    2. Antioxidants quench free radicals, suspected of being the source of many diseases.
  D. Minerals.
    1. Many transition metals are needed as micronutrients and enzyme cofactors.
    2. Other minerals are needed in greater amounts for structural and metabolic purposes.

Solutions for Chapter 19 Problems

HOC 19.1

Bubbling took place in both the uncooked meat sample and uncooked potato sample, but not in the cooked samples. Catalase was present and active in the uncooked samples, but it had been denatured by heat in the cooked samples and was no longer active.

19.1 Kinase catalyzes a maximum of 1000 (10^3) reactions per second.

19.2 D-Lactate is the lactate present in invertebrates, and the form of LDH must be able to accept it as a substrate.

HOC 19.2

Most vitamin/mineral supplements contain the listed metal ion cofactors (iron, zinc, copper, manganese, molybdenum, vanadium, chromium, cobalt, and nickel), as well as selenium and boron. All of the dietary molecules in Table 19.2 are also included in the multivitamin.

19.3 (a) NAD^+, coenzyme A, and FAD are coenzymes.
(b) Coenzymes are nonprotein organic molecules, and cofactors are minerals.

19.4 (a) Alcohol dehydrogenase catalyzes the removal of two —H from an alcohol.
(b) Aspartate transaminase catalyzes the transfer of an amino group from aspartate to a second substrate.
(c) Tyrosine–tRNA synthetase catalyzes the synthesis of tyrosine–tRNA from tyrosine and its tRNA, coupled with ATP hydrolysis.
(d) Phosphohexose isomerase catalyzes the isomerization of a phosphohexose.

19.5 (a) The enzyme is named urease. (b) The enzyme is cellulase.

19.6 Hexokinase belongs to the transferase class. It catalyzes the transfer of a phosphoryl group to a hexose.

19.7 In this lyase reaction, water adds to fumarate (substrate) to give L-malate (product).

19.8 Reaction (a) is catalyzed by a decarboxylase because it involves a loss of CO_2. Reaction (b) isn't catalyzed by a decarboxylase.

19.9 Acidic, basic, and polar side chains can take part in catalytic activity and in holding the substrate in the active site. Nonpolar side chains can take part in holding the substrate in the active site but are not likely to be involved in catalytic activity.

19.10 If an enzyme is saturated with substrate, substrate molecules are bound to all of the active sites. (a) Increasing the substrate concentration has no effect on the rate because there are no empty sites where the added substrate can bind. (b) Adding more enzyme increases the rate because more enzyme–substrate complexes can form.

19.11 The rate of the reaction shown in Figure 19.7a is greater at 35 °C than at 25 °C. The rate is greater at 35 °C than at 45 °C.

19.12 The rate of the reaction catalyzed by pepsin is much greater at pH = 2 than at pH = 4.

19.13

(a)   (b)

p-Aminobenzoate

Structure (b) is a possible competitive inhibitor for the enzyme that has *p*-aminobenzoate (PABA) as a substrate because it resembles PABA both in shape and in the type of functional groups present.

19.14 A product of an enzyme-catalyzed reaction that resembles the substrate might be a competitive inhibitor for the enzyme.

19.15 (a) L-Isoleucine most likely inhibits enzyme E1.
(b) If product A inhibited E1, no isoleucine would form. Thus, product A can't be an inhibitor in the sequence shown.

CIA 19.1
(a) A mild ACE inhibitor allows for more careful control of blood pressure. If a very potent ACE inhibitor were used to control blood pressure, blood pressure might drop drastically and cause serious medical problems.
(b) If pit viper peptide were modified so that it caused irreversible inhibition of ACE, it would be a more powerful inhibitor, but it would also be a more dangerous drug. A possible modification to pit viper peptide might be the inclusion of an $-SH$ group near proline that binds irreversibly to the zinc atom at the active site.

CIA 19.2
AZT greatly slows the synthesis of the HIV virus RNA by competitive inhibition.

CIA 19.3
Ritonavir acts on HIV protease by irreversible inhibition because it bonds to the active site of the enzyme.

19.16 (a) *Competitive inhibition* by use of a drug can regulate an overactive enzyme.
(b) *Covalent modification* by phosphorylation–dephosphorylation can regulate enzymes that are needed for quick energy. Alternatively, *feedback control* can regulate glucose production. When glucose levels are adequate, glucose inhibits the enzyme that catalyzes its formation. When glucose levels drop, the enzyme is no longer inhibited, and glucose is formed.
(c) *Covalent modification,* either by activation of a zymogen or by phosphorylation–dephosphorylation, is a type of enzyme regulation that occurs when an enzyme must be quickly activated.
(d) *Genetic control* regulates enzymes that are needed only during certain stages of development.

19.17 (a), (b) Both of these enzymes need a cofactor.
(c) The enzyme does not need K^+ as a cofactor.

19.18 (a) Vitamin B_3 provides NAD^+.
(b) Pantothenic acid (vitamin B_5) provides coenzyme A.

19.19 Vitamin A is fat-soluble because it has a long hydrocarbon chain. Vitamin C is water-soluble because it has polar hydroxyl groups.

19.20 The difference among these compounds occurs at the functional group at the end of the hydrocarbon chain. In retinal, a $-CHO$ group (aldehyde) replaces the $-CH_2OH$ group; in retinoic acid, the $-CH_2OH$ group becomes a $-COOH$ group.

19.21 Vitamins can be enzyme cofactors; some are antioxidants; they aid absorption of calcium and phosphate ions; they are used in synthesis of visual pigments; they are involved in the synthesis of blood-clotting factors.

19.22 Vitamins C and E and β-carotene (vitamin A) function as antioxidants. These vitamins scavenge damaging free radicals.

CIA 19.4

vitamin A: 0%; vitamin C: 0%; calcium: 4%; iron: 6% This food is not a good source of vitamins and minerals.

CIA 19.5

Answers will vary.

CIA 19.6

Listings for vitamin A, vitamin C, iron, and calcium are mandatory on food labels because they are considered to be of greatest importance in maintaining good health.

CIA 19.7

Other than fat, cholesterol, sodium, total carbohydrate, and protein, the trace minerals phosphorus, iodine, magnesium, zinc, and copper may also be listed.

CIA 19.8

Creatine phosphokinase (CPK) gives the earliest indication of a heart attack. The levels of lactate dehydrogenase (LDH) isoenzymes confirm a heart attack after several days.

CIA 19.9

Enzyme activity must be monitored under standard conditions because activity is affected by pH, temperature, and substrate concentration. The standard is the unit (IU), which is the amount of enzyme that converts one micromole of substrate to product at defined standard conditions of pH, temperature, and substrate concentration.

19.23 Selenium is the most toxic micronutrient, causing damage to the respiratory tract and hair loss. Nevertheless, it is an enzyme cofactor that is a necessary micronutrient in small quantities.

Understanding Key Concepts

19.24

---- hydrogen bonds

||||| salt bridges

19.25

L-Lactate   Pyruvate

(a) The enzyme that catalyzes the above reaction is an oxidoreductase.
(b) The enzyme belongs to the dehydrogenase subclass.
(c) L-Lactate is the substrate for the above enzyme.
(d) Pyruvate is the product.
(e) The enzyme is named L-lactate dehydrogenase.

19.26 The enzyme in Problem 19.25 will most likely not use D-lactate as a substrate because most enzymes are "handed" and catalyze the reaction of only one enantiomer of a pair of enantiomers. It is possible that D-lactate might act as a competitive inhibitor of the enzyme.

19.27 The coenzyme needed for the above reaction is NAD^+ (nicotinamide adenine dinucleotide), which is an oxidizing agent and includes the vitamin niacin.

19.28 (a) Increasing the substrate concentration at constant inhibitor concentration increases the rate when substrate concentration is low. Very soon, however, a maximum rate is reached, and adding more substrate does not change the rate of reaction. This maximum rate is always lower than the rate of the uninhibited reaction.
(b) Decreasing the inhibitor concentration at constant substrate concentration increases the rate of reaction. As inhibitor concentration is decreased, more inhibitor dissociates from the enzyme, leaving active enzyme that is free to react with substrate.

19.29
(a) In *covalent modification*, the activity of an enzyme is influenced by the addition or removal of a group that is covalently bonded to the enzyme.
(b) In *genetic control*, hormones control the synthesis of enzymes.
(c) *Allosteric regulation* (either positive or negative) occurs when a regulator binds to the enzyme at a site other than the active site. This binding changes the shape of the enzyme and alters the catalytic ability of the enzyme.
(d) *Feedback inhibition* takes place when the product of a series of reactions acts as an inhibitor for an earlier reaction.

19.30 (a) Feedback inhibition   (b) Irreversible inhibition   (c) Genetic control
(d) Noncompetitive inhibition

19.31 The amino acids (from left to right): aspartate, serine, glutamine, arginine, histidine. Arginine and histidine are basic, and aspartate is acidic.

Additional Problems

Enzyme Cofactors

19.32 (a) FAD is related to riboflavin (vitamin B_2).
(b) Coenzyme A is related to pantothenic acid (vitamin B_5).
(c) NAD^+ is related to vitamin B_3.

19.33 A cofactor is often a metal ion, whereas a coenzyme is a small organic molecule.
Cofactors: (a) Cu^{2+}; (d) Mg^{2+}.
Coenzymes: (b) tetrahydrofolate; (c) NAD^+.

19.34 Vitamin C (b) can serve as a cofactor.

19.35 Cofactors: (a) Fe^{2+}; (d) Ni^{2+}.
Coenzymes: (b) pyridoxal phosphate; (c) FAD.

Structure and Classification of Enzymes

19.36 (a) A *dehydrogenase* catalyzes the removal of two –H to form a double bond.
(b) A *decarboxylase* catalyzes the replacement of a carboxyl group by –H.
(c) A *lipase* catalyzes the hydrolysis of ester groups in lipids.

19.37 (a) A *kinase* catalyzes the transfer of phosphoryl groups between substrates.
(b) An *isomerase* catalyzes the isomerization of a substrate.
(c) A *synthetase* catalyzes the formation of a bond between two substrates, with the participation of a high-energy molecule.

19.38 (a) Amylase   (b) Peroxidase   (c) DNAse

19.39 (a) Lactase   (b) Protease   (c) RNAse

19.40 An enzyme is a large three-dimensional molecule with a catalytic site into which a substrate can fit. Enzymes are specific in their action because only one or a few molecules have the appropriate shape and functional groups to fit into the catalytic site.

19.41 An enzyme forms an enzyme–substrate complex with a substrate at the enzyme's active site. The substrate is in the proper orientation for functional groups at the active site to carry out a reaction. Strain is introduced into the substrate to lower the energy barrier to reaction. After the reaction, the enzyme and products separate.

19.42 (a) A *hydrolase* catalyzes this reaction, which is the hydrolysis of a peptide bond.
(b) A *lyase* (decarboxylase) catalyzes this reaction.
(c) An *oxidoreductase* catalyzes the introduction of a double bond into a molecule by removal of H_2.

19.43 (a) This reaction is catalyzed by a *transferase*, because it involves the transfer of an amino group between substrates.
(b) An *isomerase* catalyzes this reaction.
(c) Formation of oxaloacetate from pyruvate and CO_2 is catalyzed by a *ligase*, which catalyzes the formation of a new bond with participation of ATP.

19.44 (a) A *ligase* catalyzes the formation of a bond between two substrates, with the hydrolysis of ATP to provide energy.
(b) A *transmethylase* catalyzes the transfer of a methyl group between substrates.
(c) A *reductase* catalyzes the reduction of a substrate.

19.45 (a) A *dehydrase* catalyzes the loss of water from a substrate to give a double bond.
(b) A *carboxylase* catalyzes the addition of CO_2 to a substrate.
(c) A *protease* catalyzes the hydrolysis of peptide (amide) bonds in proteins.

19.46 A urease is a hydrolase because it catalyzes the hydrolysis of urea to ammonia and carbonic acid.

19.47 Alcohol dehydrogenase is an oxidoreductase because it catalyzes the oxidation of an alcohol to an aldehyde.

How Enzymes Work

19.48 In the lock-and-key model of enzyme action, the active site of an enzyme has a specific shape (lock) into which only a specific substrate (key) can fit. In the induced-fit model, the shape of the active site changes to accommodate the substrate.

19.49 The induced-fit model is a more likely model for enzyme function because it accounts for the fact that enzymes are not rigid but can change their shape to some extent in order to accommodate a substrate and to catalyze a reaction.

19.50 The amino acid residues involved in the active site can be brought close to each other by protein folding without being near each other in the polypeptide chain.

19.51 The active site of an enzyme, where catalysis takes place, is only a small part of the enzyme. Other regions of the enzyme are needed to stabilize the three-dimensional structure of the enzyme, to align the active-site functional groups that catalyze the reaction, to hold the substrate in place, and to provide sites for enzyme control and regulation.

19.52 To function in the very acidic environment of the stomach, an enzyme must be active at a low pH. To function in the intestine, an enzyme must have high catalytic activity at a pH near neutral. Each enzyme has an optimum pH for catalysis, and an enzyme that is most active at neutral pH will not be active in a strongly acidic environment.

19.53 Recall from Chapter 18 that an amino acid with a pI of less than 5 is negatively charged at physiological pH and can act as a base, and an amino acid with a pI greater than 8 is positively charged at physiological pH and can act as an acid. Thus, at physiological pH, glutamate and aspartate can accept H^+, and lysine and arginine can donate H^+. Histidine (pI = 7.6) can both accept H^+ and donate H^+.

Factors Affecting Enzyme Activity

19.54 At a high substrate concentration relative to enzyme concentration, the rate of reaction triples if the concentration of enzyme is tripled.

19.55 Since the rate of reaction increases as the substrate concentration increases, doubling substrate concentration doubles the rate if the enzyme is not saturated.

19.56 (a) Raising the temperature of the enzyme-catalyzed reaction from 37 °C to 70 °C drastically decreases the rate of reaction and possibly denatures the enzyme.
(b) Lowering the pH of the enzyme-catalyzed reaction from 7 to 3 will probably decrease the rate of reaction, since most enzymes have their optimum rates at close to neutral pH. However, if the enzyme has a rate optimum near 3, then the decrease in pH will increase the rate of reaction.
(b) Adding an organic solvent denatures the enzyme, stopping the reaction.

19.57 (a) If the enzyme has maximum activity at 37 °C, lowering the reaction temperature to 10°C decreases the rate of reaction.
(b) Adding a drop of dilute $HgCl_2$ could either slow down or stop the enzyme-catalyzed reaction because Hg^{2+} is a heavy metal that binds to the $-SH$ groups of enzymes, causing irreversible inhibition.
(c) Adding an oxidizing agent will most likely denature the enzyme and thus will stop the reaction or greatly decrease its rate.

Enzyme Regulation: Inhibition

19.58 (a) *Competitive inhibition* occurs when the structure of a second substrate closely resembles that of the normal substrate for an enzyme. The second substrate can occupy the active site and inhibit binding at the active site of the usual substrate. Competitive inhibition is usually reversible.
Uncompetitive inhibition occurs when an inhibitor binds to the enzyme at a location other than the active site. The conformation of the enzyme changes, and the enzyme is inactivated or its catalytic ability is reduced. Noncompetitive inhibition is reversible.
Irreversible inhibition occurs when a molecule covalently bonds to the active site of an enzyme. The active site is thus irreversibly inactivated, and enzyme activity stops.
(b) In competitive inhibition, the inhibitor bonds noncovalently to the active site. In uncompetitive inhibition, the inhibitor binds noncovalently with a group away from the active site. In irreversible inhibition, the inhibitor forms a covalent bond at the active site.

19.59 (a) This is an example of irreversible inhibition because penicillin binds to the enzyme and doesn't dissociate.
(b) This is an example of competitive inhibition because both alcohols can be converted to aldehydes by the same enzyme.
(c) This uncompetitive inhibitor binds to the enzyme away from the active site.

19.60 Graph B represents the activity of Eco R1 after EDTA has been added. Since the graph shows the total product accumulation over time, no more product is formed after EDTA is added.

19.61 Graph A represents the activity of lactic dehydrogenase; since the enzyme cofactor is not a metal, addition of EDTA should not affect the reaction rate.

19.62 Lead poisons enzymes by bonding to cysteine residues or by displacing an essential metal from an active site. Bonding to cysteine is irreversible and denatures the enzyme.

19.63 EDTA can form a coordinate covalent complex with lead when it has displaced a metal from an active site. This complex is water-soluble and can be excreted in the urine.

19.64 Papain is effective as a meat tenderizer because it catalyzes the hydrolysis of peptide bonds and partially digests the proteins in the meat and softens tough connective tissue.

19.65 Papain works by catalyzing hydrolysis of the peptide bonds in bee venom.

Enzyme Regulation: Allosteric Control and Feedback

19.66 Allosteric enzymes have two types of binding sites—one type of site is for catalysis of the enzymatic reaction and one type of site is for regulation of the reaction. Having two kinds of sites allows for greater regulation and control of enzyme activity.

19.67 Positive regulators change the shape of the active site so that the rate of the enzyme-catalyzed reaction increases. Negative regulators change the shape of the active site so that the enzyme is a less effective catalyst.

19.68 Feedback inhibition occurs when a product of a series of reactions is an inhibitor for an earlier reaction. When this product accumulates, it inhibits an enzyme that catalyzes an earlier step in the series, and the rate of product formation slows down. When the amount of product decreases, inhibition stops, and product formation is resumed.

19.69 Feedback inhibition regulates the quantities of substances in a cell and keeps excess products from accumulating. It also allows a cell to respond quickly to changing situations and environments.

Enzyme Regulation: Covalent Modification and Genetic Control

19.70 A zymogen is an enzyme that is synthesized in a form different from its active form and is activated when needed by removal of a covalently bonded portion of the enzyme. Some enzymes are secreted as zymogens because they would digest or otherwise injure the body if they were secreted in their active form.

19.71 Phosphorylation of a serine or threonine residue of an enzyme replaces a polar –OH group with a charged phosphate group. If this replacement occurs near the active site of an enzyme, the activity of the enzyme might be either increased or inhibited, depending on the enzyme mechanism. The activity of the enzyme may change even if the phosphorylation takes place at a distance from the active site.

19.72 These enzymes are secreted as zymogens because they would digest the pancreas if they were released in their active forms.

19.73 This regulation is an example of genetic control.

Vitamins, Antioxidants, and Minerals

19.74 Vitamins are small essential organic molecules that the body can't manufacture and that must be obtained in trace amounts from the diet.

19.75 Water-soluble vitamins act as cofactors for enzymes and make it possible for enzymes to catalyze biochemical reactions. The relationship between fat-soluble vitamins and enzymes is less evident.

19.76 Vitamin C (water-soluble) is excreted in the urine and must be replenished daily, whereas vitamin A (fat-soluble) can be stored in fatty tissue.

19.77 The four fat-soluble vitamins are vitamins A, D, E, and K. All except vitamin E are stored in the body and are toxic when ingested in large amounts.

19.78 Calcium and phosphorus are both components of bone.

19.79 Transition metals can form coordinate covalent bonds within enzymes and can transfer charge in one-electron steps.

Conceptual Problems

19.80

19.81 Two molecules of vitamin A are produced when a double bond of β-carotene is cleaved.

19.82 Blanching denatures enzymes that would cause deterioration of food quality in frozen food.

19.83 Because competitive inhibition is reversible, addition of a large amount of the normal substrate will reverse the binding of the inhibitor, and the rate of reaction of the normal substrate will return to its usual value. The effects of noncompetitive inhibition can't be reversed by addition of excess substrate.

19.84 Activation energy is the amount of energy needed for reactants to surmount the energy barrier to reaction (see Section 7.5). Energy must be supplied in order for reactions to take place.

19.85 An enzyme-mediated reaction has a lower energy of activation than an uncatalyzed reaction. An enzyme brings substrates close together, induces strain in the bonds to be broken, orients substrates, and provides groups that catalyze the reaction.

19.86 To answer this question, you need to know the optimum temperature and pH for the enzyme-catalyzed reaction. For a reaction with an optimum temperature of 37 °C and an optimum pH close to physiological pH:
(a) Lowering the temperature from 37 °C to 15 °C will slow down the rate.
(b) Raising the temperature from 37 °C to 60 °C will probably denature the protein and stop the reaction.
(c) Lowering the pH from 7 to 4 will slow down the rate of reaction.
(d) Raising the pH from 7 to 10 will slow down the rate of reaction.
(e) Doubling the amount of substrate will speed up the rate of reaction if the enzyme is not saturated.
(f) Halving the amount of substrate will slow the rate of reaction.

19.87 Irreversible enzyme inhibitors are described as poisons because they destroy an enzyme's activity, harming and possibly killing an individual.

19.88

$$1.3 \text{ mg riboflavin} \times \frac{100 \text{ mL apple juice}}{0.014 \text{ mg riboflavin}} = 9300 \text{ mL} = 9.3 \text{ L apple juice}$$

19.89 If valine replaced serine, the enzyme would be inactivated because the $-OH$ functional group of serine is essential for enzyme activity. If aspartate replaced glutamate, the effect would probably be minor because both amino acids have the same functional groups.

19.90 Look for Arg, His, and Lys (basic amino acids) in the polypeptide, and break the bond between the basic amino acid and the amino acid to its right. Three fragments are formed.

Leu–Gly–Arg⌡Ile–Met–His⌡Tyr–Trp–Ala

↓

Leu–Gly–Arg   +   Ile–Met–His   +   Tyr–Trp–Ala

19.91 (a) An apple contains more phenolase because the rate of reaction is faster for apples than it is for potatoes.
(b) A potato contains more catalase because the rate of reaction is faster for potatoes than for apples.
(c) Assuming that the reactions were carried out under identical conditions, the major variable would be the variety of apple or potato.
(d) Catalase has a higher turnover number because the time for reaction to occur is shorter for catalase than for phenolase.

Self-Test for Chapter 19

Multiple Choice

1. A subclass of enzyme that catalyzes cleavage of peptide bonds is a:
    (a) lipase   (b) dehydrase   (c) protease   (d) reductase

2. Which of the following is an enzyme cofactor?
    (a) thiamine   (b) tocopherol   (c) hexokinase   (d) trypsin

3. The ability of enzymes to bring reactants together is known as the:
    (a) orientation effect   (b) catalytic effect   (c) proximity effect   (d) energy effect

4. Fat-soluble vitamins serve all of the following functions except:
    (a) promotion of bone growth   (b) acting as an antioxidant   (c) aiding night vision
    (d) preventing anemia

5. Which of the following enzymes probably requires a coenzyme that is reduced during reaction?
    (a) transferase   (b) oxidase   (c) isomerase   (d) dehydrase

6. The most likely control mechanism for reactions that occur during embryonic development is:
    (a) feedback control   (b) genetic control   (c) zymogens   (d) allosteric control

7. Snake venom probably causes its effects by:
    (a) noncompetitive inhibition   (b) competitive inhibition   (c) allosteric control
    (d) irreversible inhibition

8. Why does the rate of enzymatic reactions decrease as temperature is lowered?
    (a) Salt bridges are disrupted.   (b) The shape of the active site is changed.   (c) Reactions are
    slower at lower temperatures.   (d) The enzyme begins to denature.

9. If an enzyme is saturated, adding more substrate will:
    (a) increase reaction rate   (b) have no effect on reaction rate   (c) decrease reaction rate
    (d) have varying effects, depending on the enzyme

10. In which of the following types of enzyme regulation can adding large amounts of substrate
    restore an enzyme's reaction rate to its maximum value?
    (a) feedback inhibition   (b) irreversible inhibition   (c) uncompetitive inhibition
    (d) competitive inhibition

Sentence Completion

1. An enzyme called a _____ catalyzes the isomerization of a chiral center.

2. _____ are enzymes that have slightly different structures but that catalyze the same reaction.

3. Vitamin __ is necessary for the synthesis of blood-clotting factors.

4. In the _____ _____ model of enzyme action, an enzyme can change its shape slightly to fit
    different substrates.

5.   In _____ inhibition, an inhibitor changes the shape of an enzyme by binding at a location other than the active site.

6.   If the temperature becomes too high, enzymes begin to _____.

7.   The catalytic activity of an enzyme is measured by its _____ _____.

8.   A type of covalent modification of an enzyme involves addition or removal of a _____ group.

9.   The _____ effect is responsible for bringing substrate and catalytic sites together.

10.   The upper limit of any enzyme's rate of reaction is ___collisions per mole per second.

11.   An _____ enzyme has more than one chain and has binding sites for both substrate and regulator.

12.   An enzyme that is synthesized in a form different from its active form is called a _____.

True or False

1.   98.6 °F is the optimum temperature for most human enzymes.

2.   In an enzyme–substrate complex, the substrate is in its lowest energy shape.

3.   Uncompetitive inhibition is irreversible.

4.   A dehydrogenase catalyzes the loss of H_2O from a substrate.

5.   At the active site, the enzyme and substrate are held together by covalent bonds.

6.   Vitamins may be small organic molecules or inorganic ions.

7.   Chymotrypsinogen is a zymogen.

8. β-Carotene is the active form of vitamin A.

9.   Copper, cobalt, and selenium are essential enzyme cofactors.

10.   Heavy metals cause irreversible inhibition of enzymes.

11.   As enzyme concentration is increased, the reaction rate eventually levels off.

12.   The side chains of serine and threonine are sites of covalent modification.

Match each entry on the left with its partner on the right.

1. Papain

2. Vitamin D

3. Competitive inhibition

4. Kinase

5. Alanine transaminase

6. Cofactor

7. Uncompetitive inhibition

8. Vitamin B_6

9. Lipase

10. Vitamin C

11. Active enzyme

12. Lyase

(a) Deficiency causes anemia

(b) Occurs away from an enzyme's active site

(c) Nonspecific enzyme

(d) Catalyzes hydrolysis of ester groups in lipids

(e) Protein part of enzyme, plus cofactor

(f) Deficiency causes scurvy

(g) Transfers a phosphate group between substrates

(h) Fat-soluble vitamin

(i) Occurs at an enzyme's active site

(j) Metal ion or small organic molecule

(k) Specific enzyme

(l) Catalyzes loss of small molecule from substrate

Chapter Outline

I.   General information about carbohydrates (Sections 20.1–20.3).
   A. Classification of carbohydrates (Section 20.1).
      1. By complexity.
         a. Monosaccharides don't yield smaller molecules when hydrolyzed.
         b. Disaccharides are composed of two monosaccharides.
         c. Polysaccharides are composed of a large number of monosaccharides.
      2. By carbonyl group.
         a. Aldoses contain an aldehyde carbonyl group.
         b. Ketoses contain a ketone carbonyl group.
      3. By number of carbons in a monosaccharide unit.
         a. Prefixes *(tri-, tetra-, penta-)* indicate the number of carbons in the monosaccharide.
   B. Handedness (Section 20.2).
      1. General characteristics of handed molecules.
         a. Compounds having a carbon bonded to four different groups are chiral.
            i.   A compound with a chiral carbon can exist in either a right-handed form (D) or a left-handed form (L).
            ii. These two forms are called enantiomers.
            iii. An *optically active* compound rotates the plane of plane-polarized light.
            iv. Enantiomers rotate the plane of plane-polarized light to the same extent but in opposite directions.
         b. Compounds having n chiral centers can have 2^n possible stereoisomers.
            i.   Some of these isomers are enantiomers.
            ii. Stereoisomers that are not mirror images are diastereomers.
            iii. Some chiral compounds have fewer than the predicted number of stereoisomers because of symmetry.
      2. Fischer projections.
         a. Stereoisomers can be drawn as Fischer projections.
            i.   In these projections, the chiral carbon is drawn as the intersection of two perpendicular lines.
            ii. Bonds that point out of the page are shown as horizontal lines.
            iii. Bonds that point back into the page are vertical lines.
            iv. In sugars, the carbonyl group is at or near the top of a Fischer projection.
         b. Monosaccharides are divided into two families.
            i.   In D sugars, the —OH group bonded to the chiral carbon farthest from the carbonyl group points to the right.
            ii. In L sugars, the —OH group bonded to the chiral carbon farthest from the carbonyl group points to the left.
            iii. Most of the carbohydrates that occur naturally are D stereoisomers.
         c. Fischer projections of molecules with more than one chiral carbon are drawn by stacking the chiral centers on top of each other.
         d. The enantiomer of a Fischer projection can be drawn by reversing the horizontal substituents on each chiral carbon.
         e. The designations D and L are unrelated to the sign of the direction of rotation.
II. Monosaccharides (Sections 20.3–20.5).
   A. Structure of monosaccharides (Section 20.3).
      1. Monosaccharides with five or more carbons usually exist as cyclic internal hemiacetals, formed between a hydroxyl group and the carbonyl group.

a. The hemiacetal carbon (C1) can have two stereoisomeric forms.
b. Stereoisomers that differ in configuration only at the hemiacetal carbon are called anomers and are diastereomers.
2. Fischer projections can be converted to cyclic Haworth projections.
a. In a Haworth projection of the β anomer, the $-CH_2OH$ group and the anomeric $-OH$ group are on the same side of the ring.
b. In a Haworth projection of the α anomer, the $-CH_2OH$ group and the anomeric $-OH$ group are on opposite sides of the ring.
c. Groups that are on the left in a Fischer projection point up in a Haworth projection.
d. Groups that are on the right in a Fischer projection point down in a Haworth projection.
e. The $-CH_2OH$ group at C6 points up in D sugars.
3. In solution, glucose exists as a mixture of open-chain glucose (0.02%), the α anomer (36%), and the β anomer (64%).
a. Any mixture of glucose anomers achieves the same percent of anomers in solution.
b. This phenomenon is called mutarotation.
B. Important monosaccharides (Section 20.4).
1. Most monosaccharides are crystalline, high-melting, and sweet.
2. Glucose is the most important monosaccharide in human metabolism.
3. Galactose.
a. Galactose differs from glucose only in the configuration at C4.
b. Galactose is a component of the disaccharide lactose.
c. Galactosemia is an inherited disorder of galactose metabolism.
4. Fructose.
a. Fructose is a component of the disaccharide sucrose.
b. Fructose is a ketohexose.
5. Ribose and 2-deoxyribose.
a. These monosaccharides are aldopentoses.
b. In 2-deoxyribose, the –OH group at C2 is replaced by –H.
c. Both of these sugars occur as components of nucleic acids.
C. Reactions of monosaccharides (Section 20.5).
1. Reaction with oxidizing agents.
a. Aldoses can be oxidized to carboxylic acids.
b. In basic solution, ketoses can also be oxidized to carboxylic acids.
i. Ketoses and aldoses are in equilibrium with an intermediate enediol.
c. Carbohydrates that react with oxidizing agents are called reducing sugars.
2. Reduction to alditols.
3. Reaction with alcohols.
a. Hemiacetals react with alcohols to form acetals that are called glycosides.
b. The bond between the anomeric carbon atom and the oxygen of the –OR group is a glycosidic bond.
4. Formation of a bond between two monosaccharides.
a. Glycosidic bonds can form between two monosaccharides or between a monosaccharide and another molecule.
i. These bonds may have either α or β orientation.
ii. Glycosides are not reducing sugars.
b. Bonds between monosaccharides in more complex carbohydrates are also glycosidic bonds.
i. The bond is described by specifying the positions of the connected carbon atoms and by using α or β to indicate the orientation.
c. Hydrolysis is the cleavage of a glycosidic bond.

5. The alcohol groups of monosaccharides can react with phosphate groups to form phosphate esters.

III. Important disaccharides (Section 20.6).
    A. Maltose.
      1. Maltose consists of two glucose molecules joined by an α-1,4 glycosidic bond.
      2. Maltose occurs in prepared foods and occurs in the body as a product of starch digestion.
      3. Maltose is a reducing sugar.
    B. Lactose.
      1. Lactose consists of a β-D-glucose molecule joined to a β-D-galactose molecule by a β-1,4 glycosidic linkage.
      2. Lactose is a reducing sugar.
      3. Lactose occurs in milk.
      4. Lactose intolerance is a metabolic disorder in which lactose can't be digested.
    C. Sucrose.
      1. Sucrose consists of an α-D-glucose molecule joined to a β-D-fructose molecule by a glycosidic bond connecting the two anomeric carbons.
      2. Sucrose is not a reducing sugar.
      3. Sucrose is the most common disaccharide.

IV. Polysaccharides (Section 20.7).
    A. Polysaccharides consist of hundreds to thousands of monosaccharides connected by glycosidic bonds.
    B. Cellulose.
      1. Cellulose consists of thousands of β-D-glucose units connected by β-1,4 glycosidic bonds.
      2. Chains of cellulose fibers can form hydrogen bonds.
      3. Humans can't digest cellulose.
    C. Starch.
      1. Starch consists of glucose units joined by α-1,4 glycosidic bonds.
      2. There are two kinds of starch.
        a. Amylose consists of hundreds of glucose units joined by α-1,4 glycosidic bonds and is soluble in hot water.
        b. Amylopectin is much larger than amylose, also contains α-1,6 branches every 25 units, and is insoluble in water.
        c. Starch is digested in the small intestine by α-amylase.
    D. Glycogen.
      1. Glycogen resembles amylopectin but is larger and has more branches.
      2. In humans and animals, glycogen is used for carbohydrate storage.
    E. Important polysaccharides based on modified glucose.
      1. Cartilage and tendons consist of protein fibers and polysaccharides composed of β-D-glucuronic acid and *N*-acetyl-D-glucosamine or galactose derivatives.
      2. Hyaluronate and chondroitin 6-sulfate are present in joints, tendons, and cartilage.
      3. Heparin is a polysaccharide that contains sulfate groups and that functions as an anticoagulant.
      4. Glycoproteins contain carbohydrates bonded to proteins and have important functions at cell surfaces.

Solutions to Chapter 20 Problems

20.1 The classification of a monosaccharide consists of three parts:
(1) the prefix *aldo-* (if the carbonyl group is an aldehyde) or *keto-* (if the carbonyl group is a ketone);
(2) a number prefix, such as *tri-, tetr-, pent-, hex-* (to show the number of carbons in the monosaccharide);
(3) the suffix *-ose* (which indicates that the compound is a sugar).

(a)

$$\underset{\substack{5\quad\ \ 4\quad\ \ 3\quad\ \ 2\quad\ \ 1}}{\text{HOCH}_2-\overset{\text{OH}}{\underset{|}{\text{CH}}}-\overset{\text{OH}}{\underset{|}{\text{CH}}}-\overset{\text{OH}}{\underset{|}{\text{CH}}}-\overset{\text{O}}{\overset{\|}{\text{CH}}}}$$

An aldopentose

(b)

$$\underset{\substack{3\qquad\ \ 2\qquad 1}}{\text{HOCH}_2-\overset{\text{O}}{\overset{\|}{\text{C}}}-\text{CH}_2\text{OH}}$$

A ketotriose

(c)

$$\underset{\substack{4\qquad 3\quad\ \ 2\quad\ \ 1}}{\text{HOCH}_2-\overset{\text{OH}}{\underset{|}{\text{CH}}}-\overset{\text{OH}}{\underset{|}{\text{CH}}}-\overset{\text{O}}{\overset{\|}{\text{CH}}}}$$

An aldotetrose

20.2

$$\text{HOCH}_2-\overset{\text{OH}}{\underset{|}{\text{CH}}}-\overset{\text{OH}}{\underset{|}{\text{CH}}}-\overset{\text{OH}}{\underset{|}{\text{CH}}}-\overset{\text{O}}{\overset{\|}{\text{CH}}}$$

An aldopentose

$$\text{HOCH}_2-\overset{\text{OH}}{\underset{|}{\text{CH}}}-\overset{\text{OH}}{\underset{|}{\text{CH}}}-\overset{\text{OH}}{\underset{|}{\text{CH}}}-\overset{\text{O}}{\overset{\|}{\text{C}}}-\text{CH}_2\text{OH}$$

A ketohexose

20.3 The bottom carbon is not chiral. The orientations of the hydroxyl groups bonded to the chiral carbons must be shown in order to indicate which stereoisomer is pictured.

20.4

(a)   (b)   (c)   (d)

If the monosaccharide shown on the left were to look in the mirror, it would see monosaccharide (d), its enantiomer.

20.5 An aldoheptose has five chiral carbon atoms. Using the fact that a molecule with n chiral carbons has 2^n stereoisomers, we calculate that an aldoheptose has a maximum of $2^5 = 32$ stereoisomers.

20.6   The enantiomers are mirror images. The –OH bonded to the chiral carbon farthest from the carbonyl group points to the right in the D enantiomer and to the left in the L enantiomer.

(a)   (b)

A D-aldopentose   An L-aldopentose   An L-ketohexose   A D-ketohexose

HOC 20.1

1. In the open-chain form, glucose has four chiral carbon atoms.
2. Glucose has five chiral carbons in the cyclic form.
3. C1 is different. In the open-chain form, C1 is an achiral aldehyde, but in the cyclic form C1 is a tetrahedral carbon with four groups bonded to it and is chiral.
4. The two cyclic glucose molecules are diastereomers, not enantiomers.
5. Anomers are cyclic monosaccharides that differ only in the configuration of substituents at C1. These two models are anomers.

20.7   First, coil D-talose into a circular shape:

Next, rotate around the indicated single bond between C4 and C5 so that the —CH$_2$OH group (C6) points up.

Finally, form a hemiacetal bond between the aldehyde and the —OH group at C5.

β-anomer

α-anomer

20.8 Reverse the series of steps shown in Problem 20.7. First, break the hemiacetal bond.

Rotate around the single bond between C4 and C5 so that the hydroxyl group on C5 points down.

Uncoil D-idose to arrive at the Fischer projection.

D-Idose

20.9

(a)

α-D-Galactose   Open-chain D-Galactose   β-D-Galactose

(b)

α-D-Fructose   Open-chain D-Fructose   β-D-Fructose

20.10 (a) Rings **1** and **4** (5 carbons) are amino sugars. (b) Ring **3** (4 carbons) is an unmodified sugar. (c) Ring **2** (6 carbons) is a nonsugar.

20.11

← anomeric carbon (2)

The drawing represents the β anomer because the —OH group bonded to the anomeric carbon is on the same side of the ring as carbon 6.

20.12 Chiral carbons are starred. Remember that anomeric carbons and all other carbons bonded to four different groups are chiral.

(a)   (b)   (c)

α-D-Fructose   α-D-Ribose   β-D-2-Deoxyribose

20.13

L-Fucose α-D-Galactose β-D-Galactose

(a) L-Fucose is an α anomer because the anomeric hydroxyl group is on the opposite side of the hemiacetal ring from carbon 6.

(b) L-Fucose is missing a hydroxyl group on carbon 6.

(c) The —OH group at carbon 2 that is below the plane of the ring in D-galactose is above the plane of the ring in L-fucose. The —OH groups that are above the plane of the ring in D-galactose are below the plane of the ring in L-fucose.

(d) 6-Deoxy-L-galactose is a correct name for L-fucose.

CIA 20.1

All blood groups have N-acetyl-D-glucosamine, D-galactose, and L-fucose as components of their antigenic determinants. Blood group O is the universal donor because it has components common to all blood groups. Blood group A has a molecule of N-acetyl-D-glucosamine linked to D-galactose, blood group B has D-galactose linked to D-galactose, and blood group AB has both. These additional components provoke an immune response if blood with incompatible groups is mixed.

CIA 20.2

People with type O blood can receive blood only from other donors that have type O blood. People with type AB blood can give blood only to other people with type AB blood because antibodies to AB blood are present in the blood of people having any of the other blood types.

CIA 20.3

N-acetyl-D-glucosamine, D-galactose, and L-fucose are found in all blood types.

20.14

α-D-Ribose   + CH₃OH $\xrightarrow[\text{catalyst}]{\text{H}^+}$ Methyl α-D-riboside +   H₂O

+

β-D-Ribose   Methyl β-D-riboside

The two products are acetals because reaction with methanol occurs at the hemiacetal hydroxyl group, producing a compound with two —OR groups.

20.15 Cellobiose has a β-1,4 glycosidic link.

20.16

Cellobiose $\xrightarrow{\text{H}_3\text{O}^+}$ β-D-Glucose   +   β-D-Glucose

20.17

(a)   (b)

α-D-Glucose   α-D-Glucose   α-D-Glucose   β-D-Fructose

Maltose — fermenting grain   Sucrose — sugar beets

(c)

β-D-Galactose   Lactose—milk

CIA 20.4
Cell walls make cells rigid, give cells shape, and protect cells from pathogens.

CIA 20.5
Most of a bacterial cell wall is made up of peptidoglycan, a polymer of alternating *N*-acetylglucosamine and *N*-acetylmuraminic acid units crosslinked by short polypeptide bridges.

CIA 20.6
Penicillin inhibits the enzyme that synthesizes peptidoglycan polypeptide bridges.

CIA 20.7
Since mammals don't contain the enzyme that synthesizes peptidoglycan polypeptide bridges, penicillin doesn't harm mammalian cells.

20.18

	Modified Glucose	*Glucose*
(a) β-D-Glucuronate	Carboxylate at C6	Alcohol at C6
(b) β-D-Glucosamine	$-NH_2$ at C2	$-OH$ at C2
(c) *N*-Acetyl-β-D-Glucosamine	*N*-Acetyl group at C2	$-OH$ at C2

20.19 Glutamine and asparagine are the two amino acids with side-chain amide groups.

HOC 20.2
(a) The three vitamins are niacin (vitamin B_3), folic acid, and ascorbic acid (vitamin C).
(b) **Triticale:** a hybrid of wheat and rye   **Millet:** a type of grain
Flax: a type of seed   **Durum:** a variety of wheat
Inulin: a fructose polymer   **Modified food starch:** starch that has been chemically modified
(c) "Shredded Wheat and Bran" contains only whole wheat (starch) and whole wheat bran.

CIA 20.8
Starch, a polysaccharide, is an example of a complex carbohydrate in the diet. Glucose and fructose are simple carbohydrates found in the diet. Soluble and insoluble fiber are complex carbohydrates that can't be hydrolyzed to simple carbohydrates, and most of them pass undigested through the body.

CIA 20.9

In a healthy diet, cellulose provides fiber, which aids in the passage of food through the digestive system and reduces the risk of cancer, hemorrhoids, diverticulosis, and cardiovascular disease.

CIA 20.10

Pectin and vegetable gum are two kinds of soluble fiber and can be found in fruits, barley, oats, and beans.

Understanding Key Concepts

20.20

$$\text{Starch} \xrightarrow{\text{Amylase}} \text{Maltose} \xrightarrow{\text{Maltase}} \text{Glucose}$$

polysaccharide   disaccharide   monosaccharide

Starch is a large polysaccharide consisting of many glucose subunits connected by α-1,4 glycosidic bonds. α-Amylase hydrolyzes these bonds to yield maltose, a disaccharide, and maltase cleaves the α-1,4 glycosidic bond of maltose to give the monosaccharide glucose.

20.21 (a) The two fructose molecules are diastereomers and anomers; they are stereoisomers that aren't mirror images. (b) D-Galactose and L-galactose are enantiomers. (c) L-Allose and D-glucose are diastereomers.

20.22 (a), (b)

α anomer β anomer β anomer

(c) The linkage between monosaccharide **A** and monosaccharide **B** is an α-1,4 link that connects carbon 1 of **A** to carbon 4 of **B**.
(d) The linkage between monosaccharide **B** and monosaccharide **C** is an β-1,4 link that connects carbon 1 of **B** to carbon 4 of **C**.

20.23 (a), (b) None of the three monosaccharides are identical, and none are enantiomers.

hydrolysis

A   B   C

L-Fucose   D-Glucose   D-Galactose

20.24 Monosaccharide **C** (D-galactose) is oxidized by an oxidizing agent because it is the only one of the three monosaccharides that has a hemiacetal in equilibrium with an open-chain aldehyde. Isolation and identification of the resulting carboxylic acid also identifies the terminal hemiacetal monosaccharide.

20.25 Monosaccharides **B** and **C** are the components of lactose, but **B** and **C** would have to be connected **C–B**, instead of **B–C**, in order for lactose to be part of the trisaccharide.

20.26

Polysaccharide	Linkage	Branching?
Cellulose	β-1,4	No
Amylose	α-1,4	No
Amylopectin	α-1,4	Yes: α-1,6 branches occur ~ every 25 units
Glycogen	α-1,4	Yes: even more α-1,6 branches than in amylopectin

20.27 In solution, an equilibrium exists between open-chain glucose and its cyclic hemiacetal forms. The open-chain form reacts with the oxidizing agent, and, according to Le Châtelier's principle, the equilibrium shifts to produce more of the open-chain form, which is also oxidized. Ultimately, all of the glucose is oxidized to gluconate.

Additional Problems

Classification and Structure of Carbohydrates

20.28 A carbohydrate is a polyhydroxylated aldehyde or ketone (or a derivative of one of these) that belongs to one of the biologically most important classes of compounds.

20.29 The family-name ending for a sugar is -ose.

20.30 An aldose contains an aldehyde carbonyl group, and a ketose contains a ketone carbonyl group.

20.31, 20.32

(a)

Threose

an aldotetrose

two chiral carbons

(b)

Ribulose

a ketopentose

two chiral carbons

(c)

Xylose

an aldopentose

three chiral carbons

(d)

Tagatose

a ketohexose

three chiral carbons

20.33 A repeating unit of heparin has nine chiral carbons.

20.34

A ketoheptose

20.35

oxygen missing here

A four-carbon deoxy sugar

20.36 *Glucose* occurs in most food, especially fruits and vegetables, and in all living organisms.
Galactose occurs in brain tissue and as a component of lactose (milk sugar).
Fructose occurs in honey and in many fruits.
Ribose occurs in nucleic acids.

20.37 *Glucose,* the most important monosaccharide in living organisms, is the principal energy source for metabolism.
Galactose can also serve as an energy source and is a component of compounds needed in brain tissue.
Fructose is a component of sucrose and can be metabolized as glucose.
Ribose occurs in nucleic acids and forms part of CoA and ATP.

Handedness in Carbohydrates

20.38 *Enantiomers* are stereoisomers that are nonsuperimposable mirror images.

20.39 L-Glucose and D-glucose are mirror images (enantiomers).

20.40

$$CH_3$$

Cl—C◄Br

Br►C◄Cl

$$CH_3$$

$$CH_3$$

Br►C◄Cl

Cl►C◄Br

$$CH_3$$

$$CH_3$$

Br►C◄Cl

-----┼------ symmetry plane

Br►C◄Cl

$$CH_3$$

Enantiomers

The third stereoisomer does not have an enantiomer because it has a symmetry plane and is not chiral.

20.41

H—C=O

HO——H

H——OH

$$CH_2OH$$

D-Threose

[H] →

$$CH_2OH$$

HO——H

H——OH

$$CH_2OH$$

Chiral

H—C=O

H——OH

H——OH

$$CH_2OH$$

D-Erythrose

[H] →

$$CH_2OH$$

H——OH

H——OH

$$CH_2OH$$

Achiral

The reduction product of D-erythrose has a symmetry plane and is achiral.

20.42 (a) Fructose rotates light to the left to a greater degree than glucose rotates light to the right. (b) The mixture is called invert sugar because the sign of rotation of the fructose–glucose mixture is inverted from (opposite to) the sign of rotation of sucrose.

20.43 Equimolar solutions of enantiomers rotate light to the same degree but in opposite directions.

Reactions of Carbohydrates

20.44 A reducing sugar is a sugar that gives a positive reaction when treated with an oxidizing agent such as Tollens' reagent or Benedict's reagent.

20.45 A sugar must have an aldehyde carbonyl group or a group that can isomerize to an aldehyde in order to be a reducing sugar.

20.46 Mutarotation occurs when either a pure anomer or a mixture of anomers is dissolved in water. In either case, if the rotation of plane-polarized light is measured, the degree of rotation and thus the ratio of anomers, changes until it reaches a constant value. At this point, an equilibrium mixture of both anomers is present in the solution. Mutarotation is not a general characteristic of all chiral compounds.

20.47 Most sugars exist in the cyclic hemiacetal form, in which a sugar hydroxyl group has added to the sugar carbonyl group in the same molecule. *Anomer* is the name given to each of the two stereoisomers formed. Anomers differ in the orientation of substituents at the anomeric carbon.

20.48 In the β form of a carbohydrate hemiacetal, the $-OH$ group attached to C1 (the hemiacetal carbon) is on the same side of the ring as the $-CH_2OH$ group. In the α form, the $-OH$ group at C1 and the $-CH_2OH$ group are on opposite sides of the ring.

A β anomer   An α anomer

20.49 The α form of D-gulose is shown, because the anomeric $-OH$ group at C1 is on the opposite side of the hemiacetal ring from the $-CH_2OH$ group at C6.

20.50

β-D-Mannose α-D-Mannose

20.51

β-D-Altrose α-D-Altrose

20.52

D-Glucose → [H] → Sorbitol

20.53

D-Fructose → [H] → Sorbitol + Mannitol

20.54

D-Glucose → Tollens' Reagent → Gluconic acid

20.55

D-Ribose → Tollens' Reagent → Ribonic acid

20.56 In a hemiacetal, an —OH group and an —OR group are bonded to a carbon atom that was formerly a carbonyl carbon. In an acetal, the carbon is bonded to two —OR groups.

A hemiacetal   An acetal

20.57 A *glycoside* is the acetal product that results from reaction of the hemiacetal —OH group of a carbohydrate with an alcohol.

20.58

20.59

The product in this problem has a cyclic hemiacetal carbon and is in equilibrium with an open-chain aldehyde that is a reducing sugar. The glycosides in Problem 20.58 are acetals and are not in equilibrium with an aldehyde.

Disaccharides and Polysaccharides

20.60

Disaccharide	Occurs in	Made from
Maltose	Fermenting grains	Glucose + glucose
Lactose	Milk	Galactose + glucose
Sucrose	Many plants	Glucose + fructose

20.61 Lactose and maltose have hemiacetal linkages that give positive Tollens' test results. Sucrose has no hemiacetal group and is thus unreactive toward Tollens' reagent.

20.62 Amylose, a major component of the human diet, consists of α-D-glucose units linked by α-1,4 glycosidic bonds. Cellulose, used by plants as a structural material, consists of β-D-glucose units linked by β-1,4 glycosidic bonds.

20.63 Amylose and amylopectin are both components of starch and both consist of long polymers of α-D-glucose linked by α-1,4 glycosidic bonds. Amylopectin is much larger and has α-1,6 branches every 25 units or so along the chain.

20.64 Cellulose is used for all of the listed products except (c).

20.65 A person with lactose intolerance can eat only french fries (b). The other foods (ice cream, chocolate milkshake) contain lactose.

20.66

Gentiobiose β-D-Glucose β-D-Glucose

20.67 Gentiobiose contains both an acetal grouping and a hemiacetal grouping. Gentiobiose is a reducing sugar because it has a hemiacetal group on the right-hand monosaccharide unit. A β-1,6 linkage connects the two monosaccharides.

20.68

Trehalose α-D-Glucose

α-D-Glucose

20.69 Trehalose is a nonreducing sugar because it contains no hemiacetal linkages. The two D-glucose monosaccharides are connected by an α-1,1 acetal link.

20.70 Amylopectin and glycogen are both polymers that are composed of α-D-glucose units linked by α-1,4 glycosidic bonds. Both amylopectin and glycogen have α-1,6 branches, but glycogen has many more branches and is much larger (1 million glucose units per molecule) than amylopectin (100,000 glucose units per molecule).

20.71 Amylopectin is the major component of starch, which is used for carbohydrate storage in plants, and glycogen is used for carbohydrate storage in animals.

20.72 **Heparin** is a polymer made up of various modified disaccharides; one of the most common is the disaccharide formed from α-D-glucuronate-2-sulfate and acetylsulfate-β-D-glucosamine-6-sulfate.
Hyaluronate is a polymer composed of repeating units of the disaccharide formed from β-D-glucuronate and N-acetyl-β-D-glucosamine.
Chondroitin-6-sulfate is a polymer composed of repeating units of the disaccharide formed from β-D-glucuronate and *N*-acetyl-β-D-glucosamine-6-sulfate.

20.73 Heparin is an anticoagulant that keeps blood from clotting.
Hyaluronate is found in synovial fluid in joints and in the vitreous humor of the eye.
Chondroitin-6-sulfate is found in tendons and cartilage.

Conceptual Problems

20.74 The α and β hemiacetal forms of monosaccharides are diastereomers because they are not mirror images.

20.75 As in the previous problem, the α and β forms are not enantiomers because they are not mirror images.

20.76

α-D-Fructose

20.77

Raffinose

20.78

1,3-Dihydroxyacetone

1,3-Dihydroxyacetone has no optical isomers because it has no chiral carbons.

20.79 1,3,4-Trihydroxy-2-butanone has one chiral carbon and exists as a pair of enantiomers.

$$HOCH_2 - \overset{\overset{\displaystyle OH}{|}}{\underset{\underset{\displaystyle H}{|}}{\overset{*}{C}}} - \overset{\overset{\displaystyle O}{\|}}{C} - CH_2OH \qquad \text{1,3,4-Trihydroxy-2-butanone}$$

20.80 Lactose intolerance is an inability to digest lactose. Symptoms include bloating, cramps, and diarrhea.

20.81 Symptoms of the disease galactosemia, which results when the body lacks an enzyme needed to digest galactose, include
   In infancy: vomiting, enlarged liver, failure to thrive
   Eventual: liver failure, mental retardation, cataracts

20.82

```
              CH2OH
              |
       H ----+---- OH
             |
      HO ----+---- H      plane of
Dulcitol -- -+- -- --     symmetry
      HO ----+---- H
             |
       H ----+---- OH
             |
            CH2OH
```

Dulcitol is optically inactive because it has a plane of symmetry and thus doesn't have an enantiomer.

20.83 A monosaccharide is the simplest carbohydrate, containing 3 to 7 carbons. A disaccharide consists of 2 monosaccharide units, and a polysaccharide consists of three or more units.

20.84

Carbohydrate	Example	Source
monosaccharide	fructose	fruit
disaccharide	sucrose	sugar cane
polysaccharide	cellulose	all plants

20.85

	Lactose intolerance	Galactosemia
Severity:	unpleasant	life-threatening
Cause:	gradual loss of activity of the enzyme lactase	inherited inability to to metabolize galactose
Treatment:	lactose-free diet enzymes that digest lactose	galactose-free diet

Group Problems

20.86 Enzymes produced by the bacteria in yogurt predigest most of the lactose, making it possible for lactose-intolerant people to eat yogurt without symptoms.

20.87

$$\frac{4 \text{ kcal}}{1 \text{ g}} \text{ x } 200 \text{ g} = 800 \text{ kcal: } \frac{800 \text{ kcal}}{2000 \text{ kcal}} \text{ x } 100\% = 40\%$$

100% of calories in a 2000-kcal daily diet would be digestible carbohydrate.

20.88

$$42 \text{ g } \text{ x } \frac{4.0 \text{ kcal}}{\text{g}} = 170 \text{ kcal; } 170 \text{ kcal } \text{ x } \frac{4.18 \text{ kJ}}{\text{kcal}} = 710 \text{ kJ}$$

20.89 Hydrogen bonds between chains stabilize cellulose and make it rigid and insoluble.

Self-Test for Chapter 20

Multiple Choice

1. One of these disaccharides is not a reducing sugar. Which one?
   (a) sucrose   (b) maltose   (c) lactose   (d) cellobiose

2. Which one of the following bonds doesn't occur in common polysaccharides?
   (a) α-1,4 bonds   (b) α-1,6 bonds   (c) β-1,4 bonds   (d) β-1,6 bonds

3. The compound shown is:
   (a) The α anomer of a ketopentose   (b) the α anomer of an aldohexose
   (c) the β anomer of an aldopentose   (d) the β anomer of a ketohexose

4. Oxidation of the aldotetroses pictured in Section 20.2 can be controlled to produce dicarboxylic acids known as tartaric acids. How many stereoisomeric tartaric acids are there?
   (a) 1   (b) 2   (c) 3   (d) 4

5. *N*-Acetyl-D-glucosamine is a component of all of the following, except:
   (a) chitin   (b) blood group determinants   (c) tendons   (d) heparin

6. Humans can't digest cellulose because:
   (a) Its molecules are too big.   (b) Humans don't produce enzymes to digest polysaccharides.
   (c) Humans can't digest branched polysaccharides.   (d) Human enzymes that digest carbohydrates can only hydrolyze α glycosidic bonds.

7. Which blood group is the "universal donor"?
   (a) A   (b) B   (c) AB   (d) O

8. Which of the following polysaccharides consists of only α-1,4 glycosidic bonds?
   (a) amylopectin   (b) cellulose   (c) amylose   (d) glycogen

9. Which of these errors of carbohydrate metabolism is not life-threatening?
   (a) galactosemia   (b) lactose intolerance   (c) diabetes mellitus   (d) blood-group incompatibility

10 How many stereoisomers does an open-chain 2-ketopentose have?
   (a) 2   (b) 4   (c) 6   (d) 8

Sentence Completion

1. An object that has handedness is said to be _____.

2. Starch molecules are digested by enzymes called _____.

3. The two mirror image forms of a chiral molecule are called _____.

4. A reaction between an aldehyde carbonyl group and an alcohol hydroxyl group in the same molecule yields a _____ _____.

5. *Levulose* is another name for _____.

6. _____ and _____ are two kinds of starch.

7. The reaction of a monosaccharide with an alcohol yields a _____.

8. _____ is used for food storage in animals.

9. D-Glucose can be classified as a _____.

10. 3-Pentanol is an _____ molecule.

11. A _____ is a stereoisomer that is not an enantiomer.

12. _____ have important functions at cell surfaces.

True or False

1. All naturally-occurring carbohydrates are chiral.

2. Glucose is a reducing sugar.

3. In an L sugar, the —OH on the carbon nearest to the carbonyl group points to the left.

4. Humans can digest polysaccharides containing β-1,4 acetal links but not α-1,4 acetal links.

5. Achiral objects possess a plane of symmetry.

6. Crystalline glucose is a mixture of α anomer, β anomer, and open-chain forms.

7. Sucrose contains both acetal and hemiacetal groups.

8. Amylopectin and glycogen contain 1,4 and 1,6 acetal links.

9. An acetal is an ester.

10. Maltose is a disaccharide composed of glucose and galactose.

11. Two diastereomers rotate plane-polarized light in equal amounts but in opposite directions.

12. Not all compounds with n chiral carbons have 2^n stereoisomers.

Match each entry on the left with its partner on the right.

1. Maltose   (a) Aldohexose

2. Cellulose   (b) Glucose

3. Methyl glucoside   (c) Aldotriose

4. Ribose   (d) Glucose + fructose

5. Fructose   (e) Glucose + glucose

6. Glycogen   (f) Polysaccharide of α-D-glucose

7. Dextrose   (g) Galactose + glucose

8. Glucose   (h) Aldopentose

9. Sucrose   (i) Ketohexose

10. Lactose   (j) Acetal

11. Glyceraldehyde   (k) Animal starch

12. Amylose   (l) Polysaccharide of β-D-glucose

Chapter 21   The Generation of Biochemical Energy

Chapter Outline

I. Energy and cells (Sections 21.1–21.2).
    A. Energy requirements for living things (Section 21.1).
        1. Energy must be released from food gradually.
        2. Energy must be stored in a readily accessible form.
        3. The rate of release of energy from storage must be finely controlled.
        4. Just enough energy must be released as heat to maintain body temperature.
        5. Energy in a form other than heat must be available to drive reactions that are not spontaneous at body temperature.
    B. Free energy and biochemical reactions.
        1. In favorable chemical reactions, free energy is released.
            a. Favorable reactions occur when heat is released and/or disorder increases.
            b. These reactions are described as exergonic.
            c. In favorable reactions, the products are more stable than the reactants.
            d. Favorable reactions have a negative value for ΔG.
        2. Free energy must be provided for unfavorable chemical reactions to take place.
            a. Unfavorable reactions are described as endergonic.
            b. Energy must be supplied in order for endergonic reactions to proceed.
            c. In unfavorable reactions, the products are less stable than the reactants.
            d. Unfavorable reactions have a positive value for ΔG.
    C. Cells and their structure (Section 21.2).
        1. Simple organisms have prokaryotic cells.
        2. More complicated organisms have eukaryotic cells.
            a. These cells have membrane-enclosed nuclei and organelles.
        3. Structure of eukaryotic cells.
            a. Everything between the cell membrane and the nuclear membrane is the cytoplasm.
            b. The fluid of the cytoplasm is the cytosol.
            c. The most important organelles are the mitochondria.
                i. Mitochondria consist of a smooth outer membrane and a folded inner membrane.
                ii. The space between the two membranes is the intermembrane space.
                iii. The space enclosed by the inner membrane is the mitochondrial matrix.
                iv. Most of the energy-producing reactions of the cell begin in the mitochondrial matrix.
II. Metabolism (Sections 21.3–21.6).
    A. Overview of metabolism (Section 21.3).
        1. A sequence of reactions in metabolism is known as a metabolic pathway.
            a. Pathways may be linear, cyclic, or spiral.
            b. Pathways that break molecules apart are known as catabolic pathways.
            c. Pathways that build molecules are known as anabolic pathways.
        2. Catabolism consists of four phases.
            a. Digestion occurs in the mouth and small intestine.
            b. Small molecules are degraded in the cell to yield acetyl-CoA.
            c. Acetyl groups are oxidized to yield CO_2, H_2O, and energy.
            d. Some of the energy is transferred to ATP and reduced coenzymes.

B. Strategies of metabolism (Sections 21.4–21.6).
1. ATP and energy transfer (Section 21.4).
   a. ATP contains two phosphoric anhydride bonds and a phosphate ester bond.
   b. Hydrolysis of one phosphate anhydride bond provides –7.3 kcal/mol (–30.5 kJ/mol) of energy.
   c. Synthesis of ATP from ADP and phosphate requires +7.3 kcal/mol (–30.5 kJ/mol) of energy.
   d. Synthesis of ATP occurs when there is energy to store, and hydrolysis of ATP occurs when energy is needed.
2. Metabolic pathways and coupled reactions (Section 21.5).
   a. In any reaction, the amount of energy consumed or released is the same, no matter what the path.
   b. The overall energy change for a series of reactions can be found by adding up the free energy changes for the individual steps.
   c. A reaction that is energetically unfavorable can take place if it is coupled with a reaction that is energetically favorable.
      i. The phosphorylation of glucose can take place by coupling with the conversion of ATP to ADP.
      ii. ATP can be synthesized by coupling with the transfer of a phosphoryl group from phosphoenolpyruvate.
3. Oxidized and reduced coenzymes (Section 21.6).
   a. Many biochemical reactions are redox reactions.
      i. Oxidation may be loss of electrons, loss of hydrogen, or addition of oxygen.
      ii. Reduction may be gain of electrons, gain of hydrogen, or removal of oxygen.
      iii. Reductions and oxidations always occur together.
   b. Common redox coenzymes are NAD^+, $NADP^+$, and FAD (written in the oxidized form).
      i. NAD^+ and $NADP^+$ are used to oxidize alcohols to carbonyl compounds.
      ii. FAD catalyzes the formation of a double bond.
   c. The energy stored in $FADH_2$, $NADH/H^+$, and $NADPH/H^+$ is passed on to ATP.

III. Metabolic pathways (Sections 21.7–21.8).
A. The citric acid cycle (Section 21.7).
1. Oxidation of two-carbon acyl groups from acetyl-CoA to CO_2 occurs in the citric acid cycle.
2. The citric acid cycle is a closed loop of eight reactions in which the product of step 8 is a reactant in step 1.
3. The cycle operates only if:
   a. Acetyl groups are available.
   b. NAD^+ and FAD are in their oxidized form.
   c. Oxygen is available.
4. A summary of steps:
   a. Steps 1, 2: Preparation.
      i. Acetyl-CoA + oxalate —> citrate.
      ii. Citrate is isomerized to isocitrate.
   b. Steps 3, 4: Oxidative decarboxylation.
      i. Isocitrate is oxidized and decarboxylated to form α-ketoglutarate, with reduction of NAD^+.
      ii. α-Ketoglutarate is decarboxylated to form succinyl-CoA and reduced NAD^+.
   c. Step 5: Phosphorylation.
      i. Succinyl-CoA + GDP —> succinate + HSCoA + GTP.
   d. Step 6: Oxidation of succinate.
      i. Succinate + FAD —> fumarate + $FADH_2$.

e. Steps 7, 8: Regeneration of oxaloacetate.
    i. Fumarate + H_2O —> malate.
    ii. Malate + NAD^+ —> oxaloacetate + $NADH/H^+$.
5. Net result of citric acid cycle:
    a. Production of four reduced coenzyme molecules.
    b. Conversion of an acetyl group to two CO_2 molecules.
    c. Production of one GTP.
B. The electron-transport chain and production of ATP (Section 21.8).
    1. The end product of the electron-transport chain is water in the reaction:
        $$O_2 + 4\ e^- + 4\ H^+ —> 2\ H_2O$$
    2. The reactions of the electron transport chain are coupled to oxidative phosphorylation.
    3. Enzymes for electron transport are embedded in the inner mitochondrial membrane.
    4. Features of the electron-transport chain:
        a. Hydrogen and electrons from NADH and $FADH_2$ enter the respiratory chain from the mitochondrial matrix.
        b. Electrons pass from weaker to stronger electron acceptors.
        c. Some of the energy released is used to transport H^+ into the intermembrane space.
        d. The H^+ concentration gradient creates a potential energy difference that is crucial for ATP synthesis.
        e. H^+ ions can return to the complex only by passing through a channel that is part of the ATP synthase complex.
            i. The energy that they release drives the phosphorylation of ATP.
    5. Various cofactors, including iron–sulfur clusters and cytochromes, are involved in electron transport.
    6. Ultimately, electrons are passed on to oxygen, which combines with H^+ to form H_2O.
    7. The energy needed for ATP synthesis is provided by the passage of H^+ ions through the ATP synthase enzyme complex.
    8. The passage of electrons from one mol NADH produces from 2.5 to 3 mol ATP.

Solutions to Chapter 21 Problems

21.1   The amount of energy produced by both processes is the same.

21.2   (a) Oxidation of glucose is exergonic, and photosynthesis is endergonic.
(b) Sunlight provides the energy for the endergonic pathway.

CIA 21.1
Chlorophyll contains a Mg^{2+} ion instead of a Fe^{2+} ion.

CIA 21.2
The light-dependent reaction of photosynthesis occurs in the chloroplasts and is used to produce O_2, NADPH, and ATP. The NADPH and ATP are used in the light-independent reaction to produce carbohydrate molecules from water and carbon dioxide.

CIA 21.3
The reaction is exergonic because ΔG is negative.

CIA 21.4
Refrigeration slows the breakdown of carbohydrates by decreasing the rate of respiration.

21.3   (a) Carbohydrates are digested in the mouth, stomach, and small intestine to yield sugars, which are degraded in the cytosol to pyruvate via glycolysis. Pyruvate is converted to acetyl-CoA, which enters the citric acid cycle and is oxidized to CO_2. The reduced coenzymes that are generated by oxidation of acetyl-CoA enter the electron transport chain and are used in the production of ATP by oxidative phosphorylation.
(b) The products of amino acid catabolism can enter the central metabolic pathway as pyruvate, as acetyl-CoA, and as citric acid cycle intermediates.

21.4

Acetyl phosphate

21.5   This strategy ensures that an exergonic reaction occurs only when the energy it releases is needed.

CIA 21.5
The reactive oxygen species(ROS) are H_2O_2 (b), $ROO\cdot$ (c), and $\cdot OH^-$ (d).

CIA 21.6
The enzyme catalase converts H_2O_2 to water. Vitamins E, C, and A can inactivate the other reactive oxygen species.

21.6

Glycerol   Glycerol 1-phosphate

21.7   If a metabolic pathway that breaks down a molecule is exergonic, the reverse of that pathway (synthesis) *must* be endergonic. The only way to synthesize the product is to couple the energetically unfavorable reactions with other reactions that are energetically favorable in a pathway that differs from the reverse of the breakdown pathway.

CIA 21.7
Basal metabolic rate is the minimum amount of energy per unit time needed for breathing, maintaining body temperature, circulating blood, and keeping all body organs functioning.

CIA 21.8
Together, the soda + hamburger contain 660 kcal. If jogging consumes 10 kcal/minute, it would take 660 kcal ÷ 10 kcal/minute = 66 minutes to burn off these calories.

CIA 21.9

The basal metabolic rate:

$$\frac{1.0 \text{ kcal}}{1 \text{ kg} \times 1 \text{ hr}} \times \frac{24 \text{ hr}}{1 \text{ day}} \times 80 \text{ kg} = \frac{1920 \text{ kcal}}{1 \text{ day}}$$

A moderately active male uses an additional 50% calories above the basal rate, or 1.5 x 1920 kcal/day = about 2900 kcal/day.

CIA 21.10

Daily activities such as walking use energy, and thus the body requires a larger caloric intake than that needed to maintain basal metabolism.

HOC 21.1

First, calculate the basal metabolic rate (in kg/day), and then adjust for activity level. For a 175 lb male:

$$175 \text{ lb} \times \frac{1 \text{ kg}}{2.2 \text{ lb}} \times \frac{1.0 \text{ kcal}}{1 \text{ kg} \times 1 \text{hr}} \times \frac{24 \text{ hr}}{1 \text{ day}} = 1900 \text{ kcal}$$

For a 115 lb female:

$$115 \text{ lb} \times \frac{1 \text{ kg}}{2.2 \text{ lb}} \times \frac{0.95 \text{ kcal}}{1 \text{ kg} \times 1\text{hr}} \times \frac{24 \text{ hr}}{1 \text{ day}} = 1200 \text{ kcal}$$

Next, estimate activity level:
A 175 lb male who is lightly active uses 50% of the basal rate, 950 kcal.
A 115 lb female who is very active uses 100% of the basal rate, 1200 kcal.

Finally, add the activity level to the basal metabolic rate:
For a 175 lb male: 1900 kcal + 950 kcal = 2850 kcal
For a 115 lb female: 1200 kcal + 1200 kcal = 2400 kcal

If you were to perform a more detailed calculation, you would probably find that your actual caloric requirement was higher because you would be calculating your activity level more carefully.

21.8   Acetyl phosphate + H_2O $\longrightarrow$ Acetate + $HOPO_3^{2-}$ + H^+ $\Delta G = -10.3$ kcal/mol
$(= -43.1$ kJ/mol$)$

ADP + $HOPO_3^{2-}$ + H^+ $\longrightarrow$ ATP + H_2O $\Delta G = +7.3$ kcal/mol
$(= +30.5$ kJ/mol$)$

_____

Acetyl phosphate + ADP $\longrightarrow$ Acetate + ATP $\Delta G = -3.0$ kcal/mol
$(= -12.6$ kJ/mol$)$

The reaction is favorable because ΔG is negative.

21.9   All of these are found in FAD. The part of the structure on the right is ADP (b), which contains a phosphate anhydride bond (d). The part of the structure on the left contains a substituted benzene ring [(c), top]. There are five heterocyclic rings (a)—the three in ADP and the two rings of FAD that are not benzene rings.

21.10 The hydrogen atoms removed are circled. The reactions are catalyzed by oxidoreductases.

Step 3:

Step 6:

Step 8:

21.11 Citrate and isocitrate are the anions of tricarboxylic acids.

21.12 Reduced coenzymes are produced in steps 3, 4, 6, and 8.

21.13 FAD is the coenzyme involved with the formation of a carbon–carbon double bond; NAD^+ is involved with the formation of a carbonyl double bond.

21.14 Citrate is a tertiary alcohol, and isocitrate and malate are secondary alcohols.

21.15 Isocitrate has two chiral carbon atoms.

21.16 Steps 1–4 correspond to the first stage, and steps 5–8 correspond to the second stage.

21.17 The pH is higher in the mitochondrial matrix because the $[H^+]$ is higher in the intermembrane space.

21.18 *Similarities*: both involve the reaction of glucose, oxygen, carbon dioxide, and water; both take place in organelles (chloroplasts, mitochondria); both involve large, metal-ion-containing molecules (chlorophyll, heme); both involve electron transfer; both involve similar coenzymes.

Differences: photosynthesis captures energy, whereas electron transport releases energy; photosynthesis requires light, whereas oxidative phosphorylation does not.

21.19 O_2 is the final electron acceptor. When H^+ ions move from a region of high concentration to a region of low concentration, energy is released that is used in ATP synthesis.

CIA 21.11

Although 2,4-dinitrophenol decouples electron transport (leading to weight loss), the dose needed to slow ATP synthesis is close to the dose that *stops* ATP synthesis, resulting in death.

CIA 21.12

A blocker of respiration interferes with the flow of electrons through the electron-transport chain (ETC). Blockers may act as reversible inhibitors, or they may bind to the metal ions in cytochromes and heme, causing irreversible inhibition.

CIA 21.13

An uncoupler of respiration allows electron transport but prevents the conversion of ADP to ATP by ATP synthase.

Understanding Key Concepts

21.20 (a) Exergonic reaction:

Succinyl phosphate + $H_2O \longrightarrow$ Succinate + $HOPO_3^{2-}$ + H^+

(b) Endergonic reaction:

ADP + $HOPO_3^{2-}$ + $H^+ \longrightarrow$ ATP + H_2O $\Delta G = +7.3$ kcal/mol (+30.5 kJ/mol)

21.21 (a) Hydrolysis of starch to glucose occurs in Stage 1 of food metabolism.
(b) The synthesis of ATP takes place in Stage 4 of digestion (electron-transport chain).
(c) Production of acetyl-CoA from glucose is a result of Stage 2 of food metabolism.
(d) The reactions shown represent the citric acid cycle and occur in Stage 3 of food metabolism.

21.22 When ATP "drives" a reaction, energy from ATP hydrolysis is used to allow another reaction to proceed. In this case, the reaction that forms fatty acid-CoA from a fatty acid and coenzyme A is endergonic and is unfavorable in the absence of ATP. When the energy from hydrolysis of ATP is used to drive the reaction, the formation of fatty acid-CoA can take place. Coupling is the metabolic strategy in which the energy from an energetically favorable reaction can be used to allow an energetically unfavorable reaction to take place.

21.23 In the steps of the citric acid cycle in which acetyl groups are oxidized to CO_2, NAD^+ or FAD are acceptors of hydride ions, forming NADH or $FADH_2$. The hydrogen ions that are removed in an oxidation using NAD^+ or FAD are released to the mitochondrial matrix. NADH and $FADH_2$ transfer electrons to the enzymes and coenzymes of the electron transport chain in the fourth stage of metabolism. Ultimately, the hydrogen ions combine with reduced oxygen to form H_2O.

21.24

Isocitrate   α-Ketoglutarate

(a) The coenzyme NAD$^+$ is involved with reaction **A**.
(b) CO$_2$ is evolved and an H$^+$ is added in reaction **B**.
(c) The product of step **A** is a β–keto acid because its ketone group is two carbons away from the carboxylic acid group bonded to the middle carbon.
(d) Isocitrate dehydrogenase is an oxidoreductase.

21.25

Step	Reaction Type	Name of Enzyme	Class of Enzyme
1	Addition	Citrate synthase	Lyase
2	Isomerization	Aconitase	Isomerase
3	Oxidation, loss of CO$_2$	Isocitrate dehydrogenase complex	Oxidoreductase
4	Oxidation, loss of CO$_2$	α-Ketoglutarate dehydrogenase complex	Oxidoreductase; lyase
5	Hydrolysis, phosphorylation	Succinyl CoA synthetase	Ligase
6	Oxidation	Succinate dehydrogenase	Oxidoreductase
7	Addition of H$_2$O	Fumarase	Lyase
8	Oxidation	Malate dehydrogenase	Oxidoreductase

21.26 Metals are better oxidizing and reducing agents. Also, they can accept and donate electrons in one-electron increments.

Additional Reactions

Free Energy and Biochemical Reactions

21.27 For a reaction to be favorable, it must release energy and thus have a negative ΔG.

21.28 An endergonic reaction requires energy, and an exergonic reaction releases energy.

21.29 The sign of ΔG predicts whether a reaction is favorable (negative ΔG) or unfavorable (positive ΔG).

21.30 Enzymes only increase the rate of reaction and have no effect on either the magnitude or the sign of ΔG.

21.31 Reactions (a) and (b) are exergonic, and reaction (c) is endergonic. Reaction (a) produces a phosphate (GTP) that can later yield energy by giving up a phosphoryl group.

21.32 Reactions (a) and (b) are exergonic; reaction (c) is endergonic. Reaction (b) proceeds farthest toward products at equilibrium because it has the largest negative value for ΔG.

Cells and Their Structure

21.33 The streptococcus bacterium (b) is a prokaryotic cell, characteristic of bacteria and algae. Eukaryotic cells are found in all other organisms, including humans (a), carrots (c), and brewer's yeast (d).

21.34 All of the traits listed can be characteristic of eukaryotic cells; (a), (c), and (d) are specifically eukaryotic characteristics. Traits (b) and (e) are characteristic of prokaryotes.

21.35 The cytoplasm consists of everything between the cell membrane and the nuclear membrane. The cytosol is the medium that fills the interior of the cell and contains electrolytes, nutrients, and many enzymes, in aqueous solution.

21.36 Organelles are subcellular structures that perform specialized tasks within the cell.

21.37 A mitochondrion is egg-shaped and consists of a smooth, outer membrane and a folded inner membrane. The intermembrane space lies between the outer and inner membranes, and the space enclosed by the inner membrane is called the *mitochondrial matrix*.

21.38 Cristae, the folds of the inner mitochondrial membrane, provide extra surface area for electron transport and ATP production to take place.

Metabolism

21.39 Metabolic processes that break down molecules are called *catabolism*. Metabolic processes that assemble larger molecules from smaller molecules are known as *anabolism*.

21.40 Metabolism refers to all reactions that take place inside cells. Digestion is a part of metabolism in which food is broken down into small organic molecules.

21.41 *First* —————————————————————————> *Last*
digestion, citric acid cycle, electron transport, oxidative phosphorylation

21.42 Acetyl-CoA is formed from the catabolism of all three major classes of food.

21.43 ATP is called a high-energy molecule because energy is released when ATP transfers a phosphoryl group to other molecules.

21.44 An ATP molecule transfers a phosphoryl group to another molecule in exergonic reactions.

21.45 A reaction that is energetically unfavorable can be combined with a reaction that is energetically favorable so that the overall reaction is energetically favorable. The two reactions are then said to be coupled.

21.46

1,3-Bisphosphoglycerate + H_2O $\longrightarrow$
3-Phosphoglycerate + $HOPO_3^{2-}$ + H^+ $\Delta G = -11.8$ kcal/mol
         $(-49.4$ kJ/mol$)$

$ADP + HOPO_3^{2-} + H^+$ $\longrightarrow$ $ATP + H_2O$ $\Delta G = +7.3$ kcal/mol
         $(+30.5$ kJ/mol$)$

- -

1,3-Bisphosphoglycerate + ADP $\longrightarrow$
3-Phosphoglycerate + ATP $\Delta G = -4.5$ kcal/mol
         $(-18.8$ kJ/mol$)$

The reaction is favorable because ΔG is negative.

21.47

1,3-Bisphosphoglycerate   (ADP → ATP)   3-Phosphoglycerate

21.48 The hydrolysis of fructose 6-phosphate ($\Delta G = -3.3$ kcal/mol) is not favorable for phosphorylating ADP ($\Delta G = +7.3$ kcal/mol) because the overall value of ΔG ($+4.0$ kcal/mol) is positive. Any of the reactions in Table 21.1 that have a value of ΔG greater than -7.3 kcal/mol can be coupled with the phosphorylation of ADP to produce a favorable reaction.

21.49 (a) FAD is reduced when a molecule is dehydrogenated.
(b) FAD is an oxidizing agent.
(c) FAD oxidizes $-CH_2CH_2-$ to $-CH{=}CH-$.
(d) Dehydrogenation converts FAD to $FADH_2$.
(e)

$-CH_2CH_2-$ (FAD → FADH$_2$) $-CH{=}CH-$

21.50 (a) NAD^+ is reduced when a molecule is dehydrogenated.
(b) NAD^+ is an oxidizing agent because it oxidizes substrates and is itself reduced.
(c) NAD^+ oxidizes secondary alcohols to ketones.
(d) After dehydrogenation, NAD^+ becomes NADH/$H^{+\cdot}$
(e)

The Citric Acid Cycle

21.51 The purpose of the citric acid cycle is to break down acetyl groups to yield CO_2 and reduced coenzymes.

21.52 The citric acid cycle takes place in cellular mitochondria.

21.53 Oxaloacetate is the starting point for the citric acid cycle.

Oxaloacetate

21.54 The two acetyl-CoA carbons are oxidized to CO_2 in the citric acid cycle.

21.55 (a) Oxidations occur in the steps where NAD^+ and FAD are reduced (steps 3, 4, 6, and 8).
(b) Decarboxylations occur in steps 3 and 4.
(c) A hydration occurs in step 7.

21.56 Three molecules of $NADH/H^+$ and one molecule of $FADH_2$ are produced in each turn of the citric acid cycle.

21.57 Step 6 (succinate → fumarate) stores energy as $FADH_2$.

21.58 Step 3 (isocitrate → α-ketoglutarate), step 4 (α-ketoglutarate → succinyl-CoA), and step 8 (malate → oxaloacetate) store energy as NADH.

The Electron-Transport Chain/Oxidative Phosphorylation

21.59 The two primary functions of the electron-transport chain are the conversion of ADP to ATP and the oxidation of the coenzymes NADH and $FADH_2$.

21.60 One complete citric acid cycle produces four reduced coenzymes, which enter the electron-transport chain and ultimately generate ATP.

21.61 The coenzymes NADH and $FADH_2$ initiate the events of the electron-transport chain.

21.62 The ultimate products of the electron-transport chain are water, energy in the form of ATP, and the oxidized coenzymes FAD and NAD^+.

21.63 In the electron transport chain, $NADH/H^+$ and FAD are found in the mitochondrial matrix, and Cytochrome c and CoQ are found in the inner mitochondrial membrane. In other circumstances, $NADH/H^+$ and FAD can be found in other locations.

21.64 (a) FAD = flavin adenine dinucleotide
(b) CoQ = coenzyme Q
(c) $NADH/H^+$ = reduced nicotinamide adenine dinucleotide, plus hydrogen ion
(d) Cyt c = Cytochrome c

21.65 The iron atoms of the cytochromes are oxidized and reduced in the electron transport chain. The –OH and =O atoms in the CoQ ring undergo oxidation and reduction.

21.66 *First* —————————————————> *Last*
                NADH,   coenzyme Q,   cytochrome *c*

21.67

FAD → CoQH$_2$ → 2 Fe^{3+}
FADH$_2$ → CoQ → 2 Fe^{2+}

21.68 If FADH$_2$ and NADH weren't reoxidized, the body's supply of the oxidized forms of these coenzymes would be depleted, and the citric acid cycle would stop.

21.69 Oxidative phosphorylation refers to formation of ATP from the reactions of reduced coenzymes as they are oxidized in the electron-transport chain. Substrate-level phosphorylation refers to the transfer of a phosphoryl group between substrates and does not involve oxidized and reduced coenzymes. The two processes are not the same.

21.70 In oxidative phosphorylation, reduced coenzymes are oxidized, and ADP is phosphorylated.

21.71 (a) ATP is the energy-carrying product.
(b) H$_2$O and oxidized coenzymes are the other two products.

21.72 The energy to drive oxidative phosphorylation is supplied by the movement of H$^+$ ions from the intermembrane space to the mitochondrial matrix through channels that are part of the ATP synthase enzyme complex embedded in the inner mitochondrial membrane.

21.73 Piericidin, after crossing the mitochondrial membrane, might be oxidized in place of CoQ in the electron transport chain. If this occurred, all steps of electron transfer occurring after CoQ oxidation, including ATP synthesis, would stop.

21.74 When ATP production is decoupled, oxygen consumption increases because the proton gradient disappears, more water is formed, and more oxygen is used.

21.75 A seal has more brown fat because it needs to generate more body heat than a cat does.

Conceptual Problems

21.76 The breakdown of molecules for energy must occur in several steps to avoid the production of large amounts of heat, to allow for storage of energy, and to control the rate of metabolism. Stepwise breakdown also allows energetically favorable steps to be coupled with other energetically unfavorable reactions.

21.77

21.78 Evidently, the isomer with a cis double bond cannot act as a substrate for the enzyme that is responsible for the next step in the citric acid cycle.

21.79 NAD$^+$ and FAD are associated with oxidoreductases.

21.80 Electrons from the oxidation of reduced coenzymes pass through the electron-transport chain. Ultimately, the electrons are used to reduce O_2, which combines with H$^+$ ions to form H_2O.

21.81 This reaction requires FAD because it involves the removal of two hydrogen atoms to introduce a double bond. The enzyme involved is an oxidoreductase (dehydrogenase).

21.82 Energy from combustion is released to the surroundings as heat and is wasted. Energy from metabolic oxidation is released in several steps and is stored in each step so that it is available for use in other metabolic processes.

21.83 The intermembrane space is more acidic because it has a higher [H$^+$] concentration.

21.84 Technically, no. Step 5 of the citric acid cycle produces GTP, which is immediately converted to ATP.

21.85 *Least oxidized* $-------> $ *Most oxidized*
        succinate, fumarate = malate, oxaloacetate

21.86 Cells store energy as polymers, which can be hydrolyzed as needed for energy.

Group Problems

21.87 Fluoroacetate reacts with acetyl-CoA to form fluoroacetyl-CoA, which reacts with oxaloacetate, yielding fluorocitrate. Fluorocitrate binds to aconitase, inhibiting it and stopping the citric acid cycle.

21.88 Oxygen debt occurs because the increased metabolic rate due to running consumes oxygen in the electron-transport chain. Panting is the body's attempt to resupply tissues with oxygen.

21.89 The cells that need the most energy have the most mitochondria. In order, from fewest to most mitochondria per cell: adipose tissue, skin cells, brain, skeletal muscle, heart muscle.

Self-Test for Chapter 21

Multiple Choice

1. Which of the following oxygen by-products is not produced in the human body?
   (a) $\cdot O_2^-$ (b) O_3 (c) $HO\cdot$ (d) H_2O_2

2. The hydrolysis of which of the following substrates can be coupled with the reaction ADP + phosphate $\longrightarrow$ ATP to produce an energetically favorable reaction?
   (a) glucose 6-phosphate (b) glucose 1-phosphate (c) creatine phosphate
   (d) fructose 6-phosphate

3. Which of the following cofactors is not held in a fixed position in electron transport?
   (a) coenzyme Q (b) FMN (c) iron–sulfur clusters (d) cytochrome a

4. Which of the following reactions is an oxidation?
   (a) $Fe^{3+} \longrightarrow Fe^{2+}$ (b) $CH_3COOH \longrightarrow CH_3CHO$
   (c) $CH_3CH_2CH_2COOH \longrightarrow CH_3CH=CHCOOH$ (d) $CH_3COOCH_3 \longrightarrow CH_3COOH + CH_3OH$

5. All of the following statements about electron transport are true except:
   (a) A concentration gradient is established between the inner membrane and the mitochondrial matrix. (b) Electrons are passed from weaker electron acceptors to stronger electron acceptors. (c) A greater amount of energy is produced from oxidation of NADH than from oxidation of $FADH_2$. (d) Some of the enzyme cofactors in the electron transport chain are mobile.

6. All of the following reactions are oxidations except:
   (a) isocitrate $\longrightarrow \alpha$-ketoglutarate (b) α-ketoglutarate $\longrightarrow$ succinyl-SCoA
   (c) succinate $\longrightarrow$ fumarate (d) fumarate $\longrightarrow$ malate

7. Which of the following citric acid cycle enzymes is an isomerase?
   (a) citrate synthetase (b) aconitase (c) succinyl-CoA synthetase (d) fumarase

8. Which coenzyme is not an oxidizing agent?
   (a) FAD (b) coenzyme Q (c) acetyl-CoA (d) NAD^+

9. All of the following reactions have positive ΔG except:
   (a) photosynthesis (b) hydrolysis of acetyl-CoA (c) formation of ATP from ADP
   (d) formation of glucose 6-phosphate

10. Which of the following metabolic pathways does not occur in mitochondria?
    (a) oxidative phosphorylation (b) ATP synthesis (c) citric acid cycle
    (d) degradation of carbohydrates

Sentence Completion

1. _____ cells are found in higher organisms.

2. A → B → C is known as a _____ _____.

3. ATP is synthesized from _____ and _____.

4. A reaction that requires energy in order to take place is said to be energetically _____.

5. The citric acid cycle is also known as the _____ and as the _____ - _____ cycle.

6. α-Ketoglutaric acid reacts with acetyl-CoA to yield _____ and _____.

7. Reactions that use the coenzyme _____ remove two hydrogens from an alcohol to yield a ketone.

8. A complete citric acid cycle yields _____ molecules of ATP.

9. _____ is the material that fills the interior of a cell.

10. _____ is a fast-acting enzyme that catalyzes the decomposition of reactive oxygen species.

11. The reaction succinate —> fumarate is catalyzed by _____ _____.

12. _____ _____ is the minimal amount of energy required to stay alive.

True or False

1. ATP is a higher-energy molecule than ADP.

2. Both prokaryotic and eukaryotic cells are found in higher organisms.

3. Anabolism is the breakdown of high-energy molecules.

4. In a reaction that is energetically unfavorable, the energy of the products is less than the energy of the reactants.

5. The third and fourth stages of digestion take place in the mitochondria.

6. A reaction that is energetically unfavorable can use ATP as an energy source.

7. $FADH_2$ donates two electrons to cytochrome c in the respiratory chain.

8. FAD removes two hydrogens from a carbon chain to yield a double bond.

9. Release of heat and increase of order contribute to making a reaction favorable.

10. ATP is synthesized only in the respiratory chain.

11. In the synthesis of ATP from ADP, a phosphate ester bond is formed.

12. The reactions of the citric acid cycle may be exergonic or endergonic.

Match each entry on the left with its partner on the right.

1. Mitochondrion   (a) Uses energy

2. Oxidative phosphorylation   (b) Yields one molecule of H_2O

3. Isocitrate   (c) Releases energy

4. Anabolic reaction   (d) Site of energy production in cell

5. Succinyl-CoA → succinate   (e) Reduces NAD^+ to $NADH/H^+$

6. Citric acid cycle   (f) Tricarboxylic acid

7. Malate → oxaloacetate   (g) Contains an iron atom

8. Catabolic reaction   (h) Subcellular structure

9. Respiratory chain   (i) Reduces FAD to $FADH_2$

10. Cytochrome   (j) ATP synthesis

11. Succinate → fumarate   (k) Yields 2 molecules of CO_2

12. Organelle   (l) Yields acetyl-CoA plus ATP

Chapter Outline

I. Introduction to carbohydrate metabolism (Sections 22.1–22.2).
  A. Digestion of carbohydrates (Section 22.1).
    1. Digestion is the breakdown of food into small molecules.
    2. The products of digestion are absorbed from the intestinal tract.
    3. Digestion of carbohydrates.
       a. α–Amylase in the mouth breaks starch into smaller polysaccharides and maltose.
       b. α–Amylase in the small intestine further breaks down polysaccharides.
       c. Disaccharides are broken down to monosaccharides in the small intestine.
  B. Overview of glucose metabolism (Section 22.2).
    1. *Glycolysis* is the conversion of glucose to pyruvate; it occurs when energy is needed.
    2. *Glycogenesis* is the synthesis of glycogen from glucose; it occurs when excess glucose is present.
    3. The *pentose phosphate pathway* supplies other monosaccharides; it occurs when these are in short supply.
    4. *Glycogenolysis* is the breakdown of glycogen to glucose units; it occurs when the body needs glucose.
    5. Pyruvate has several fates.
       a. Under normal circumstances, pyruvate is converted to acetyl-CoA.
       b. When oxygen is lacking, pyruvate is converted to lactate.
       c. When the body is starved for glucose, pyruvate and other noncarbohydrates are converted to glucose via *gluconeogenesis*.
II. Glycolysis and pyruvate metabolism (Sections 22.3–22.6).
  A. Glycolysis is the breakdown of glucose to two pyruvates (Section 22.3).
    1. Glycolysis occurs in the cytosol.
  B. Steps in glycolysis:
    1. Steps 1–3: Phosphorylation.
       a. Step 1: Glucose moves into the cell and is converted to glucose 6-phosphate, with the expenditure of one ATP.
       b. Step 2: Glucose 6-phosphate is isomerized to fructose 6-phosphate.
       c. Step 3: Fructose 6-phosphate is converted to fructose 1,6-bisphosphate, with the expenditure of one ATP.
    2. Steps 4, 5: Cleavage and isomerization.
       a. Step 4: Fructose 1,6-bisphosphate is cleaved to glyceraldehyde 3-phosphate and dihydroxyacetone phosphate.
       b. Step 5: Dihydroxyacetone phosphate is isomerized to glyceraldehyde 3-phosphate.
    3. Steps 6–10: Energy generation.
       a. Step 6: Glyceraldehyde 3-phosphate is phosphorylated and oxidized to 1,3-bisphosphoglycerate: $NAD^+ \longrightarrow NADH/H^+$.
       b. Step 7: 1,3-Bisphosphoglycerate transfers a phosphate to ADP to yield 3-phosphoglycerate and ATP.
       c. Step 8: 3-Phosphoglycerate is isomerized to 2-phosphoglycerate.
       d. Step 9: 2-Phosphoglycerate is dehydrated to phosphoenolpyruvate.
       e. Step 10: Phosphoenolpyruvate transfers a phosphate group to ADP to yield ATP and pyruvate.
  C. End results of glycolysis.
    1. Conversion of glucose to two pyruvates.
    2. Production of two ATP.
    3. Production of two $NADH/H^+$.

D. Entry of other sugars into glycolysis (Section 22.4).
   1. Fructose is converted to fructose 6-phosphate and enters glycolysis.
   2. Mannose is converted to fructose 6-phosphate and enters glycolysis.
   3. Galactose is converted to glucose 6-phosphate and enters glycolysis.
E. Pyruvate metabolism (Section 22.5).
   1. When oxygen is available, pyruvate is converted to acetyl-CoA, with the conversion of NAD^+ to $NADH/H^+$ and the production of CO_2.
   2. Under anaerobic conditions, pyruvate is reduced to lactate, and $NADH/H^+$ is converted to NAD^+.
   3. When glucose is in short supply, pyruvate can be converted back to glucose by gluconeogenesis.
   4. In yeast, pyruvate is converted to ethanol and CO_2.
F. Energy output in complete catabolism of glucose (Section 22.6).
   1. The total energy output from glucose metabolism is the combined result of:
      a. Glycolysis.
      b. Pyruvate —> acetyl-CoA.
      c. Citric acid cycle of two acetyl-CoA molecules (and passage of the resulting reduced coenzymes through the respiratory chain).
      d. Passage of two reduced coenzymes from glycolysis through the respiratory chain.
   2. The number of ATP produced depends on:
      a. The number of electrons that enter the respiratory chain from reduced coenzymes.
      b. The number of ATP produced per reduced coenzyme.
      c. A maximum of 38 ATP are produced per mol glucose catabolized.
III. Regulation of glucose metabolism (Section 22.7).
   A. Regulation of glucose concentration in the blood (Section 22.7).
      1. Hypoglycemia: low blood sugar.
         a. The hormone glucagon stimulates breakdown of glycogen to glucose to raise blood sugar.
      2. Hyperglycemia: high blood sugar.
         a. The hormone insulin stimulates passage of glucose into cells to be used in energy production.
   B. Fasting and starvation.
      1. First, glucose is released from glycogen.
      2. Then, glucose is synthesized from proteins via gluconeogenesis.
      3. Then, fats are metabolized to acetyl-CoA, which enters the citric acid cycle.
      4. As the citric acid cycle is overloaded, acetyl-CoA is removed by formation of ketone bodies.
      5. Ultimately, the body produces 50% of ATP from ketone bodies.
      6. The body can endure this state for months.
IV. Other pathways of carbohydrate metabolism (Sections 22.8–22.9).
   A. Glycogen metabolism (Section 22.8).
      1. Glycogenesis.
         a. Glycogenesis occurs when glucose levels are high.
         b. The pathway to glycogen synthesis:
            i. Glucose —> Glucose 6-phosphate —> Glucose 1-phosphate —> Glucose-UDP —> Glycogen
      2. Glycogenolysis.
         a. Glycogenolysis occurs when there is an immediate need for energy.
         b. The pathway of glycogenolysis:
            i. Glycogen —> Glucose 1-phosphate —> Glucose 6-phosphate —> Glucose

B. Gluconeogenesis (Section 22.9).
   1. Gluconeogenesis is the synthesis of glucose from any of several small molecules, including lactate, pyruvate, amino acids, and glycerol.
   2. Gluconeogenesis occurs when the body is starved for glucose.
   3. Seven of the steps of gluconeogenesis are reversals of glycolysis steps.
       a. The other three steps are too endergonic, and alternate pathways must be used.
       b. Steps 1 and 2 of gluconeogenesis are endergonic:
           i. Pyruvate —> Oxaloacetate —> Phosphoenolpyruvate
       c. Oxaloacetate is an intermediate that allows amino acids to enter gluconeogenesis.
       d. Gluconeogenesis takes place in the cytosol.

Solutions to Chapter 22 Problems

22.1 (a) *Glycogenesis* is the synthesis of glycogen from glucose.
(b) *Glycogenolysis* is the release of glucose units from glycogen.
(c) *Gluconeogenesis* is the synthesis of glucose from lactate or other noncarbohydrates.

22.2 The following synthetic pathways have glucose 6-phosphate as their first reactant:

Glycogenesis: synthesis of glycogen from glucose
Pentose phosphate pathway: synthesis of five-carbon sugars from glucose
Glycolysis: conversion of glucose to pyruvate

22.3 (1) In step 6, 1,3-bisphosphoglycerate is synthesized. In step 7, the energy is harvested in ATP as 3-phosphoglycerate is formed.

(2) In step 9, phosphoenolpyruvate is synthesized. In step 10, the energy is harvested in ATP as pyruvate is formed.

22.4 The following steps of glycolysis are isomerizations:
Step 2: Glucose 6-phosphate $\rightarrow$ Fructose 6-phosphate
Step 5: Dihydroxyacetone phosphate $\rightarrow$ Glyceraldehyde 3-phosphate
Step 8: 3-Phosphoglycerate $\rightarrow$ 2-Phosphoglycerate

22.5

Glucose 6-phosphate   Fructose 6-phosphate

22.6 (a) Pyruvate is oxidized to a greater extent than glucose because it has two carbon–oxygen double bonds.
(b) Oxidation occurs in step 6, in which glyceraldehyde 3-phosphate is oxidized and phosphorylated to give 1,3-bisphosphoglycerate. NAD^+ is the oxidizing agent.

CIA 22.1
Dextran adheres to teeth and allows bacterial colonies to stick to teeth without being washed away by water or saliva.

CIA 22.2
Plaque is composed of glycoproteins from saliva, bacteria, dextran, and polysaccharide storage granules released by the bacteria.

CIA 22.3
The bacteria associated with plaque secrete products that are toxic to the gums.

CIA 22.4
In an environment rich in sucrose, bacteria associated with tooth decay secrete an enzyme that transfers glucose units from digested sucrose to the dextran polymer. The residual fructose is metabolized to lactate, which lowers pH. The resulting acidic environment in the mouth dissolves minerals in teeth, leading to cavities.

CIA 22.5
Sucrose promotes the growth of decay-producing bacteria. Honey, which contains glucose and fructose, is better for tooth health.

22.7

Fructose 6-phosphate enters glycolysis at step 3.

22.8

Glucose and galactose differ in configuration at carbon 4.

22.9 (a) The remaining energy escapes as heat.
(b) The exergonic nature of the reaction makes it very favorable in the forward direction and unfavorable in the reverse direction. Also, escape of the gaseous product drives the reaction to completion in the forward direction.

22.10 Humans have used fermentation to produce beer and wine, to leaven bread, and to make yogurt and cheese.

22.11 (1) Pyruvate is converted to acetyl-CoA when oxygen is in good supply. (2) Pyruvate is converted to lactate when oxygen is in short supply. (3) In yeast, pyruvate is fermented to ethanol.

HOC 22.1
Bread rises because fermentation by yeast causes bubbles of carbon dioxide to form. The aroma of rising bread is "yeasty", but it is actually an alcohol-based smell. If the bread rises too long, you may notice the smell of alcohol, too.

22.12 Glycolysis of 10 molecules of glucose produces 80 ATP molecules.

22.13 Complete catabolism of 1 mole of glucose produces 38 moles of ATP.

22.14 As starvation begins, insulin starts to decrease as glucagon starts to increase, while liver glycogen begins a rapid decrease. After several hours, the levels of insulin and glucagon continue to change, and blood glucose drops significantly. As glycogen is used up, fatty acids and ketone bodies appear in the bloodstream. At the end of 24 hours, liver glycogen is depleted, the levels of glucose and insulin are reduced, and the levels of glucagon, free fatty acids, and ketone bodies are elevated.

22.15

$$
\begin{array}{c}
CH_2OH \\
| \\
H-C-OH \\
| \\
HO-C-H \\
| \\
H-C-OH \\
| \\
H-C-OH \\
| \\
CH_2OH
\end{array}
$$

Sorbitol

Sorbitol can't form a cyclic hemiacetal because it doesn't have a carbonyl group.

22.16 (a) The increase in $[H^+]$ drives the equilibrium shown in Section 22.7 to the right, causing the production of CO_2.
(b) This effect is an example of Le Châtelier's principle.

HOC 22.2
When you begin to breathe heavily, your muscles are in a state of oxygen debt due to lactic acid accumulation. When you reduce your exertion level, lactic acid can be converted back to pyruvate.

CIA 22.6

The distance a person can sprint is limited by readily available energy that comes from creatine phosphate and glycogen, both of which are quickly used up.

CIA 22.7

First used –> *Last used*
ATP, creatine phosphate, glucose, glycogen, fatty acids from triacylglycerols

CIA 22.8

Reaction of creatine phosphate with ADP produces ATP in one step. Production of ATP from glucose or glycogen requires several steps.

22.17 Glycogenesis is the synthesis of glycogen from glucose 6-phosphate when glucose concentration is high. Glycogenolysis is the breakdown of glycogen to glucose when glucose is needed.

22.18 Glycogenesis is necessary for storage of excess glucose. Glycogenesis is necessary when blood glucose levels are low.

22.19

$$
\begin{array}{ccc}
\underset{\displaystyle \text{Glycerol}}{
\begin{array}{c}
CH_2OH \\
| \\
H-C-OH \\
| \\
CH_2OH
\end{array}}
& \xrightarrow{\text{phosphorylation}} &
\underset{\displaystyle \text{Glycerol 3-phosphate}}{
\begin{array}{c}
CH_2OPO_3^{2-} \\
| \\
H-C-OH \\
| \\
CH_2OH
\end{array}}
& \xrightarrow{\text{oxidation}} &
\underset{\displaystyle \begin{array}{c}\text{Dihydroxyacetone}\\\text{phosphate}\end{array}}{
\begin{array}{c}
CH_2OPO_3^{2-} \\
| \\
C=O \\
| \\
CH_2OH
\end{array}}
\end{array}
$$

22.20 In the Cori cycle, lactate from muscles travels through the bloodstream to the liver, where it undergoes gluconeogenesis to produce glucose. The glucose formed travels through the bloodstream to muscles, where it can be used for energy or stored as glycogen.

22.21 Gluconeogenesis is needed in situations where blood glucose is low and glycogen supplies have been depleted.

CIA 22.9

Glucose is oxidized by glucose oxidase to yield gluconate and H_2O_2, which reacts with a dye in a peroxidase-catalyzed reaction to yield a colored oxidized dye (and H_2O). The color change, which can be measured, is specific for glucose.

CIA 22.10

The fasting level of glucose in a diabetic is 140 mg/dL or greater, as compared to an average level of 90 mg/dL for a nondiabetic.

CIA 22.11

After drinking a glucose solution, both a diabetic person and a nondiabetic person experience a great increase in blood sugar level in the first hour. After two hours, the glucose level in a nondiabetic person returns to near fasting level, while the level in a diabetic person remains high.

CIA 22.12
To confirm a diagnosis of diabetes, a doctor would order a fasting blood glucose test and a glucose tolerance test.

Understanding Key Concepts

22.22 Most of the enzymes involved with digestion are hydrolases, which catalyze the hydrolysis of bonds in large molecules to produce smaller molecules.

22.23 (a) *Glycolysis* occurs when the body needs energy and the supply of glucose is adequate.
(b) *Hydrolysis to free glucose* takes place in the liver when free glucose is needed in cells in other parts of the body. (Only free glucose can be transported in the blood.)
(c) The *pentose phosphate pathway* operates when either ribose 5-phosphate or the reduced coenzyme NADPH is needed.
(d) *Glycogenesis* occurs when the concentration of glucose in the blood is high and the body doesn't need glucose for energy production.

22.24 The energy investments made when glucose is converted to glucose 6-phosphate (step 1) and when fructose 1-phosphate is converted to fructose 1,6-bisphosphate (step 3) produce the phosphorylated intermediates needed in later stages of glycolysis. These intermediates take part in the reactions that ultimately repay the initial energy investment. Cleavage of fructose 1,6-bisphosphate at step 4 generates the two three-carbon molecules whose transformations result in the formation of pyruvate.

22.25 (a) Pyruvate enters into the *citric acid cycle* when the body needs energy. To enter the cycle, pyruvate is first converted to acetyl-CoA in mitochondria.
(b) In yeast, pyruvate yields *ethanol and CO_2* in the absence of oxygen.
(c) Pyruvate is *converted to lactate* in muscles and in red blood cells under anaerobic conditions.
(d) Pyruvate is directed to *gluconeogenesis* when the body needs glucose and both glucose and glycogen are in short supply. Gluconeogenesis takes place in the liver.

22.26

Step	Enzyme	Class
1	Hexokinase	Transferase
2	Glucose 6-phosphate isomerase	Isomerase
3	Phosphofructokinase	Transferase
4	Aldolase	Lyase
5	Triose phosphate isomerase	Isomerase
6	Glyceraldehyde 3-phosphate dehydrogenase	Oxidoreductase, transferase
7	Phosphoglycerate kinase	Transferase
8	Phosphoglycerate mutase	Isomerase
9	Enolase	Lyase
10	Pyruvate kinase	Transferase

Transferases are the most common class of enzymes in glycolysis because several reactions involve phosphate transfers to ADP and from ATP. Ligases are not represented among the enzymes of glycolysis because ligases are involved with the synthesis of larger molecules from smaller molecules, with the expenditure of ATP. Glycolysis is a pathway for the breakdown of larger molecules to smaller molecules, with the eventual generation of ATP.

22.27 (1) Insulin levels rise (g).
    (2) Glucose is absorbed by cells (c).
    (3) Glycolysis replenishes ATP supplies (b).
    (4) Glycogen synthesis (glycogenesis) occurs if there is excess glucose (e).
    (5) Blood levels of glucose pass through normal to below normal (hypoglycemic) (f).
    (6) Glucagon is secreted (a).
    (7) The liver releases glucose into the bloodstream (d).

22.28 Compounds available for gluconeogenesis include pyruvate, lactate, many amino acids, glycerol, and intermediates of the citric acid cycle, especially oxaloacetate. Lactate, formed in muscle tissue, is carried in the bloodstream to the liver, where, in the cytosol of liver cells, it is converted to pyruvate. Pyruvate is transported to liver mitochondria, where gluconeogenesis takes place. Gluconeogenesis takes place when the body's supply of glucose is low.

22.29 The conversion of fatty acids to glucose occurs in plants, where fatty acids stored in seeds can be used as a source of carbohydrates after germination. Unlike plants, animals don't need carbohydrates as structural materials and are able to obtain all the carbohydrates they need from food.

22.30 (a) No. Molecular oxygen is not involved in any of the steps of glycolysis.
    (b) Molecular oxygen is used in electron transport, where it combines with hydrogen ions and with electrons from coenzymes (reduced in the citric acid cycle) to produce water. The reoxidized coenzymes can reenter the citric acid cycle or can be used in glycolysis.

Additional Problems

Digestion and Metabolism

22.31 Digestion occurs primarily in the mouth, stomach, and small intestine; it involves the enzyme-catalyzed hydrolysis of food components into small molecules.

22.32 Lactose + H_2O $\xrightarrow{\text{Lactase}}$ Glucose + Galactose

This process occurs in the mucous lining of the small intestine.

22.33 The major monosaccharide products of digestion are glucose, fructose, and galactose.

22.34

Type of Food Molecules	Products of Digestion
Proteins	Amino acids
Triacylglycerols	Glycerol and fatty acids
Maltose	Glucose
Sucrose	Glucose and fructose
Lactose	Glucose and galactose
Starch	Glucose

22.35 The word *aerobic* describes a condition in which oxygen is plentiful; the word *anaerobic* describes a condition where oxygen is absent.

22.36 Under aerobic conditions, pyruvate forms acetyl-CoA.
Under anaerobic conditions, pyruvate forms lactate.
Under fermentation conditions, pyruvate forms ethanol and CO_2.

22.37

	Substrate(s)	Product
(a) Glycolysis	Glucose	Pyruvate
(b) Gluconeogenesis	Pyruvate, lactate, amino acids, glycerol	Glucose

22.38

	Substrate(s)	Product
(a) Glycogenesis	Glucose	Glycogen
(b) Glycogenolysis	Glycogen	Glucose + chain-shortened glycogen

22.39 The pentose phosphate pathway provides the coenzyme NADPH and produces ribose 5-phosphate, a sugar used in nucleic acid synthesis.

22.40

If the Body Needs	Fate of Intermediates in Pentose Phosphate Pathway
NADPH	Pentose phosphate intermediates are recycled to glucose 6-phosphate for further production of NADPH.
ATP	Fructose 6-phosphate and glyceraldehyde 3-phosphate (pentose phosphate intermediates) enter glycolysis.
Nucleic acid synthesis	Ribose 5-phosphate is the major product.

Glycolysis

22.41 All of the pathways take place in the cytosol; some steps of gluconeogenesis occur in mitochondria.

22.42 (a) Glycolysis takes place in all of these types of cells.
(b) Gluconeogenesis takes place in the liver.
(c), (d) Glycogenesis and glycogenolysis are carried out in liver and muscle cells.

22.43 An initial energy investment is needed to prepare glucose for glycolysis; this energy is repaid in later, energy-producing steps.

22.44 None of the ATP produced in glycolysis results from oxidative phosphorylation; all of the phosphorylations are substrate-level.

22.45

Enzyme	Catalyzes
(a) Pyruvate kinase	Phosphoenolpyruvate $\rightarrow$ Pyruvate
(b) Glyceraldehyde 3-phosphate dehydrogenase	Glyceraldehyde 3-phosphate $\rightarrow$ 1,3-Bisphosphoglycerate
(c) Hexokinase	Glucose $\rightarrow$ Glucose 6-phosphate
(d) Phosphoglycerate mutase	3-Phosphoglycerate $\rightarrow$ 2-Phosphoglycerate
(e) Aldolase	Fructose 1,6-bisphosphate $\rightarrow$ Glyceraldehyde 3-phosphate + Dihydroxyacetone phosphate

22.46 (a) Phosphorylation/dephosphorylation:

> *Step 1:*   Glucose →   Glucose 6-phosphate
>
> *Step 3:*   Fructose 6-phosphate → Fructose 1,6-bisphosphate
>
> *Step 6:*   Glyceraldehyde 3-phosphate → 1,3-Bisphosphoglycerate
>
> *Step 7:*   1,3-Bisphosphoglycerate →   3-Phosphoglycerate
>
> *Step 10:*   Phosphoenolpyruvate → Pyruvate

(b) Oxidation:

> *Step 6:*   Glyceraldehyde 3-phosphate → 1,3-Bisphosphoglycerate

(c) Dehydration:

> *Step 9:*   2-Phosphoglycerate →   Phosphoenolpyruvate
>
> The last two phosphorylations involve phosphate transfers *to* ADP.

22.47 (a) As shown in Section 22.6, glycolysis of 1 mol glucose yields 2 ATP and 2 NADH, each of which can form a maximum of 3 ATP, giving a maximum of 8 ATP.
(b) Aerobic conversion of 1 mol pyruvate to acetyl-CoA yields 1 mol NADH, giving a maximum of 3 ATP.
(c) Catabolism of 1 mol acetyl-CoA yields 3 NADH (forming a maximum of 9 ATP), 1 $FADH_2$ (forming a maximum of 2 ATP), and 1 ATP, giving a maximum of 12 ATP. Note: The values given for ATP yield from NADH and $FADH_2$ are the ideal yields. The actual yield is probably smaller.

22.48 Note: Substrate-level phosphorylation is formation of ATP as a byproduct of a reaction, as opposed to formation of ATP as a by-product of electron transport.
(a) Glycolysis of 1 mol glucose produces 2 mol ATP as a result of substrate-level phosphorylation, and a maximum of 6 mol ATP formed by oxidative phosphorylation. See the note in the previous problem. (Actually, 4 mol ATP are produced by phosphorylation, but 2 mol ATP are consumed in steps 1 and 3 of glycolysis.)
(b) Aerobic conversion of pyruvate to acetyl-CoA doesn't produce ATP by substrate-level phosphorylation but produces a maximum of 3 mol ATP by oxidative phosphorylation.
(c) In the citric acid cycle, 1 mol of acetyl-CoA produces 1 mol ATP by substrate-level phosphorylation (step 5) and a maximum of 11 mol ATP from oxidative phosphorylation.

Most of the ATP produced in the citric acid cycle results from passage of the reduced coenzymes NADH and $FADH_2$ through the electron–transport chain.

22.49 Under anaerobic conditions, NADH can't enter the respiratory chain to be oxidized. Instead, in order to replenish the body's supply of NAD^+, NADH reacts with pyruvate to form lactate and NAD^+.

22.50

22.51 The complete catabolism of 1 mole of sucrose yields 12 moles of CO_2.

22.52 Catabolism of 1 mole of sucrose produces 4 moles of acetyl-CoA.

Regulation of Glucose Metabolism and Metabolism During Stress

22.53 Insulin is released when blood glucose levels are high and results in passage of glucose into cells, where it is metabolized. Glucagon is released when blood glucose levels are low and results in breakdown of glycogen in the liver to produce glucose.

22.54

Condition	Blood Sugar Level	Symptoms
Hypoglycemia	Low	Weakness, sweating, rapid heartbeat, confusion, coma, death
Hyperglycemia	High	Increased urine flow, low blood pressure, coma, death

22.55 Glucose is initially formed from glycogen during starvation or fasting. When glycogen is used up, glucose is formed from pyruvate, lactate, amino acids, and other small molecules via gluconeogenesis.

22.56 As starvation continues, acetyl-CoA is converted to *ketone bodies* to prevent buildup in the cells.

Glycogen Catabolism and Anabolism

22.57 Most of the glycogen in the body is stored in muscle cells and in the liver.

22.58 Muscle cells, unlike liver cells, lack an enzyme that converts glucose 6-phosphate to glucose, which can pass into the bloodstream. Consequently, the glucose 6-phosphate product of glycogenolysis in muscle cells enters directly into glycolysis.

22.59 UTP reacts with glucose to form the high-energy compound UDP-glucose, which carries glucose to a growing glycogen chain.

22.60 The exact reverse of an energetically favorable reaction must be energetically unfavorable. Since glycogenolysis is an energetically favorable hydrolysis, its exact reverse must be energetically unfavorable. Thus, glycogenesis must occur by an alternate pathway, which involves more steps.

Glucose from Noncarbohydrates

22.61 *Gluconeogenesis* is the anabolic pathway for making glucose.

22.62 Lactate and pyruvate are two molecules that can serve as starting materials for glucose synthesis. Other small molecules may also enter into gluconeogenesis.

22.63 In glucose anabolism, pyruvate is initially converted to *oxaloacetate*, which in turn is converted to *phosphoenolpyruvate*.

22.64 Several steps of the exact reverse of the energetically favorable conversion of glucose to pyruvate are energetically unfavorable.

22.65 The steps of glycolysis and of its reverse in gluconeogenesis are shown below:

Glycolysis: Phosphoenolpyruvate → Pyruvate

Gluconeogenesis: Pyruvate → Oxaloacetate → Phosphoenolpyruvate

Glycolysis: Glucose + ATP → Glucose 6-phosphate + ADP

Gluconeogenesis: Glucose 6-phosphate → Glucose + P_i

Glycolysis: Fructose 6-phosphate + ATP → Fructose 1,6-bisphosphate + ADP

Gluconeogenesis: Fructose 1,6-bisphosphate → Fructose 6-phosphate + P_i

22.66 Three steps of gluconeogenesis are not the exact reverse of steps of glycolysis–steps 1, 3, and 10. All of these steps involve phosphate transfers.

22.67 The Cori cycle takes place in the muscles and liver and involves the conversions glucose (in muscle) → pyruvate → lactate (in muscles) → lactate (in liver) → pyruvate → glucose (in liver) → glucose (in muscle).

22.68 The Cori cycle is needed when muscle glucose is depleted and when oxygen is in short supply.

Conceptual Problems

22.69 Pyruvate can cross the mitochondrial membrane because it is the only molecule in glycolysis that is not a phosphate. (Phosphates can't cross the mitochondrial membrane.)

22.70 Kinase enzymes are associated with phosphoryl group transfers.

22.71 Glucose obtained from the hydrolysis of glycogen is phosphorylated by reaction with inorganic phosphate ion and enters the glycolysis pathway as glucose 6-phosphate. Thus, one fewer ATP is needed (at step 1), and one more ATP is produced.

22.72 Glycolysis must be tightly controlled because the body must avoid extreme fluctuations in glucose concentration.

22.73 In both cases, 2 moles of ATP are generated.

22.74 Reaction (a) consumes energy, and reaction (b) releases energy.

22.75 (1) Lactate in the bloodstream is transported to the liver, where it can be converted to pyruvate, which can be converted to glucose via gluconeogenesis. The glucose can be returned to the muscles for energy. (2) Buildup of lactate in muscle tissue lowers pH, which causes physiological problems.

22.76 (a) Glycolysis predominates when glucose is in good supply and energy is needed.
(b) Gluconeogenesis predominates when glucose is in short supply and during fasting and early stages of starvation.

22.77 The NAD$^+$ produced when pyruvate is converted to lactate is used in glycolysis.

22.78 Characteristics of Type I diabetes include excessive thirst, frequent urination, high concentrations of glucose in the urine and blood, and weight loss.

22.79 In Type II diabetes, insulin is in good supply, but cell membrane receptors fail to recognize insulin and to allow the passage of glucose into cells.

22.80 Metabolic syndrome resembles a prediabetic state in which blood sugar level is slightly elevated, blood pressure is slightly high, and glucose tolerance is slightly impaired, and is a predictor for diabetes.

22.81 In diabetes, excess glucose in the eyes and extremities is converted to sorbitol, which can't be transported out of cells. This elevated sorbitol level changes the osmolarity of tissues and causes cataracts.

Group Problems

22.82 Muscle tissue needs a steady supply of glucose, and the compounds needed for glucose synthesis (gluconeogenesis) are present in the liver.

22.83 People with galactosemia must avoid dairy products.

22.84 In the absence of oxygen, the pyruvate product of catabolism of glucose in wine is fermented by yeast enzymes to ethanol and CO_2, which increased the pressure in the bottle and popped the lid.

22.85 Fructose and glucose have the same net ATP production because fructose can be phosphorylated to form fructose 6-phosphate, which can enter glycolysis directly as a glycolysis intermediate.

Self-Test for Chapter 22

Multiple Choice

1. The synthesis of glucose from small molecules is called:
    (a) glycolysis   (b) glucogenesis   (c) gluconeogenesis   (d) glycogenesis

2. Which one of the following substances can cross cell membranes?
    (a) glucose 6-phosphate   (b) ATP   (c) glucose   (d) acetyl-CoA

3. The synthesis of glycogen occurs in the:
    (a) liver and bloodstream   (b) liver and muscles
    (c) liver and pancreas   (d) muscles and pancreas

4. The exact reverse of which of the following steps of glycolysis also occurs in gluconeogenesis?
    (a) glucose —> glucose 6-phosphate   (b) phosphoenolpyruvate —> pyruvate
    (c) fructose 6-phosphate —> fructose 1,6-bisphosphate
    (d) 3-phosphoglycerate —> 2-phosphoglycerate

5. When the body needs to synthesize nucleic acids:
(a) The oxidative phase of the pentose phosphate path is bypassed.
(b) The synthesis of NADPH is increased. (c) Glucose 6-phosphate enters glycolysis.
(d) Three-, four-, or seven-carbon sugars are produced.

6. After several weeks of starvation, the body's principal energy source is:
(a) protein (b) fat (c) carbohydrate (d) all three

7. Type I *diabetes mellitus* is a disorder of:
(a) insufficient insulin production (b) faulty glucose metabolism (c) autoimmune response
(d) all three

8. Which of the following is not a route for pyruvate metabolism in the human body?
(a) conversion to acetyl-CoA (b) reduction to lactate (c) fermentation to alcohol
(d) reformation of glucose

9. Glycogenesis resembles glycogenolysis in that:
(a) Both require the coenzyme UTP. (b) Both occur when glucose is in good supply.
(c) Both are stimulated by the hormone epinephrine.
(d) Both involve glucose 6-phosphate as an intermediate.

10. Which of the following monosaccharides doesn't enter glycolysis?
(a) ribose (b) mannose (c) galactose (d) fructose

Sentence Completion

1. The _____–_____ pathway is another name for glycolysis.

2. Step __ of glycolysis involves an oxidation and a phosphorylation.

3. _____ _____ _____ catalyzes the formation of acetyl-CoA from pyruvate.

4. _____ aids in the passage of blood glucose across cell membranes.

5. The first step of glycogenolysis is the formation of ____ ____.

6. Steps ___ and _____ are known as the energy investment part of glycolysis.

7. The metabolic response to _____ _____ resembles starvation.

8. _____ and _____ supply the energy for gluconeogenesis.

9. When pyruvate is converted to acetyl-CoA, one molecule of _____ is lost.

10. The isomerization of glucose 6-phosphate to fructose 6-phosphate is catalyzed by ___ ___ __.

11. The hormone _____ stimulates breakdown of glycogen.

12. _____ may be due to an overdose of insulin.

True or False

1. The pentose phosphate pathway is important because it produces NADH.

2. Fructose, galactose, and mannose can all enter the glycolysis pathway.

3. Conversion of glucose 6-phosphate to fructose 6-phosphate requires ATP.

4. Glycogenesis is the exact reverse of glycogenolysis.

5. Hyperglycemia is a high level of blood glucose.

6. The reactions of glycolysis occur in the mitochondria.

7. Thirty-six ATP molecules are produced for each glucose that passes through the glycolysis pathway.

8. Glucagon is the storage form of glucose.

9. All four transformations of pyruvate occur in the human body.

10. Cleavage of fructose 1,6-diphosphate yields two molecules of glyceraldehyde 3-phosphate.

11. Pyruvate is converted to lactate under anaerobic conditions.

12. Each conversion of glyceraldehyde 3-phosphate to pyruvate requires two ATPs.

Match each entry on the left with its partner on the right.

1. Glycolysis	(a)	Formed by fermentation of pyruvate
2. Hyperglycemia	(b)	Formed from pyruvate under anaerobic conditions
3. Enolase	(c)	Intermediate in glycogen synthesis
4. Glycogenolysis	(d)	Synthesis of glucose from pyruvate
5. UDP-glucose	(e)	Catalyzes the cleavage of fructose 1,6-diphosphate
6. Ribose 5-phosphate	(f)	Low blood sugar
7. Gluconeogenesis	(g)	Synthesis of glycogen from glucose
8. Ethanol	(h)	Product of pentose phosphate pathway
9. Hypoglycemia	(i)	Catalyzes loss of water from 2-phosphoglyceric acid
10. Glycogenesis	(j)	Breakdown of glycogen to glucose
11. Lactate	(k)	High blood sugar
12. Aldolase	(l)	Breakdown of glucose to pyruvate

Chapter Outline

I. Structure and classification of lipids (Section 23.1).
   A. Lipids are defined by solubility in nonpolar solvents.
   B. Classification of lipids.
      1. Waxes.
      2. Triacylglycerols.
      3. Glycerophospholipids.
      4. Sphingolipids.
      5. Sterols and eicosanoids.
II. Fatty acids and their esters (Sections 23.2–23.4).
   A. Fatty acids (Section 23.2).
      1. Fatty acids are long-chain carboxylic acids.
         a. A shorthand notation indicates the length of the chain (C_n).
      2. Fatty acids are classified by the number of double bonds they have.
         a. Fatty acids with no double bonds are saturated ($C_n:0$).
         b. Fatty acids with one double bond are monounsaturated ($C_n:1$).
         c. Fatty acids with more than one double bond are polyunsaturated ($C_n:>1$).
   B. Waxes are esters of long-chain alcohols with fatty acids.
   C. Fats and oils.
      1. Fats and oils are mixtures of esters of glycerol with three fatty acids and are known as triacylglycerols.
      2. Properties of fats and oils (Section 23.3).
         a. Hydrophobic.
         b. Uncharged.
         c. Melting point is determined by the length of the fatty acid side chains and the number of double bonds.
            i. Solid fats have saturated fatty acid side chains and are higher melting.
            ii. Liquid fats have unsaturated fatty acid side chains and are lower melting.
      3. Reactions of triacylglycerols (Section 23.4).
         a. Hydrogenation.
            i. Control of hydrogenation determines the consistency of a fat.
         b. Hydrolysis.
            i. In the body, enzymes catalyze hydrolysis of triacylglycerols.
            ii. In the laboratory, strong aqueous base is used.
            iii. The resulting fatty acid salts are called soaps.
            iv. Soaps clean because the polar end dissolves in water and the nonpolar end dissolves in grease.
            v. Soap molecules cluster together in water to form micelles; grease is trapped in the middle of a micelle.
III. Cell membrane lipids (Sections 23.5–23.6).
   A. Phospholipids (Section 23.5).
      1. Glycerophospholipids.
         a. Glycerophospholipids are esters of glycerol with two fatty acids and a phosphate group that is bonded to one of a number of different compounds containing an –OH group.
         b. The phospholipid with choline as a phosphate ester is a lecithin.
            i. Lecithins are components of cell membranes.
            ii. Lecithin is often used as an emulsifying agent.

    2. Sphingolipids and other membrane lipids.
       a. Sphingolipids are derivatives of the amino alcohol sphingosine.
       b. Sphingolipids consist of:
          i. An amide bond between a fatty acid and sphingosine.
          ii. A phosphate ester also bonded to choline.
       c. Sphingomyelins are sphingolipids that are constituents of the coating around nerve fibers.
  B. Glycolipids resemble sphingolipids, except that the phosphate group at C1 is replaced by a carbohydrate.
    1. Cerebrosides, found in cell membranes, have a monosaccharide at C1.
    2. Gangliosides have a small polysaccharide at C1.
       a. Gangliosides are components of neurotransmitters and receptors.
  C. Sterols have a central structure consisting of four fused rings (Section 23.6).
    1. Cholesterol.
       a. Cholesterol serves as a starting material for steroid hormones.
       b. Cholesterol is a component of cell membranes that helps maintain the structure of the membrane.
    2. Bile acids are synthesized from cholesterol and aid in digestion of fats.
    3. Steroid hormones.
       a. Mineralcorticoids regulate fluid balance.
       b. Glucocorticoids regulate glucose metabolism and inflammation.
       c. Sex hormones regulate reproduction and secondary sex characteristics.
IV. Cell membranes (Sections 23.7).
  A. Structure of cell membranes.
    1. Cell membranes are composed of two parallel layers of phospholipids, called a *lipid bilayer*.
    2. The nonpolar tails are clustered in the middle, and the polar heads point inward and outward.
    3. The fluid-mosaic model of a cell membrane explains the more complex structural details.
       a. Glycolipids and cholesterol are in the lipid part and provide membrane structure.
       b. Glycoproteins (20%) mediate the action of cell contents with the outer environment.
          i. Some proteins form channels to allow specific molecules to enter or leave.
          ii. Glycoproteins that extend through the bilayer are called integral proteins.
          iii. Glycoproteins that are partially embedded are peripheral proteins.
          iv. The carbohydrate portions act as receptors for enzymes, proteins, and neurotransmitters.
       c. The membrane is fluid and does not rupture.
          i. Fluidity of the membrane increases with the amounts of saturated and unsaturated fatty acids in the lipid portion.
  B. Transport across cell membranes.
    1. Passive transport.
       a. Simple diffusion.
          i. Gases and small nonpolar molecules diffuse through the bilayer.
          ii. Small hydrophilic molecules pass through channels formed by integral proteins.
       b. Facilitated diffusion.
          i. Facilitated diffusion is passive transport, but proteins help solutes to pass across the membrane.
    2. Active transport.
       a. Ions and other substances that maintain concentration gradients inside and outside the cell move by active transport.
       b. Active transport requires an expenditure of energy and involves changing the shape of an integral protein.

Solutions to Chapter 23 Problems

23.1 The circled groups can be used to identify lipid families.

(a)

An eicosanoid

(b)

$$CH_2OCCH_2(CH_2)_{15}CH_3$$

$$CHOCCH_2(CH_2)_{13}CH_3$$

$$CH_2O-P-O^-$$

A glycerophospholipid

(c)

$$CH_3(CH_2)_{16}C-OCH_2(CH_2)_6CH=CH(CH_2)_6CH_3$$

A wax

23.2

$$CH_3(CH_2)_{18}\overset{O}{\overset{\|}{C}}-OCH_2(CH_2)_{30}CH_3$$

From C_{20} carboxylic acid   From C_{32} alcohol

23.3

$$CH_2-O-\overset{O}{\overset{\|}{C}}-CH_2CH_2CH_2CH_2CH_2CH_2CH_2CH=CHCH_2CH_2CH_2CH_2CH_2CH_2CH_2CH_3$$

$$CH-O-\overset{O}{\overset{\|}{C}}-CH_2CH_2CH_2CH_2CH_2CH_2CH_2CH=CHCH_2CH_2CH_2CH_2CH_2CH_2CH_2CH_3$$

$$CH_2-O-\overset{O}{\overset{\|}{C}}-CH_2CH_2CH_2CH_2CH_2CH_2CH_2CH=CHCH_2CH_2CH_2CH_2CH_2CH_2CH_2CH_3$$

Glyceryl trioleate

23.4 (a) Butter has the largest percentage of saturated fatty acids.
(b) Soybean oil has the largest percentage of polyunsaturated fatty acids.
(c) Soybean oil has the largest percentage of linoleic acid.

CIA 23.1

   In addition to fats and oils, other foods with high lipid content are meat, fish, poultry, dairy products, nuts, seeds, whole-grain cereals, and prepared foods.

CIA 23.2

   According to the FDA, no more than 35% of daily caloric intake should come from fats and oils.

CIA 23.3

   Although the two treats contain a similar number of calories, the ice cream cone contains a higher percentage of both saturated and unsaturated fat. The cookies (36% fat) also have a percentage of fat closer to the nutritional guidelines (35% fat) and are the better choice.

HOC 23.1

   Observations depend on which food was selected. A potato chip makes a greasy spot almost immediately; a cookie takes 15 minutes or so; breakfast cereal can take a while, but a spot may show up after a half hour or so.

23.5

Arachidonic acid

23.6   The starred carbon of the triacylglycerol pictured below is bonded to four different groups and is chiral. (Whenever two different fatty acids are bonded to carbon 1 and carbon 3 of glycerol, carbon 2 is chiral.)

23.7   London forces hold the nonpolar segments of lipids together. These forces are generally weak, but when there are many such interactions, their total contribution is relatively large. Lipids don't mix with water because the hydrogen bonds between water molecules are stronger than the forces between lipid molecules or between water molecules and lipid molecules.

23.8

Glyceryl trioleate (Problem 23.3)

$\downarrow$ 3 H$_2$, catalyst

$$CH_2-O-\overset{\overset{\displaystyle O}{\|}}{C}-CH_2CH_2CH_2CH_2CH_2CH_2CH_2CH_2CH_2CH_2CH_2CH_2CH_2CH_2CH_2CH_2CH_3$$
$$CH-O-\overset{\overset{\displaystyle O}{\|}}{C}-CH_2CH_2CH_2CH_2CH_2CH_2CH_2CH_2CH_2CH_2CH_2CH_2CH_2CH_2CH_2CH_2CH_3$$
$$CH_2-O-\overset{\overset{\displaystyle O}{\|}}{C}-CH_2CH_2CH_2CH_2CH_2CH_2CH_2CH_2CH_2CH_2CH_2CH_2CH_2CH_2CH_2CH_2CH_3$$

Glyceryl tristearate

The acyl groups of the triacylglycerol shown are derived from stearic acid.

23.9 Butter contains cholesterol (unhealthy), and solid margarine may contain trans fatty acids (unhealthy).

23.10

Hydrophobic tail

$$Na^+ \quad {}^-O-\overset{\overset{\displaystyle O}{\|}}{C}-CH_2CH_2CH_2CH_2CH_2CH_2CH_2CH_2CH_2CH_2CH_2CH_2CH_2CH_2CH_2CH_3$$

Hydrophilic head   Sodium stearate

The hydrophilic head is on the exterior of the micelle, where it can interact with water. The hydrophobic tail is on the interior, where it can avoid water.

23.11

$$CH_2-O-\overset{\overset{\displaystyle O}{\|}}{C}-(CH_2)_{16}CH_3$$
$$CH-O-\overset{\overset{\displaystyle O}{\|}}{C}-(CH_2)_{16}CH_3 \qquad or$$
$$CH_2-O-\overset{\overset{\displaystyle O}{\|}}{C}-(CH_2)_7CH{=}CH(CH_2)_7CH_3$$

$$CH_2-O-\overset{\overset{\displaystyle O}{\|}}{C}-(CH_2)_{16}CH_3$$
$$CH-O-\overset{\overset{\displaystyle O}{\|}}{C}-(CH_2)_7CH{=}CH(CH_2)_7CH_3$$
$$CH_2-O-\overset{\overset{\displaystyle O}{\|}}{C}-(CH_2)_{16}CH_3$$

$\downarrow$ NaOH, H$_2$O

$$CH_2OH$$
$$CHOH \quad + \quad 2\ CH_3(CH_2)_{16}-\overset{\overset{\displaystyle O}{\|}}{C}-O^-\ Na^+ \quad + \quad CH_3(CH_2)_7CH{=}CH(CH_2)_7-\overset{\overset{\displaystyle O}{\|}}{C}-O^-\ Na^+$$
$$CH_2OH$$

Glycerol   Sodium stearate   Sodium oleate

23.12 Lecithins emulsify fats in the same way that soaps dissolve grease: The fats are coated by the nonpolar part of a lecithin, and the polar part of lecithins allows fats to be suspended in aqueous solution.

23.13

(a)

R′ is an unsaturated hydrocarbon group, and R is a saturated hydrocarbon group.

(b)

23.14

A sphingomyelin

23.15

Stearic acid acyl group

$CH_2-O-\overset{\overset{O}{\|}}{C}-CH_2CH_2CH_2CH_2CH_2CH_2CH_2CH_2CH_2CH_2CH_2CH_2CH_2CH_2CH_2CH_2CH_2CH_3$

$CH-O-\overset{\overset{O}{\|}}{C}-CH_2CH_2CH_2CH_2CH_2CH_2CH_2CH=CHCH_2CH_2CH_2CH_2CH_2CH_2CH_2CH_3$

Oleic acid acyl group

$CH_2-O-\overset{\overset{}{\underset{O^-}{P}}}{}-OCH_2CH_2NH_3^+$

Phosphate   Ethanolamine

A glycerophospholipid

23.16

$CH_3(CH_2)_{12}-CH=CH-CH-OH$

$R-\overset{\overset{O}{\|}}{C}-NH-CH$

$CH_2-O-\overset{\overset{O}{\|}}{\underset{O^-}{P}}-OCH_2-\overset{\overset{NH_3^+}{|}}{\underset{H}{C}}-COO^-$

The illustrated structure is a phospholipid (a) because it contains a phosphate group. It is a sphingolipid (c) because it has a sphingosine backbone. It is a lipid (e) because it is soluble in nonpolar solvents. It contains a phosphate ester group (f).

23.17 Integral membrane proteins must be hydrophobic to remain in the hydrophobic environment of a cell membrane. They must contain many amino acids with nonpolar side chains and must be folded so that the hydrophilic regions face outward.

CIA 23.4

COOH ← carboxylic acid (most acidic)

C-C double bond

acetal

CH_3

OH ← alcohol

The carboxylic acid group and the hydroxyl group are capable of forming hydrogen bonds. The molecule has both polar and nonpolar regions.

CIA 23.5

Thromboxane A$_2$ Arachidonic acid

Thromboxane A$_2$ is an eicosanoid. Arachidonic acid is the precursor of thromboxane A$_2$, just as it is the precursor of all other eicosanoids.

CIA 23.6

By transferring its acetyl group, aspirin inhibits the enzyme that is responsible for the first step in the conversion of arachidonic acid to prostaglandins.

CIA 23.7

Eicosanoids are called "local hormones" because they are synthesized near their site of action.

CIA 23.8

Prostaglandins can lower blood pressure, assist in blood clotting, stimulate uterine contractions, lower gastric secretions, and cause some of the pain and swelling associated with inflammation.

CIA 23.9

An itchy bump from a mosquito bite (a), and a stuffy nose after working in the rose garden (d) involve prostaglandin responses.

HOC 23.2

A lipid bilayer consists of two parallel layers of lipids. In each layer, the polar heads point outward, and the nonpolar tails congregate in the middle. The membrane is fluid, not rigid, and is not easily ruptured. The bilayer also contains transmembrane proteins, glycolipids, cholesterol, and gated channels.

23.18 You would predict that NO would cross a lipid bilayer by simple diffusion because it is small and relatively nonpolar.

23.19 Glucose 6-phosphate has a charged phosphate group and can't pass through the hydrophobic lipid bilayer.

23.20 The inner and outer surfaces of the lipid bilayer are in different environments and serve different functions. For example, the outer surface must receive signals from the cell's surroundings and must contain groups that allow the cell to be identified. The inner surface must include groups needed to pass messages to other parts of the cell.

Understanding Key Concepts

23.21 The triacylglycerol that has the highest melting point has the greatest percent of saturated fatty acids. Of the four fatty acids listed, palmitic acid and stearic acid are saturated, and oleic acid and linoleic acid are unsaturated. Thus, if you add the percentages in the table, you arrive at a new table.

Triacylglycerol	Percent Saturated Fatty Acids	Percent Unsaturated Fatty Acids
Triacylglycerol A	49.2%	47.5%
Triacylglycerol B	28.9%	70.8%
Triacylglycerol C	19.5%	76.8%

This table shows that triacylglycerol A has the highest melting point. Triacylglycerols B and C are probably liquids at room temperature because their fatty acid composition more closely resembles that of the oils in Table 23.2 than that of the animal fats.

23.22 Three of the fatty acids in the table shown in the previous problem are C_{18} fatty acids (stearic acid, oleic acid, and linoleic acid). When the triacylglycerols are hydrogenated, the unsaturated fatty acids are converted to stearic acid. The composition of the triacylglycerols after hydrogenation can be found by adding the percentages of C_{18} fatty acids.

Triacylglycerol	Percent C_{16} Fatty Acids	Percent C_{18} Fatty Acids
Triacylglycerol A	21.4%	75.3%
Triacylglycerol B	12.2%	87.5%
Triacylglycerol C	11.2%	85.1%

After hydrogenation, triacylglycerol B would be composed of 12.2% palmitic acid and 87.5% stearic acid. The hydrogenation product of triacylglycerol B closely resembles the hydrogenation product of triacylglycerol C because they have similar percentages of C_{16} fatty acids and C_{18} fatty acids.

23.23

The membrane lipid is a glycerophospholipid that is a phosphatidylethanolamine.

23.24 Because the membrane is fluid, rather than rigid, it is able to flow together after an injury instead of shattering.

23.25

$$CH_2-O-\overset{\displaystyle O}{\overset{\|}{C}}-(CH_2)_{14}CH_3$$
$$CH-O-\overset{\displaystyle O}{\overset{\|}{C}}-(CH_2)_{14}CH_3 \qquad \text{DPPC}$$
$$CH_2-O-\overset{\displaystyle O}{\overset{\|}{P}}-O-CH_2CH_2\overset{+}{N}(CH_3)_3$$
$$\overset{\displaystyle |}{\underset{O^-}{}}$$

The two fatty acids are 16-carbon saturated straight-chain fatty acids. The polar head is situated in lung tissue, and the nonpolar groups protrude into the air space within the alveoli.

Additional Problems

Waxes, Fats, and Oils

23.26 A lipid is a naturally occurring organic molecule that dissolves in nonpolar solvents.

23.27 Many kinds of lipids are found in the body; these include glycolipids, phospholipids, sterols, and eicosanoids.

Fatty Acids and Their Esters

23.28 A C_{18} fatty acid is a straight-chain carboxylic acid.

$$CH_3CH_2CH_2CH_2CH_2CH_2CH_2CH_2CH_2CH_2CH_2CH_2CH_2CH_2CH_2CH_2CH_2\overset{\displaystyle O}{\overset{\|}{C}}\diagdown_{OH} \qquad \text{Stearic acid}$$

23.29 The double bonds in this fatty acid cause it to be bent.

$$CH_3CH_2CH_2CH_2CH_2CH_2CH_2CH \quad CH_2 \quad CH_2CH_2CH_2CH_2\overset{\displaystyle O}{\overset{\|}{C}}\diagdown_{OH}$$
$$\underset{H}{\overset{|}{C}}=\underset{H}{\overset{|}{C}} \quad \underset{H}{\overset{|}{C}}=\underset{H}{\overset{|}{C}}$$

23.30 *Saturated fatty acids* are long-chain carboxylic acids that contain no carbon–carbon double bonds.
Monounsaturated fatty acids contain one carbon–carbon double bond.
Polyunsaturated fatty acids contain two or more carbon–carbon double bonds.

23.31 The double bonds in naturally occurring fatty acids are cis.

23.32 An essential fatty acid can't be synthesized by the human body and must be part of the diet.

23.33 Linoleic acid and linolenic acid are two essential fatty acids. Vegetable oils and nuts are good sources of these acids.

23.34 The unsaturated fatty acid (linoleic acid) is lower melting because its double bonds introduce kinks that prevent it from solidifying.

23.35 The reasoning in the previous problem applies here. Stearic acid has a higher melting point because its molecules form a more closely packed solid, and more heat is required to liquify it.

Fats and Oils

23.36 Fats are composed of triacylglycerols containing both saturated and unsaturated fatty acids and are solids at room temperature. Oils are made up of triacylglycerols containing mainly unsaturated fatty acids and are liquids at room temperature.

23.37 Fats are usually obtained from animal sources (butter, lard), and oils are usually obtained from plant sources (peanuts and olives).

23.38

$$CH_2-O-\overset{\overset{\displaystyle O}{\|}}{C}-CH_2CH_2CH_2CH_2CH_2CH_2CH_2CH_2CH_2CH_2CH_3$$
$$CH-O-\overset{\overset{\displaystyle O}{\|}}{C}-CH_2CH_2CH_2CH_2CH_2CH_2CH_2CH_2CH_2CH_2CH_3$$
$$CH_2-O-\overset{\overset{\displaystyle O}{\|}}{C}-CH_2CH_2CH_2CH_2CH_2CH_2CH_2CH_2CH_2CH_2CH_3$$

Glyceryl trilaurate

23.39

$$CH_2-O-\overset{\overset{\textstyle O}{\|}}{C}-CH_2CH_2CH_2CH_2CH_2CH_2CH_2CH_2CH_2CH_2CH_2CH_2CH_2CH_2CH_2CH_3 \quad \longleftarrow \text{Stearic acid}$$

$$CH-O-\overset{\overset{\textstyle O}{\|}}{C}-CH_2CH_2CH_2CH_2CH_2CH_2CH_2CH_2CH_2CH_2CH_2CH_2CH_2CH_2CH_2CH_3 \quad \longleftarrow \text{Stearic acid}$$

$$CH_2-O-\overset{\overset{\textstyle O}{\|}}{C}-CH_2CH_2CH_2CH_2CH_2CH_2CH_2CH_2CH_2CH_2CH_2CH_2CH_2CH_3 \quad \longleftarrow \text{Palmitic acid}$$

$$CH_2-O-\overset{\overset{\textstyle O}{\|}}{C}-CH_2CH_2CH_2CH_2CH_2CH_2CH_2CH_2CH_2CH_2CH_2CH_2CH_2CH_2CH_2CH_3 \quad \longleftarrow \text{Stearic acid}$$

$$CH-O-\overset{\overset{\textstyle O}{\|}}{C}-CH_2CH_2CH_2CH_2CH_2CH_2CH_2CH_2CH_2CH_2CH_2CH_2CH_2CH_3 \quad \longleftarrow \text{Palmitic acid}$$

$$CH_2-O-\overset{\overset{\textstyle O}{\|}}{C}-CH_2CH_2CH_2CH_2CH_2CH_2CH_2CH_2CH_2CH_2CH_2CH_2CH_2CH_2CH_2CH_3 \quad \longleftarrow \text{Stearic acid}$$

In one isomer, palmitic acid forms an ester with a terminal hydroxyl group of glycerol; in the other, palmitic acid forms an ester with the middle hydroxyl group. (The first isomer is chiral.)

23.40 A wax serves as a protective coating for fruits, berries, leaves, animal fur, and feathers.

23.41 In animals, fats are used for energy storage, as components of cell membranes, and as precursors for steroid hormones.

23.42

$$CH_3(CH_2)_{13}CH_2\overset{\overset{\textstyle O}{\|}}{C}-OCH_2(CH_2)_{14}CH_3$$
Cetyl palmitate

23.43 Spermaceti is a wax.

23.44

$$CH_2-O-\overset{\overset{\textstyle O}{\|}}{C}-(CH_2)_nCH_3$$

$$CH-O-\overset{\overset{\textstyle O}{\|}}{C}-(CH_2)_nCH=CH(CH_2)_nCH_3$$

$$CH_2-O-\overset{\overset{\textstyle O}{\|}}{P}-OCH_2CH_2\overset{+}{N}(CH_3)_3 \quad \longleftarrow \text{Choline}$$
$$\underset{O^-}{|}$$

Glycerol   Phosphate   a lecithin
            group

23.45 A lecithin is a glycerophospholipid.

Chemical Reactions of Lipids

23.46 Hydrogenation converts unsaturated fatty acids to saturated fatty acids.

23.47 The partially hydrogenated product contains both cis and trans double bonds, which are not found in nature.

23.48 The reaction is a saponification.

23.49 All double bonds are cis.

$$
\begin{array}{l}
\text{CH}_2\text{-O}-\overset{\displaystyle O}{\overset{\|}{C}}-(\text{CH}_2)_{16}\text{CH}_3 \\[4pt]
\text{CH}-\text{O}-\overset{\displaystyle O}{\overset{\|}{C}}-(\text{CH}_2)_7\text{CH}=\text{CH}(\text{CH}_2)_7\text{CH}_3 \qquad \xrightarrow{\ \text{KOH, H}_2\text{O}\ } \\[4pt]
\text{CH}_2\text{-O}-\overset{\displaystyle O}{\overset{\|}{C}}-(\text{CH}_2)_7\text{CH}=\text{CHCH}_2\text{CH}=\text{CHCH}_2\text{CH}=\text{CHCH}_2\text{CH}_3
\end{array}
$$

CH$_2$OH   K$^+$ $^-$O$-$C(=O)$-$(CH$_2$)$_{16}$CH$_3$ Potassium stearate

CHOH + K$^+$ $^-$O$-$C(=O)$-$(CH$_2$)$_7$CH=CH(CH$_2$)$_7$CH$_3$ Potassium oleate

CH$_2$OH   K$^+$ $^-$O$-$C(=O)$-$(CH$_2$)$_7$CH=CHCH$_2$CH=CHCH$_2$CH=CHCH$_2$CH$_3$

Glycerol   Potassium linolenate

23.50

$$
\begin{array}{l}
\text{CH}_2\text{-O}-\overset{\displaystyle O}{\overset{\|}{C}}-(\text{CH}_2)_{16}\text{CH}_3 \\[4pt]
\text{CH}-\text{O}-\overset{\displaystyle O}{\overset{\|}{C}}-(\text{CH}_2)_7\text{CH}=\text{CH}(\text{CH}_2)_7\text{CH}_3 \qquad \xrightarrow{\ \text{H}_2,\ \text{Pd catalyst}\ } \\[4pt]
\text{CH}_2\text{-O}-\overset{\displaystyle O}{\overset{\|}{C}}-(\text{CH}_2)_7\text{CH}=\text{CHCH}_2\text{CH}=\text{CHCH}_2\text{CH}=\text{CHCH}_2\text{CH}_3
\end{array}
$$

$$
\begin{array}{l}
\text{CH}_2\text{-O}-\overset{\displaystyle O}{\overset{\|}{C}}-(\text{CH}_2)_{16}\text{CH}_3 \\[4pt]
\text{CH}-\text{O}-\overset{\displaystyle O}{\overset{\|}{C}}-(\text{CH}_2)_{16}\text{CH}_3 \qquad \text{Glyceryl tristearate} \\[4pt]
\text{CH}_2\text{-O}-\overset{\displaystyle O}{\overset{\|}{C}}-(\text{CH}_2)_{16}\text{CH}_3
\end{array}
$$

Glyceryl tristearate is higher melting than the original lipid because it has no double bonds and can solidify more easily.

23.51 (a) If one double bond were converted to a single bond, four different products would result (since there are four double bonds).
(b) There are six different products that can result when two double bonds are hydrogenated.
(c) If three double bonds were converted to single bonds, four different products would result.
(d) Glyceryl tristearate (pictured above) is the sole product of hydrogenation of all double bonds.

23.52 (a), (b) Margarine contains more monounsaturated and polyunsaturated fatty acids.
(c) Butter contains fewer trans fatty acids.

23.53 Peanut oil and olive oil are the best sources of monounsaturated fatty acids.

Phospholipids and Glycolipids

23.54 A triacylglycerol is an ester of glycerol and three fatty acids; a phospholipid contains a phosphate group. There are two main kinds of phospholipids. A *glycerophospholipid* is an ester of glycerol 3-phosphate, two fatty acids, and an amino alcohol bonded to the phosphate group. A *sphingolipid* is an amide of sphingosine and a fatty acid, and has an amino alcohol bonded to a phosphate group, which is attached to C1 of sphingosine.

23.55 Glycerophospholipids have an ionic part (the *head*) and a nonpolar part (the *tail*). The ionic head protrudes outward toward the aqueous environment of the cell or inward toward the cell contents, and the nonpolar tails cluster together to form the membrane. Triacylglycerols don't have an ionic head and thus cannot function as membrane components.

23.56 A sphingomyelin and a cerebroside are similar in that both have a sphingosine backbone. The difference between the two occurs at C1 of sphingosine. A sphingomyelin has a phosphate group bonded to an amino alcohol at C1; a cerebroside has a glycosidic link to a monosaccharide at C1.

23.57 Sphingomyelins and glycolipids are two different kinds of sphingosine-based lipids.

23.58 Glycerophospholipids are more soluble in water than triacylglycerols because they have an ionic phosphate group that is solvated by water.

23.59 Glycerophospholipids are components of cell membranes. Stored fats in the body are triacylglycerols.

23.60

23.61

$$CH_2-O-\overset{\overset{O}{\|}}{P}-OCH_2CH_2\overset{+}{N}(CH_3)_3$$

Phosphate   Choline

$\overset{|}{O^-}$

Stearic acid

$CH-NH-\overset{\overset{\|}{O}}{C}-CH_2CH_2CH_2CH_2CH_2CH_2CH_2CH_2CH_2CH_2CH_2CH_2CH_2CH_2CH_2CH_2CH_3$

$CH-OH$

$CH=CHCH_2CH_2CH_2CH_2CH_2CH_2CH_2CH_2CH_2CH_2CH_2CH_2CH_3$

Sphingosine   A sphingomyelin

23.62

$$CH_2-O-\overset{\overset{O}{\|}}{C}-(CH_2)_{14}CH_3 \longleftarrow \text{Palmitic acid}$$

$$CH-O-\overset{\overset{O}{\|}}{C}-(CH_2)_7CH=CH(CH_2)_7CH_3 \longleftarrow \text{Oleic acid}$$

$$CH_2-O-\overset{\overset{O}{\|}}{P}-OCH_2CH_2CH_2NH_3^+ \longleftarrow \text{Propanolamine}$$

$$\overset{|}{O^-}$$

Glycerol   Phosphate
group

A glycerophospholipid

23.63 Aqueous NaOH cleaves all ester bonds, including phosphate esters.

Cardiolipin

$$\downarrow \text{NaOH, H}_2\text{O}$$

$$RC\overset{O}{\|}-O^-Na^+ + R'C\overset{O}{\|}-O^-Na^+ + 3\ HOCH_2CHCH_2OH + 2\ Na^+\ ^-O-\overset{O^-Na^+}{\underset{O^-Na^+}{P}}-O^-Na^+$$

$$+$$

$$R''C\overset{O}{\|}-O^-Na^+ + R'''C\overset{O}{\|}-O^-Na^+$$

Sterols

23.64 Cholesterol is a component of cell membranes and is the starting material for the synthesis of steroid hormones and bile acids.

23.65 The function of bile acids is to emulsify triacylglycerols during digestion.

23.66 Male sex hormones: androsterone, testosterone.
Female sex hormones: estradiol, estrone.

23.67 The fused tetracyclic ring system is the same, but the functional groups on the rings differ.

Cell Membranes

23.68 In a micelle, the polar hydrophilic heads are on the exterior, and the hydrophobic tails cluster in the center. In a membrane bilayer, hydrophilic heads are on both the exterior and interior surfaces of the membrane, and the region between the two surfaces is occupied by hydrophobic tails.

23.69 Both liposomes and micelles are spherical clusters of lipids. A liposome resembles a spherical lipid bilayer, in which polar heads cluster both inside and outside of the sphere. Consequently, water-soluble substances can be trapped in the center. Liposomes can be used for drug delivery; a water-soluble drug can be carried to its target site without interfering with surrounding tissue. The liposome can fuse with the cell membrane, emptying its contents into the target cell.

23.70 Glycolipids, cholesterol, proteins, and glycoproteins are present in cell membranes in addition to phospholipids.

23.71 If cell membranes were freely permeable, the concentrations of all substances would be the same on both sides of the cell membrane, and it would be impossible for cells to maintain concentration gradients.

23.72 Active transport requires energy because it is a process in which substances are transported across a membrane in a direction opposite to their tendency to diffuse.

23.73 In both simple diffusion and facilitated diffusion, substances cross cell membranes from areas of high concentration to areas of low concentration. In facilitated diffusion, proteins help specific substances cross cell membranes.

23.74 (a) NO moves into cells by simple diffusion.
(b) Fructose, like glucose, is transported into cells by facilitated diffusion.
(c) Ca^{2+} probably crosses a cell membrane through protein pores. (It is also possible that active transport is used.)

23.75 (a) Galactose is transported by facilitated diffusion.
(b) CO moves into cells by simple diffusion.
(c) Mg^{2+} crosses a cell membrane by active transport.

Conceptual Problems

23.76 Glyceryl trioleate (b), a sphingomyelin (c), a cerebroside (e), and a lecithin (f) are saponifiable lipids.

23.77

Lipid	Backbone	Groups Attached
(b) Glyceryl trioleate	Glycerol	Three oleic acids
(c) A sphingomyelin	Sphingosine	A fatty acid, a phosphate group, choline, or ethanolamine
(e) A cerebroside	Sphingosine	A fatty acid, a monosaccharide
(f) A lecithin	Glycerol	A saturated fatty acid, an unsaturated fatty acid, a phosphate group, choline

23.78

$$\begin{array}{l} CH_2-O-\overset{\overset{\displaystyle O}{\parallel}}{C}-(CH_2)_{12}CH_3 \\ | \qquad\qquad \overset{\displaystyle O}{\parallel} \\ CH-O-C-(CH_2)_7CH=CHCH_2CH=CHCH_2CH=CHCH_2CH_3 \\ | \qquad\qquad \overset{\displaystyle O}{\parallel} \\ CH_2-O-C-(CH_2)_{12}CH_3 \end{array}$$

or

$$\begin{array}{l} CH_2-O-\overset{\overset{\displaystyle O}{\parallel}}{C}-(CH_2)_{12}CH_3 \\ | \qquad\qquad \overset{\displaystyle O}{\parallel} \\ CH-O-C-(CH_2)_{12}CH_3 \\ | \qquad\qquad \overset{\displaystyle O}{\parallel} \\ CH_2-O-C-(CH_2)_7CH=CHCH_2CH=CHCH_2CH=CHCH_2CH_3 \end{array}$$

23.79 The fat described in the previous problem is slightly lower melting than a fat composed of linolenic, myristic, and stearic acids, because myristic acid is slightly lower melting than stearic acid.

23.80 (a) tallow = beef fat   (b) cooking oil = plant oil   (c) lard = pork fat

23.81 Cholesterol is not saponifiable because it contains no ester linkages.

23.82 Cholesterol acetate is saponifiable because it has an ester bond.

Cholesterol acetate

23.83 Sphingomyelins, cerebrosides, and gangliosides are abundant in brain tissue.

23.84 Sphingomyelins serve as coatings for nerve fibers and are present in brain tissue.

23.85 A decrease in sphingomyelin is observed in multiple sclerosis.

23.86

Molar mass of fat = 1500 g
Molar mass of NaOH = 40 g

$$5.0 \text{ g oil} \times \frac{1 \text{ mol oil}}{1500 \text{ g}} \times \frac{3 \text{ mol NaOH}}{1 \text{ mol oil}} \times \frac{40 \text{ g}}{1 \text{ mol NaOH}} = 0.40 \text{ g NaOH}$$

23.87

$$\frac{200 \text{ mg cholesterol}}{1 \text{ dL}} \times \frac{10 \text{ dL}}{1 \text{ L}} \times 5.75 \text{ L} \times \frac{1 \text{ g}}{1000 \text{ mg}} = 11.5 \text{ g cholesterol}$$

An average person has approximately 11–12 grams of cholesterol in his or her blood.

23.88

$CH_3(CH_2)_{16}C$ — $OCH_2(CH_2)_{20}CH_3$ Jojoba wax
From stearic acid   From a C_{22} alcohol

Jojoba wax is the ester formed by a C_{22} alcohol and a C_{18} carboxylic acid. Spermaceti is the ester formed by a C_{18} alcohol and a C_{16} carboxylic acid. It might be possible to substitute jojoba wax for spermaceti in the cosmetic industry as long as its greater molar mass and resulting higher melting point were unimportant.

Self-Test for Chapter 23

Multiple Choice

1. Small polar molecules pass through the lipid bilayer by:
    (a) simple diffusion through the lipid bilayer (b) facilitated transfer
    (c) active transport   (d) simple diffusion through channels

2. Cholesterol is:
    (a) a male sex hormone (b) a saponifiable lipid (c) a component of cell membranes
    (d) common to plants and animals

3. Which of the following is not an effect of prostaglandins?
    (a) stimulation of allergic response   (b) lowering of blood pressure
    (c) reduction of gastric secretions   (d) stimulation of uterine contractions

4. Active transport requires all of the following except:
    (a) the solutes to be nonpolar organic molecules   (b) the expenditure of ATP
    (c) the need for cells to maintain a concentration gradient between the inside and the outside
    (d) the assistance of integral proteins

5. Which of the following contain a glycosidic bond?
    (a) triacylglycerols (b) sphingomyelins (c) lecithins (d) cerebrosides

6. The basic hydrolysis of which of the following triacylglycerols produces only polyunsaturated fatty acids?
    (a) glyceryl trioleate (b) glyceryl tristearate (c) glyceryl linoleate dioleate
    (d) glyceryl trilinolenate

7. How many different triacylglycerols (including enantiomers!) can be formed from glycerol, two stearic acids, and one oleic acid?
    (a) 1 (b) 2 (c) 3 (d) 4

8. Which of the following molecules aren't components of glycolipids?
    (a) sphingosine (b) phosphate groups (c) fatty acids (d) sugars

9. A cell membrane is composed of all of the following except:
    (a) triacylglycerols (b) phosphoglycerides (c) sphingomyelins (d) glycolipids

10. Waxes:
    (a) are triacylglycerols   (b) can be hydrolyzed by base
    (c) contain no more than 30 carbons   (d) are only produced by plants

Sentence Completion

1. A glycolipid contains an _____ link between sphingosine and a sugar.

2. _____ is a mixture of long-chain fatty acid salts.

3. The carboxylate end of a fatty acid is _____ and the organic chain end is _____.

4. _____ and _____ are two components of cell membranes.

5. The common model of a cell membrane is called the _____ _____ model.

6. Steroid structures are based on a _____ ring system.

7. Prostaglandins are synthesized in the body from a fatty acid called _____ acid.

8. Clusters of soap molecules in water are called _____.

9. _____ are phospholipids that are abundant in brain tissue.

10. Phospholipids aggregate in a closed, sheetlike membrane called a _____ _____.

11. _____ are a group of compounds derived from 20-carbon unsaturated fatty acids.

12. _____ proteins are involved with active transport across cell membranes.

True or False

1. Lipids are defined by their physical properties, not by their structure.

2. Saturated fats are lower melting than unsaturated fats.

3. Sphingolipids contain a carboxylic acid ester group.

4. Cholesterol is synthesized by the human body.

5. Sphingosine is a component of both phospholipids and glycolipids.

6. When soap molecules are dissolved in water, they form a lipid bilayer.

7. Cholesterol is a hormone.

8. The main difference between fats and oil is in their melting points.

9. Cerebrosides are a major constituent of the coating around nerve fibers.

10. Waxes and fats are both carboxylic acid esters.

11. Leukotrienes are synthesized in the cells where they act.

12. Facilitated diffusion requires an energy investment.

Match each entry on the left with its partner on the right.

1. Decyl stearate

2. Prostaglandin

3. Choline

4. Stearic acid

5. Integral protein

6. Glyceryl trilaurate

7. Cerebroside

8. Oleic acid

9. Lecithin

10. Linoleic acid

11. Sphingomyelin

12. Peripheral protein

(a) Extends completely through the cell membrane

(b) Monounsaturated fatty acid

(c) Partially embedded in the cell membrane

(d) Phosphoglyceride

(e) Amino alcohol

(f) Saturated fatty acid

(g) Polyunsaturated fatty acid

(h) C_{20} acid with a five-membered ring

(i) Sphingolipid

(j) Fat

(k) Glycolipid

(l) Wax

Chapter 24   Lipid Metabolism

Chapter Outline

I. Triacylglycerols (Sections 24.1–24.4).
  A. Digestion of triacylglycerols (Section 24.1).
    1. Triacylglycerols (TAGs) pass through the mouth unchanged.
    2. In the stomach, TAGs are broken down into small droplets.
    3. As TAGs pass into the intestines, lipases and bile are produced.
      a. Bile acids solubilize lipid droplets.
      b. Lipases hydrolyze lipids to mono- and diacylglycerols, fatty acids, and glycerol.
    4. Products of TAG digestion are absorbed through the intestinal wall.
      a. Smaller molecules are transported to the liver.
      b. Free fatty acids and acylglycerols are reconverted to triacylglycerols and packaged into lipoproteins (chylomicrons).
  B. Lipoproteins for lipid transport (Section 24.2).
    1. Triacylglycerols and fatty acids enter metabolism from three sources:
      a. Diet.
      b. Storage in adipose tissue.
      c. Synthesis in the liver.
    2. These lipids must be made soluble by association with lipoproteins.
    3. A lipoprotein is a globule of triacylglycerols and fatty acids surrounded by a layer of phospholipids.
    4. Five kinds of lipoproteins are important.
      a. Chylomicrons carry lipids in the diet through the lymphatic system into the blood.
      b. VLDLs carry TAGs from the liver to tissues for storage or energy production.
      c. IDLs carry remnants of VLDLs from peripheral tissues back to the liver for use in synthesis.
      d. LDLs carry cholesterol from the liver to tissues.
      e. HDLs carry cholesterol from cells to the liver, where it is converted to bile acids.
  C. Metabolism of triacylglycerols (Section 24.3–24.4).
    1. Overview (Section 24.3).
      a. Triacylglycerols are hydrolyzed.
        i.   TAGs in the diet are hydrolyzed to glycerol and fatty acids by lipoprotein lipase in capillary walls in adipose tissue.
        ii. TAGs in adipocytes are hydrolyzed within the cells, and the fatty acids travel into the bloodstream with albumins.
      b. Glycerol is carried to the liver or kidneys.
        i.   Glycerol is converted to dihydroxyacetone phosphate, which enters glycolysis or gluconeogenesis.
      c. Fatty acids have two fates.
        i.   When energy is in good supply, fatty acids are converted to TAGs for storage.
        ii. When energy is needed, fatty acids are metabolized to acetyl-CoA.
      d. Acetyl-CoA is used for:
        i.   Generation of energy via the citric acid cycle and oxidative phosphorylation.
        ii. Biosynthesis of fatty acids.
        iii. Production of ketone bodies.
        iv. Synthesis of sterols.

2. Storage and mobilization of triacylglycerols (Section 24.4).
   a. When blood glucose levels are high, TAGs are synthesized for storage.
      i. The glycerol 3-phosphate needed is made from dihydroxyacetone phosphate, which comes from glycolysis.
      ii. The fatty acyl-CoAs needed come from digestion or biosynthesis.
      iii. Two fatty acyl-CoA are added to glycerol 3-phosphate.
      iv. After removal of phosphate from phosphatidic acid, the third fatty acyl-CoA is added.
   b. When blood glucose levels are low, fatty acids and glycerol are released from adipocytes and enter the bloodstream.

II. Fatty acid oxidation (Sections 24.5–24.6).
   A. Steps in fatty acid oxidation (Section 24.5).
      1. Activation.
         a. Fatty acids react with HSCoA to form fatty acyl-CoA.
         b. At the same time, ATP is converted to AMP + 2 $HOPO_3^{2-}$.
      2. Transport.
         a. Fatty acyl-CoAs react with carnitine to form fatty acyl-carnitine, which is transported across the mitochondrial membrane.
         b. Fatty acyl-CoA is regenerated.
      3. Oxidation (β oxidation).
         a.

         b.

         c.

         d.

         e. An n-carbon fatty acid produces $n/2$ acetyl-CoA.
   B. Energy from fatty acid oxidation.
      1. The number of molecules of ATP produced from a fatty acid that yields n acetyl-CoA molecules is $14n - 6$.
      2. Fats produce more than twice the amount of energy per unit weight than carbohydrates do.
   C. Production of ketone bodies (ketogenesis) (Section 24.6).
      1. Ketone bodies are formed when the body produces excess acetyl-CoA.
      2. Synthesis of ketone bodies.
         a. 2 Acetyl-CoA —> Acetoacetyl-CoA + ⁻SCoA.
         b. Acetoacetyl-CoA + Acetyl-CoA —> 3-Hydroxy-3-methylglutaryl-CoA + HSCoA.
         c. 3-Hydroxy-3-methylglutaryl-CoA —> Acetyl-CoA + Acetoacetate.
         d. Acetoacetate + NADH/H⁺ —> 3-Hydroxybutyrate + NAD⁺.
         e. Acetoacetate —> Acetone + CO_2.

3. Catabolism of ketone bodies produces much of the energy that the body needs during starvation.
4. When ketone bodies are formed faster than the body can utilize them, ketoacidosis occurs.
   a. In ketoacidosis, blood pH drops, ketone bodies are excreted in the urine, and coma and death may result.

III. Fatty acid biosynthesis (lipogenesis) (Section 24.7).
A. Lipogenesis occurs when amino acids and carbohydrates are in good supply.
B. Lipogenesis resembles fatty acid oxidation but is not its exact reverse.
   1. Lipogenesis occurs in the cytosol, and intermediates are carried by ACP.
C. Mechanism of lipogenesis.
   1. Condensation.
      a. Acetyl-CoA + HCO_3^- —> H_2O + malonyl-CoA.
      b. Malonyl-CoA + HSACP —> malonyl-ACP + HSCoA.
      c. Acetyl-CoA + HSACP —> acetyl-ACP + HSCoA.
      d. Acetyl-ACP + malonyl-ACP —> acetoacetyl-ACP + SACP + CO_2.
   2. Reduction, using NADPH/H^+.
   3. Dehydration.
   4. Reduction, using the coenzyme NADPH/H^+.
   5. Addition of acetyl-ACP units, followed by Steps 2–5, occurs until the chain is of the correct length.

Solutions to Chapter 24 Problems

24.1 Both cholate and cholesterol have the tetracyclic ring structure of a sterol. Cholate has more polar functional groups (three —OH groups and one carboxylate group) and is better able to emulsify fats than cholesterol (one —OH group). The hydrophobic side of cholate interacts with TAGs (triacylglycerols), and the hydrophilic side interacts with the aqueous environment. The less polar cholesterol is more suitable for incorporation into cell membranes. Thus, the two molecules can't change roles.

24.2 The enzyme *triose phosphate isomerase* catalyzes the isomerization of dihydroxyacetone phosphate to glyceraldehyde 3-phosphate, which proceeds through the last five steps of glycolysis to yield pyruvate.

24.3 Fatty acids released from triacylglycerols have two fates: (1) When energy from their metabolism is not needed, they are reformed into triacylglycerols and transported by VLDLs for storage in adipose tissue; and (2) If energy is needed, they are converted to fatty acyl-CoA and oxidized to acetyl-CoA. Fatty acids from storage in adipose tissue are transported by serum albumin.

24.4 (a), (b) In step 1, a C=C double bond is introduced, using FAD as the oxidizing agent. In step 3, an alcohol is oxidized to a ketone, using NAD$^+$ as the oxidizing agent.
(c) In step 2, water is added to a carbon–carbon double bond.
(d) In step 4, HSCoA displaces acetyl-CoA, producing a chain-shortened acyl-CoA fatty acid.

24.5

(a)

Seven β oxidations are needed.
(b) Since lignoceric acid has 24 carbons, 12 acetyl-CoAs are produced as a result of 11 β oxidations.

24.6 In the citric acid cycle, steps 6 (dehydrogenation of succinate to produce fumarate), 7 (hydration of fumarate to produce malate), and 8 (oxidation of malate to produce oxaloacetate) are similar to the first three reactions of the β oxidation of a fatty acid.

24.7 Stearic acid (a C_{18} fatty acid) produces 9 acetyl-CoA from 8 β oxidations.
Each β oxidation produces 5 ATPs:

$$8 \ \beta \ \text{oxidations} \ \times \ 5 \ \text{ATP}/\beta \ \text{oxidation} \ = \ 40 \ \text{ATP}$$

The 9 acetyl-CoAs each yield 12 ATP from passage through the citric acid cycle:

$$9 \ \text{acetyl-CoA} \ \times \ 12 \ \text{ATP/cycle} = 108 \ \text{ATP}$$

Two ATP are consumed when stearic acid is activated: $\qquad = \ \underline{-2 \ \text{ATP}}$
$$146 \ \text{ATP}$$

A total of 146 ATPs are produced from the complete oxidation of stearic acid.

CIA 24.1
Important functions of the liver include:
(a) Synthesis of glycogen, glucose, triacylglycerols, fatty acids, cholesterol, bile acids, plasma proteins, and blood clotting factors.
(b) Catabolism of glucose, fatty acids, and amino acids.
(c) Storage of glycogen, lipids, amino acids, iron, and fat-soluble vitamins.
(d) Inactivation of toxic substances.

CIA 24.2
Cirrhosis is the development of fibrous tissue in the liver and is due to excessive deposits of triacylglycerols as a result of alcoholism or metabolic disorders.

CIA 24.3
The liver stores nutrients, synthesizes important biomolecules, regulates the concentration of substances in the blood, and inactivates toxins.

24.8 The formation of 3-hydroxybutyrate from acetoacetate is a reduction (d).

24.9 (a) Acetyl-CoA provides the acetyl groups used in the synthesis of the ketone bodies. (b) Three acetyl-CoA molecules are used. (c) During prolonged starvation, the body, including the brain, uses ketone bodies as an energy source.

24.10 Seven more malonyl-ACP molecules are needed, and eight CO_2 are released.

HOC 24.1

1. A heart attack takes place when blood flow to a part of the heart stops, causing damage to heart muscle.
2. Symptoms in men: chest discomfort; pain in one or both arms, the back, neck, jaw, or stomach; shortness of breath; cold sweat; nausea; lightheadedness.
   Symptoms in women: similar to men but more likely to involve shortness of breath, nausea or vomiting, and back or jaw pain.
3. Heart attacks can cause the heart to stop beating; damage to heart muscles can prevent the heart from supplying enough blood to tissues; heart muscles, walls, or valves can rupture. All of these problems can lead to death.
4. Coronary bypass surgery is used to redirect blood flow from areas of the heart where arteries have narrowed. A blood vessel is harvested from another area of the body, one end is attached to the aorta, and the other end is attached to the heart near the blocked artery.

CIA 24.4

Obese people are in danger of developing Type II diabetes, colon cancer, heart attacks, and stroke.

CIA 24.5

Calorie-dense food and lack of exercise also lead to fat accumulation.

CIA 24.6

Total cholesterol: 200 mg/dL or lower.
            LDL: 100 mg/dL or lower.
           HDL: 60 mg/dL or higher.

CIA 24.7

Atherosclerosis is the formation of deposits composed of cholesterol and other lipid-containing material in the arteries of the brain and heart.

CIA 24.8

Arterial plaque is a deposit of cholesterol and other lipid-containing material that forms within larger arteries. A high HDL value indicates that cholesterol is being removed to the liver, where it is converted to bile acids. A high LDL indicates that an excess of cholesterol is being transported to tissues, where it can form arterial plaque.

Understanding Key Concepts

24.11 The NADH and $FADH_2$ produced in β oxidation must be reoxidized in the electron transport system in order for β oxidation to continue. Since oxygen is required to reoxidize these reduced coenzymes in the electron transport system, β oxidation is an aerobic pathway even though oxygen isn't directly involved.

24.12 (a) Chylomicrons have the lowest density because they have the highest percentage of lipids, and lipids are less dense than protein.
(b) Chylomicrons carry triacylglycerols from the diet.
(c) HDL removes cholesterol from circulation and transports it to the liver, where it is converted to bile acids and excreted.
(d) LDL contains "bad" cholesterol, which is deposited in tissues and contributes to atherosclerosis.
(e) HDL has the highest ratio of protein to lipid.
(f) VLDL carries triacylglycerols from the liver to peripheral tissue, where they are used either for storage or for energy generation.
(g) LDL carries cholesterol from the liver to peripheral tissue.

24.13

A	B	C

high blood glucose ⤬ high glucagon/low insulin ⤬ fatty acid and TAG synthesis

low blood glucose ↘ high insulin/low glucagon ↗ TAG hydrolysis and fatty acid oxidation

24.14 Fatty acid + HSCoA + ATP $\longrightarrow$ Fatty acyl-CoA + AMP + $P_2O_7^{4-}$
The above reaction, which is used to activate fatty acids for β oxidation, requires the energy expenditure of the equivalent of two ATPs. This energy is later recaptured in β oxidation of the fatty acid.

24.15 When oxaloacetate in liver is used for gluconeogenesis, less oxaloacetate is available for the citric acid cycle, and less acetyl-CoA is able to enter the cycle. Instead, there is a buildup of acetyl-CoA in the bloodstream. Heart and muscle use the excess acetyl-CoA for synthesis of ketone bodies. Buildup of ketone bodies leads to ketoacidosis, in which blood pH is lowered and the odor of acetone is noticeable in urine and on the breath. Ketoacidosis can lead to dehydration, labored breathing, and death.

24.16 Catabolism of one gram of stored triacylglycerols provides more energy than does catabolism of one gram of glycogen.

24.17 During starvation, glucose is in short supply, and the body responds by producing ketone bodies, which can be metabolized to acetyl-CoA, as a source of energy.

24.18 Table 24.1 summarizes the two pathways, which differ in several ways. (1) Intermediates of oxidation are carried by coenzyme A, whereas intermediates of lipogenesis are carried by acyl carrier protein. (2) The two processes use different enzymes and coenzymes. (3) Oxidation occurs in mitochondria, and synthesis occurs in cytosol.

Additional Problems

Digestion of Lipids

24.19 Lipids make you feel full because they slow the rate of movement of food through the stomach.

24.20 Digestion of triacylglycerols occurs in the small intestine, where pancreatic lipase partially hydrolyzes emulsified triacylglycerols. Short-chain fatty acids are absorbed into the bloodstream. In the intestinal lining, longer-chain fatty acids and acylglycerols are recombined to form triacylglycerols, which are bound in chylomicrons that travel through the lymphatic system to enter the bloodstream.

Hydrolysis of stored triacylglycerols occurs in fat cells. Hydrolysis of triacylglycerols from food occurs at capillary walls of adipose tissue, muscles, and liver.

24.21 Bile emulsifies lipid droplets so that they can be attacked by enzymes.

24.22 Bile acids are synthesized in the liver from cholesterol.

24.23

24.24 The products of pancreatic lipase hydrolysis of triacylglycerols (TAGs) are mono- and diacylglycerols, plus fatty acids and glycerol. The products of lipoprotein lipase (in adipocytes) are fatty acids and glycerol.

Lipid Transport

24.25 Acylglycerols, fatty acids, and protein are combined to form *chylomicrons,* which are lipoproteins used to transport lipids from the diet into the bloodstream.

24.26 Very-low-density lipoproteins originate in the liver and are used to transport triacylglycerols that have been synthesized in the liver to tissues where they will be used or stored.

24.27 Fatty acids from adipose tissue are carried in the bloodstream by albumins.

24.28 Cholesterol is transported by LDLs to peripheral tissue, where it is used in cell membranes and to synthesize sterols.

Overview of Triacylglycerol Storage and Metabolism

24.29 Steps 6–10 of the glycolysis pathway are necessary for converting glyceraldehyde 3-phosphate to pyruvate. These steps include an oxidation (step 6), a phosphorylation (step 6), two phosphate transfers to ADP (steps 7 and 10), an isomerization (step 8), and a dehydration (step 9).

24.30 For each molecule of glycerol, six molecules of ATP are released in forming pyruvate. One ATP results from the conversion of glycerol to glyceraldehyde-3-phosphate, and five more arise from the conversion of glyceraldehyde-3-phosphate to pyruvate (two substrate-level phosphorylations, which produce two ATP, and production of one NADH, which enters the electron transport chain to produce three more ATP).

24.31 Nine molecules of ATP are formed in the catabolism of glycerol to yield acetyl-CoA—the six mentioned in Problem 24.30 and an additional three in the conversion of pyruvate to acetyl-CoA (Section 22.5).
A maximum of 21 molecules of ATP are released in the complete catabolism of glycerol to CO_2 and H_2O – the nine molecules mentioned above and an additional 12 molecules formed when acetyl-CoA is metabolized in the citric acid cycle.

24.32

Glyceryl trilaurate

Glycerol is converted to glyceraldehyde 3-phosphate, which enters glycolysis, is converted to pyruvate and yields one acetyl-CoA. Each laurate produces 6 acetyl-CoA molecules; three laurates yield 18 acetyl-CoA molecules. A total of 19 acetyl-CoA molecules are thus formed.

24.33 An adipocyte is a cell, almost entirely filled with fat globules, in which triacylglycerols are stored and mobilized.

24.34 Adipose tissue is located under the skin and in the abdominal cavity and is used for storage and mobilization of triacylglycerols.

Oxidation of Fatty Acids

24.35 Fatty acid oxidation occurs primarily in heart, liver, and muscle cells.

24.36 β Oxidation occurs in the mitochondrial matrix of cells.

24.37 A fatty acid is converted to its fatty acyl-CoA in order to activate it for catabolism.

24.38 The activated fatty acid must be transported from the cytosol into the mitochondrial matrix.

24.39 Stepwise oxidation of fatty acids is known as β oxidation because the carbon β to the thioester group (two carbons away from the thioester group) is oxidized in the process.

24.40 The sequence is a spiral because the same reaction series is repeated on a two-carbon-shortened fatty acid until the original acid is consumed. In a cycle, the product of the final step is a reactant in the first step.

24.41 The coenzymes NAD^+ and FAD are required for β oxidation.

24.42 No. The coenzyme NADPH replaces NAD^+ and FAD in fatty acid synthesis.

24.43 Each cycle of β oxidation produces five molecules of ATP (three ATPs from the formation of the reduced coenzyme NADH and two ATPs from the formation of the reduced coenzyme $FADH_2$), plus one molecule of acetyl-CoA, which enters into the citric acid cycle to yield 12 more molecules of ATP, for a total of 17.

24.44 In the previous problem, we calculated that each cycle of β oxidation produced 17 ATP. Since the last acetyl-CoA doesn't undergo β oxidation, five fewer ATP are produced from that cycle. Two ATP are also used for activation, and when they are subtracted, we arrive at the formula:
ATP produced per fatty acid = $17n - 7$, where n = the number of acetyl-CoA produced.

For a 14-carbon acid, $n = 7$; thus, the energy content of $CH_3(CH_2)_{12}COOH$ (lauric acid) is a maximum of 112 ATP per mole.

24.45

Least energy/mole			⟶	*Greatest energy/mole*
Molecule	Glucose	Sucrose	$CH_3(CH_2)_8COOH$	$CH_3(CH_2)_{12}COOH$
ATP content/mole	38 ATP	76 ATP	78 ATP	112 ATP

24.46

Least energy/mole			⟶	*Greatest energy/mole*
Molecule	Mannose	Fructose	$CH_3(CH_2)_{14}COOH$	$CH_3(CH_2)_{16}COOH$
ATP content/mole	38 ATP	38 ATP	129 ATP	146 ATP

Depending on the tissue and point of entry into glycolysis, the values for mannose and fructose may be smaller than this.

24.47

(a)

$$CH_3CH_2CH_2CH_2CH_2\overset{\overset{\displaystyle O}{\|}}{C}SCoA \xrightarrow[\substack{\text{Acyl-CoA} \\ \text{dehydrogenase}}]{\text{FAD} \quad \text{FADH}_2} CH_3CH_2CH_2CH=CH\overset{\overset{\displaystyle O}{\|}}{C}SCoA$$

(b)

$$CH_3CH_2CH_2CH=CH\overset{\overset{\displaystyle O}{\|}}{C}SCoA + H_2O \xrightarrow[\text{hydratase}]{\text{Enoyl-CoA}} CH_3CH_2CH_2\overset{\overset{\displaystyle OH}{|}}{C}HCH_2\overset{\overset{\displaystyle O}{\|}}{C}SCoA$$

(c)

$$CH_3CH_2CH_2\overset{\overset{\displaystyle OH}{|}}{C}HCH_2\overset{\overset{\displaystyle O}{\|}}{C}SCoA \xrightarrow[\substack{\beta\text{-Hydroxyacyl-CoA} \\ \text{dehydrogenase}}]{\text{NAD}^+ \quad \text{NADH/H}^+} CH_3CH_2CH_2\overset{\overset{\displaystyle O}{\|}}{C}CH_2\overset{\overset{\displaystyle O}{\|}}{C}SCoA$$

(d)

$$CH_3CH_2CH_2\overset{\overset{\displaystyle O}{\|}}{C}CH_2\overset{\overset{\displaystyle O}{\|}}{C}SCoA \xrightarrow[\substack{\text{Acetyl-CoA} \\ \text{transferase}}]{\text{HSCoA}} CH_3CH_2CH_2\overset{\overset{\displaystyle O}{\|}}{C}SCoA + CH_3\overset{\overset{\displaystyle O}{\|}}{C}SCoA$$

24.48

$$CH_3\overset{\overset{\displaystyle O}{\|}}{C}CH_2\overset{\overset{\displaystyle O}{\|}}{C}SCoA \xrightarrow[\substack{\text{Acetyl-CoA} \\ \text{transferase}}]{\text{HSCoA}} 2\ CH_3\overset{\overset{\displaystyle O}{\|}}{C}SCoA$$

24.49 The number of molecules of acetyl-CoA produced from one molecule of fatty acid is half the number of carbons in the acid. The number of cycles of β oxidation is one fewer than the number of molecules of acetyl-CoA produced.

Acid	Acetyl-SCoA Produced	Number of Cycles
(a) $CH_3(CH_2)_{12}COOH$	7	6
(b) $CH_3(CH_2)_6COOH$	4	3

24.50 Three cycles of β oxidation catabolize caprylic acid (a C_8 acid), and six cycles of β oxidation catabolize myristic acid (a C_{14} acid).

Ketone Body Production

24.51 Acetone, acetoacetate, and 3-hydroxybutyrate are ketone bodies, which are produced when blood sugar is low and the body metabolizes fats. If more acetyl-CoA is produced than can enter the citric acid cycle, it is converted to ketone bodies, in the process known as ketogenesis, to be used as an alternate energy source. Ketone bodies are so named because they either contain the ketone functional group or are directly derived from ketones.

24.52 Ketosis is a condition in which ketone bodies accumulate in the blood faster than they can be metabolized. Since two of the ketone bodies are carboxylic acids, they lower the pH of the blood, producing the condition known as ketoacidosis. Symptoms of ketoacidosis include dehydration, labored breathing, and depression; prolonged ketoacidosis may lead to coma and death.

24.53 In uncontrolled diabetes, ketone bodies are produced faster than they can be used. The odor of acetone, a volatile ketone body, appears on the breath.

24.54 Ketones have little effect on pH, but the two other ketone bodies are acidic, and they lower the pH of urine.

24.55 When carbohydrates are restricted, the body metabolizes fats for energy, resulting in a quantity of acetyl-CoA that can't be metabolized in the citric acid cycle due to the reduced quantity of oxaloacetate, which is used in gluconeogenesis, The body, instead, uses ketogenesis for energy, and ketosis may result.

Fatty Acid Anabolism

24.56 The anabolic pathway for synthesizing fatty acids is called *lipogenesis*.

24.57 Since β oxidation is an energetically favorable reaction, its exact reverse is energetically unfavorable and thus doesn't occur.

24.58 Acetyl-CoA is the starting material for fatty acid synthesis.

24.59 Fatty acids have an even number of carbons because they are synthesized from a starting material (acetyl-CoA) that has an even number of carbons.

24.60 Stearic acid has 18 carbons and is synthesized from 9 acetyl-CoA units through 8 rounds of the lipogenesis spiral.

24.61 Each cycle of lipogenesis uses two NADPH molecules. For the synthesis of stearic acid, which requires 8 cycles of lipogenesis, 16 molecules of NADPH are needed.

24.62 Lipogenesis takes place in the cytosol, whereas β oxidation takes place in mitochondria.

24.63 The two processes require different coenzymes; oxidation releases energy, whereas synthesis requires energy.

Conceptual Problems

24.64 An excess of carbohydrates produces an excess of acetyl-CoA. If more acetyl-CoA is produced than is needed in the respiratory chain, it is used to synthesize fatty acids, which are used to produce the triglycerides that are deposited in adipose tissue. Once carbohydrates have been catabolized to acetyl-CoA, it is not possible to resynthesize them, because animals don't have enzymes to synthesize carbohydrates from acetyl-CoA.

24.65 Calories are stored as fat because fat is a more efficient and energy-dense means of storage.

24.66 In β oxidation, the alcohol intermediate is chiral because the hydroxyl carbon is bonded to four different groups.

24.67 Consider a carbohydrate and a fatty acid with similar molar masses. Glucose (molar mass = 180 g) yields a maximum of 38 ATP/mole; $CH_3(CH_2)_8COOH$ (molar mass = 172 g) yields a maximum of 78 ATP/mole. Thus, the energy yield per mole from a fat is about twice the yield from a sugar. There are two principal reasons for this difference. (1) A carbohydrate has fewer carbons than a fatty acid of similar molar mass and thus produces fewer moles of acetyl-CoA per mole of carbohydrate. (2) Only two-thirds of a carbohydrate's carbons are used to form acetyl-CoA; the other carbons are released as CO_2 in the transformation pyruvate —> acetyl-CoA.

24.68 (a) LDL, which transports cholesterol from the liver to peripheral tissue, is endogenous.
(b) Chylomicrons, which carry dietary triacylglycerols in the bloodstream, are exogenous.

24.69 (a) Eleven acetyl-CoAs
(b) Eleven acetyl-CoAs are produced from 10 cycles of β oxidation, each of which yields five ATPs, resulting in 50 ATP from β oxidation. Two ATP are used to activate the fatty acid, reducing the total to 48 ATP.
(c) Twenty-two molecules of CO_2 are produced from the C_{22} fatty acid.
(d) Each acetyl-CoA can enter the citric acid cycle, in which 12 ATPs are formed per acetyl CoA, to yield 11 x 12 = 132 ATP.
(e) The total ATP from (b) + (d) = 180 ATP

24.70 The body synthesizes cholesterol when no cholesterol is present in the diet. Thus, eliminating exogenous sources of cholesterol won't eliminate cholesterol in the bloodstream. A certain amount of cholesterol is essential for membrane function and as the starting material for synthesis of steroid hormones.

24.71

$$CH_3$$
$$|$$
$$H_2C{=}CH-C{=}CH_2 \quad \text{2-Methyl-1,3-butadiene}$$

Since cholesterol contains 27 carbon atoms, at least 6 molecules of 2-methyl-1,3-butadiene are needed for its synthesis.

24.72 The excess carbohydrate passes through glycolysis and ends up as acetyl-CoA. Since the body doesn't need extra energy, acetyl-CoA instead enters lipogenesis to form fatty acids, which are stored as triacylglycerols in adipocytes, leading to weight gain.

Self-Test for Chapter 24

Multiple Choice

1. ATP is ultimately produced in all of the following steps of fatty acid catabolism except:
   (a) fatty acid activation   (b) introduction of a double bond into a fatty acid
   (c) oxidation of an alcohol to a ketone   (d) entry of acetyl-SCoA into the citric acid cycle

2. An excess of ketone bodies in the blood is known as:
   (a) ketosis   (b) ketonuria   (c) ketoacidosis   (d) ketonemia

3. The reactants for TAG synthesis are:
   (a) glycerol and fatty acids   (b) glycerol 3-phosphate and fatty acids
   (c) glyceraldehyde 3-phosphate and fatty acyl-SCoA   (d) glycerol and fatty acyl-SCoA

4. Production of ketone bodies occurs in:
   (a) the liver   (b) adipose tissue   (c) muscle   (d) all of the above

5. Which of the following statements about lipogenesis is false?
   (a) Synthesis occurs two carbons at a time.   (b) Formation of acyl-SACP intermediates requires ATP.   (c) The coenzyme NADPH is needed.   (d) Lipogenesis occurs when acetyl-CoA is in abundance.

6. Chylomicrons transport all of the following except:
   (a) acylglycerols   (b) small-chain fatty acids   (c) phospholipids   (d) cholesterol

7. Fatty acids from storage in adipose tissue are transported:
   (a) by chylomicrons   (b) by VLDL   (c) by serum albumin   (d) free in the bloodstream

8. Acetyl-CoA is the starting material for synthesis of all of the following except:
   (a) steroids   (b) fatty acids   (c) ketone bodies   (d) glycerol

9. β Oxidation of $CH_3(CH_2)_8COOH$ produces a maximum of how many ATPs?
   (a) 78   (b) 72   (c) 64   (d) 60

10. When glucose is in good supply, the body:
    (a) synthesizes ketone bodies from acetyl-SCoA   (b) hydrolyzes TAGs from adipocytes
    (c) breaks down fatty acids via β-oxidation   (d) synthesizes TAGs for storage

Sentence Completion

1. One _____ and one _____ are needed for each β oxidation cycle.

2. Bile acids solubilize lipids by forming _____.

3. Triacylglycerols are stored in _____.

4. TAG synthesis begins with the conversion of glycerol to _____ _____.

5. An alternate name for β oxidation is the ___ - ____ ____.

6. The first step in β oxidation involves introduction of a _____ _____.

7. _____ is the presence of ketone bodies in the urine.

8. _____ is a coenzyme for fatty-acid biosynthesis.

9. _____ is a ketone body containing a hydroxyl group.

10. Chylomicrons are transported in the _____ ____ and in the bloodstream.

11. TAGs that are released from storage are said to be _____.

12. All fatty acids are synthesized from _____.

True or False

1. Bile acids are steroids.

2. TAGs are hydrolyzed by pancreatic lipases to glycerol and fatty acids.

3. Activation of fatty acids for fatty acid oxidation requires two phosphates to be cleaved.

4. A chylomicron is a type of lipoprotein.

5. Chylomicrons carry fatty acid-CoA across the mitochondrial membrane.

6. A fatty acid with n carbons requires $n/2 - 1$ cycles of β-oxidation for complete breakdown.

7. TAGs are synthesized from glycerol and fatty acyl-CoAs.

8. HDL transports cholesterol to tissues, where it can accumulate.

9. A fatty acid with a molecular weight similar to that of glucose yields more ATPs per molecule than glucose.

10. Fatty acid synthesis occurs in mitochondria.

11. The terminal $-CH_3$ group of a fatty acid comes from malonyl-SCoA.

12. Most naturally-occurring fatty acids have an even number of carbon atoms.

Match each entry on the left with its partner on the right.

1. Chylomicron   (a) Lipoprotein rich in cholesterol

2. Lipogenesis   (b) Protein that carries fatty acids in the bloodstream

3. LDL   (c) Solubilizes fats

4. Acetone   (d) Hydrolyzes TAG in adipose tissue

5. Malonyl-CoA   (e) Lipoprotein that aids in lipid transport of fats in the diet

6. Fatty acid spiral   (f) Lipoprotein rich in protein

7. Micelle   (g) Starting material in lipogenesis

8. Lipoprotein lipase   (h) Acidic ketone body

9. HDL   (i) Fatty acid biosynthesis

10. Albumin   (j) Breakdown of fatty acids into acetyl-CoA

11. Bile   (k) Formed from bile and lipids

12. Acetoacetate   (l) Nonacidic ketone body

Chapter Outline

I. Introduction to protein and amino acid metabolism (Sections 25.1–25.2).
    A. Protein digestion (Section 25.1).
        1. Proteins are denatured in the stomach.
        2. In the stomach, the enzyme pepsin hydrolyzes proteins to polypeptides.
        3. In the intestine, other enzymes hydrolyze the polypeptides to amino acids.
        4. Amino acids cross the cell membrane of the intestine and are absorbed directly into the bloodstream.
            a. Active transport is required for amino acids to cross the intestinal lining and to move into cells.
    B. Overview of amino acid metabolism (Section 25.2).
        1. The amino acid pool is the collection of all free amino acids in the body.
            a. The amino acid pool is the source of amino acids for protein synthesis and of nitrogen for the synthesis of nitrogen-containing biomolecules.
        2. Amino acid catabolism occurs in two steps.
            a. Removal of nitrogen, which can be:
                i.   Used for synthesis of nitrogen-containing biomolecules.
                ii. Excreted as urea.
            b. Entry of the remaining carbon atoms into metabolic pathways.
                i.   Citric acid cycle.
                ii. Gluconeogenesis.
                iii. Ketogenesis.
                iv. Lipogenesis.
II. Amino acid catabolism (Sections 25.3–26.5).
    A. Removal of the amino group (Section 25.3).
        1. Transamination.
            a. In transamination, the amino group of an amino acid and the keto group of an α-keto acid change places.
            b. Transaminase enzymes are usually specific for α-ketoglutarate as the amino group acceptor.
            c. Transamination reactions are equilibria that regulate amino acid concentration.
        2. Deamination.
            a. If the glutamate from transamination is in excess, the $-NH_3^+$ group can be removed and excreted and α-ketoglutarate regenerated.
            b. This reaction is known as oxidative deamination.
            c. Either NAD^+ or $NADP^+$ is required.
            d. The reverse reaction, reductive amination, occurs in biosynthesis.
    B. The urea cycle (Section 25.4).
        1. Because ammonia is toxic, the human body converts it to urea in order to excrete it.
        2. The reactions of the urea cycle take place in the liver.
        3. Ammonia is first converted to carbamoyl phosphate.
            $$NH_4^+ + HCO_3^- + 2\ ATP \longrightarrow \text{carbamoyl phosphate} + 2\ ADP + HOPO_3^{2-} + H_2O.$$
        4. Steps of the cycle:
            a. Step 1: Carbamoyl phosphate + ornithine $\longrightarrow$ citrulline + $HOPO_3^{2-}$.
            b. Step 2: Citrulline + aspartate + ATP $\longrightarrow$ argininosuccinate + AMP + $P_2O_7^{4-}$.
            c. Step 3: Argininosuccinate $\longrightarrow$ fumarate + arginine.
            d. Step 4: Arginine + H_2O $\longrightarrow$ urea + ornithine.

  5. Results of the cycle.
     a. Urea is eliminated.
        i. The urea carbon comes from bicarbonate, one nitrogen comes from aspartate, and the other nitrogen comes from ammonia.
     b. Four phosphate bonds are broken.
     c. Fumarate, a citric acid cycle intermediate, is produced.
        i. Fumarate is converted to oxaloacetate, which undergoes transamination to yield aspartate, which reenters the cycle. (Fumarate can also enter gluconeogenesis.)
  C. Fate of the carbon atoms (Section 25.5).
     1. All amino acid carbon skeletons are either citric acid cycle intermediates or fatty acid metabolism intermediates.
     2. Glucogenic amino acids.
        a. Amino acids that are converted to pyruvate or enter the citric acid cycle as α-ketoglutarate, succinyl-CoA, fumarate, or oxaloacetate are glucogenic.
        b. These amino acids can be converted to glucose.
     3. Ketogenic amino acids.
        a. Amino acids that are converted to acetyl-CoA or acetoacetyl-CoA are known as ketogenic amino acids.
        b. These amino acids can enter either ketogenesis to form ketone bodies or fatty acid biosynthesis.
     4. Some amino acids are both glucogenic and ketogenic.
III. Biosynthesis of nonessential amino acids (Section 25.6).
  A. The human body can synthesize 11 of the 20 common amino acids.
  B. All 11 amino acids derive their amino group from glutamate.
  C. Pyruvate, oxaloacetate, α-ketoglutarate, and 3-phosphoglycerate are the precursors of all nonessential amino acids.
  D. Tyrosine is formed from phenylalanine.
     1. An inability of the body to perform this conversion is the metabolic error in phenylketonuria.

Solutions to Chapter 25 Problems

HOC 25.1
(a) For a person weighing 135 lb:

$$135 \text{ lb } \times \frac{1 \text{ kg}}{2.2 \text{ lb}} \times \frac{0.8 \text{ g}}{1 \text{ kg}} = 49 \text{ g}$$

(b) For a fairly active person:

$$135 \text{ lb } \times \frac{1 \text{ kg}}{2.2 \text{ lb}} \times \frac{1.1 \text{ g}}{1 \text{ kg}} = 68 \text{ g}$$

(c) (no answer)

(d) Quality protein is protein that contains essential amino acids, doesn't contain excessive amounts of fat, and is easily digested. Good sources of quality protein are lean beef, chicken, fish, milk, beans, yogurt, eggs, and nuts.

25.1 (a) False. The amino acid pool is the collection of free amino acids in the body. (b) False. Our bodies do not store nitrogen-containing compounds.   (c) True.

25.2

Tryptophan → (oxidation, oxidoreductase) → hydroxylated tryptophan

hydroxylated tryptophan → (decarboxylation, lyase) → Serotonin + CO_2

25.3

Phenylalanine + α-Ketoglutarate → product + Glutamate

25.4

$CH_3SCH_2CH_2CHCOO^-$ (with NH_3^+) → (α-Ketoglutarate → Glutamate, α-Aminotransferase) → $CH_3SCH_2CH_2CCOO^-$ (with =O)

25.5 The conversion of alanine to pyruvate can be identified as an oxidation by noting the loss of –H from the α-carbon and the replacement of $-NH_3^+$ by =O.

25.6 The three branched-chain amino acids are valine, leucine, and isoleucine.

$CH_3CHCHCOO^-$ (Valine) + α-Ketoglutarate → α-Keto-3-methylbutanoate + Glutamate

CIA 25.1

Adenosine

The changes include deamination (1), hydrolysis (2), and oxidation (3).

CIA 25.2

(a) Gout is a painful inflammation of the joints caused by deposits of insoluble salts of uric acid. These deposits occur because of an excess of purines in the body or because of an impaired ability to excrete the products of purine metabolism. Although the symptoms of gout are often due to metabolic problems that are genetic, some cases of gout are caused by injury or excessive alcohol consumption.

(b) Simple ways to prevent attacks of gout that are not genetic are to limit consumption of alcohol and purine-rich foods.

CIA 25.3

Allopurinol   Hypoxanthine   Xanthine   Uric acid

Allopurinol is identical to hypoxanthine in all but one respect: The nitrogen at position 7 of the hypoxanthine ring is at position 8 of the allopurinol ring. Just as hypoxanthine can be oxidized to xanthine, allopurinol can be oxidized to a compound similar to xanthine (alloxanthine). The product of allopurinol oxidation, however, is an inhibitor of the enzyme that converts xanthine to uric acid. Instead of producing uric acid, purine catabolism ends with the formation of the more soluble hypoxanthine and xanthine, which can be easily excreted.

Oxidation of xanthine to uric acid occurs at position 8 of the purine ring system. In allopurinol, this site is occupied by nitrogen.

25.7

H_2N^+
$\|$
$HN-C-NH_2$
$|$
CH_2 H_2O H_2NCNH_2 (O)
$|$
CH_2
$|$
CH_2
$|$
$HC-NH_3^+$
$|$
COO^-

Arginine
(a)

NH_3^+
$|$
$H-C-H$ α-Keto-glutarate   Glutamate
$|$
CH_2
$|$
CH_2
$|$
$HC-NH_3^+$
$|$
COO^-

Ornithine
(b)

H
$\diagdown$
$C=O$
$|$
CH_2
$|$
CH_2
$|$
$HC-NH_3^+$
$|$
COO^-

(c)

$\xrightarrow{[O]}$

COO^-
$|$
CH_2
$|$
CH_2
$|$
$HC-NH_3^+$
$|$
COO^-

Glutamate
(d)

$NAD^+,$ H_2O $NADH, NH_4^+$

Glutamate dehydrogenase

COO^-
$|$
CH_2
$|$
CH_2
$|$
$C=O$
$|$
COO^-

α-Ketoglutarate
(e)

25.8

H COO^-
$\diagdown / $
C
$\|$
C
$/ \diagdown$
^-OOC H

Fumarate

H_2O $\xrightarrow[\text{(5)}]{\text{Addition}}$

COO^-
$|$
$HO-CH$
$|$
CH_2
$|$
COO^-

Malate

NAD^+ $NADH/H^+$ $\xrightarrow[\text{(1)}]{\text{Oxidation}}$

COO^-
$|$
$C=O$
$|$
CH_2
$|$
COO^-

Oxaloacetate

Gluta-mate α-Keto-glutarate

$\xrightarrow[\text{(3)}]{\text{Transamination}}$

COO^-
$|$
$^+H_3N-CH$
$|$
CH_2
$|$
COO^-

Aspartate

25.9 Table 25.1 shows which amino acids are ketogenic and which are glucogenic. Several amino acids are listed as both ketogenic and glucogenic.
Glucogenic: histidine, methionine, valine, threonine
Ketogenic: leucine
Both glucogenic and ketogenic: lysine, isoleucine, tryptophan, phenylalanine

25.10

3-Phosphoglycerate $\xrightarrow[\text{(c)}]{\text{oxidation}}$ 3-Phosphohydroxypyruvate $\xrightarrow[\text{(a)}]{\text{transamination}}$

3-Phosphoserine $\xrightarrow[\text{(b)}]{\text{hydrolysis}}$ Serine $+ \ HOPO_3{}^{2-}$

CIA 25.4

Conditional amino acids are usually produced in sufficient quantity, but it may become necessary to increase intake in periods of stress or illness.

CIA 25.5

A deficiency of methionine may lead to an accumulation of cholesterol, liver cirrhosis, kidney inflammation, and chronic rheumatic fever.

CIA 25.6

A deficiency of tryptophan, "nature's Prozac," may lead to serotonin deficiency syndrome, which causes behavioral and emotional problems.

Understanding Key Concepts

25.11

| Dietary protein | Pyruvate, acetoacetate, acetyl-CoA, citric acid cycle intermediates | Tissue protein |

Digestion   Synthesis   Turnover

Amino acid pool

25.12 Catabolism of an amino acid begins with a transamination reaction that removes the amino acid nitrogen.

α-amino acid α-Ketoglutarate $\xrightarrow{\alpha\text{-Aminotransferase}}$ α-keto acid   Glutamate

The carbon atoms: The resulting α-keto acid, which contains the carbons from the original amino acid, undergoes reactions that convert it to a common metabolic intermediate. This intermediate may be a citric acid cycle intermediate, pyruvate, acetyl-CoA, or acetoacetyl-CoA.

The nitrogen atoms: Glutamate, which contains the amino group from the original amino acid, undergoes an oxidative deamination reaction that converts the amino group to ammonium ion and that regenerates α-ketoglutarate. Either NAD^+ or $NADP^+$ can be used as the enzyme cofactor.

$$\underset{\text{Glutamate}}{{}^-OOCCH_2CH_2\overset{\overset{+}{N}H_3}{\underset{|}{C}}HCOO^-} + H_2O \xrightarrow[\text{dehydrogenase}]{\overset{\overset{NAD^+ \qquad NADH/H^+}{\curvearrowright}}{\text{Glutamate}}} \underset{\alpha\text{-Ketoglutarate}}{{}^-OOCCH_2CH_2\overset{\overset{O}{\|}}{C}COO^-} + NH_4^+$$

Ammonium ion enters the urea cycle, where it is transformed to urea and is excreted.

$$NH_4^+ + HCO_3^- + \underset{\text{Aspartate}}{{}^-OOCCH_2-\overset{\overset{+}{N}H_3}{\underset{|}{C}}HCOO^-}$$

$$\downarrow \begin{array}{l} \curvearrowleft \text{3 ATP} \\ \curvearrowright \text{2 ADP} + AMP + 2 HOPO_3^{2-} + P_2O_7^{4-} \end{array}$$

$$\underset{\text{Urea}}{H_2N-\overset{\overset{O}{\|}}{C}-NH_2} + \underset{\text{Fumarate}}{{}^-OOCCH=CHCOO^-}$$

25.13

$$NH_4^+ + \underset{\alpha\text{-Ketoglutarate}}{{}^-OOCCH_2CH_2\overset{\overset{O}{\|}}{C}COO^-} \xrightarrow[\text{dehydrogenase}]{\overset{\overset{NADH/H^+ \qquad NAD^+}{\curvearrowright}}{\text{Glutamate}}} \underset{\text{Glutamate}}{{}^-OOCCH_2CH_2\overset{\overset{+}{N}H_3}{\underset{|}{C}}HCOO^-} + H_2O$$

$NADPH/H^+$ can also be used as an enzyme cofactor.

$$\underset{\text{Glutamate}}{{}^-OOCCH_2CH_2\overset{\overset{+}{N}H_3}{\underset{|}{C}}HCOO^-} + \underset{\text{Pyruvate}}{CH_3\overset{\overset{O}{\|}}{C}COO^-} \underset{\underset{\text{aminotransferase}}{\text{Alanine}}}{\rightleftharpoons} \underset{\alpha\text{-Ketoglutarate}}{{}^-OOCCH_2CH_2\overset{\overset{O}{\|}}{C}COO^-} + \underset{\text{Alanine}}{CH_3\overset{\overset{+}{N}H_3}{\underset{|}{C}}HCOO^-}$$

The enzymes for both reactions can catalyze reaction in both directions. The product is the amino acid alanine.

25.14 The carbon atoms from ketogenic amino acids can be converted to ketone bodies or to acetyl-CoA. The carbon atoms from glucogenic amino acids can be converted to compounds that can enter gluconeogenesis and can form glucose, which can enter glycolysis and also yield acetyl-CoA.

25.15 An essential nutrient can't be synthesized by the body and must be provided in the diet. For example, all amino acids are "essential" in the sense that all of them are necessary for protein synthesis. The body can synthesize half of them from non-amino acid precursors; these are described as "nonessential." The other amino acids must be part of the diet.

25.16 It is important for ornithine transcarbamoylase to have a high activity in order to remove NH_4^+ from the body quickly. It is also important that arginase activity be high in order to release urea for excretion by the kidneys and to provide the ornithine needed to react with carbamoyl phosphate. If arginase activity were low, hyperammonemia might result.

Additional Problems

Amino Acid Pool

25.17 The body's amino acid pool, which is the collection of free amino acids that result from diet or tissue breakdown, occurs throughout the body.

25.18 The chemical digestion of proteins begins in the stomach.

25.19 Pyruvate and 3-phosphoglycerate are glycolytic precursors of amino acids.

25.20 Oxaloacetate and α-ketoglutarate are citric acid cycle precursors of amino acids.

Amino Acid Catabolism

25.21 In a transamination reaction, a keto group of an α-keto acid and an amino group of an α-amino acid change places.

25.22

25.23

α-Ketoglutarate   Glutamate

Amino Acid →[α-Aminotransferase]→ *α-Keto Acid*

(a)

$$^-OOCCH_2CH_2\overset{\overset{\displaystyle NH_3^+}{|}}{C}H-COO^-$$
Glutamate

$$^-OOCCH_2CH_2\overset{\overset{\displaystyle O}{||}}{C}-COO^-$$

(b)

$$CH_3\overset{\overset{\displaystyle NH_3^+}{|}}{C}H-COO^-$$
Alanine

$$CH_3\overset{\overset{\displaystyle O}{||}}{C}-COO^-$$

25.24

α-Ketoglutarate   Glutamate

Amino Acid →[α-Aminotransferase]→ *α-Keto Acid*

(a)

$$CH_3CH_2\overset{\overset{\displaystyle CH_3}{|}}{C}H-\overset{\overset{\displaystyle NH_3^+}{|}}{C}H-COO^-$$
Isoleucine

$$CH_3CH_2\overset{\overset{\displaystyle CH_3}{|}}{C}H-\overset{\overset{\displaystyle O}{||}}{C}-COO^-$$

(b)

$$H_3C-\overset{\overset{\displaystyle CH_3}{|}}{C}H-\overset{\overset{\displaystyle NH_3^+}{|}}{C}H-COO^-$$
Valine

$$H_3C-\overset{\overset{\displaystyle CH_3}{|}}{C}H-\overset{\overset{\displaystyle O}{||}}{C}-COO^-$$

25.25 In an oxidative deamination reaction, an $-NH_3^+$ group of an amino acid (usually glutamate) is replaced by a carbonyl oxygen, and ammonium ion is eliminated. In a transamination reaction, a keto group of an α-keto acid and an amino group of an α-amino acid change places. An oxidative deamination is a redox reaction.

25.26 Either NAD^+ or $NADP^+$ is associated with oxidative deamination.

25.27

	Amino Acid	*α-Keto Acid*

(a)

$$CH_3CHCH_2-CH-COO^-$$

with CH_3 and NH_3^+ substituents

Leucine

$$CH_3CHCH_2-C-COO^-$$

with CH_3 and O substituents

(b)

$$CH_2-CH-COO^-$$

with NH_3^+ substituent, attached to indole ring

Tryptophan

$$CH_2-C-COO^-$$

with O substituent, attached to indole ring

25.28 Ammonium ion is the other product of oxidative deamination.

25.29 A ketogenic amino acid is catabolized to acetoacetyl-CoA or acetyl-CoA, which can be used either in the synthesis of ketone bodies or in fatty acid biosynthesis. Examples: leucine, isoleucine, lysine. (The last two are both glucogenic and ketogenic.)

25.30 A glucogenic amino acid is an amino acid that is catabolized to pyruvate or to citric acid cycle intermediates and can enter gluconeogenesis. Examples: alanine, glycine, serine.

The Urea Cycle

25.31 Ammonia is toxic and must be eliminated as nontoxic urea, which is water-soluble.

25.32 The *urea carbon* comes from carbamoyl phosphate, which is synthesized from bicarbonate produced from CO_2 in the citric acid cycle.

25.33 One nitrogen comes from carbamoyl phosphate, which is synthesized from ammonium ion produced from oxidative deamination. One nitrogen comes from aspartate.

25.34 Aspartate enters the urea cycle at step 2 (reaction with citrulline) and leaves in step 3 as fumarate, which can enter the citric acid cycle.

Amino Acid Biosynthesis

25.35 Nonessential amino acids can be synthesized in one to three steps by most organisms. Plants and microorganisms synthesize essential amino acids in many more steps.

25.36 Glutamate is the source of nitrogen for amino acid anabolism because all other amino acids acquire their amino group from glutamate.

25.37 Meat, fish, and dairy products are all rich in lysine. Legumes are a good plant-based source of lysine.

25.38 In the body, tyrosine is biosynthesized by hydroxylation of phenylalanine. Tyrosine is an essential amino acid for phenylketonurics, who are unable to synthesize tyrosine from phenylalanine.

25.39 Phenylketonuria (PKU) is caused by a genetic inability to convert phenylalanine to tyrosine, leading to the accumulation of phenylalanine and its metabolites in the body. PKU causes mental retardation if not detected early in life. Treatment consists of a diet restricted in phenylalanine.

25.40 Aspartame is a dipeptide that contains phenylalanine, which must be severely restricted in the diet of phenylketonurics, who can't metabolize phenylalanine.

25.41 All biomolecules that contain nitrogen acquire it from amino acids. These include (b) nitric oxide, (c) collagen, and (d) epinephrine.

Conceptual Problems

25.42 Three molecules of ATP are consumed in producing each molecule of urea, making the process energy-intensive.

25.43

$$C_6H_5CH_2-\overset{\overset{+}{N}H_3}{\underset{|}{C}}HCOO^- + CH_3\overset{\overset{O}{||}}{C}COO^- \overset{ALT}{\underset{}{\rightleftharpoons}} C_6H_5CH_2-\overset{\overset{O}{||}}{C}COO^- + CH_3\overset{\overset{+}{N}H_3}{\underset{|}{C}}HCOO^-$$

Phenylalanine   Pyruvate   Alanine

The enzyme that catalyzes the transamination is alanine aminotransferase.

25.44 (a) Amino acids can enter the citric acid cycle as α-ketoglutarate, succinyl-CoA, fumarate, and oxaloacetate. α-Ketoglutarate enters the cycle at step 4, succinyl-CoA enters at step 5, fumarate enters at step 7, and oxaloacetate enters at step 1.
(b) Fumarate arises from the catabolism of phenylalanine and tyrosine.

25.45 An amino acid can be both glucogenic and ketogenic. After catabolism, certain amino acids yield products that can enter the citric acid cycle and other products that can be intermediates of fatty acid metabolism.

25.46 The conversion of ammonium ion to urea takes place in the liver. Urea is transported to the kidneys, where it is excreted in urine.

25.47 Tissue is dynamic because its components are constantly being broken down and reformed. Dynamic relationships are common in protein and amino acid metabolism, where proteins are digested and resynthesized, and where amino acids are converted to other amino acids and to carbohydrate intermediates.

25.48 *Storage:* Unlike fats and carbohydrates, amino acids are not stored in the body. Instead, when excess amino acids accumulate, amino acid nitrogen is excreted and the resulting compounds are converted to either fats or carbohydrates.

Energy: Fats and carbohydrates that are not stored are catabolized. Surplus amino acids must be converted to either fats or carbohydrates in order to be an energy source.

25.49 The basic scheme of the conversion of glutamate to glycogen is deamination —>
gluconeogenesis —> glycogenesis.
Step 1: Glutamate (b) —> α-Ketoglutarate (e) (deamination or transamination).
Step 2: α-Ketoglutarate (e) —> Oxaloacetate (d) (citric acid cycle).
Step 3: Oxaloacetate (d) —> Phosphoenolpyruvate (f) (gluconeogenesis).
Step 4: Phosphoenolpyruvate (f) —> Glucose (a) (gluconeogenesis).
Step 5: Glucose (a) —> Glycogen (c) (glycogenolysis).
The complete sequence: (b) —> (e) —> (d) —> (f) —> (a) —> (c).

25.50 The activated forms of the proteases would hydrolyze the proteins in the lining of the
pancreas if they were stored there until needed in digestion. They can be stored safely in
their inactive form and still be readily available when needed for digestion.

25.51 (1) The amino group is removed. (2) The amino nitrogen is either used in synthesis of new
nitrogen-containing compounds or excreted as urea. (3) The remaining carbon atoms are
converted into compounds that can either enter the citric acid cycle or enter ketogenesis.

25.52 One of the ATPs in the urea cycle is hydrolyzed to AMP, which is the equivalent of
spending two ATPs.

25.53 Active transport of amino acids across cell membranes is achieved by several transport
systems, each of which is specific for several different amino acids. An excess of one
amino acid might overwhelm its transport system, making it unavailable to transport other
amino acids and causing an amino acid deficiency.

Group Problems

25.54 The answer depends on the foods that were eaten. The essential amino acids (histidine,
lysine, threonine, isoleucine, methionine, tryptophan, leucine, phenylalanine, valine) are
found in complete protein sources such as meat and dairy products. Incomplete protein
sources include grains, nuts, seeds, legumes, and corn.

25.55 Breakfast: yogurt, two scrambled eggs, peanut butter on toast.
Lunch: cheese and bean burrito, glass of milk.
Dinner: roast chicken, corn on the cob, brown rice, whole wheat roll.
Don't forget the fruit and vegetables!

25.56 Use all the steps of protein and amino acid metabolism that we studied in this chapter to
answer this intriguing question.
(1) Hydrolysis of the tetrapeptide neither produces nor consumes ATP.
(2) Transamination does not involve ATP. However, oxidative deamination to produce the
α-keto acid produces one reduced coenzyme (NADH), which yields 3 ATP through
oxidative phosphorylation.
(3) Unfortunately, excretion of the NH_4^+ produced in the previous step, consumes 3 ATP,
giving a net yield of zero ATP.

At this point, we are left with four α-keto acids, each of which enters the citric acid cycle at a different point after undergoing a series of transformations too complicated to outline here. ATP may be produced during these pathways, but we won't include it.

Lysine and leucine enter the citric acid cycle as acetyl-CoA after deamination and yield 12 ATPs for each acetyl-CoA; each lysine is catabolized to two acetyl-CoA, and each leucine is catabolized to 3 acetyl-CoA.

After deamination and several transformations, histidine enters the citric acid cycle as α-ketoglutarate and produces 9 ATP as it proceeds through the steps of the cycle. (Remember, that α-ketoglutarate is part way through the cycle, and that one NADH has already been produced.)

Valine enters the cycle as succinyl-CoA and produces 6 ATP.

Self-Test for Chapter 25

Multiple Choice

1.   Which of the following amino acids is strictly ketogenic?
(a) proline   (b) threonine   (c) leucine   (d) isoleucine

2.   The product of transamination of valine is:
(a) α-keto-3-methylbutanoic acid   (b) α-ketobutanoic acid
(c) α-keto-4-methylpentanoic acid   (d) α-ketopentanoic acid

3.   The amino acid proline is probably synthesized by cyclization of:
(a) arginine   (b) glutamate   (c) leucine   (d) threonine

4.   The urea cycle takes place in the:
(a) muscles   (b) kidneys   (c) bloodstream   (d) liver

5.   Which of the following proteases is necessary for the activation of other proteases?
(a) pepsin   (b) chymotrypsin   (c) carboxypeptidase   (d) trypsin

6.   How many grams of protein tissue are broken down in an adult body per day?
(a) none   (b) 50 g   (c) 300 g   (d) 1000 g

7.   Ornithine is:
(a) a urea cycle intermediate   (b) an amino acid   (c) an acceptor of a carbamoyl group
(d) all three

8.   Which of the following nitrogen-containing compounds is used to treat gout?
(a) hypoxanthine   (b) xanthine   (c) allopurinol   (d) uric acid

9.   Which of the following is not a precursor in nonessential amino acid biosynthesis?
(a) fumarate   (b) pyruvate   (c) α-ketoglutarate   (d) 3-phosphoglycerate

10.   Glucogenic amino acids enter gluconeogenesis via:
(a) acetyl-CoA   (b) oxaloacetate   (c) α-ketoglutarate   (d) fumarate

Sentence Completion

1.   In the transamination reaction of alanine and α-ketoglutarate, _____ and _____ are formed.

2.   _____ is a primary metabolic source of sulfur.

3.   Ketogenic amino acids are catabolized to _____ _____ or _____ _____.

4.   The first step in the urea cycle is the formation of _____ _____ from ammonia, CO_2, and ATP.

5.   A _____ is an enzyme that carries out peptide hydrolysis.

6.   The amino acid serine enters the citric acid cycle as _____.

7.   _____ is the key reaction in amino acid anabolism.

8.   The essential amino acid _____ is part of the urea cycle.

9.   PKU is a metabolic error in converting _____ to _____.

10.   The pain of gout is caused by an inflammatory response to _____ _____ in tissues.

11.   Liver damage is associated with a high level of the enzyme _____.

12.   The carbon in urea comes from _____.

True or False

1.   3-Phosphoglycerate is a precursor in the synthesis of nonessential amino acids.

2.   Protein digestion starts in the mouth.

3.   All ketogenic amino acids except lysine and isoleucine are also glucogenic amino acids.

4.   Essential amino acids can neither be synthesized nor catabolized in the human body.

5.   Some organisms can excrete nitrogen as ammonia without undergoing any harm.

6.   Amino acid catabolism occurs in the intestinal lining.

7.   Some intermediates in the urea cycle also occur in the citric acid cycle.

8.   The urea cycle is an endergonic process.

9.   Both nitrogens of urea come from ammonium ion.

10.   Proline is a nonessential amino acid.

11.   Each amino acid is associated with its own transaminase enzyme.

12.   Oxidative deamination is the reverse of reductive amination.

Match each entry on the left with its partner on the right.

1. Oxidative deamination

2. Ornithine

3. α-Ketoglutarate

4. Serine

5. Trypsin

6. Reductive amination

7. Leucine

8. Phenylketonuria

9. Pepsin

10. Uric acid

11. Transamination

12. Threonine

(a) Hydrolyzes proteins in the small intestine

(b) Process for synthesizing glutamate from α-ketoglutarate and NH_4^+

(c) Ketogenic amino acid

(d) End product of purine catabolism

(e) Process that transfers an amino group from an amino acid to α-ketoglutarate

(f) Nonessential amino acid

(g) Intermediate in the urea cycle

(h) Process that regenerates a-ketoglutarate from glutamate

(i) Glucogenic amino acid

(j) Intermediate in both the citric acid cycle and in amino acid metabolism

(k) Hydrolyzes proteins in the stomach

(l) Metabolic disease

Chapter 26   Nucleic Acids and Protein Synthesis

Chapter Outline

I. Nucleic acids (Sections 26.1–26.4).
   A. Introduction (Section 26.1).
      1. When a cell is not dividing, its nucleus consists of chromatin, a tangle of fibers composed of protein and DNA.
      2. Just before dividing, the DNA is duplicated.
      3. During division, chromatin organizes itself into chromosomes, each of which is a huge DNA molecule.
      4. The chromosomes are made up of genes, individual segments of DNA that contain instructions for synthesizing a single protein.
      5. The human body has 23 pairs of chromosomes.
   B. Composition of nucleic acids (Section 26.2).
      1. Nucleic acids are polymers of nucleotides.
         a. DNA and RNA are the two types of nucleic acids.
      2. Nucleotides consist of:
         a. A pentose.
            i. For RNA, the pentose is D-ribose.
            ii. For DNA, the pentose is D-2-deoxyribose.
            iii. The pentose is connected to the base by a *β-N*-glycosidic bond to the anomeric carbon.
         b. A heterocyclic base.
            i. The bases for DNA are the purines adenine and guanine, and the pyrimidines cytosine and thymine.
            ii. For RNA, uracil replaces thymine.
         c. A phosphate group.
            i. The phosphate group is attached at C5 of the pentose.
      3. The combination of sugar + base is known as a nucleoside.
         a. Nucleosides are named by replacing the *-ine* of the base name by *-osine* for purines and *-idine* for pyrimidines.
         b. If the pentose is 2-deoxyribose, the prefix *deoxy* is added to the name.
         c. To indicate position, numbers without primes are used for bases, and numbers with primes are used for the pentose.
         d. Nucleosides have important biochemical functions other than as components of nucleic acids.
      4. Nucleotides.
         a. Nucleotides are named by adding "5'-monophosphate" to the nucleoside name.
         b. Nucleotides can form di- and triphosphates.
            i. ATP plays an important role in biochemical energetics.
   C. The structure of nucleic acid chains (Section 26.3).
      1. Nucleotides are joined by phosphate ester bonds between the 5' phosphate and the 3' —OH group of the sugar of the next nucleotide.
      2. Vast numbers of nucleotides can be connected by these bonds to form immense polynucleotides.
      3. The structure and function of nucleic acids depends on the sequence of nucleotides.
      4. The sequence is described by starting at the 5' end and identifying the bases by one-letter abbreviations.

    D. Base pairing in DNA (Section 26.4).
        1. In DNA, two polynucleotide strands coil around each other in a double helix.
            a. The sugar–phosphate backbone is on the outside.
            b. The bases are inside.
        2. The two strands run in opposite directions.
        3. The two strands are held together by hydrogen bonding between bases.
            a. A and T form two hydrogen bonds to each other.
            b. C and G form three hydrogen bonds to each other.
        4. The two strands are complementary.
            a. A and T always occur in DNA in the same percentage.
            b. C and G always occur in DNA in the same percentage.
  II. The function of nucleic acids (Sections 26.5–26.10).
    A. The central dogma of molecular genetics (Section 26.5).
        1. The function of DNA is to store genetic information.
        2. The function of RNA is to read and decode DNA and use the information to synthesize proteins.
        3. Nucleic acids take part in three major processes.
            a. Replication is the process in which DNA makes a copy of itself.
            b. Transcription is the process in which RNA reads the genetic message.
            c. Translation is the process in which the genetic message is used to make proteins.
    B. Replication of DNA (Section 26.6).
        1. Mechanism of replication.
            a. The double helix partially unwinds, with the assistance of helicase enzymes.
               i. Unwinding occurs simultaneously at many locations.
            b. Bases line up and form hydrogen bonds with their complements.
            c. DNA polymerase catalyzes bond formation between the 5' phosphate group on the arriving nucleotide and the 3' —OH of the old strand.
        2. Two identical new copies of the double helix are formed.
        3. Because strand growth occurs in one direction (5' —> 3'), only one strand of DNA is synthesized continuously (the leading strand).
            a. The other strand (the lagging strand) is synthesized in segments (Okazaki fragments), beginning at the replication fork.
            b. DNA ligase attaches the segments of the lagging strand.
        4. The copying of DNA requires several hours, involves many replication forks, and proceeds virtually without error.
    C. RNA (Section 26.7).
        1. Structure of RNA.
            a. RNA resembles DNA but has the following differences:
               i. The pentose is ribose.
               ii. Uracil replaces thymine.
            b. RNA chains are shorter than DNA chains.
            c. RNA chains may be single stranded or double stranded.
        2. Types of RNA.
            a. Ribosomal RNA has a molecular mass of 5×10^6 amu and occurs in ribosomes, which are the site of protein synthesis.
            b. Messenger RNA reads DNA and carries the message to the ribosomes.
            c. Transfer RNA delivers amino acids to the growing protein chain.
    D. Transcription (Section 26.8).
        1. mRNA is synthesized by transcription of DNA.
            a. RNA polymerase recognizes a control segment of DNA that precedes the nucleotides to be transcribed.
            b. A small section of DNA unwinds.
            c. Complementary bases are attached.

      d. Transcription stops when RNA polymerase recognizes a "stop" nucleotide sequence.

      e. All RNA is synthesized in the same way, although we will focus on mRNA in this description.

  2. Only one DNA strand is transcribed—the template strand.

      a. The other strand is the information strand.

      b. mRNA is a copy of the information strand (with U replacing T).

  3. Not all genes are continuous strands of DNA.

      a. The coding sections of DNA (exons) are interrupted by noncoding sections (introns).

      b. In the final mRNA, the intron sections are cut out.

E. Translation (Sections 26.9–26.10).

  1. The genetic code (Section 26.9).

      a. A sequence of three mRNA bases codes for a specific amino acid.

      b. The codes are written in the 5' —> 3' direction.

      c. Each amino acid is represented by one or more codes.

      d. There are 64 codons; 61 code for amino acids, and 3 are "stop" codons.

  2. Transfer RNA (Section 26.10).

      a. The message carried by mRNA is decoded by tRNA.

      b. Each amino acid is transported by its own tRNA to the ribosome.

      c. Amino acids are bonded to tRNA by an ester bond between the —COOH group of an amino acid and C3' of ribose at the end of the tRNA.

      d. Each tRNA has an anticodon that corresponds to a codon sequence on mRNA.

  3. Protein synthesis.

      a. Initiation.

          i. The first AUG codon on the 5' end of mRNA is a "start" codon (which also codes for methionine).

         ii. When the Met tRNA comes together with the mRNA codon and occupies one of the two binding sites in the ribosome, elongation can start.

        iii. If Met isn't the first amino acid in the protein, it is removed after protein synthesis.

      b. Elongation.

          i. A tRNA with an anticodon complementary to the second mRNA codon occupies the second binding site.

         ii. An enzyme in the ribosome catalyzes peptide bond formation and breaks the bond between the first amino acid and its tRNA.

        iii. This tRNA leaves the ribosome.

        iv. The ribosome shifts one position along the mRNA chain (translocation).

         v. The second site is free to accept the next tRNA.

        vi. A single mRNA can be read by many ribosomes.

      c. Termination.

          i. A "stop" codon signals the end of translation.

         ii. An enzyme called "releasing factor" frees the polypeptide chain from the last tRNA, and the mRNA is released from the ribosome.

Solutions to Chapter 26 Problems

26.1

2'-Deoxythymidine

26.2 D-Ribose has one more oxygen atom than 2-deoxy-D-ribose. This added oxygen allows D-ribose to form an additional hydrogen bond. The hydroxyl hydrogen can also take part in hydrogen bonding.

β-D-Ribose
($C_5H_{10}O_5$)

β-2-Deoxy-D-ribose
($C_5H_{10}O_4$)

26.3

2'-Deoxyadenosine 5'-monophosphate

26.4

Guanosine 5'-triphosphate (GTP)

26.5 dUMP—2'-Deoxyuridine 5'-monophosphate
UMP—Uridine 5'-monophosphate
CDP—Cytidine 5'-diphosphate
AMP—Adenosine 5'-monophosphate
ATP—Adenosine 5'-triphosphate

26.6 G-A-U-C-A = guanine-adenine-uracil-cytosine-adenine. The pentanucleotide comes from RNA because it contains uracil.

26.7 The 5' end may also be a triphosphate.

26.8 (a) *Original strand:* 5' T-A-T-A-C-T-G 3'
⋮ ⋮ ⋮ ⋮ ⋮ ⋮ ⋮
Complement: 3' A-T-A-T-G-A-C 5'

(b) *Original strand:* 5' G-A-T-C-G-C-T-C-T 3'
⋮ ⋮ ⋮ ⋮ ⋮ ⋮ ⋮ ⋮ ⋮
Complement: 3' C-T-A-G-C-G-A-G-A 5'

26.9

26.10 The phosphate groups cause the DNA molecule to be negatively charged.

26.11 (a) A longer strand of DNA has a higher "melting temperature" because larger molecules have greater numbers of hydrogen bonds.
(b) A strand of DNA containing a higher percentage of C/G base pairs has a higher "melting temperature." A C/G base pair has three hydrogen bonds, whereas an A/T pair has only two hydrogen bonds. A DNA strand with a higher percent of C/G base pairs has more hydrogen bonds, and more heat is required to separate the two strands.

26.12 Okazaki fragments are segments of DNA synthesized by using the lagging strand as a template. The fragments are later joined by a DNA ligase enzyme.

26.13 DNA polymerase catalyzes the addition of nucleoside triphosphates to a growing strand of DNA. DNA ligase connects Okazaki fragments formed from replication of DNA that forms the lagging strand.

26.14 A spliceosome is a protein-RNA complex that removes introns from heterogeneous nuclear RNA (hnRNA) and reconnects the exons.

26.15 (a) *DNA template strand:* 5' C-A-T–G-C-T-C-T--A-C-A-G 3'

mRNA complement: 3' G-U-A-C-G-A-G-A-U-G-U-C 5'

(b) *DNA template strand:* 3' T-A-T-T-A–G–C-G-A-C-C-G 5'

mRNA complement: 5' A-U-A-A-U-C-G-C-U-G-G-C 3'

HOC 26.1

(1)–(3)	Informational strand: (5' —> 3')	CTC--ATG--CCG--AGC--TAA
(4)	mRNA: (5' —> 3')	CUC--AUG-CCG--AGC--UAA
	Peptide:	Leu Met---Pro----Ser (stop)
		control
		segment

The peptide Met--Pro--Ser is a tripeptide.

(5a)	Informational strand: (5' —> 3')	GTA---CGA--GAT--GTC
	mRNA: (5' —> 3')	GUA--CGA--GAU--GUC
	Peptide:	Val-----Arg----Asp---Val
(5b)	Informational strand: (5' —> 3')	ATA--ATC---GCT--GGC
	mRNA: (5' —> 3')	AUA--AUC--GCU--GGC
	Peptide:	Ile-----Ile------Ala----Gly

CIA 26.2

Viruses consist of a strand of nucleic acid wrapped in a protein coat. Unlike higher organisms, viruses cannot replicate or manufacture protein independent of a host cell.

CIA 26.2

Type A influenza is the most severe flu, and type C influenza is the mildest.
Symptoms of influenza A: cough, fever, muscle aches, headache, sore throat, conjunctivitis; may lead to pneumonia.
Symptoms of influenza B: fever, respiratory problems, digestive problems.
Symptoms of influenza C: mild respiratory symptoms.

CIA 26.3

The influenza vaccination for 2015–2016 includes the inactivated forms of H1N1 and H3H2 strains of influenza A and the Phuket strain of influenza B.

CIA 26.4

Rapid variation in virus antigens make it difficult to develop a universal influenza vaccine.

26.16 *Amino Acid*   *Possible Codons*   (5' –> 3')

(a) Val	GUU GUC GUA GUG
(b) Phe	UUU UUC
(c) Asn	AAU AAC
(d) Gly	GGU GGC GGA GGC
(e) Met	AUG

26.17 The sequence guanine-adenine-guanine (GAG) codes for the amino acid glutamic acid.

26.18 *Codon*   *Amino Acid*

(a) AUC	Ile
(b) GCU	Ala
(c) CGA	Arg
(d) AAG	Lys

26.19 Six mRNA base triplets code for leucine: UUA, UUG, CUU, CUC, CUA, CUG. Many possible mRNA strands could code for this tripeptide if different codons were used for each Leu. Six of them are listed below:

5' UUAUUGCUU 3'   5' UUAUUGCUC 3'   5' UUAUUGCUA 3'
5' UUAUUGCUG 3'   5' UUACUUCUC 3'   5' UUACUUCUA 3'

26.20,   26.21

mRNA sequence:	5' CUC—AUU—CCA—UGC—GAC—-GUA 3'
Amino-acid sequence:	Leu——Ile——Pro——-Cys—-Asp——Val
tRNA anticodons:	3' GAG UAA GGU ACG CUG CAU 5'

Understanding Key Concepts

26.22

Guanosine 5'-monophosphate

26.23

The sequence of the left chain: 5' –A–G–T–C– 3'.
The sequence of the right chain: 3' –T–C–A–G– 5'.

26.24

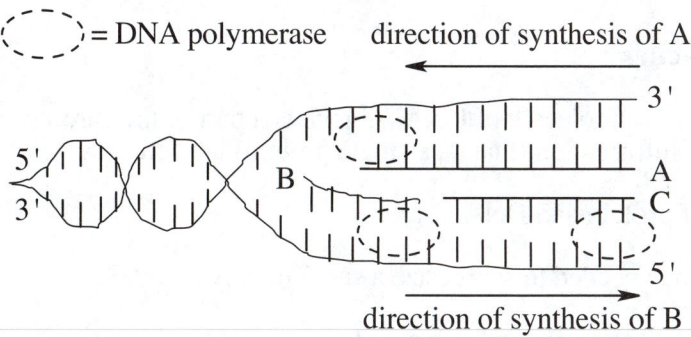

Segments B and C are joined by the action of a DNA ligase enzyme.

26.25 The sugar–phosphate backbone is found on the exterior of the DNA double helix. Since the phosphate groups are negatively charged, histone proteins should have a large number of positively charged amino acids, such as lysine, arginine, and histidine, on their surfaces.

26.26

26.27 Each amino acid has more than one mRNA codon.
Codons for Gln: CAA, CAG
Codons for His: CAC, CAU
Codons for Pro: CCC, CCU, CCA, CCG
Codons for Gly: GGG, GGU, GGA, GGC
Using the first codon listed:

(a) 5′ C A A C A C C C C G G G 3′ mRNA

(b) 3′ G T T G T G G G G C C C 5′ DNA template strand

(c) 5′ C A A C A C C C C G G G 3′ DNA informational strand

The upper strand of DNA is the template for the synthesis of mRNA, and its sequence of bases is the complement of the sequence of bases that make up the pro-TRH gene (with T replacing U). The lower strand is the informational strand.
(d) To find the number of DNA sequences, multiply the numbers of different codons for each amino acid.
2 x 2 x 4 x 4 = 64 possible DNA sequences code for the first four amino acids in pro-TRH.

Additional Problems

DNA, Chromosomes, and Genes

26.28 A chromosome is an enormous molecule of DNA. A gene is a part of the chromosome that codes for a single piece of information (that is, a single protein) needed by a cell.

26.29 Chromatin is composed of protein and DNA.

26.30 A gene carries the DNA code needed to synthesize a specific polypeptide.

26.31 There are 46 chromosomes (23 pairs) in a human cell.

Structure and Function of Nucleic Acids

26.32

26.33 Ribose is the sugar in RNA, and deoxyribose is the sugar in DNA. The —OH group at carbon 2 of ribose is replaced by —H in deoxyribose.

26.34 (a) Adenine, cytosine, guanine, and thymine are the four heterocyclic bases in DNA.
(b) Adenine, cytosine, guanine, and uracil are the four major heterocyclic bases in RNA.
(c) Adenine, cytosine, and guanine are common to both DNA and RNA. Thymine differs from uracil in having a methyl group at position 5 of the pyrimidine ring.

26.35 The purine bases (two fused heterocyclic rings) are adenine and guanine. The pyrimidine bases (one heterocyclic ring) are cytosine, thymine (in DNA), and uracil (in RNA).

26.36, 26.37, 26.40

Bond between phosphate group and C5 of sugar

Phosphate group

Ribose

Uracil

Bond between uracil and C1 of ribose

Uridine 5'-monophosphate

A phosphate ester bond is formed between the phosphate group and the sugar. An *N*-glycosidic bond is formed between the base and the sugar. Water is removed in the formation of both the sugar–phosphate linkage and the sugar–base linkage.

26.38 The 5' end of a nucleotide is a phosphate group bonded to carbon 5' of ribose. The 3' end is an —OH group bonded to carbon 3' of ribose.

26.39 Polynucleotides are synthesized from 5' to 3'.

26.40 See Problem 26.36.

26.41

5' end

Uridine

Cytidine

3' end

Base Pairing

26.42 (a) *Base pairing* is the hydrogen-bonded pairing of two complementary heterocyclic bases in the double helix of DNA and during replication, transcription, and translation.
(b) Adenine pairs with thymine (or uracil, in RNA), and guanine pairs with cytosine.
(c) Adenine and thymine (or uracil) form two hydrogen bonds with each other. Cytosine and guanine form three hydrogen bonds with each other.

26.43 The DNA double helix is held together by hydrogen bonds between base pairs and by hydrophobic interactions between the bases.

26.44 Bases that are complementary form hydrogen bonds with each other and thus always occur in pairs. For example, adenine is always hydrogen-bonded to thymine or uracil, and guanine is always hydrogen-bonded to cytosine.

26.45 The percent of A always equals the percent of T, since A and T are complementary. The same is true for G and C. Thus, sea urchin DNA contains about 32% each of A and T, and 18% each of G and C.

26.46 If a sample of DNA is 22% G, the percent of C is also 22%, since G and C are complementary. The sample contains 28% A and 28% T, because A and T are also complementary.

Nucleic Acids, Replication of DNA, and Structure and Replication of RNA

26.47 *Similarities:* Replication, transcription, and translation are all polymerizations in which a nucleic acid is used as a template for the synthesis of another biopolymer. In all of these processes, hydrogen bonding is used to bring the subunits into the correct position for bond formation and to determine the order of assembly.
Differences: In replication, DNA makes a copy of itself. In transcription, DNA is used as a template for the synthesis of mRNA. In translation, mRNA is used as a template for the synthesis of proteins. Replication and transcription take place in the nucleus of cells, and translation takes place in ribosomes.

26.48 A number of replication forks are needed when DNA is duplicated in order to increase the speed of replication of the enormous DNA chain.

26.49 To say that DNA replication is semiconservative means that each of the two new copies of DNA has a strand of DNA that was the original template and one strand that is newly synthesized.

26.50 *Messenger RNA (mRNA)* carries the genetic message from DNA to ribosomes.
Ribosomal RNA (rRNA) complexes with protein to form ribosomes, where protein synthesis takes place.
Transfer RNA (tRNA) transports specific amino acids to the ribosomes, where they are incorporated into proteins.

26.51 DNA is the largest nucleic acid, mRNA is intermediate in size, and tRNA is the smallest of the three nucleic acids.

Transcription: RNA Synthesis

26.52 Exons are sequences of DNA that code for part of a specific protein. Introns are sequences of DNA that are found between exons and whose function is unclear; they are spliced from mRNA before protein synthesis.

26.53 Introns may serve a regulatory purpose, they may separate exons that can be mixed or matched, or they may merely be left over from an earlier stage of development.

26.54 Transcribed RNA is complementary to the template strand of DNA.

26.55 A codon is a sequence of three nucleotides on mRNA that codes for a specific amino acid in protein synthesis.

The Genetic Code and Translation

26.56 An anticodon is a sequence of three nucleotides that is complementary to a sequence on a codon. The anticodon occurs on tRNA and matches the codon sequence on mRNA when it brings the correct amino acid into position for protein synthesis.

26.57 Ser, Arg, and Leu each have six codons; Ala, Gly, Pro, Val, and Thr each have four. Met and Trp each have only one codon. To an extent, these numbers are correlated with the relative abundance of these amino acids in proteins.

26.58 *Amino Acid*   *Codons (5' –> 3')*

(a) Val	GUU	GUC	GUA	GUG		
(b) Arg	CGU	CGC	CGA	CGG	AGA	AGG
(c) Ser	UCU	UCC	UCA	UCG	AGU	AGC

26.59, 26.60

Codon (5' –> 3')	*Amino Acid*	*tRNA Anticodon (3' –> 5')*
(a) CCC	Pro	GGG
(b) GCG	Ala	CGC
(c) UUA	Leu	AAU

26.61 Codons are written (5' –> 3'), and anticodons are written (3' –>5')

(a) Val	Codons:	GUU	GUC	GUA	GUG		
	Anticodons:	CAA	CAG	CAU	CAC		
(b) Arg	Codons:	CGU	CGC	CGA	CGG	AGA	AGG
	Anticodons:	GCA	GCG	GCU	GCC	UCU	UCC
(c) Ser	Codons:	UCU	UCC	UCA	UCG	AGU	AGC
	Anticodons:	AGA	AGG	AGU	AGC	UCA	UCG

26.62–26.64

The DNA sequence of the template strand is complementary to the DNA sequence of the informational strand. The mRNA sequence is identical to the DNA sequence on the informational strand, except that U in mRNA replaces T in DNA.

Informational strand: (5' —> 3')	TAC--CCT
Template strand: (3' —> 5')	ATG--GGA
mRNA: (5' —> 3')	UAC--CCU
Dipeptide:	Tyr----Pro

26.65

Informational strand: (5' —> 3')	GTC--AGT--ACG--TTA
mRNA: (5' —> 3')	GUC--AGU-ACG--UUA
Tetrapeptide:	Val----Ser----Thr----Leu

26.66 Metenkephalin:
mRNA (5' —> 3'):

Tyr—Gly—Gly—Phe—Met   Stop
UAU–GGU–GGU–UUU–AUG–UAA
UAC GGC GGC UUC   UAG
          GGG GGG   UGA
          GGA GGA

26.67 Using the first set of base pairs in Problem 26.66 to solve this problem:

Informational strand (5' —> 3'): TAT–GGT–GGT–TTT—ATG–TAA
Template strand (3' —> 5'):   ATA–CCA–CCA–AAA—TAC–ATT

26.68 A tRNA molecule is cloverleaf-shaped and contains 70–100 nucleotides. The tRNA anticodon triplet is on one "leaf", and an amino acid bonds covalently to the 3' end.

26.69 The tRNAs for each amino acid differ in their anticodon sequences.

Conceptual Problems

26.70 The two mRNA codons for glu are GAA and GAG. Of the four codons for val (GUU, GUC, GUA, and GUG), the last two differ from the glu codons by one base. Thus, a change in one base can completely alter the structure and function of a protein.

26.71 To code for preproinsulin, 81 x 3 = 243 bases are needed to code for the protein's 81 amino acids. In addition, a three-base "start" codon and a three-base "stop" codon are needed, for a total of 249 bases.

26.72 *Position 9:*

Horse amino acid = Gly   Human amino acid = Ser
mRNA codons (5' —> 3'):
GGU GGC GGA GGG   UCU UCC UCA UCG AGU AGC

DNA bases (template strand 3' —> 5'):
<u>CCA</u> <u>CCG</u> CCT CCC   AGA AGG AGT AGC <u>TCA</u> <u>TCG</u>

The underlined horse DNA base triplets differ from their human counterparts (also underlined) by only one base.

Position 30:

Horse amino acid = Ala   Human amino acid = Thr
mRNA codons (5' —> 3'):
GCU GCC GCA GCG   ACU ACC ACA ACG

DNA bases (template strand 3' —> 5'):
CGA CGG CGT CGC   TGA TGG TGT TGC

Each group of three DNA bases from horse insulin has a counterpart in human insulin that differs from it by only one base. It is possible that horse insulin DNA differs from human insulin DNA by only two bases out of 159!

26.73 If a protein does not have methionine as its first amino acid, the methionine is removed after protein synthesis is complete.

26.74 Recall from Problems 26.45 and 26.46 that the percent of A equals the percent of T, the percent of C equals the percent of G, and all percents add to 100%. If a DNA molecule contains 22% ATP, it will also contain 22% TTP, 28% CTP, and 28% GTP. If you check the list of available nucleoside triphosphates, you should notice that only 22% dCTP is available for replication. dCTP is the nucleoside triphosphate that is limiting to the replication, and dTTP is in excess.

Group Problems

26.75 AIDS is caused by a retrovirus—an RNA virus that enters a host cell and forms single-stranded DNA from its own RNA by using an enzyme, reverse transcriptase, provided by the virus. This DNA makes a copy of itself and inserts itself into the host's DNA, where it produces more RNA to assemble into new virus particles.

There are number of challenges for treatment of AIDS. First, it is difficult to find a drug that is effective against the AIDS virus without harming the host. Currently, AIDS drugs consist of synthetic nucleosides, which, when incorporated into the DNA of the virus, slow down production of new virus particles. Also, it would be desirable to develop an AIDS vaccine, but the rapid rate of mutation of the virus makes that difficult.

26.76 Avian flu viruses may be transmitted to humans from domesticated birds, which have been infected by migratory waterfowl. The virus can also be transmitted from an intermediate host, such as swine.

26.77 Influenza A viruses are described by a code that describes the hemagglutinins (H) and the neuraminidases (N) in the virus. A list of viruses:

H1N1: humans, pigs	H2N2: humans
H3N2: humans, pigs	H5N1: (humans), birds
H7N7: (humans), poultry	H1N2: humans, pigs
H3N8: birds, horses, dogs	H7N7: horses
H9N2: (humans), poultry	H5N2: (humans), poultry

Most of the viruses are endemic to the animal or bird species listed. At present, only H1N1, H1N2, and H3N2 are of concern for humans. The viruses in parentheses have affected humans in contact with the endemic species but have not caused major outbreaks.

26.78 The H1N1 virus was responsible for the 1918 influenza pandemic, and the H5N1 virus is present in avian flu. Since these viruses can undergo antigenic shift in host animals, there is concern when infected birds and animals harbor influenza viruses.

Self-Test for Chapter 26
Multiple Choice

1. The name for a nucleoside formed from cytosine and 2'-deoxyribose is:
   (a) deoxycytidine 5'-monophosphate (b) 2'-deoxycytidine (c) 2'-deoxycytosine
   (d) 2'-deoxycytosine 5'-monophosphate

2. Adenine and thymine are complementary because:
   (a) They are both purines. (b) They are both pyrimidines. (c) They both occur in DNA.
   (d) They form two hydrogen bonds to each other.

3. If the base sequence CAC-TTA-GGT appears on the informational strand of DNA, which sequence occurs in mRNA?
   (a) CAC-TTA-GGT (b) CAC-UUA-GGU (c) GTG-AAT-CCA (d) GUG-AAU-CCA

4. Which statement about the RNA base sequence CGG-AAA-GUU is true?
   (a) It codes for a basic tripeptide. (b) It contains a "stop" codon.
   (c) Changing the third codon from GUU to GUC changes the amino acids in the tripeptide.
   (d) The 5' end is on the right, and the 3' end is on the left.

5. Which of the following completes the initiation phase of protein synthesis?
   (a) the small subunit of a ribosome (b) the large subunit of a ribosome (c) mRNA (d) tRNA

6. The process by which the genetic message contained in DNA is read is called:
   (a) replication (b) transcription (c) translation (d) translocation

7. Which of the following heterocyclic bases contains a methyl group?
   (a) guanine (b) cytosine (c) thymine (d) uracil

8. An enzyme involved in translation is:
   (a) DNA polymerase (b) restriction endonuclease (c) RNA polymerase (d) releasing factor

9. Which of the following statements about DNA replication is untrue?
   (a) Replication is semiconservative.   (b) Each DNA strand has many replication forks.
   (c) DNA ligase catalyzes bond formation between each arriving nucleotide.
   (d) Replication proceeds in the 5'—>3' direction.

10. Which heterocyclic base contains only carbon, hydrogen, and nitrogen?
    (a) adenine   (b) guanine   (c) cytosine   (d) thymine

Sentence Completion

1. _____ is the process by which an identical copy of DNA is made.

2. The two strands of DNA have base sequences that are _____.

3. When a cell is not actively dividing, its nucleus is occupied by _____.

4. Another name for an RNA virus is a _____.

5. An _____ is a segment of a gene that does not code for protein synthesis.

6. tRNA binds to an amino acid by a(n) _____ linkage.

7. A _____ consists of a heterocyclic amine base bonded to an aldopentose.

8. An enzyme called a _____ _____ frees the polypeptide chain from the last tRNA during protein synthesis.

9. The point of replication of a DNA chain is called the _____ _____.

10. The sugar _____ has an —H instead of an —OH at its 2' position.

11. The _____-_____ model describes DNA as two polynucleotide strands coiled about each other in a double helix.

12. The enzyme _____ _____ catalyzes the bonding of nucleotides to form new strands of DNA.

True or False

1. There are a total of 64 codons for the 20 amino acids.

2. The amount of adenine in an organism's DNA is equal to the amount of guanine.

3. In transcription, mRNA is a copy of the informational strand of DNA.

4. An amino acid codon occurs on tRNA.

5. The codon AUG initiates protein synthesis.

6. Synthetase enzymes catalyze the formation of ester bonds between amino acids and their tRNAs.

7. The individual units of RNA and DNA are called nucleotides.

8. The two strands of the DNA double helix run in the same direction.

9. Transcription is the process by which the genetic message is decoded and used to make proteins.

10. Most human DNA contains genetic instructions.

11. Each amino acid is specified by more than one codon.

12. The correct name for the RNA nucleotide containing G is guanidine 5'-monophosphate.

Match each entry on the left with its partner on the right.

1. 5' end   (a) Enzyme used in replication

2. A-C-G   (b) Cysteine codon

3. U-G-C   (c) Forms three hydrogen bonds

4. Adenine   (d) Coded for by six codons

5. DNA ligase   (e) RNA containing both exons and introns

6. 3' end   (f) "Stop" codon

7. Leu   (g) Purine

8. Cytosine   (h) Enzyme used in transcription

9. hnRNA   (i) —OH group

10. Uracil   (j) Cysteine anticodon

11. U-G-A   (k) Pyrimidine

12. RNA polymerase   (l) Phosphate group

Chapter 27   Genomics

Chapter Outline

I.   Mapping the human genome (Section 27.1).
A. Two research groups have collectively decoded 99% of the human genome.
  1. The Human Genome Project.
     a. The Human Genome Project is made up of 20 international not-for-profit groups.
     b. The technique.
        i. For each chromosome, a genetic map is developed; the map shows the location of markers determined by inheritance.
        ii. A map of finer resolution (physical map) shows markers determined by other methods.
        iii. The chromosome is cut into large segments, and the resulting clones are organized into an overlapping sequence.
        iv. Each clone is fragmented into hundreds of pieces, which are labeled and separated by electrophoresis.
        v. The sequences are assembled into a nucleotide map.
  2. Celera.
     a. Celera is a for-profit research group.
     b. The technique.
        i. The entire genome is broken into fragments, without identifying their origin.
        ii. The fragments are copied many times, sequenced and separated.
        iii. The sequences are reassembled, using a supercomputer.
II. DNA and chromosomes (Sections 27.2–27.3).
A. Chromosome structure (Section 27.2).
  1. Telomeres.
     a. Telomeres are long series of repeating DNA that are found at the ends of each chromosome.
     b. Telomeres protect the ends of chromosomes from accidents.
     c. Telomeres also prevent the ends of chromosomes from pairing with other DNA.
     d. The enzyme telomerase catalyzes the addition of telomeres to DNA.
        i. Telomerase is more active in young cells than in older cells.
  2. Centromeres.
     a. Centromeres are constrictions that determine the shapes of chromosomes during cell division.
     b. Centromeres contain large repetitive base sequences.
  3. Other noncoding DNA.
     a. Not all of the functions of noncoding DNA are known.
        i. Some segments are regulatory.
        ii. Other proposed functions: accommodation of DNA folding, evolutionary roles.
  4. Genes.
     a. The coding sections of a gene (exons) are interspersed with noncoding introns.
B. Mutations and polymorphisms (Section 27.3).
  1. Mutations.
     a. An error in mRNA transcription is not serious.

b. An error in DNA replication is a mutation and is passed on each time DNA replicates.
   i. Substances that cause mutations are mutagens.
   ii. Each time mRNA is transcribed, the resulting protein will have an error.
   iii. The consequences of a mutation range from insignificant to lethal.
c. If the error occurs in a somatic cell, cancer may result.
d. If the error occurs in a germ cell, a genetic defect may be passed on to offspring.

2. Polymorphisms.
   a. Polymorphisms are variations in nucleotide sequences that are common within a given population.
   b. Replacement of one nucleotide by another in the same location is a single-nucleotide polymorphism (SNP).
      i. SNPs are responsible for most of the variation among human beings.
      ii. SNPs are responsible for many genetic diseases.
      iii. A catalog of SNPs has been developed; it lists one SNP for every 300 base pairs.
      iv. This catalog has been used to identify SNPs responsible for abnormalities.

III. Recombinant DNA and genomics (Section 27.4–27.5).
   A. Recombinant DNA technology is used to cut a specific gene out of one organism and insert it into the DNA of another organism (Section 26.4).
      1. The second organism is usually a bacterium.
      2. Bacterial DNA is contained in plasmids, which are easy to isolate.
   B. Technique of recombinant DNA.
      1. The plasmid is cleaved by a restriction endonuclease.
         a. There are more than 200 restriction endonucleases.
         b. Cleavage occurs at a specific site in a base sequence.
         c. Each DNA strand is left with a few unpaired bases, called "sticky ends."
      2. The gene to be inserted is also cleaved by the same endonucleases and thus has complementary sticky ends.
      3. The gene and the plasmid are mixed in the presence of a DNA ligase enzyme and the gene is inserted into the plasmid.
      4. The plasmid can be reinserted into a bacterial cell, and the bacteria, which multiply rapidly, manufacture the protein.
   C. Uses of recombinant DNA technology—genomics (Section 26.5).
      1. Agriculture.
         a. Transgenic plants and animals have been bred; their DNA includes single genes that code for desirable traits.
      2. Medicine.
         a. Gene therapy might cure inherited diseases by replacing a disease-causing gene with a healthy copy of that gene.
         b. A personal genomic survey might be helpful in drug and therapeutic choices.
         c. All of this work has social and ethical implications.

Solutions to Chapter 27 Problems

27.1   uouo**the**dtttrr**fat**naedigo**pred**sldjflsjfxx**rat**ponxbv**ate**ugfaqq**the**nqeut**bad**pagf**rat**meabrrx
(The fat red rat ate the bad rat).

HOC 27.1
(a) Let us use as an example Turner Syndrome (not on the list).
(b) Check the Internet for information; a good resource is
http://www.genome.gov/10001204.
Characteristics of Turner Syndrome:
   Other names: XO Syndrome, 45,X.
   Gene mutation: Occurs on the X chromosome, but the exact gene has not been
   identified.
   Prevalence: Occurs in approximately 1 in 2500 births.
   Symptoms and treatment: Affects only females; females are shorter than average, and
   their growth slows after age 3; they do not undergo puberty without hormone therapy;
   usually infertile; have frequent ear infections during childhood.
   Research: None at present.

(c) "Your daughter has been diagnosed with a rare genetic disease commonly known as
Turner Syndrome, which affects ~1 out of 2500 people and is not usually inherited. In
other words, there is nothing you could have done to prevent this.

"Turner Syndrome affects development in females, resulting in slower growth rates. In
other words, your daughter will be shorter than normal. Other symptoms include an early
loss of ovarian function, so that your daughter will not undergo puberty without hormone
therapy, and a reduced likelihood of bearing children. Here is the good news: with
hormone therapy, your daughter will go through the normal stages of puberty. Therefore,
we recommend that around the age in which she would normally undergo puberty, we
place her on a hormone therapy program."

CIA 27.1
A polymerase chain reaction is used to produce a large number of copies of a specific DNA
chain.

CIA 27.2
An outline of a polymerase chain reaction:
(1) A solution of DNA that is to be copied is heated so that the DNA separates into two
   strands.
(2) Two oligonucleotide primers, which are complementary to the ends of the segment of
   DNA to be copied, are attached to the DNA strands.
(3) DNA polymerase and nucleotide triphosphates are added, and the segment of DNA
   between the primers is copied.
The process is repeated many times until a large amount of the desired DNA is produced.
The steps can be carried out at an elevated temperature because a heat-stable enzyme *(Taq
polymerase)* is used.

CIA 27.3
Taq polymerase can withstand high temperatures and does not need to be replaced for each
PCR cycle.

27.2   Look back to Table 26.3 and note that UGU codes for cysteine, and UGG codes for tryptophan. In the synthesized protein, a tryptophan would replace a cysteine. Since cysteine forms disulfide bridges, the protein's three-dimensional structure would possibly be disrupted and its function might be affected.

27.3   The complementary DNA strand is drawn by pairing the bases and by reversing the labels of the two ends of the strands. The restriction enzyme cleaves the complementary strand at the same location.

$$5' \quad -A- // -G-A-T-C-T- \quad 3'$$
$$3' \quad -T-C-T-A-G- // -A- \quad 5'$$

27.4   The explanation in the previous problem applies here.

$$5' \quad -G- // -A-A-T-T-C- \quad 3'$$
$$3' \quad -C-T-T-A-A- // -G- \quad 5'$$

27.5   To be sticky, base sequences must be complementary. Draw the complement of the first sequence. If the complement you draw is identical to the second sequence, the two sequences are sticky.
(a) The pairs are sticky.
(b), (c) The pairs aren't sticky.

27.6   (a) The field of *comparative genomics* deals with the identification of genes that perform identical functions in mice and humans.
(b) A person working in *genetic engineering* might be involved with creating a variety of wheat that is not harmed by a herbicide that kills weeds that interfere with wheat crops.
(c) *Pharmacogenetics* is used for screening an individual's genome to choose the most appropriate pain-killing medication for that person.
(d) *Bioinformatics* is the computer analysis of base-sequence information from groups of people with and without a certain disease to discover where the disease-causing polymorphism lies.

CIA 27.4

If the chance of two people having an identical fingerprint match is one in 1.5 billion, around four people might have an identical match.

CIA 27.5

The five steps of DNA fingerprinting using the RFLP method.
(1) A sample of DNA is digested with a restriction endonuclease.
(2) The resulting fragments are separated by size using gel electrophoresis.
(3) The DNA fragments are transferred to a nylon membrane.
(4) The DNA blot is treated with a radioactive DNA probe, which binds to the sticky ends of the DNA fragments.
(5) The radioactive RFLPs can be identified by exposing an X-ray film to the blot.

The PCR method is favored because it produces a greater quantity of DNA in less time.

CIA 27.6

A VNTR (variable number tandem repeat) is a region of noncoding DNA that contains repeating nucleotide sequences. The number of repeats of a given VNTR varies among individuals. Comparing the VNTRs on several genes of an individual allows scientists to create a DNA fingerprint of that person.

Understanding Key Concepts

27.7   (1) A genetic map, which shows the location of markers 1 million nucleotides apart, is created. (2) Next comes a physical map, which refines the distance between markers to 100,000 base pairs. (3) The chromosome is cleaved into large segments of overlapping clones. (4) The clones are fragmented into 500 base pieces, which are sequenced.

27.8   The variations in DNA sequences occupy only a small part of the genome; the rest is identical among humans. A diverse group of individuals contributed DNA to the project.

27.9   *Telomeres* protect the chromosome from mutations, prevent DNA from bonding to other DNA, and are believed to play a role in human aging. *Centromeres* form the constrictions that determine the shapes of chromosomes during cell division. *Promoter sequences* regulate which genes will be turned on. *Introns* are noncoding segments that are interspersed with gene-encoding exons and whose function is unknown.

27.10 Both mutations and polymorphisms are variations in base sequences. A mutation is an error in a base sequence that is transferred during DNA replication, occurs in a very small number of individuals, and is caused by a random error or a mutagen. A polymorphism is a variation in DNA sequence that is common within a given population.

27.11 Recombinant DNA contains two or more DNA segments that do not occur together in nature. The DNA that codes for a specific human protein can be incorporated into a bacterial plasmid using recombinant DNA technology. The plasmid is then reinserted into a bacterial cell, where its protein-synthesizing machinery makes the desired protein.

27.12 Major benefits of genomics include creation of disease-resistant and nutrient-rich crops, gene therapy to cure or prevent genetic diseases, and genetic screening that leads to more effective health care. Major negative outcomes are associated with the misuse of an individual's genetic information, and with prediction of a genetic disease for which there is no cure.

Additional Problems

The Human Genome Map

27.13 Genomics is the study of entire sets of genes and of their functions.

27.14 Celera broke the genome into many unidentified fragments. The fragments were multiplied and cut into 500 base pieces, which were sequenced. A supercomputer was used to determine the order of the bases. An advantage was that the order of the bases in each fragment could be found quickly, although time was needed to put the fragments in the correct order.

27.15 Competition caused the working draft of the Human Genome Project to be completed in 15 months, rather than in the 4 years originally anticipated.

27.16 Fifty percent of the human genome is composed of repeat sequences.

27.17 Approximately 3 billion base pairs were identified.

27.18 (a) About 200 genes are identical to bacterial genes.
(b) A single gene may produce several proteins.

27.19 The most surprising result is that humans have about 19,000 genes, a number much smaller than the previous estimate of 100,000 genes.

27.20 The clones used in DNA mapping are multiple identical copies of DNA segments from a single individual. Clone samples large enough for experimental manipulation are essential for mapping.

Chromosomes, Mutations, and Polymorphisms

27.21 Telomeres protect the ends of chromosomes from changes that might alter the coding sequence of DNA.

27.22 The youngest cells have long telomeric sequences of DNA, and the telomeres shorten as cells age. Thus, cell age can be predicted by telomere length.

27.23 Telomerase is responsible for adding telomeres to DNA. Telomerase is most active in young cells and least active in older cells.

27.24 The centromere is the constriction that determines the shape of a chromosome during cell division. It is composed of noncoding and repetitive DNA.

27.25 A mutagen is an external agent that can cause an error in base sequence in DNA.

27.26 An error in the sequence of RNA affects only one molecule of RNA; other intact molecules of RNA can still carry out protein synthesis. A mutation in DNA is much more damaging, however, because there is only one molecule of DNA per gene, and any error will be copied into all subsequent DNA molecules during replication.

27.27 Some mutations occur as point mutations, and others occur as frameshift mutations.

27.28 A single-nucleotide polymorphism (SNP) is the replacement of one nucleotide by a different nucleotide at the same location in a strand of DNA.

27.29 An SNP can result in the change in identity of an amino acid inserted into a protein at a particular location in a polypeptide chain. The effect of an SNP depends on the function of the protein and the nature of the SNP. Most of the variations between individual human beings are due to SNPs.

27.30 The effects of some SNPs are negligible. Other SNPs are responsible for differences in hair and eye color and for genetic diseases such as sickle-cell anemia, a type of epilepsy, and total color blindness. Some SNPs are associated with the risk of developing diseases such as Alzheimer's disease and breast cancer, whereas other SNPs impart resistance to diseases, including AIDS.

27.31 Analysis of an individual's DNA for SNPs would allow a physician to predict the age at which inherited diseases might become active, their severity, and the response of the individual to various types of treatment.

27.32 Not necessarily. Since many amino acids are coded by several different base sequences, some substitutions cause no change in the amino acid composition of the protein being synthesized.

27.33 A change in the nature of an amino acid's side chain determines whether an amino acid will create a major or a minor alteration in the protein's shape. Similar side chains have a lesser effect than very different side chains.

27.34 *Normal Codon*　　　　*Codes for:*　　　　*Mutated Codon*　　　　*Codes for:*

	Normal Codon	Codes for:	Mutated Codon	Codes for:
(a)	UCA	Ser	UCG	Ser
(b)	UAA	Stop	UAU	Tyr

In (a), the mutated mRNA codes for the same amino acid as the nonmutated mRNA, and thus the mutation has no effect. In (b), the mutated mRNA replaces a "stop" codon with a Tyr codon; instead of stopping protein synthesis, mRNA continues adding amino acids to the polypeptide chain. Mutation (b) is much more serious that mutation (a).

27.35 *Normal Codon*　　　　*Codes for:*　　　　*Mutated Codon*　　　　*Codes for:*

	Normal Codon	Codes for:	Mutated Codon	Codes for:
(a)	GCU	Ala	GCC	Ala
(b)	ACU	Thr	AUU	Ile

In (a), the normal mRNA and the mutated mRNA code for the same amino acid (Ala). In (b), the mutated mRNA codes for an amino acid that has a hydrophobic side chain, whereas the normal amino acid has a hydrophilic side chain. The second mutation is more serious because the mutation changes the type of amino acid and would probably affect the shape and function of the protein.

Recombinant DNA

27.36 The DNA of bacterial cells occurs in plasmids, each of which carries only a few genes. The plasmids are easy to isolate, and several copies of each plasmid are present in a bacterial cell. The DNA of plasmids replicates rapidly by normal base pairing, and it is possible for foreign DNA to enter bacterial plasmids.

27.37 Proteins can be produced in greater quantity than otherwise possible.

27.38 Electrophoresis is used to separate DNA fragments by size.

27.39 Sticky ends are unpaired bases on each end of the cleaved DNA. Recombinant DNA is formed when the sticky ends of the DNA of interest and of the DNA of the plasmid have complementary base pairs and can be joined by a DNA ligase. If the sticky ends are not complementary, the DNA segments can't be joined.

27.40 *Unpaired bases:*　(a) GGTAC　　(b) ACCCA　　(c) GTGTC
Sticky sequence:　　CCATG　　　TGGGT　　　CACAG

27.41 Write out the sticky sequence for the first of the two base sequences. If the sticky sequence agrees with the second sequence shown, the two sequences are sticky.
(a) The sequence TTAGC (5' to 3') is sticky with AATCG (3' to 5'), which is the same as GCTAA (5' to 3'). Thus, the two sequences are sticky.
(b) CGTACG isn't sticky with CCTTCG.

Genomics

27.42 Pharmacogenomics is the study of the genetic basis of responses to drug treatment. Pharmacogenomics helps doctors prescribe the most effective medicine for a patient, based on the patient's genetic makeup.

27.43 Both genetic engineering and gene therapy are involved with altering the genetic makeup of an organism. Gene therapy is used on humans, with the goal of curing or preventing disease. Genetic engineering is used for plants or other nonhuman organisms, with the goal of producing a new substance or of changing a genetic trait.

27.44 Corn and soybeans have been genetically altered and improved to provide disease and herbicide resistance.

27.45 Possible advantages include treatment of a genetic defect by gene therapy, choice of the most effective drug for treatment of a disease, and lifestyle adjustments to avoid development of a disease. Major drawbacks include identification of a disease for which there is no cure and discrimination against a person on the basis of genotype.

27.46 The field of bioethics studies the ethical issues involved with advances in genomics. Among the topics studied by bioethicists: ownership of, and access to, genetic information; implications of genetic testing; prevention of genetic disabilities; use of gene therapy.

Conceptual Problems

27.47 A monogenic disease is caused by the variation in just one gene.

27.48 A vector is the agent used to carry therapeutic quantities of DNA directly into cell nuclei.

27.49 TATGACT is sticky with ATACTGA.

27.50

	Normal	*Mutated*
DNA:	5' A-T-T-G-G-C-C-T-A 3'	5' A-C-T-G-G-C-C-T-A 3'
mRNA:	5' A-U-U-G-G-C-C-U-A 3'	5' A-C-U-G-G-C-C-U-A 3'
Amino acids:	Ile------Gly-----Leu	Thr-----Gly-----Leu

The mutation would substitute a Thr (hydrophilic side chain) for an Ile (hydrophobic side chain) and would affect the tertiary structure of the protein.

27.51 A restriction endonuclease is an enzyme that recognizes a specific DNA sequence and cleaves the DNA between two particular nucleotides in that sequence.

27.52 A mutation must occur in a germ cell (sperm or egg) in order for it to be passed down to future generations.

Group Problems

27.53 Advantages: knowing your genetic strengths and weaknesses; being able to make lifestyle adjustments to counteract predicted genetic problems; determining if you are a carrier of genetic diseases.

Disadvantages: misuse of genetic data; fatalism; knowing that you are predicted to have a genetically-related disease for which there's no cure.

27.54 Much research is currently being done on gene therapy, both for genetic diseases and in cancer therapy, although few results have reached a clinical stage. Among the research projects:
Parkinson's Disease: Gene therapy may reprogram brain cells to produce dopamine, which helps control motor function.
Huntington's Disease: It may be possible to switch off the gene mutation that causes Huntington's Disease.
Prostate cancer: Gene therapy may stimulate the immune system to attack the tumor.
Muscular Dystrophy: After identification of the gene sequence that helps muscle tissue function, it may be possible to develop a drug that counteracts the mutated gene.

27.55 "Unlocking Life's Code" is an exhibit created by the Smithsonian Institute that will travel to selected museums from 2014 to 2018. It includes a timeline of decoding the DNA of prehistoric animals and humans, starting from 1984. Use the above phrase to perform a web search that will tell you more.

Self-Test for Chapter 27

Multiple Choice

1. A field of study that deals with the use of computers to manage and interpret genetic information is:
(a) genetic engineering   (b) proteomics   (c) bioinformatics   (d) gene therapy

2. Approximately how many genes are in the human genome?
(a) 3.1 billion   (b) 1.4 million   (c) 100,000   (d) 20,000

3. Which of the following is not a characteristic of a telomere?
(a) It has a repeating series of nucleotides.   (b) Each chromosome has one telomere.
(c) The nucleotide groups of telomeres are noncoding.
(d) A telomere protects the end of a chromosome from accidental damage.

4. Which of the following mutations is least likely to cause genetic damage?
(a) AUC to AUA   (b) AUC to AUG   (c) GUU to GGU   (d) CAA to CGA

5. Which of the following mutations is most likely to cause genetic damage?
(a) AUC to AUA   (b) AUC to AUG   (c) GUU to GGU   (d) CAA to CGA

6. All of the following are composed of noncoding DNA except:
   (a) telomere   (b) centromere   (c) promoter sequence   (d) exon

7. Which of the following is not a characteristic of plasmids?
   (a) They are easy to isolate.   (b) They replicate in the same way as human DNA does.
   (c) They are circular.   (d) Each cell has one plasmid.

8. A hereditary disease caused by a defect in hemoglobin is:
   (a) cystic fibrosis   (b) sickle-cell anemia   (c) Tay–Sachs disease   (d) albinism

9. In which type of map are marker locations determined by inheritance patterns?
   (a) clone map   (b) physical map   (c) chromosome   (d) genetic map

10. DNA fingerprinting uses all of the following except:
    (a) a radioactive DNA probe   (b) restriction endonuclease   (c) a fluorescent marker
    (d) gel electrophoresis

Sentence Completion

1. Groups of unpaired bases at the end of a DNA chain treated with a restriction endonuclease are known as _____ ____.

2. Repeat sequences in noncoding DNA are often referred to as ____ DNA.

3. In recombinant DNA technology, plasmids are cut open with a(n) _____ _____.

4. The heat-stable enzyme used in the polymerase chain reaction is _____.

5. The ELSI program of the National Human Genome Research Institute deals with _____.

6. Mutations may be caused by exposure to _____ _____ or to chemicals called _____.

7. A(n) _____ is a segment of a gene that does not code for protein synthesis.

8. The replacement of one nucleotide by another in the same location along the polynucleotide chain is known as a(n) ____ _____ _____.

9. The field of _____ is directed to the study of the complete set of proteins coded by a genome or synthesized within a given type of cell.

10. A(n) _____ is a constriction that determines the shape of a chromosome during cell division.

11. A point mutation that doesn't change the amino acid incorporated into a protein is a(n) _____ mutation.

12. The film result of DNA fingerprinting is a(n) _____.

True or False

1. In a frameshift mutation, the number of inserted or deleted nucleotides is not a multiple of 3.

2. "SNP" stands for single-nucleoside polymorphism.

3. The DNA used for the Human Genome Project came from a single donor.

4. By reorganization of exons during mRNA synthesis, a gene can produce several different proteins.

5. A person working in the field of pharmacogenomics is likely to work at a company that manufactures drugs.

6. Mapping of each chromosome starts at one end and works to the other end.

7. Telomeres consist of a long series of repeats of the nucleotide sequence TTAGGG.

8. The nucleotide sequences -GAACT- and -CTGGA- are "sticky."

9. The mutation of GAU to GAG is a nonsense mutation.

10. A mutation in RNA causes little damage.

11. Telomerase is responsible for removing telomeres from DNA.

12. Human insulin can be manufactured by recombinant DNA technology.

Match each entry on the left with its partner on the right.

1. Silent mutation	a. UUG —> UAG
2. Exon	b. UUA —> CUA
3. Human Genome Project	c. Shows little telomerase activity
4. Mature somatic cell	d. Addition or loss of a nucleotide
5. Missense mutation	e. Occurs in a very small number of individuals
6. Plasmid	f. Noncoding DNA
7. Mutation	g. Uses a series of "maps" for genome mapping
8. Nonsense mutation	h. Occurs in at least 1% of population
9. Celera	i. Coding DNA
10. SNP	j. GCU —> GGU
11. Intron	k. "Shotgun" approach to DNA mapping
12. Frameshift mutation	l. Bacterial DNA

Chapter 28 Chemical Messengers

Chapter Outline

I. Introduction to chemical messengers (Section 28.1).
    A. Control of vital functions is accomplished by chemical messengers.
        1. The messengers interact with the cells of a target tissue.
        2. The message is delivered by interaction between the chemical messengers and receptors.
        3. Noncovalent interactions draw messengers and receptors together.
        4. The results of the interaction are chemical changes within the target cell.
    B. Two kinds of chemical messengers are important in the body.
        1. Hormones are the chemical messengers of the endocrine system.
            a. Hormones travel through the bloodstream.
            b. The responses produced by hormones are slow but long-lasting.
        2. The chemical messengers of the nervous system are neurotransmitters.
            a. Signals travel very quickly along nerve fibers.
            b. Neurotransmitters carry the message of the signal across the gap that separates one nerve fiber from the next fiber.
            c. The effects of neurotransmitters are very short-lived.
II. Hormones (Sections 28.2–28.4).
    A. Hormones and the endocrine system (Section 28.2).
        1. The endocrine system consists of all the glands that secrete hormones.
        2. The endocrine glands are managed by the hypothalamus in three different ways.
            a. Direct neural control by a nervous system pathway from the hypothalamus.
            b. Direct release of hormones from the hypothalamus.
            c. Indirect control by release of regulatory hormones.
        3. There are three types of hormones.
            a. Polypeptide hormones.
            b. Steroid hormones.
            c. Amino acid derivatives.
        4. The signal of a hormone enters cells in either of two ways.
            a. Steroid hormones pass through the hydrophobic cell membrane.
            b. Other hormones bind with receptors in the cell wall.
    B. How hormones work: epinephrine (Section 28.3).
        1. Epinephrine carried in the bloodstream binds to a receptor on the surface of a cell.
        2. The hormone–receptor complex interacts with a nearby G protein, causing it to bind GDP.
        3. GDP from the G protein is exchanged for GTP in the cytosol.
        4. This complex activates the enzyme adenylate cyclase, which resides in the cell membrane.
        5. Adenylate cyclase catalyzes production within the cell of the second messenger, cyclic AMP.
        6. Cyclic AMP initiates reactions that activate glycogen phosphorylase, which catalyzes release of glucose from storage.
        7. After the emergency, a phosphodiesterase catalyzes hydrolysis of cyclic AMP to AMP.

C. Types of hormones (Section 28.4).
   1. Amino acid derivatives.
      a. Many amino acid derivatives are also neurotransmitters.
         i.   Examples include epinephrine, dopamine, and norepinephrine.
      b. Thyroxine is an iodine-containing hormone that regulates the synthesis of various enzymes.
   2. Polypeptide hormones.
      a. TRH and TSH control the thyroid gland.
      b. Vasopressin and oxytocin are small cyclic polypeptides.
      c. Insulin regulates glucose metabolism.
   3. Steroid hormones.
      a. Mineralocorticoids (such as aldosterone) regulate ionic balance in cellular fluid.
      b. Glucocorticoids (such as cortisone) regulate inflammation.
      c. Sex hormones.
         i.   Androgens (male hormones) include testosterone and androsterone.
         ii. Female hormones include estrogens and progestins.
      d. There are also hundreds of synthetic hormones of all types.
III. Neurotransmitters (Sections 28.5–28.6).
   A. General features of neurotransmitters (Section 28.5).
      1. A nerve impulse is transmitted along a neuron by variations in electric potential.
      2. Neurotransmitters convey the impulse across the synaptic cleft from a presynaptic neuron to a postsynaptic neuron.
      3. The impulse is transmitted down the next neuron, and the preceding process is repeated.
      4. Neurotransmitters are synthesized and stored in vesicles in the presynaptic neuron.
      5. After transmission, the neurotransmitter is inactivated in one of two ways.
         a. A chemical change inactivates the neurotransmitter.
         b. The neurotransmitter is returned to the presynaptic neuron.
      6. Most neurotransmitters are amines.
      7. Some neurotransmitters act directly, and others rely on second messengers.
   B. How neurotransmitters work: acetylcholine (ACh) (Section 28.6).
      1. Events of neurotransmission.
         a. An impulse arrives at the presynaptic neuron.
         b. The vesicles move to the cell membrane, fuse with it, and release ACh.
         c. ACh crosses the synaptic cleft and binds to receptors on the postsynaptic neuron.
         d. A change in the permeability of the postsynaptic neuron initiates the nerve impulse in that neuron.
         e. Acetylcholinesterase in the synaptic cleft catalyzes the decomposition of ACh.
         f.   Choline is reabsorbed into the presynaptic neuron, and new ACh is synthesized.
      2. Interactions of drugs at ACh synapses.
         a. To have an effect, a drug must connect with a receptor.
         b. Drugs that are agonists prolong the biochemical response of a receptor.
            i.   Nicotine at a low dose is an example of an agonist.
         c. Drugs that are antagonists block the normal response of a receptor.
            i.   Antagonists can be either competitive (tubocurarine) or irreversible (nerve gas).
IV. Histamines, antihistamines, and important neurotransmitters (Section 28.7).
   A. Histamine and antihistamines.
      1. Histamine is the neurotransmitter responsible for the symptoms of allergic reactions.
      2. Antihistamines are drugs that are histamine receptor antagonists.

B. Monoamines.
1. Serotonin, norepinephrine, and dopamine are neurotransmitters active in the brain.
2. Since these neurotransmitters affect mood, drugs have been created to affect the concentration of these neurotransmitters at synapses.
   a. Tricyclic antidepressants (Elavil) prevent the reuptake of serotonin and norepinephrine from the synaptic cleft.
   b. MAO inhibitors (Nardil) inhibit the enzyme that breaks down monoamine neurotransmitters.
   c. SSRI antidepressants (Prozac) inhibit only the reuptake of serotonin.
3. Dopamine.
   a. Dopamine plays a role in emotion, thought, and behavior through interactions with five different kinds of receptors in the brain.
   b. Dopamine levels are responsible for many types of behaviors.
      i.   An oversupply of dopamine is associated with schizophrenia.
      ii. An undersupply of dopamine is responsible for Parkinson's disease.
      iii. An ample supply of dopamine is responsible for a feeling of well-being.
   c. Drugs such as heroin, cocaine, marijuana, and alcohol increase dopamine levels.
E. Neuropeptides.
1. The brain has receptors for small polypeptides (enkephalins) that also act as receptors for opiates.

Solutions to Chapter 28 Problems

28.1 Noncovalent interactions may include hydrogen bonding, ionic interactions, and hydrophobic interactions.

28.2 Epinephrine is an amino acid derivative.

28.3 Thyroxine is related to tyrosine.

28.4 TRH contains a histidine residue.

28.5

If you redraw the reaction shown in Figure 28.3, you can see the anion that is a by-product of the conversion of ATP to cyclic AMP. The anion (PP_i) is $P_2O_7^{4-}$ (b).

28.6 The heterocyclic rings of caffeine and theobromine resemble the heterocyclic rings of cyclic AMP. It is possible that these two molecules could act as inhibitors to one of the enzymes that inactivates cAMP.

28.7

Glu   His   Pro

The tripeptide is Glu–His–Pro. Glutamic acid has formed a cyclic amide (lactam).

28.8

Thyroxine is hydrophobic because its aromatic, hydrophobic part is larger than its zwitterionic, hydrophilic part.

CIA 28.1
Homeostasis is the maintenance of a constant internal environment in the body.

CIA 28.2
Clinical chemistry analyzes the concentrations of significant ions and compounds in tissues and bodily fluids to determine which bodily systems are failing to maintain homeostasis.

CIA 28.3
Blood glucose and triacylglycerols are important metabolites whose levels reflect homeostasis. Since the concentrations of both compounds are regulated by hormones, these values also indicate the state of regulatory genes.

28.9

Testosterone

Nandrolone

Nandrolone and testosterone are identical except for the presence in testosterone of a methyl group between the first two rings that is absent in nandrolone.

28.10

5-Hydroxytryptophan

Serotonin

N-Acetylserotonin

Melatonin

(a) decarboxylation (3)
(b) acetylation (1)
(c) methylation (2)

28.11

Propanolol   Epinephrine

Both structures have aromatic rings and have secondary alcohol and secondary amine functional groups. Propranolol has a naphthalene aromatic ring system and has an ether group. Epinephrine has an aromatic ring with two phenol groups. The compounds have different carbon skeletons in their side chains.

28.12 (a)

Parathion   Malathion   Diazinon

(b) Malathion, with an LD_{50} of 1000–1375 mg/kg, is the best choice because its lethal dose is highest, and it is thus the least harmful pesticide of the three.
(c) Parathion, with an LD_{50} of 3–13 mg/kg, is the deadliest insecticide because its lethal dose per kilogram of body weight is smallest.

28.13 (a) Sumatripan, an agonist, prolongs the biochemical response at the serotonin receptor.
(b) Ondansetron, an antagonist, blocks or inhibits the normal response at the serotonin receptor.

28.14

Tetrahydrocannabinol (THC)

THC is likely to be hydrophobic (because the phenol hydroxyl group is the only polar functional group in the molecule), and it might be expected to accumulate in fatty tissue.

28.15 (a) This structure represents an antihistamine because it has the general antihistamine structure pictured in Section 28.7.
(b) This structure represents an antidepressant. It is similar to Prozac, an SSRI, but there are noticeable differences; the aromatic rings have different substituents, and a third ring is present in the compound pictured.

HOC 28.1
A list of possible hormones and neurotransmitters:
Concentration: Epinephrine, norepinephrine, and acetylcholine increase concentration.
Heart rate: Epinephrine and norepinephrine increase heart rate.
Cold hands: Cortisol affects cold hands.
Quality of sleep: Melatonin and serotonin regulate day–night cycle.
        Acetylcholine may affect sleep–wake cycle.
Frustration: Acetylcholine may affect mood.
        Melatonin, serotonin, and dopamine decrease frustration and improve mood.
        Enkephalins may decrease frustration.
Headache: Epinephrine and norepinephrine decrease headache.
Change of appetite: Cortisol, cortisone, epinephrine, and norepinephrine suppress appetite.
Respiration: Epinephrine and norepinephrine increase respiration rate.

Concentration levels are affected by sleep, as the same neurotransmitters affect quality of sleep and concentration. Mood should improve after taking a walk, as released enkephalins decrease frustration and improve mood. Heart rate increases during frustration, as similar neurotransmitters affect these two functions.

Understanding Key Concepts

28.16 (a) Luteinizing hormone (LH) is a polypeptide hormone that is produced in the anterior pituitary gland.
(b) Progesterone is a steroid hormone that is produced in the ovaries.
(c) LH doesn't enter progesterone-producing cells because it is a polypeptide hormone. Instead, LH interacts on the cell surface with LH receptors that generate second messengers, which carry out the function of LH.
(d) Progesterone enters the cell directly because it is a steroid hormone and can pass through the hydrophobic cell membrane.

28.17 Since a small amount of epinephrine produces a massive response, at least one step in the sequence that results in the release of glucose must be amplified many fold. Although amplification occurs in all steps of the sequence of events, two steps have the greatest amplification. (1) Adenylate cyclase is capable of catalyzing the production of a great number of cAMP molecules. (2) The kinase enzymes phosphorylated as a result of cAMP activation (Section 19.8) quickly catalyze the breakdown of glycogen to release large amounts of glucose.

28.18 (a) Insulin, a polypeptide hormone, is involved in type I diabetes.
(b) In unaffected individuals, insulin is released by the pancreas. The pancreas fails to produce insulin in individuals with type I diabetes.
(c) Insulin is transported through the bloodstream to cells that need it.
(d) Insulin doesn't enter cells because it is a polypeptide hormone and can't pass through the hydrophobic cell membrane. Instead, it binds with a cell surface receptor, which generates a second messenger within the cell.

28.19 Neurotransmitters can act either by binding to receptors or by activating second messengers.

28.20 The transmission of a nerve impulse may be terminated by an enzyme that converts the neurotransmitter to an inactive form. It is also possible for the neurotransmitter to be returned to the presynaptic neuron.

28.21 Elevated concentrations of dopamine in the brain result in a "high" feeling. All of these drugs increase the concentration of dopamine in the brain. The brain attempts to return to its normal state by decreasing the number of dopamine receptors and by decreasing their sensitivity. Consequently, more of the drug is required to achieve the same effect, leading to addiction.

Additional Problems

Chemical Messengers

28.22 A chemical messenger is a molecule that travels from one part of the body to another location, where it delivers a signal or acts to control metabolism. The target tissue is the cell or group of cells whose activity is regulated by the messenger. A hormone receptor is the molecule with which the chemical messenger interacts if it is a hormone.

28.23 Hormones are molecules of different sizes and types that travel through the bloodstream and regulate the rates of biochemical reactions without directly taking part in the reactions. A hormone is detected by a receptor, either at a cell surface or within the cell.

28.24 A vitamin is usually an enzyme cofactor, whereas a hormone regulates enzyme activity.

28.25 A hormone transmits a chemical message from an endocrine gland to a target tissue. A neurotransmitter carries an impulse between neighboring nerve cells. Neurotransmitters act rapidly and with short duration, whereas hormones act more slowly and produce effects of longer duration.

28.26 Neither a hormone nor its receptor is changed as a result of binding to each other. The binding forces between hormone and receptor are noncovalent.

28.27 Like allosteric regulation, hormone binding is noncovalent and serves to control the rate of a reaction, rather than to take part in a reaction.

Hormones and the Endocrine System

28.28 The body's endocrine system manufactures and secretes hormones, which regulate the functioning of biochemical reaction pathways.

28.29 Endocrine glands include the pituitary, the adrenals, the pancreas, ovaries, testes, thyroid, and parathyroid glands.

28.30 The three major classes of hormones are polypeptide hormones, steroid hormones, and hormones that are derivatives of amino acids.

28.31 *Polypeptide hormones:* insulin, oxytocin, vasopressin, thyrotropin-releasing hormone, thyroid-stimulating hormone.
Steroid hormones: aldosterone, cortisone, hydrocortisone, estrone, estradiol, progesterone, testosterone, androsterone.
Amino acid derivatives: epinephrine, dopamine, norepinephrine, thyroxine.

28.32 Enzymes are proteins, whereas hormones vary greatly in structure. Some hormones are polypeptides, some are proteins, some are steroids, and some are derivatives of amino acids.

28.33 The same type of specificity exists for both an enzyme–substrate complex and a hormone–receptor complex.

28.34 Polypeptide hormones travel through the bloodstream and bind to cell receptors, which are on the outside of a cell. The receptors cause production within cells of second messengers that activate enzymes.

28.35 Steroid hormones travel through the bloodstream to target tissues. Because they are lipids, they are able to pass through the cell membrane. Receptors for steroid hormones occur either in intracellular fluid or in the nucleus.

How Hormones Work: Epinephrine

28.36 Epinephrine is produced and released in the adrenal medulla.

28.37 Ephinephrine is released as a result of a nervous system signal from the hypothalamus. It is produced when the body needs an instant response to stress.

28.38 Ephinephrine travels through the bloodstream to reach its target tissues.

28.39 At target tissues, epinephrine stimulates the production of glucose as a source of energy to deal with whatever stress is at hand.

28.40 The three membrane-bound proteins that transmit the epinephrine message are, in order of involvement, the hormone receptor, G protein, and adenylate cyclase.

28.41 Adenylate cyclase produces cyclic AMP, the second messenger. The ratio of epinephrine molecules to cyclic AMP molecules is less than 1:1 because each epinephrine molecule stimulates the production of more than one cyclic AMP molecule.

28.42 The second messenger initiates reactions that activate glycogen phosphorylase, the enzyme responsible for releasing glucose from storage.

28.43 When the message is ready to be terminated, the enzyme phosphodiesterase converts cyclic AMP to AMP, which is recycled to ATP.

28.44 Epinephrine is used to treat anaphylaxis, a life-threatening allergic response whose symptoms include lowered heart rate and blood pressure and difficulty in breathing.

28.45 People susceptible to anaphylactic shock carry epinephrine with them, usually in injectable form.

Hormones

28.46 Insulin is an example of a polypeptide hormone and contains 51 amino acids in two chains that are linked by disulfide bridges. It is stored in the pancreas and is released in response to high glucose levels in the blood. The hormone stimulates cells to take up glucose for use or storage.

28.47 Estradiol, an example of a steroid hormone, has a tetracyclic ring structure. The first ring is a phenol, and the fourth ring has a hydroxyl group. Estradiol is released from the ovaries and acts on most cells. It is responsible for the development of female secondary sex characteristics.

Estradiol

28.48 All steroid hormones have a characteristic tetracyclic ring skeleton.

28.49 All steroid hormones are derived from cholesterol and are lipids, unlike other classes of hormones, which are water-soluble.

28.50 *Male sex hormones:* androsterone, testosterone.

28.51 *Female sex hormones:* estrone, estradiol, progesterone.

28.52 A person who takes androgens becomes more muscular faster than a person who doesn't take androgens.

28.53 Side effects of androgen use include increased susceptibility to various cancers, aggressive behavior, liver damage, and high blood pressure.

28.54 Epinephrine, norepinephrine, and dopamine can function as both hormones and as neurotransmitters.

28.55 Epinephrine's mode of action is to stimulate the production of the second messenger cyclic AMP. It performs this function both as a neurotransmitter and as a hormone.

28.56 *Hormone*   *Class*

(a)   amino acid derivative

Dopamine

(b) Insulin   polypeptide hormone

(c)   steroid hormone

Testosterone

28.57 *Hormone*   *Class*

(a) Glucagon   polypeptide hormone

(b)   amino acid derivative

Thyroxine

(c)   steroid hormone

Estradiol

Neurotransmitters

28.58 A synapse is the gap between two nerve cells (neurons). Neurotransmitters released by one neuron cross the synapse to receptors on a second neuron and transmit the nerve impulse.

28.59 An axon is the long, thin part of a nerve call (neuron) along which a nerve impulse travels.

28.60 The signal from a neurotransmitter might be received by another nerve cell, a muscle cell, or an endocrine cell.

28.61 Neurotransmitters transmit nerve impulses, they may cause changes in adjacent cells, or they may cause the release of second messengers.

28.62 A nerve impulse arrives at the presynaptic end of a neuron. The nerve impulse stimulates the movement of a vesicle, containing neurotransmitter molecules, to the cell membrane. The vesicle fuses with the cell membrane and releases the neurotransmitter, which crosses the synaptic cleft to a receptor site on the postsynaptic end of a second neuron. After reception, the cell transmits an electrical signal down its axon and passes on the impulse. Enzymes then deactivate the neurotransmitter so that the neuron can receive the next impulse. Alternatively, the neurotransmitter may be returned to the presynaptic neuron.

28.63 (1) The neurotransmitter can be enzymatically inactivated. (2) The neurotransmitter may be returned to the presynaptic neuron and stored until it is needed again.

28.64 (1) Neurotransmitter molecules are released from a presynaptic neuron in response to an impulse. (2) The neurotransmitter molecules bind to receptors on the target cell. (3) The neurotransmitter is deactivated after transmitting the message.

28.65

$$CH_3\overset{\overset{\displaystyle O}{\|}}{C}{-}OCH_2CH_2\overset{+}{N}(CH_3)_3 \xrightarrow[\text{H}_2\text{O}]{\text{Acetylcholinesterase}} CH_3\overset{\overset{\displaystyle O}{\|}}{C}{-}O^- \ + \ HOCH_2CH_2\overset{+}{N}(CH_3)_3$$

Acetylcholine   Acetate   Choline

28.66 Enkephalins are called neurohormones because they are secreted in the central nervous system and have receptors in brain tissue.

28.67 (1) An impulse arrives at the presynaptic neuron.
(2) Vesicles containing acetylcholine (ACh) move to the cell membrane, fuse with it, and release ACh.
(3) ACh crosses the synaptic cleft and binds to receptors on the postsynaptic neuron.
(4) The resulting change in the permeability of the postsynaptic neuron to ions initiates the nerve impulse in that neuron.
(5) After the message is delivered, acetylcholinesterase catalyzes the breakdown of acetylcholine.
(6) Choline is reabsorbed into the presynaptic neuron, where new ACh is synthesized by reaction with acetyl coenzyme A.

Chemical Messengers and Drugs

28.68 Drugs that are agonists interact with receptors to produce or prolong the normal response of the receptor. Antagonists block or inhibit the normal response of a receptor.

28.69 Black widow spider venom acts as an agonist for acetylcholine, and botulism toxin acts as an antagonist.

28.70 Nicotine is an agonist, and atropine is an antagonist.

Histamines, Antihistamines, and Important Neurotransmitters

28.71 Antihistamines such as doxylamine counteract allergic responses caused by histamine by blocking histamine receptors in mucous membranes. Antihistamines such as cimetidine block receptors for histamine that stimulate production of stomach acid.

28.72 Three families of drugs used to treat depression are tricyclic antidepressants, monoamine oxidase inhibitors (MAO inhibitors), and selective serotonin reuptake inhibitors (SSRI).

28.73 The three major monoamine neurotransmitters are serotonin, norepinephrine, and dopamine.

28.74 Cocaine increases dopamine levels in the brain by blocking reuptake of dopamine.

28.75 Amphetamines accelerate release of dopamine in the brain.

28.76 Tetrahydrocannabinol (THC) increases dopamine levels in the same brain areas where dopamine levels increase after administration of heroin and cocaine.

28.77 When it was discovered that the brain had receptors for neurotransmitters that came from plants, scientists reasoned that there must also be animal neurotransmitters that acted on the same receptors. It seems to be a coincidence that the brain receptors that respond to animal neurotransmitters also respond to plant neurotransmitters.

28.78 Chlorpromazine acts as an antagonist because it blocks the dopamine signal.

28.79 Methamphetamine acts as an agonist because it produces an overstimulated state that resembles schizophrenia.

28.80 Endorphins are polypeptides with morphine-like activity. They are produced by the pituitary gland and have receptors in the brain.

28.81 Enkephalins and endorphins act at the same receptors as opiates.

28.82 If an animal produced its own pain-suppressing molecule, it would have an advantage in escaping from prey while injured.

Conceptual Problems

28.83 The hormone epinephrine is responsible for the response to imminent danger.

28.84 Curare, the mixture of toxins applied to the tips of arrows, acts as an antagonist to acetylcholine at receptors.

28.85 Both thyroxine and steroid hormones have large nonpolar regions and can cross a cell membrane to activate the synthesis of enzymes.

28.86 The *hormone receptor* recognizes the hormone and sets into motion the series of reactions that result in the response of the cell to hormonal stimulation.
The hormone–receptor complex interacts with the *G protein* and causes it to bind GTP. The G protein mediates the reaction between the receptor and adenylate cyclase.
The G protein–GTP complex activates *adenylate cyclase,* which catalyzes the formation of the second messenger, cyclic AMP. Cyclic AMP initiates the reactions that the hormone is designed to stimulate.

28.87 It is important for the second messenger (cyclic AMP) to break down rapidly after synthesis for two reasons. First, once the message has been delivered, cyclic AMP is no longer needed to activate reactions. Second, the precursor to cyclic AMP must be ready for the next signal.

28.88 Signal amplification is the process in which a small signal induces a response much larger in magnitude than the original signal. For hormones, this amplification begins with the activation of the G-protein; one hormone–receptor complex can activate many G-protein–GTP complexes. Each G-protein–GTP complex, in turn, can activate many molecules of adenylate cyclase, which can stimulate production of many molecules of cyclic AMP. The importance of signal amplification is that a small amount of hormone can cause a very large response.

28.89 With the exception of the circled groups, testosterone and progesterone are identical.

Testosterone   Progesterone

28.90

Ethynyl estradiol   Norethindrone

Ethynyl estradiol and norethindrone differ only in the ring on the left. Ethynyl estradiol has a phenolic ring, and norethindrone has a ketone group and a double bond in that ring.

Estradiol   Progesterone

The only difference between ethynyl estradiol and estradiol is in the five-membered ring, where a –C≡CH group that is present in ethynyl estradiol is absent in estradiol. Norethindrone and progesterone differ in two respects. (1) The five-membered ring of norethindrone contains a –C≡CH group and an —OH group, whereas the five-membered group of progesterone contains a methyl ketone (acetyl) group. (2) A methyl group between the first two rings is present in progesterone and absent in norethindrone.

28.91 The first step in the conversion of tyrosine to epinephrine (substitution of $-OH$ for $-H$ on the aromatic ring) is catalyzed by tyrosine hydroxylase, an oxidoreductase (subclass: hydroxylase). The second step (loss of CO_2) is catalyzed by dopa decarboxylase, a lyase (subclass: decarboxylase).

28.92 Testosterone can be converted to androsterone by reduction of the ketone group and the double bond in the first ring, and by oxidation of the hydroxyl group in the five-membered ring. These reactions are oxidations and reductions.

28.93 Estradiol can be converted to estrone by oxidation of the $-OH$ group in the five-membered ring.

Group Problems

28.94 The craving for chocolate might be explained by the stimulation of dopamine receptors by anandamides, producing feelings of satisfaction similar to those produced by THC. The effect of chocolate consumption may be a milder version of marijuana's effects.

28.95 When caffeine inhibits the enzymatic hydrolysis of cyclic AMP, response to a signal is prolonged.

Self–Test for Chapter 28

Multiple Choice

1. Which of the following compounds is an acetylcholine agonist?
   (a) nicotine   (b) succinylcholine   (c) tubocurarine   (d) atropine

2. Which of the following hormones regulates the balance of Na^+ and K^+ ions in cellular fluid?
   (a) androsterone   (b) progesterone   (c) aldosterone   (d) cortisone

3. Cocaine affects the brain because it:
   (a) accelerates the release of dopamine   (b) blocks reuptake of dopamine
   (c) raises the level of endorphins   (d) all of the above

4. An effective antihistamine has all of the structural elements listed except:
   (a) an ethylamine skeleton   (b) a tertiary amine   (c) an ether oxygen
   (d) two bulky $-R$ groups.

5. Which of the following compounds is a neurotransmitter but not a hormone?
   (a) dopamine   (b) norepinephrine   (c) epinephrine   (d) acetylcholine

6. Which of the following compounds does not reside in the cell membrane?
   (a) cyclic AMP   (b) epinephrine receptor   (c) adenylate cyclase   (d) G protein

7. Hormones and neurotransmitters have all of the following in common except:
   (a) They both regulate reaction rates.   (b) They both can be amino acid derivatives.
   (c) They both interact with receptors at target tissues.
   (d) They both travel through the bloodstream.

8. Which of the following events doesn't occur in "fight or flight"?
    (a) Epinephrine crosses the cell membrane.   (b) Epinephrine stimulates production of cyclic AMP.   (c) Glucose is released from storage.   (d) Cyclic AMP is converted to ATP.

9. Control of the endocrine system occurs in the:
    (a) pancreas   (b) pituitary gland   (c) adrenal glands   (d) hypothalamus

10. Which of the following isn't a site for hormone receptors?
    (a) outer cell surface   (b) cytoplasm within the cell   (c) cell nucleus   (d) bloodstream

Sentence Completion

1. _____ is a life-threatening allergic response.

2. In direct neural control, the hypothalamus controls the release of hormones by the ____ gland.

3. Neurotransmitter molecules are stored in _____.

4. _____ is the neurotransmitter responsible for the symptoms of allergic reactions.

5. Nerves that rely on acetylcholine as their neurotransmitter are referred to as _____ nerves.

6. The hormone _____ prepares the uterus to receive a fertilized egg.

7. A(n) _____ is a neuropeptide that acts at opiate receptors.

8. A person who studies the healing powers of plants from remote regions is a(n) _____.

9. One neuron is separated from another by a gap called a(n) _____ _____.

10. Heroin, alcohol, and cocaine all produce an elevated level of the neurotransmitter _____.

11. The hormone _____ contains iodine atoms.

12. After an emergency, the enzyme _____ catalyzes hydrolysis of cyclic AMP to ATP.

True or False

1. Only steroid hormones can cross the cell membrane to activate enzymes.

2. Cocaine accelerates the release of dopamine.

3   In direct release of hormones, hormones move through the bloodstream to the posterior pituitary gland, where they are stored until needed.

4. Some neurotransmitters can act directly; others use second messengers.

5. Acetylcholine is synthesized at postsynaptic neurons.

6. Choline is a quaternary ammonium ion.

7. Epinephrine reduces blood pressure and slows the heart rate.

8. An enlarged thyroid gland is a symptom of excess iodine in the diet.

9. Polypeptide hormones vary greatly in size.

10. Succinylcholine is a competitive antagonist for acetylcholine.

Match each entry on the left with its partner on the right.

1.	Androgen	(a)	Polypeptide hormone
2.	Tricyclic antidepressant	(b)	Progesterone
3.	Acetylcholine agonist	(c)	Cortisone
4.	Natural estrogen	(d)	Growth hormone
5.	Second messenger	(e)	Testosterone
6.	Oxytocin	(f)	Norethindrone
7.	Regulatory hormone	(g)	Cyclic AMP
8.	MAO inhibitor antidepressant	(h)	Amitriptyline (Elavil)
9.	Synthetic estrogen	(i)	Fluoxetine (Prozac)
10	Acetylcholine antagonist	(j)	Phenelzine (Nardil)
11.	Glucocorticoid	(k)	Nicotine
12	SSRI inhibitor antidepressant	(l)	Atropine

Chapter Outline

I.   Water (Sections 29.1–29.2).
   A. Body water and its solutes (Section 29.1).
       1. Types of body fluids.
           a. Intracellular fluid is within cells.
           b. Extracellular fluid is outside of cells.
               i.   Blood plasma, which consists of water-soluble proteins.
               ii. Interstitial fluid, which fills the spaces between cells.
       2. Components of body fluids.
           a. Electrolytes (inorganic ions).
           b. Gases—O_2 and CO_2.
           c. Small organic molecules.
           d. Ionized biomolecules.
       3. All body fluids have the same osmolarity, even if they differ in composition.
       4. Movement of body fluids.
           a. Blood travels through peripheral tissues in capillaries.
               i.   At the arterial end, blood pressure is higher and pushes solutes into interstitial fluid.
               ii. At the venous end, blood pressure is lower, and solutes reenter blood plasma.
               iii. Blood plasma and interstitial fluid have similar composition.
           b. Lymph capillaries collect excess interstitial fluid and other molecules too large to pass through capillary walls; this combination is called lymph.
               i.   Lymph can't return to surrounding tissue.
               ii. Lymph enters the blood stream at the thoracic duct.
           c. Solutes in interstitial and intracellular fluid are exchanged by crossing cell membranes.
               i.   Often, active transport is needed to maintain concentration differences.
   B. Maintenance of fluid balance (Section 29.2).
       1. The output of water and electrolytes is controlled by hormones.
       2. Receptors in the hypothalamus monitor concentration of blood solutes.
       3. If the concentration of solutes is too high, secretion of antidiuretic hormone increases.
           a. In SIADH, too much hormone keeps kidneys from excreting enough water.
           b. In *diabetes insipidus,* excessive amounts of dilute urine are excreted.
II. Blood (Sections 29.3–29.6).
   A. Components of blood (Section 29.3).
       1. Whole blood contains:
           a. Plasma.
           b. Blood cells.
       2. Blood serum is the fluid portion that remains after blood has clotted.
   B. Functions of blood.
       1. Transport.
       2. Regulation: redistribution of heat and solutes, and buffering of tissues.
       3. Defense.
           a. Immune response.
           b. Clotting.
   C. Immune response (Section 29.4).
       1. A foreign invader (antigen) is a substance that the body does not recognize as part of itself.

2. The body responds to antigens in three ways.
    a. Inflammatory responses (nonspecific).
       i. Inflammation produces swelling, warmth, and pain.
       ii. The chemical messenger histamine is released at the site of injury.
       iii. Histamine dilates capillaries and increases blood flow to the injured area, allowing clotting factors and white blood cells to pass.
       iv. Bacteria are destroyed by phagocytes (white blood cells), which can also initiate specific immune responses.
       v. Inflammation disappears when infectious agents have been removed.
    b. Cell-mediated immune response (specific).
       i. The cell-mediated immune response is under the control of T lymphocytes.
       ii. The cell-mediated immune response guards against abnormal cells due to bacteria, viruses, or cancer.
       iii. When a T cell recognizes an invader, it produces killer T cells (that destroy the invader), helper T cells (that enhance defenses), and memory T cells (that remain on guard).
    c. Antibody-mediated immune response (specific).
       i. The antibody-mediated immune response is under the control of B lymphocytes (with the assistance of T lymphocytes).
       ii. A B cell is activated when it binds to an antigen and encounters a helper T cell.
       iii. The B cells divide to produce plasma cells, which form non–cell-bound antibodies called immunoglobulins.
       iv. The antibodies find their antigens and inactivate them by one of several methods.
       v. Memory cells produce more antibodies if the same antigen reappears.
       vi. The body contains thousands of different immunoglobulins, and many have been identified.

D. Blood clotting (Section 29.5).
  1. A blood clot consists of blood cells trapped in a mesh of the protein fibrin.
    a. The process of clot formation requires many steps.
  2. The body's mechanism for halting blood loss is called hemostasis.
    a. The first event is constriction of blood vessels and formation of a plug of platelets.
    b. Then, blood clotting occurs by either of two pathways.
       i. In the intrinsic pathway, blood makes contact with the negatively charged surface of collagen.
       ii. In the extrinsic pathway, damaged tissue releases a glycoprotein called tissue factor.
    c. In either pathway, several zymogen clotting factors are released.
    d. After several steps, the enzyme thrombin catalyzes the formation of insoluble fibrin, which forms the clot.
  3. After the injury has healed, the clot is broken down by hydrolysis of its peptide bonds.

E. Red blood cells and blood gases (Section 29.6).
  1. The purpose of red blood cells (erythrocytes) is to transport blood gases.
  2. Hemoglobin is responsible for transporting O_2 and CO_2.
    a. Hemoglobin consists of four protein chains and four heme molecules.
  3. Oxygen transport.
    a. Oxygen is transported via bonding with Fe^{2+} through an unshared electron pair.
    b. The percent of heme molecules that carry oxygen is the percent saturation.
    c. The binding curve is s-shaped.
       i. The uptake of the first molecule of oxygen is more difficult than the uptake of the remaining molecules.
       ii. The release of the first O_2 is more difficult than the release of the other three.

4. CO_2 transport.
    a. CO_2 can be transported in three ways.
        i.   Dissolved (about 7%).
        ii. Bonded to Hb.
        iii. As HCO_3^- in solution.
    b. Carbonic anhydrase in erythrocytes catalyzes the formation of HCO_3^- from CO_2.
    c. The HCO_3^- can be transported in blood to the lungs, where it is exhaled as CO_2.
    d. Electrolyte balance in erythrocytes is maintained by entry of one Cl^- for every HCO_3^- removed.
    e. Excess acidity is controlled by binding of H^+ to hemoglobin, with release of O_2.
5. The effects of changes in the oxygen saturation curve with changing pCO_2 and $[H^+]$ are illustrated in Figure 29.12.
6. Acidosis and alkalosis result from disruption of the body's mechanism for maintaining pH and by imbalances in CO_2 metabolism.

III. Urine (Sections 29.7–29.8).
 A. The kidney and urine formation (Section 29.7).
    1. The kidney is composed of nephrons.
    2. Blood enters the nephrons at the glomerulus, where filtration occurs.
        a. Plasma and smaller solutes are filtered into the surrounding fluid.
    3. Reabsorption recaptures water and smaller solutes.
    4. In secretion, some solutes are excreted in higher concentration than occurs in the filtrate.
    5. Some substances move by passive diffusion; others move by active transport.
    6. Hormones influence urine composition.
 B. Urine composition and function (Section 29.8).
    1. Composition of urine.
        a. The components of urine are mainly electrolytes and nitrogen-containing wastes.
        b. The concentration of these substances varies with water intake, temperature, exercise, and state of health.
    2. Acid–base balance.
        a. Of the 50–100 mEq of H^+ produced in metabolism, very little appears in the urine.
        b. H^+ is produced from CO_2 in the reaction:
            i. $CO_2 + H_2O \longrightarrow H^+ + HCO_3^-$
        c. The resulting H^+ combines with either NH_3 or HPO_4^{2-} and is excreted.
        d. H^+ also combines with HCO_3^- in urine filtrate to produce CO_2, which enters the bloodstream.
    3. Fluid and Na^+ balance.
        a. The amount of H_2O reabsorbed depends on:
            i.   The osmolarity of the fluid passing through the kidney.
            ii. Antidiuretic hormone-controlled membrane permeability.
            iii. The amount of Na^+ reabsorbed.
        b. Reabsorption of Na^+ is under the control of aldosterone.

Solutions to Chapter 29 Problems

29.1   When chloride concentrations are low, the reaction should proceed farther to the right, according to Le Châtelier's principle. When chloride concentrations are high, the reaction proceeds to the left. In order to move passively into a cell, a molecule cannot be charged because a cell membrane is nonpolar. Therefore, the form of cisplatin shown on the left enters the cell more readily. Once inside, the equilibrium shifts to the right and forms the charged version of cisplatin, which does not cross back through the cell membrane.

HOC 29.1

(a) An energy drink contains a blend of carbohydrates (glucose, sucrose, fructose, galactose, and/or maltodextrin) and electrolytes dissolved in water. The ingredients provide energy and help maintain electrolyte balance. Considerations: An energy drink must provide 13–19 g of carbohydrate per 8 oz. serving, and drinking 1 1/2 to 4 cups during exercise, as well as drinking before and after exercise, will provide the nutrients and electrolytes needed. Water would prevent dehydration but would not supply energy.
(b) There are high protein bars and high carbohydrate bars. The high carbohydrate bars contain sugars in the form of brown rice syrup and/or sucrose and grains in the form of rice crisps or oats. Considerations: A sports bar should have 25–40 g of carbohydrates and less than 15 g of protein, and a limited amount of fat, as large amounts of fat can slow digestion. Eat an energy bar one hour before exercising, and drink plenty of water with it. A high glycemic index refers to how quickly carbohydrates circulate in the blood stream. A high glycemic index bar is best during a workout, as it releases carbohydrates rapidly.
(c) Since sports drinks and energy bars have a significant number of calories, consuming them on a regular basis might lead to weight gain.

CIA 29.1

In an asymmetric transport system, substances can either be transported into a cell or be transported out of a cell, but not both. For example, there exists a system for transport of glycine out of brain cells, but no system exists for transport of glycine into brain cells.

CIA 29.2

Substances that are soluble in membrane lipids can cross the blood–brain barrier. Ethanol crosses this barrier because it's soluble in membrane lipids.

CIA 29.3

The metabolic blood–brain barrier is a pathway in which a substance is converted in an epithelial cell to a metabolite that cannot enter the brain.

CIA 29.4

Heroin has two acetyl groups in the same location as the two hydroxyl groups of morphine. These relatively nonpolar acetyl groups make heroin soluble in the membrane lipids of endothelial cells and more able to pass into these cells.

29.2 (a) Interstitial fluid—(iii) Fluid that fills spaces between cells
(b) Whole blood—(ii) Fluid, solutes, and cells that together flow through veins and arteries
(c) Blood serum—(iv) Fluid that remains after clotting agents are removed from plasma
(d) Intracellular fluid—(v) Fluid within cells
(e) Blood plasma—(i) Fluid that remains when blood cells are removed

29.3 (a) The value of pH goes down because of the formation of carbonic acid from dissolved CO_2, which increases $[H_3O]^+$ and thus increases acidity.
(b) A blood gas analysis measures $[O_2]$, $[CO_2]$, and pH.

29.4 (a) Emphysema is a cause of *respiratory acidosis*. In pneumonia, the lungs have difficulty exhaling CO_2, and increased blood acidity results.
(b) Failure of the kidneys to excrete H_3O^+ results in *metabolic acidosis*.
(c) Overdose of an antacid leads to overneutralization of H_3O^+, which raises pH and leads to *metabolic alkalosis*.

29.5 (a) A panic attack causes hyperventilation, which removes CO_2 from the body, raising pH and resulting in *respiratory alkalosis*.
(b) Congestive heart failure is a cause of *respiratory acidosis*.
(c) Running a marathon can cause *metabolic acidosis*.

CIA 29.5
Electrolyte tests might show if a patient was suffering from dehydration. A red blood cell (RBC) count or a mean corpuscular volume (MCV) test might show if a patient was suffering from anemia.

CIA 29.6
Blood tests show the levels of substances from many metabolic pathways and allow a doctor to assess health in organs not directly accessible.

CIA 29.7
The Vertical Auto Profile (VAP) provides levels of 15 lipid components of blood, including lipoproteins, rather than the 4 usually measured in conventional blood tests.

Understanding Key Concepts

29.6 (a) The body fluid found inside cells is called intracellular fluid.
(b) The body fluid found outside cells is called extracellular fluid.
(c) Blood plasma and interstitial fluid are the major body fluids found outside of cells.
(d) K^+, Mg^{2+}, and HPO_4^{2-} are the major electrolytes found inside cells.
(e) The major electrolytes found outside cells are Na^+ and Cl^-.

29.7

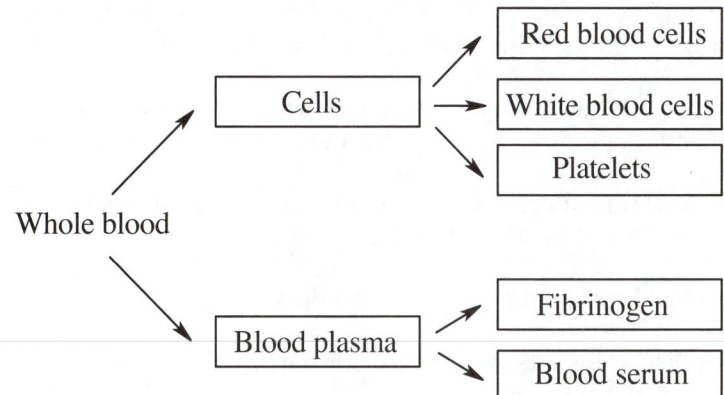

29.8 (a) Blood carries O_2 from lungs to tissues.
(b) Blood carries CO_2 (dissolved and as HCO_3^-) from the tissues to the lungs.
(c) Blood transports nutrients from the digestive system to the tissues.
(d) Blood carries waste products from the tissues to the site of excretion.
(e) Blood transports hormones from endocrine glands to their site of binding.
(f) Blood transports defensive agents such as white blood cells (to destroy foreign material) and platelets (to prevent blood loss).

29.9 Symptoms of inflammation include swelling, redness, warmth, and pain.

29.10 Histamine, produced by the enzymatic decarboxylation of histidine, causes the symptoms of inflammation by dilating capillaries, causing an increased blood flow that reddens and warms the skin. Pain and swelling occur when blood-clotting factors and defensive proteins enter the intercellular space.

29.11 *Cell-mediated immune response,* which is under the control of T cells, arises when abnormal cells, bacteria, or viruses enter cells. The invading cells are killed by T cells that destroy the invader.
Antibody-mediated immune response, which is under the control of B cells, assisted by T cells, occurs when antigenic substances enter cells. When an antigen is recognized, B cells divide to produce plasma cells, which form antibodies to the antigen. These antibodies form an antibody–antigen complex that inactivates the antigen.

29.12 Excess hydrogen ions are excreted by reaction with NH_3 or HPO_4^{2-}. H^+ ions also combine with bicarbonate, producing CO_2 that returns to the bloodstream and is again converted to bicarbonate.

Additional Problems

Body Fluids

29.13 The three principal body fluids are intracellular fluid (64%), interstitial fluid (25%), and plasma (8%).

29.14 To be soluble in bodily fluids, a substance must be an ion, a gas, a small molecule, or a molecule with many polar or ionic groups on its surface.

29.15 Substances not soluble in blood, such as lipids, are transported by blood proteins.

29.16 The difference in blood pressure between arterial capillaries and interstitial fluid pushes solutes and water into interstitial fluid. The difference in blood pressure between interstitial fluid and venous capillaries draws solutes and water into venous capillaries.

29.17 Blood pressure in arterial capillaries is higher than interstitial fluid pressure, and blood pressure in venous capillaries is lower than interstitial fluid pressure.

29.18 The lymphatic system collects excess interstitial fluid, cellular debris, proteins, and lipid droplets and ultimately returns them to the bloodstream.

29.19 Lymph enters the bloodstream at the thoracic duct.

29.20 Vasopressin is also known as antidiuretic hormone.

29.21 An excess of antidiuretic hormone (ADH) causes a decrease in the water content of the urine. This excess may be caused by pulmonary or central nervous system disorders.

29.22 Blood plasma is the fluid portion of blood that contains water-soluble solutes. Blood serum is the fluid that remains after blood has completely clotted.

29.23 At 7% body mass loss, collapse is likely to occur.

29.24 The three main types of blood cells are erythrocytes (red blood cells), platelets, and white blood cells.

29.25 *Red blood cells* transport blood gases.
White blood cells protect the body from foreign substances.
Platelets assist in blood clotting.

29.26 Electrolytes are inorganic ions.

29.27 Sodium cation is the major cation found in interstitial fluid.

29.28 Potassium cation and magnesium cation are the major cations found in intracellular fluid.

29.29 An antigen is a foreign substance that the body identifies as an invader.

29.30 Three types of bodily responses to antigens are inflammation, cell-mediated immune response, and antibody-mediated immune response.

29.31 Specific immune responses, like enzyme–substrate interactions, involve a noncovalent interaction between an antigen and a defender (antibody) specific to that antigen.

29.32 Immunoglobulins are the plasma proteins involved in the antibody-mediated immune response.

29.33 The antibody-directed immune response involves B lymphocytes, a type of white blood cell. These B lymphocytes identify antigens and divide into plasma cells that produce antibodies specific to the antigen present.

29.34 When a T cell recognizes an antigen on the surface of an invader, three kinds of T cells are produced. *Killer* T cells destroy the invader, *helper* T cells enhance defenses against the invader, and *memory* T cells can produce new killer T cells if the invader reappears.

29.35 When T cells recognize an antigen on the surface of an invading cell, they produce killer T cells that destroy the invader, often by releasing a toxic protein that perforates cell membranes.

29.36 Memory cells, which are produced from B cells, can quickly produce more plasma cells if the same antigen appears in the future. Memory cells "remember" the antigen and are capable of producing antibodies to it for a long time.

29.37 A blood clot is a mass of blood cells trapped in a fibrin mesh.

29.38 Vitamin K and calcium ion are needed for blood clotting.

29.39 Blood clotting may be triggered by the intrinsic pathway, which occurs when blood is exposed to the negatively charged fibrous protein collagen.

29.40 Blood-clotting enzymes are released as zymogens in order to avoid undesirable clotting in noninjured tissues.

29.41 Each hemoglobin tetramer can bind four O_2 molecules.

29.42 Hemoglobin iron must be in the +2 oxidation state in order to bind with oxygen.

29.43 Deoxyhemoglobin is dark red-purple. This color change is due to the loss of oxygen, which makes oxyhemoglobin bright red.

29.44 When the partial pressure of oxygen drops below 10 mmHg, hemoglobin is unsaturated. When the partial pressure of oxygen is greater than 100 mmHg, hemoglobin is completely saturated. Between these pressures, hemoglobin is partially saturated.

29.45 Because of oxygen's allosteric interaction with hemoglobin, uptake of the first oxygen facilitates the uptake of the remaining oxygens. In the reverse direction, release of the first oxygen facilitates the release of the other three oxygens.

29.46 CO_2 can be transported as a dissolved gas, as bicarbonate ion, or as covalently bonded to hemoglobin.

29.47 Hemoglobin is 50% saturated with oxygen at 30 mmHg.
Under normal conditions at sea level, the partial pressure of O_2 is:
    760 mmHg x 0.21 = 160 mmHg.
According to Figure 29.11, hemoglobin is 100% saturated with O_2 under normal conditions.

29.48 CO_2 in the blood raises $[H^+]$ in two types of reactions:

(1) Reaction of CO_2 with nonionized $-NH_2$ groups of hemoglobin amino acids produces H^+.

$$Hb-NH_2 + CO_2 \longrightarrow Hb-NHCOO^- + H^+$$

(2) Carbonic anhydrase catalyzes the reaction of CO_2 with H_2O, producing bicarbonate and H^+.

$$CO_2 + H_2O \underset{\text{anhydrase}}{\overset{\text{Carbonic}}{\rightleftharpoons}} HCO_3^- + H^+$$

29.49 (a) Increasing the temperature causes hemoglobin to release more O_2 to tissues.
(b), (c) Production of CO_2 brings about an increase of $[H^+]$, which causes the release of O_2.

29.50 Respiratory acidosis occurs when CO_2 generation and CO_2 removal by the lungs are out of balance. Metabolic acidosis results from abnormally high acid generation or from the inability of the body's buffering system to regulate bicarbonate concentration.

29.51 Ketoacidosis due to diabetes is classified as metabolic acidosis because it is caused by increased production of acid as a result of metabolic dysfunction.

29.52 Respiratory alkalosis is due to excessive loss of CO_2 due to hyperventilation. Metabolic alkalosis is due to elevated plasma bicarbonate concentration.

29.53 In addition to filtration, kidneys also recapture water and essential solutes (reabsorption) and excrete excess solutes (secretion).

29.54

$$H^+ + HPO_4^{2-} \rightleftharpoons H_2PO_4^-$$

$$H^+ + HCO_3^- \rightleftharpoons CO_2 + H_2O$$

General Questions and Problems

29.55 Ethanol is soluble in blood because it is a small, polar molecule.

29.56 A nursing mother's antibodies can be passed to her baby in breast milk.

29.57 When the concentration of sodium in the blood is high, the secretion of antidiuretic hormone (ADH) increases. ADH causes the water content of the urine to decrease, and causes the amount of water retained by the body to increase, causing swelling.

29.58 Active transport is the movement of solutes from regions of low concentration to regions of high concentration, a process that requires energy. Osmosis is the movement of water through a semipermeable membrane from a dilute solution to a more concentrated solution, a process that requires no energy.

29.59 Active transport is necessary when a cell needs a substance that has a higher concentration inside the cell than outside, or when a cell needs to secrete a substance that has a higher concentration outside the cell than inside.

29.60 In the blood, CO_2 from metabolism reacts to form $HCO_3^- + H^+$. The H^+ is bound to hemoglobin, which releases O_2, and is carried to the lungs. There, the H^+ is released and O_2 is bound to hemoglobin.

In the urine, CO_2 reacts to form HCO_3^- and H^+. The HCO_3^- returns to the bloodstream, and the H^+ is neutralized by reaction with HPO_4^{2-} or NH_3.

Whenever excess HCO_3^- accumulates in blood or urine, it can react with H^+ to form $H_2O + CO_2$.

29.61 *Homeostasis* is the maintenance of a constant internal environment in the body. *Hemostasis* is the body's mechanism for preventing blood loss and could be considered a part of homeostasis.

29.62 When blood CO_2 levels drop, the following reaction occurs to restore CO_2 supply:

$$H^+ + HCO_3^- \longrightarrow H_2CO_3 \longrightarrow CO_2 + H_2O$$

This reaction uses up H^+ ions and leads to alkalosis. Breathing into a paper bag recaptures the expired CO_2 and restores the blood CO_2 level.

Group Problems

29.63 For Lucozade energy drink (104 kcal/8 oz.):
0.2 g fat, 26 g carbohydrate, 0.6 g protein, 200 mg Na, 7 mg K.

29.64 Two factors determine a compound's ability to cross the blood–brain barrier—size and polarity. Of the alkaloids in Table 16.2, you would expect heroin and coniine to cross the BBB the easiest. Atropine and codeine are intermediate in size and polarity. Reserpine and solanine cross the BBB with more difficulty because they are large. Morphine has two hydroxyl groups and crosses the BBB with the most difficulty.

29.65 A partial list of medications that can affect liver function includes Tylenol, statins, methotrexate, Corderon, some antibiotics, NSAIDs. Methotrexate is used for cancer treatment, for autoimmune diseases, and for ectopic pregnancies.

Self-Test for Chapter 29

Multiple Choice

1. Which oxygen is taken up by heme with the most difficulty?
(a) the first oxygen   (b) the second oxygen   (c) the last oxygen   (d) All bind with equal ease.

2. Cardiac arrest may result in:
(a) metabolic alkalosis   (b) metabolic acidosis
(c) respiratory alkalosis   (d) respiratory acidosis

3. Cell-mediated immune response guards against all of the following except:
(a) cancer cells   (b) transplanted organs   (c) allergens   (d) bacteria

4. Potassium ion is most abundant in:
   (a) interstitial fluid   (b) intracellular fluid   (c) plasma   (d) extracellular fluid

5. All of the following are involved in blood clotting except:
   (a) tissue factor   (b) zymogens   (c) thrombin   (d) memory cells

6. Which of the following disorders is not an autoimmune disease?
   (a) hemophilia   (b) arthritis   (c) lupus erythematosus   (d) diabetes mellitus

7. The amount of water reabsorbed in kidney tubules depends on:
   (a) the amount of sodium reabsorbed   (b) control of membrane permeability by antidiuretic hormone   (c) aldosterone   (d) all three

8. Blood is at its lowest percent saturation with oxygen in:
   (a) peripheral tissue   (b) muscles   (c) veins   (d) arteries

9. All of the following decrease the oxygen affinity of hemoglobin except:
   (a) increased $[H^+]$ (b) increased $[O_2]$ (c) increased $[CO_2]$ (d) increased temperature

10. B cells divide to produce all of the following except:
    (a) memory cells   (b) antibodies   (c) phagocytes   (d) plasma cells

Sentence Completion

1. The percentage of heme molecules that carry oxygen is dependent on the _____ _____ of O_2.

2. Antigens can be small molecules known as _____.

3. In the process of blood clotting, _____ are activated to give active clotting factors.

4. Inorganic ions are the major contributors to the _____ of body fluids.

5. Histamine is synthesized from _____.

6. _____ ion and vitamin _____ assist in blood clotting.

7. In _____ _____, substances move from regions of low concentration to regions of high concentration.

8. The three functions of blood are _____, _____, and _____.

9. Bacteria at the site of inflammation are attacked by _____.

10. Arthritis and allergies are known as _____ diseases.

11. _____ is caused by the absence of one or more clotting factors.

12. Fibrin molecules are bound into fibers by crosslinks between the side chains of the amino acids _____ and _____.

True or False

1. The walls of lymph capillaries are constricted so that lymph can't return to surrounding tissues.

2. Homeostasis is the body's mechanism for halting blood loss.

3. O_2, H^+, and CO_2 all bond to heme.

4. Secreted H^+ is eliminated in the urine as NH_4^+ or $H_2PO_4^-$.

5. When the partial pressure of CO_2 increases, the percent saturation of hemoglobin decreases.

6. Allergies and asthma are caused by an underproduction of immunoglobulin E.

7. Blood plasma is the fluid remaining after blood has completely clotted.

8. Blood clotting begins with the release of tissue factor from injured tissue.

9. Reabsorption is the movement of solutes and water out of the kidney.

10. The metabolism of food contributes to one's water intake.

11. Lymphocytes surround and destroy bacteria.

12. Caffeine causes increased output of urine.

Match each entry on the left with its partner on the right.

1. B cells

2. Active transport

3. Fibrin

4. Carbonic anhydrase

5. Erythrocytes

6. T cells

7. Antigen

8. Histidine decarboxylase

9. Osmosis

10. Antidiuretic hormone

11. Antibody

12. Aldosterone

(a) Foreign substance

(b) Enzyme that catalyzes the formation of histamine

(c) Movement of solute against a concentration gradient

(d) White blood cells involved in antibody-directed immune response

(e) Controls reabsorption of Na^+

(f) Blood protein responsible for clotting

(g) Enzyme that catalyzes reaction between H_2O and CO_2

(h) Movement of water in response to a concentration difference

(i) White blood cells involved in cell-directed immune response

(j) Red blood cells

(k) Protein molecule that identifies a foreign substance

(l) Causes a decrease in the water content of urine

Answers to Self-Tests

Chapter One

Number questions: 1. (a) micrometer   (b) deciliter   (c) megagram   (d) liter   (e) nanogram
2. (a) kL   (b) pg   (c) cm   (d) hL   3. (a) 5   (b) 2   (c) 1   (d) exact   (e) 3, 4, 5, or 6
4. (a) 7.03×10^{-5} g; 3 sig. fig.   (b) $1.371\,00 \times 10^5$ m; 4, 5, or 6 sig. fig.   (c) 1.1×10^{-2} L; 2 sig.
fig.   (d) $1.837\,100\,8 \times 10^7$ mm; 8 sig. fig.   5. (a) 8.07×10^2 L   (b) 4.77×10^6 people
(c) 1.27×10^{-3} g   (d) 1.04×10^4 μm   6. (a) 0.564 lb   (b) 0.417 m   (c) 7.6 L   (d) 85.4 in.
(e) 95 °F   (f) 25 °C   (g) 5.92 fl oz   (h) 6.17×10^{-3} oz   7. 31 cal   8. 217 g
Multiple choice: 1. b   2. d   3. b   4. a   5. c   6. d   7. b   8. b   9. c   10. a
Sentence completion: 1. kilogram, meter, cubic meter, Kelvin   2. number, unit   3. specific
heat   4. Bi   5. 3   6. 1 lb/454 g   7. malleable   8. factor-label   9. Specific gravity   10. Celsius,
Kelvin
True/false: 1. F (specific gravity is unitless)   2. T   3. F (three sig. fig.)   4. F (0.9464 L/1 qt)   5.
T   6. F   7. F   8. T   9. F   10. T
Matching: 1. e   2. j   3. g   4. i   5. h   6. l   7. f   8. k   9. c   10. a   11. d   12. b

Chapter Two

Multiple choice: 1. c   2. b   3. a   4. c   5. d   6. d   7. a   8. b   9. d   10. d   11. a   12. b
Sentence completion: 1. p 2. Dalton   3. protons, neutrons   4. atomic number   5. shells
6. group   7. 13   8. subatomic particles   9. isotopes   10. 18   11. quantized
True/false: 1. T   2. F (5.486×10^{-4} amu)   3. F   4. T   5. F (atomic number indicates only
protons)   6. F (They're less reactive.)   7. F (in the same group)   8. T   9. T   10. F (Subshells can
contain 2, 6, 10 ·· electrons.)   11. T   12. F
Matching: 1. h   2. e   3. k   4. a   5. j   6. l   7. c   8. b   9. i   10. d   11. g   12. f

Chapter Three

Multiple choice: 1. d   2. a   3. d   4. c   5. a   6. b   7. b   8. d   9. b   10. d
Sentence completion: 1. Ionization energy   2. potassium phosphate   3. 2   4. polyatomic
5. simplest   6. 8   7. base   8. Hydroxyapatite   9. crystalline   10. 4A or 5A
True/false: 1. F (It forms only Zn^{2+}.) 2. T   3. F (energy needed to form an ion) 4. T   5. F
(Ions formed may be H^+, $H_2PO_4^-$, and HPO_4^{2-}.) 6. F (Solutions of ions conduct electricity.)   7.
F   8. T   9. T   10. F (group 8A) 11. T   12. F
Matching: 1. e   2. l   3. k   4. a   5. h   6. j   7. i   8. b   9. f   10 g   11. d   12. c

Chapter Four

Multiple choice: 1. c   2. d   3. b   4. c   5. a   6. a   7. d   8. b   9. d   10. a
Sentence completion: 1. electronegative   2. biomolecule   3. covalent   4. C≡C, C≡N
5. diatomic   6. trigonal planar   7. d 8. triple   9. Binary   10. organic   11. condensed
True/false: 1. T   2. T   3. F   4. F (valence electrons) 5. F (It's planar.) 6. F (covalent
compounds) 7. T   8. F   9. F (Some bent molecules have bond angles that are approx. 109.5°.)
10. T   11. T
Matching: 1. h   2. j   3. f   4. a   5. k   6. i   7. e   8. c   9. b   10. g   11. d   12. l

Chapter Five

Multiple choice: 1. a   2. a   3. c   4. b   5. d   6. c   7. c   8. b   9. a   10. d
Sentence completion: 1. coefficients   2. precipitate   3. spectator   4. aqueous   5. carbonate, bicarbonate 6. Metallurgy   7. reactants   8. +3   9. reducing   10. neutralization
True/false: 1. F (a lithium-iodine battery) 2. T   3. T   4. F   5. T   6. F 7. F   8. F (in blood) 9. T   10. T
Matching: 1. h   2. f   3. k   4. a   5. i   6. c   7. j   8. b   9. g   10. l   11. d   12. e

Chapter Six

Multiple choice: 1. c   2. d   3. b   4. b   5. d
True/false: 1. F (same number of atoms) 2. T   3. F   4. F (Molar mass is the conversion factor.)   5. F (The mole ratio is 1/2.)
Matching: 1. e   2. a   3. b   4. c   5. d

Chapter Seven

Multiple choice: 1. c   2. b   3. d   4. d   5. b   6. a   7. c   8. a   9. d   10. d
Sentence completion: 1. 1   2. conservation of energy   3. catalyst   4. reversible
5. endothermic 6. equilibrium-constant expression   7. temperature, concentration, catalyst
8. metabolism   9. Reaction rate   10. collide
True/false: 1. F (produced) 2. F (At equilibrium, the ratio of concentrations is constant.) 3. T
4. T   5. F (The size of E_{act} and ΔH are unrelated.) 6. F (not always true for biochemical reactions) 7. T   8. F (Catalysts lower the height of the energy barrier.) 9. F (At equilibrium, the rate of the forward reaction equals the rate of the reverse reaction.) 10. T
Matching: 1. c   2. j   3. l   4. i   5. a   6. k   7. b   8. f   9. d   10. e   11. g   12. h

Chapter Eight

Multiple choice: 1. b   2. c   3. d   4. a   5. d   6. c   7. a   8. b   9. c   10. d
Sentence completion: 1. pressure, volume   2. definite, indefinite   3. Dalton's   4. standard temperature and pressure   5. Avogadro's   6. equilibrium   7. atmosphere, Pascal, mm Hg, pounds per square inch, torr   8. heat of fusion   9. number of atoms, pressure   10. temperature
11. change of state   12. volatile
True/false: 1. T   2. F (Only ideal gases have similar physical behavior.) 3. T   4. F (Doubling pressure halves the volume.) 5. T   6. F (only true in crystals) 7. F (273 K) 8. T   9. F (higher)
10. F (Vapor pressure doesn't depend on quantity of liquid.)   11. F (viscosity)
Matching: 1. g   2. e   3. k   4. f   5. h   6. a   7. j   8. l   9. b   10. d   11. c   12. i

Chapter Nine

Multiple choice: 1. b   2. d   3. a   4. b   5. c   6. d   7. c   8. c   9. a   10. b
Sentence completion: 1. miscible   2. gram equivalent   3. volumetric flask   4. solubility, partial pressure   5. isotonic   6. hygroscopic   7. $M_1 \times V_1 = M_2 \times V_2$ 8. colloid   9. solvent, small solute molecules   10. saturated   11. glucose   12. lower
True/false: 1. F (solvent)   2. T   3. F (A quantity of the second liquid is added to make 100 mL of solution.) 4. F   5. T   6. F (Weight/weight percent is rarely used.) 7. T   8. F (Effect of temperature is unpredictable.) 9. T (It also differs with respect to particle size.) 10. F (hemolysis)
11. F (raise) 12. T
Matching: 1. h   2. f   3. d   4. j   5. a   6. k   7. b   8. i   9. g   10. l   11. e   12. c

Chapter Ten

Multiple choice: 1. b   2. a   3. c   4. d   5. c   6. a   7. b   8. a   9. b   10. d
Sentence completion: 1. equivalent   2. red   3. diprotic   4. strong   5. Dissociation   6. OH^- ions   7. Neutralization   8. pH   9. 30.0 mL   10. conjugate   11. carbonate/bicarbonate   12. amphoteric
True/false: 1. F (60 mL of 0.1 M NaOH)   2. T   3. F (H_2SO_4 is a strong acid.) 4. F   5. F   6. T   7. T   8. F (According to this definition, an acid donates H^+.)   9. F (Ammonia yields ammonium ion, plus the anion of the acid.)   10. T   11. F   12. T
Matching: 1. d   2. e   3. g   4. l   5. a   6. b   7. i   8. j   9. c   10. f   11. k   12. h

Chapter Eleven

Multiple choice: 1. c   2. d   3. c   4. a   5. c   6. d   7. a   8. d   9. b   10. c
Sentence completion: 1. transmutation   2. Gamma radiation   3. tracer   4. background   5. nucleons   6. ^{14}C 7. curie   8. nuclear fission   9. film badges   10. body imaging   11. nuclear fusion   12. critical mass
True/false: 1. F (3/4 will have decayed after 24 days.)   2. T   3. T   4. F (Becquerel)   5. T   6. F   7. F (Atomic number increases by 1.)   8. F (Hazards are due to emissions of radiation.)   9. F (The rem is more common.)   10. T   11. F (Most are synthetic.)   12. F
Matching: 1. f   2. i   3. l   4. k   5. c   6. j   7. d   8. h   9. b   10. a   11. e   12. g

Chapter Twelve

Multiple choice: 1. a   2. c   3. d   4. c   5. b   6. d   7. c   8. a   9. c   10. a
Sentence completion: 1. paraffins   2. triple   3. oxygen   4. distillation   5. condensed structure   6. quaternary   7. lower   8. ring   9. IUPAC   10. isomers   11. Combustion   12. higher
True/false: 1. F (They have different molecular formulas.)   2. T   3. F (It's puckered.)   4. F (3-Methylhexane)   5. T   6. T   7. F (It's polar covalent.)   8. T   9. T   10. F (Reactivity depends on the nature of the functional groups.)   11. T   12. F
Matching: 1. i   2. j   3. f   4. l   5. h   6. c   7. a   8. k   9. e   10. b   11. g   12. d

Chapter Thirteen

Multiple choice: 1. d   2. b   3. a   4. c   5. a   6. c   7. c   8. c   9. d   10. d
Sentence completion: 1. 2-heptyne   2. hydrogenation   3. *para*   4. more, fewer   5. hydration   6. catalyst   7. carbocation   8. unsaturated   9. Toluene   10. substitution   11. Polycyclic aromatic compounds   12. thermal cracking
True/false: 1. F   2. T   3. T   4. F (2-bromobutane)   5. F (Halogenation is the reaction of X_2 with an alkene, where X is a halogen.)   6. T   7. F   8. F (Hydration requires a strong acid as a catalyst.)   9. T   10. F   11. F   12. T
Matching: 1. e   2. k   3. i   4. j   5. b   6. l   7. d   8. c   9. a   10. f   11. g   12. h

Chapter Fourteen

Multiple choice: 1. c 2. b 3. b 4. a 5. d 6. d 7. a 8. b 9. c 10. d
Sentence completion: 1. wood alcohol 2. glycols 3. phenols 4. alkene 5. thiols
6. $KMnO_4$ or $K_2Cr_2O_7$ 7. lower 8. BHT 9. greater 10. free radical 11. peroxide 12. CFC
True/false: 1. T 2. F (A ketone is formed from the oxidation of a secondary alcohol.) 3. T
4. F (carbolic acid) 5. T 6. T 7. F (Ethyl chloride is an anesthetic.) 8. F (two hydrogen atoms)
9. T 10. T 11. F 12. T
Matching: 1. j 2. g 3. f 4. h 5. l 6. b 7. e 8. a 9. k 10. d 11. i 12. c

Chapter Fifteen

Multiple choice: 1. c 2. b 3. a 4. d 5. c 6. b 7. b 8. a 9. d 10. b
Sentence completion: 1. carbonyl 2. silver 3. Hydrolysis 4. positive, negative 5. Acetone
6. acetaldehyde 7. more 8. reduction 9. disinfectant or preservative 10. hemiacetal
11. Benedict's 12. Acetone
True/False: 1. F (a hemiacetal link) 2. T 3. F (Ketones don't undergo Tollens' reaction.) 4. F
($NaBH_4$ is used for reduction.) 5. F 6. F (Both aldehydes and ketones form acetals.) 7. T 8. T
9. F (NADH) 10. T 11. T 12. T
Matching: 1. e 2. i 3. d (or l) 4. a 5. h 6. j 7. k 8. b 9. l 10. f 11. g 12. c

Chapter Sixteen

Multiple choice: 1. c 2. a 3. d 4. c 5. c 6. a 7. b 8. d 9. a 10. b
Sentence completion: 1. *amino-* 2. quaternary 3. DNA 4. Toxicology 5. heterocycle
6. Alkaloids 7. hydrogen bonding 8. ammonium 9. free radical 10. nonaromatic amines
True/false: 1. T 2. T 3. F (They're lower boiling.) 4. T 5. F 6. T 7. F (It's synthetic.)
8. T 9. F (NO lowers blood pressure.) 10. F (*N*-Methylbutylamine)
Matching: 1. d 2. l 3. h 4. i 5. a 6. j 7. a 8. k 9. f 10. g 11. e 12. c

Chapter Seventeen

Multiple choice: 1. d 2. c 3. a 4. d 5. b 6. d 7. b 8. a 9. d 10. b
Sentence completion: 1. carbonyl group substitution 2. polyamide 3. carboxylate salt
4. esterification 5. attracts 6. *-oic* acid 7. propyl propanoate 8. esters 9. Phosphorylation
10. fats 11. hydrolysis or saponification 12. propanedioic acid
True/false: 1. T 2. F (basic hydrolysis) 3. F (Amides are nonbasic.) 4. T 5. F (The products
are acetate ion and ethanol.) 6. F (*N,N*-dimethylformamide) 7. F 8. T 9. T 10. F
(Esterification is acid-catalyzed.) 11. F (It's an acid anhydride.) 12. T
Matching: 1. j 2. f 3. k 4. b 5. i 6. d 7. a 8. e 9. l 10. h 11. c 12. g

Chapter Eighteen

Multiple choice: 1. a 2. d 3. c 4. b 5. b 6. a 7. d 8. c 9. c 10. d
Sentence completion: 1. 5.0–6.3 2. enantiomers 3. zwitterions 4. backbone 5. loop
6. glycoprotein 7. Tertiary 8. β-sheet 9. polar 10. sulfur 11. polypeptide 12. Hydrophobic
True/false: 1. F (Glycine is achiral.) 2. T 3. F (Denaturation disrupts all structural elements
except primary structure.) 4. F (They're different.) 5. T 6. F 7. T 8. T 9. T 10. F (Fibrous
proteins are insoluble.) 11. T 12. T
Matching: 1. e 2. g 3. j 4. a 5. b 6. k 7. c 8. l 9. f 10. h 11. d 12. i

Chapter Nineteen

Multiple choice: 1. c 2. a 3. c 4. d 5. b 6. b 7. d 8. c 9. b 10. d
Sentence completion: 1. isomerase 2. Isoenzymes 3. K 4. induced-fit 5. uncompetitive
6. denature 7. turnover number 8. phosphoryl 9. proximity 10. 10^7 11. allosteric
12. zymogen
True/false: 1. T 2. F 3. F 4. F (a dehydrase) 5. F (Enzyme–substrate interactions are noncovalent.) 6. F (Minerals, not vitamins, are inorganic ions.) 7. T 8. F (Retinol, retinal, retinoic acid are all active forms of vitamin A.) 9. T 10. T 11. F (Reaction rate increases.) 12. T
Matching: 1. c 2. h 3. i 4. g 5. k 6. j 7. b 8. a 9. d 10. f 11. e 12. l

Chapter Twenty

Multiple choice: 1. a 2. d 3. c 4. c 5. d 6. d 7. d 8. c 9. b 10. a
Sentence completion: 1. chiral 2. amylases 3. enantiomers 4. cyclic hemiacetal
5. fructose 6. amylose, amylopectin 7. glycoside 8. Glycogen 9. aldohexose 10. achiral
11. diastereomer 12. Glycoproteins
True/false: 1. T 2. T 3. F (The hydroxyl group on the chiral carbon farthest from the carbonyl group is on the left.) 4. F (α, not β) 5. T 6. F (Crystalline glucose is the α-anomer.) 7. F (Sucrose contains only an acetal bond.) 8. T 9. F (An acetal resembles an ether.) 10. F (Maltose is composed of two glucose molecules.) 11. F (The relationship between the rotations of diastereomers is not predictable.) 12. T
Matching: 1. e 2. l 3. j 4. h 5. i 6. k 7. b 8. a 9. d 10. g 11. c 12. f

Chapter Twenty-one

Multiple choice: 1. b 2. c 3. a 4. c 5. a 6. d 7. b 8. c 9. b 10. d
Sentence completion: 1. Eukaryotic 2. metabolic pathway or linear sequence 3. ADP and phosphate 4. unfavorable 5. Krebs cycle, tricarboxylic acid cycle 6. succinyl-CoA, CO_2
7. NAD^+ 8. one GDP (plus 4 reduced coenzymes) 9. Cytoplasm 10. Catalase 11. succinate dehydrogenase 12. Basal metabolism
True/false: 1. T 2. F (Only eukaryotic cells are found in higher organisms.) 3. F (Anabolism is the synthesis of complicated molecules from simpler molecules.) 4. F 5. T 6. T 7. F ($FADH_2$ donates electrons to coenzyme Q.) 8. T 9. F (decrease in order) 10. F (ATP can be formed in the citric acid cycle) 11. F (a phosphate anhydride) 12. T
Matching: 1. d 2. j 3. f 4. a 5. l 6. k 7. e 8. c 9. b 10. g 11. i 12. h

Chapter Twenty-two

Multiple choice: 1. c 2. c 3. b 4. d 5. a 6. b 7. d 8. c 9. d 10. a
Sentence completion: 1. Embden–Meyerhof 2. 6 3. Pyruvate dehydrogenase complex 4. Insulin 5. glucose 1-phosphate 6. 1 and 3 7. diabetes mellitus 8. GTP, ATP 9. CO_2
10. glucose 6-phosphate isomerase 11. glucagon 12. Hypoglycemia
True/false: 1. F (It produces NADPH.) 2. T 3. F 4. F 5. T 6. F (Glycolysis occurs in the cytosol.) 7. F (The yield of ATP was formerly considered to be 38 ATP/1 mol glucose. The new estimate is probably less.) 8. F (glycogen) 9. F (Fermentation occurs only in yeasts.) 10. T (but one results from isomerization of dihydroxyacetone phosphate) 11. T 12. F (It produces two ATPs.)
Matching: 1. l 2. k 3. i 4. j 5. c 6. h 7. d 8. a 9. f 10. g 11. b 12. e

Chapter Twenty-three

Multiple choice: 1. d 2. c 3. a 4. a 5. d 6. d 7. c 8. b 9. a 10. b
Sentence completion: 1. glycoside 2. Soap 3. hydrophilic, hydrophobic 4. Glycolipids, cholesterol, glycoproteins 5. fluid-mosaic 6. tetracyclic 7. arachidonic 8. micelles 9. Sphingomyelins 10. lipid bilayer 11. Eicosanoids 12. Integral
True/false: 1. T 2. F 3. F (They contain an amide group.) 4. T 5. T 6. F (They form a micelle.) 7. F (Cholesterol is a steroid component of cell membranes.) 8. F (The major difference is in the quantity of unsaturated fatty acids they contain.) 9. F (Cerebrosides are components of nerve cell membranes in the brain.) 10. T 11. T 12. F (In facilitated diffusion, solutes are transported across cell membranes by proteins, but no energy investment is required.)
Matching: 1. 1 2. h 3. e 4. f 5. a 6. j 7. k 8. b 9. d 10. g 11. i 12. c

Chapter Twenty-four

Multiple choice: 1. a 2. d 3. c 4. a 5. b 6. b 7. c 8. d 9. a 10. d
Sentence completion: 1. NAD^+, FAD 2. micelles 3. adipocytes 4. dihydroxyacetone phosphate 5. fatty-acid spiral 6. double bond 7. Ketonuria 8. NADPH 9. 3-Hydroxybutyrate 10. lymphatic system 11. mobilized 12. acetyl-CoA
True/false: 1. T 2. F (The products are mainly mono- and diacylglycerols.) 3. T 4. T 5. F 6. T 7. F (Dihydroxyacetone phosphate, not glycerol) 8. F (LDL transports cholesterol.) 9. T 10. F (It occurs in the cytosol.) 11. F (It comes from acetyl-CoA.) 12. T.
Matching: 1. e 2. i 3. a 4. 1 5. g 6. j 7. k 8. d 9. f 10. b 11. c 12. h

Chapter Twenty-five

Multiple choice: 1. c 2. a 3. b 4. d 5. a 6. c 7. d 8. c 9. a 10. b
Sentence completion: 1. pyruvate, glutamate 2. Methionine 3. acetyl-CoA, acetoacetyl-CoA 4. carbamoyl phosphate 5. protease 6. pyruvate 7. Transamination 8. arginine 9. phenylalanine, tyrosine 10. urate crystals 11. ALT 12. bicarbonate
True/false: 1. T 2. F (It starts in the stomach.) 3. F (lysine and leucine) 4. F (The human body can catabolize essential amino acids.) 5. T 6. F (Catabolism occurs in the liver and in the cytosol.) 7. T 8. T 9. F (One is from aspartate; the other is from ammonium ion.) 10. T 11. F (Some transaminases have several amino acid donors.) 12. T
Matching: 1. h 2. g 3. j 4. f 5. a 6. b 7. c 8. 1 9. k 10. d 11. e 12. i

Chapter Twenty-six

Multiple choice: 1. c 2. d 3. b 4. a 5. b 6. b 7. c 8. d 9. c 10. a
Sentence completion: 1. Replication 2. complementary 3. chromatin 4. retrovirus 5. intron 6. ester 7. nucleoside 8. releasing factor 9. replication fork 10. 2'-deoxyribose 11. Watson–Crick 12. DNA polymerase
True/false: 1. F (61 codons code for the 20 amino acids; the other three are "stop" codons.) 2. F (The amount of adenine equals the amount of thymine.) 3. T 4. F (An anticodon occurs on tRNA.) 5. T 6. T 7. T 8. F (They run in opposite directions.) 9. F (translation) 10. F (Most DNA consists of introns.) 11. F (Two have only one codon.) 12. F (guanosine 5'-monophosphate)
Matching: 1. 1 2. j 3. b 4. g 5. a 6. i 7. d 8. c 9. e 10. k 11. f 12. h

Chapter Twenty-seven

Multiple choice: 1. c 2. d 3. b 4. a 5. d 6. d 7. d 8. b 9. c 10. c
Sentence completion: 1. "sticky" ends 2. "junk" 3. restriction endonuclease 4. *Taq* polymerase 5. bioethics 6. ionizing radiation, mutagens 7. intron 8. single-nucleotide polymorphism 9. proteomics 10. centromere 11. silent 12. autoradiogram
True/false: 1. T 2. F (nucleotide) 3. F (from several people) 4. T 5. T 6. F (starts from base–base sequences at known locations) 7. T 8. F 9. F (missense) 10. T 11. F (adding) 12. T
Matching: 1. b 2. i 3. g 4. c 5. j 6. l 7. e 8. a 9. k 10. h 11. f 12. d

Chapter Twenty-eight

Multiple choice: 1. a 2. c 3. b 4. c 5. d 6. a 7. d 8. a 9. d 10. d
Sentence completion: 1. Anaphylaxis 2. adrenal 3. vesicles 4. Histamine 5. cholinergic 6. progesterone 7. enkephalin 8. ethnobotanist 9. synaptic cleft 10. dopamine 11. thyroxine 12. phosphodiesterase
True/false: 1. F (Thyroxine can cross the cell membrane.) 2. F (It blocks reuptake.) 3. T 4. T 5. F (presynaptic neurons) 6. T 7. F (It increases blood pressure and heart rate.) 8. F (It's a symptom of iodine deficiency.) 9. T 10. T
Matching: 1. e 2. h 3. k 4. b 5. g 6. a 7. d 8. j 9. f 10. l 11. c 12. i

Chapter Twenty-nine

Multiple choice: 1. a 2. d 3. c 4. b 5. d 6. a 7. d 8. b 9. b 10. c
Sentence completion: 1. partial pressure 2. haptens 3. zymogens 4. osmolarity 5. histidine 6. calcium, K 7. active transport 8. transport, regulation, defense 9. phagocytes 10. autoimmune 11. Hemophilia 12. lysine, glutamine
True/false: 1. F (constructed) 2. F (hemostasis) 3. F (CO_2 bonds to hemoglobin) 4. T 5. T 6. F (overproduction) 7. F (blood serum) 8. T 9. F (tubule) 10. T 11. F (phagocytes) 12. T
Matching: 1. d 2. c 3. f 4. g 5. j 6. i 7. a 8. b 9. h 10. l 11. k 12. e